CROSSING A CHASM
IN SMALL STEPS?

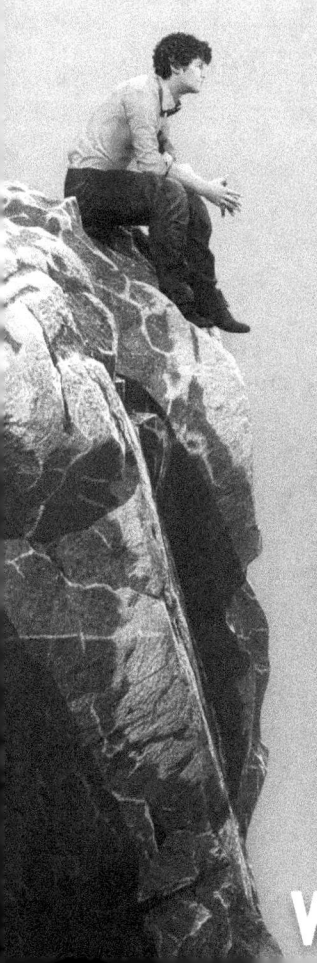
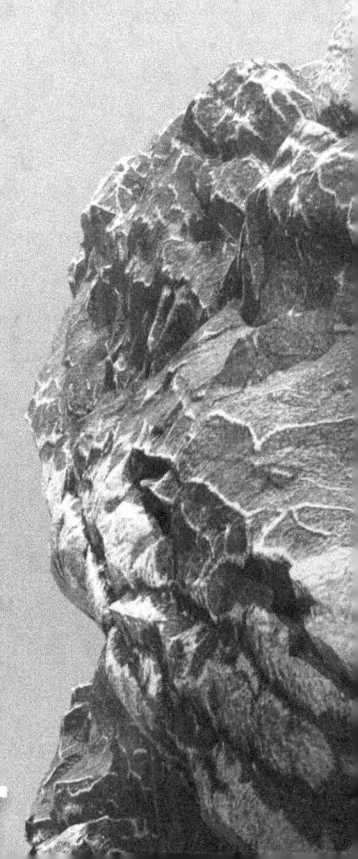

WAYNE TALBOT

Copyright @2021 by Wayne Talbot

All rights reserved. No part of this book may be reproduced in any form or by any electronic or mechanical means, including information storage and retrieval systems, without permission in writing from the publisher, except by reviewers, who may quote brief passages in a review.

This publication contains the opinions and ideas of its author. It is intended to provide helpful and informative material on the subjects addressed in the publication. The author and publisher specifically disclaim all responsibility for any liability, loss or risk, personal or otherwise, which is incurred as a consequence, directly or indirectly, of the use and application of any of the contents of this book.

WORKBOOK PRESS LLC
187 E Warm Springs Rd,
Suite B285, Las Vegas, NV 89119, USA

Website: https://workbookpress.com/
Hotline: 1-888-818-4856
Email: admin@workbookpress.com

Ordering Information:
Quantity sales. Special discounts are available on quantity purchases by corporations, associations, and others.
For details, contact the publisher at the address above.

ISBN-13: 978-1-956017-53-3 (Paperback Version)
978-1-956017-54-0 (Digital Version)

REV. DATE: 09/07/2021

Crossing a Chasm

In Small Steps?

The Elephant in the Room of Evolution

Wayne Talbot

Table of Contents

Acknowledgements...1
Author's Note..3
What Piqued My Interest?...11
Prologue..16
Introduction..22
In Defence of Philosophy..36
Foundational Propositions..38
Part 1: Primary Axioms...54
Chapter 1-1: Primary Axiom #1...55
Chapter 1-2: Primary Axiom #2...59
Chapter 1-3: Primary Axiom #3...68
Chapter 1-4: Primary Axiom #4...71
Chapter 1-5: Primary Axiom #5...74
Chapter 1-6: Primary Axiom #6...79
Chapter 1-7: Primary Axiom #7...82
Chapter 1-8: Primary Axiom #8...88
Chapter 1-9: Primary Axiom #9...92
Chapter 1-10: Primary Axiom #10......................................107
Part 2: The Exceptional Human Condition..........................110
Chapter 2-1: Sex...113
Chapter 2-2: An Unfortunate Experience...........................117
Chapter 2-3: Official Theories of Being...............................119
Chapter 2-4: Philosophy..129
Chapter 2-5: Curiosity & Imagination.................................134
Chapter 2-6: Knowledge...140
Chapter 2-7: Processing versus Being................................146
Chapter 2-8: Sentience versus Consciousness....................155
Chapter 2-9: Objective vs Subjective..................................165

Chapter 2-10: Rationality & Reason..169
Chapter 2-11: Intentionality & Causation...172
Chapter 2-12: Volition & Free Will...176
Chapter 2-13: Intelligence & Intellect...187
Chapter 2-14: Morality..192
Chapter 2-15: The Human Spirit...197
Chapter 2-16: Memory...199
Chapter 2-17: Music...204
Chapter 2-18: To Reiterate...207
Part 3: Establishing Foundations of Knowledge..............................210
Chapter 3-1: Thesis..214
Chapter 3-2: The Theory of Knowledge..236
Chapter 3-3: Knowledge versus Evolution Theory..........................239
Chapter 3-4: The Morality Conundrum...248
Chapter 3-5: The Sensory Conundrum..257
Chapter 3-6: The Information Blind Spot..266
Part 4: Principles of Information Processing...................................279
Chapter 4-1: Coding Systems...281
Chapter 4-2: Functional Requirements..296
Chapter 4-3: Understanding Information.......................................299
Chapter 4-4: Understanding Communications...............................310
Chapter 4-5: Data Processing Principles...318
Chapter 4-6: Information Processing in the Mind..........................334
Part 5: Science and Evolution...346
Chapter 5-1: The Language of Science..351
Chapter 5-2: The Truth of Science..368
Chapter 5-3: Descent with Modification..402
Chapter 5-4: Cells as Computers..410
Chapter 5-5: Determinism..426
Chapter 5-6: Self-Organization...437
Chapter 5-7: Emergence...446
Chapter 5-8: Doubting the Science of Evolution.............................456

Chapter 5-9: Is Darwinism Broken?......474
Chapter 5-10: An Ancient View......478
Part 6: The Building Blocks......486
Chapter 6-1: The Genome......489
Chapter 6-2: Neuron Communications......499
Chapter 6-3: The Nervous System......522
Chapter 6-4: Neural Network Modelling......526
Part 7: The Sensory Messaging System......537
Chapter 7-1: Human Telemetry System......542
Chapter 7-2: The Auditory System......548
Chapter 7-3: Echo Location & Ranging......554
Chapter 7-4: Evolution of the Eye......562
Chapter 7-5: Irreducibility of Sight......575
Chapter 7-6: The BOLT......601
Part 8: The Mind-Brain Complex – What Is It?......605
Chapter 8-1: What is the Brain?......615
Chapter 8-2: Neuroanatomy......625
Chapter 8-3: How the Brain Works......632
Chapter 8-4: The Concept of Mind......650
Chapter 8-5: What IS the Mind?......657
Chapter 8-6: How the Mind Works......666
Chapter 8-7: The Mind's Eye......685
Part 9: Language......688
Part 10: Postscript......694
Chapter 10-1: A Law Hypothesis......696
Chapter 10-2: Reflections on Science......714
Chapter 10-3: A Final Word on Evolution......720
Bibliography......723

Knowing That You Can

The Elephant In the Evolutionists Room

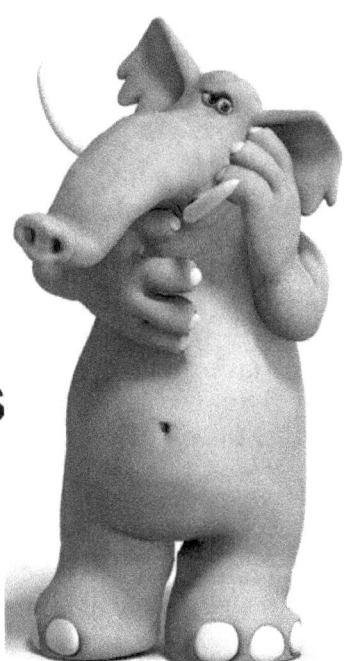

Wayne Talbot

Sequel to
Information, Knowledge, Evolution and Self

Acknowledgements

"And for all the people waiting for permission to level up enough before they start working on something big and scary--just go in. Don't be like me."

~ Mary H.K. Choi, Emergency Contacts ~

I wish to acknowledge the serious scientific enquiry of the many scientists whom I reference in this book. Without their published works, I would have little to discuss other than my own musings and conjecture. I note how fields associated with intelligence, and in particular, language and the processing of information, continue to intrigue scientists as they struggle to understand how such capabilities arose through evolution, and to what degree humans can be differentiated from their supposed evolutionary predecessors.

Acknowledged giants in these fields have long contended with each other, at times acrimoniously. This fact evidences how far each of them may be from the truth. I thank them for their contributions to the material in this, my own study, on these and related subjects. The amount of published literature is simply overwhelming, so much so that I seem to spend much of my time choosing which book to acquire, or which scientific paper I should study, as some delve into far more detail than I can comprehend, and more than is necessary for that which I am attempting to accomplish. If you ask, "Why did you not mention [fill in your expert]?", it is because there are far too many for me to survey or study.

I have done the best that I can to review contradictory positions, taking as my guide, the wisdom of the ancient Roman poet, Ovid: *"fas est et ab hoste doceri"* (Latin) – *it is right to learn even from an enemy.* Not that they are my enemies, but you know what I mean (I hope).

You may also note that many of my sources are philosophers of various religious persuasions. We discuss philosophy a little later, but to my mind, the very existence of philosophy, as a product of the human mind, argues against the contention that "the mind is an emergent property of the brain". Not beholden to the philosophical materialism of atheists and evolutionists, these philosophers dare to think beyond the materialistic paradigm, a task antithetical to those who do not believe in a higher power than man. I offer no opinion on the validity of the latter, but have strong opinions regarding the existence of the immaterial, especially the human mind. For this reason, I find the thinking of these philosophers to be invaluable in seeking the truth, irrespective of the imperatives behind their thinking.

So ... let me start working on something truly big and scary!

Author's Note

"A story is a letter that the author writes to himself, to tell himself things that he would be unable to discover otherwise."

~ Carlos Ruiz Zafón, *The Shadow of the Wind* ~

I concur, for the primary reason for any of my writings is to explain difficult subjects to myself. It is only when I attempt to express concepts in my own words that I get closer to understanding them. It is for this reason that I have chosen to follow the concept of *slow reading*, although in my advancing age, I cannot be sure whether it is deliberate or a function of age. Nevertheless, reading with the intention of learning involves far more than achieving simple comprehension. Comprehension is not equivalent to knowledge acquisition. You must employ metacognition. You have to read slowly in order to enable and activate the processes that support knowledge acquisition as you read. If you're reading to learn, you need to engage with the content and associate the new concepts with your existing knowledge. Only then can you install new knowledge in your mind and be able to utilize this knowledge in the future. You have to do the work to learn, and "the work" has to be the right work done correctly. This leads into my Primary Axiom #2, which we will get to shortly (depending on how fast, or slow, you read).

One of the goals of this study is to demonstrate that if evolution is your only method of transport, then - *you cannot get there from here*, "here" being Planet Earth before there was any life form of any type, and "there" being humanity as we know it. In short, humanity is inexplicable by any evolutionary hypotheses. Whilst I am confident with that assertion, I cannot offer any alternative explanation – it is all a mystery to me, but I can at least demonstrate why I have concluded as I have. As for alternative explanations, they are outside the scope of this study. Many people seem to believe in evolution, not because they understand and believe the pseudo-scientific explanations, but because no alternative explanation is acceptable to them. I prefer to start from the beginning, evaluating whether our understanding of the relevant sciences supports the evolution hypothesis. If they do not, then logically, one should seek alternative explanations.

The *central theme* in this study concerns the mind-brain complex, for therein lies the greatest mystery of all – *conceptualisation*. Conceptualisation is itself, a *concept*, absent of physical properties, and one must ask: Is the material capable of deriving or otherwise being aware of concepts? The human body contains an amazing telemetry system, comprised of sensors that monitor our internal biological workings (*interoceptors*), and sensors which provide us with our experience of the outside world (*exteroceptors*).

In this study, I have only a passing interest in the former, for whilst the mind is aware of some internal conditions, it plays little or no part in any autonomous response to them – that is the function of the brain. On the other hand, the mind is the primary agent for interpreting and responding to inputs from the latter (the outside world). When we "see", light waves are not conducted from our eyes to the brain, nor in hearing are sound waves conducted from our ears. Technically, exteroceptors are *transducers* which convert energy from one form to another, whilst faithfully conveying the *semantic* layer (meaning). Between the sensors and the brain are neuronal pathways (nerves) which convey electrical impulses, resulting in encoded symbolic representations of the source energy. Along the way, signal processing occurs before neural patterns are formed in the brain. Thus, the conundrum: what performs the orchestration of these neural network arrangements, and how can they perceive themselves conceptually?

$$--- \int --- \int ---$$

Let me begin with my working definition of the General Theory of Evolution (GTE), as defined by noted British zoologist and physiologist, Professor G.A. Kerkut (1927-2004), then Dean of Science, Chairman of the School of Biochemical and Physiological Sciences, and Head of the Department of Neurophysiology at Cambridge University: *"the theory that all the living forms in the world have arisen from a single source which itself came from an inorganic form"*[1], or as Charles Darwin expressed it, "all the organic beings which have ever lived on the earth have descended from some single primordial form." I prefer these definitions because they best represent what the general population believe about evolution, rather than as it is often more narrowly defined by specialists in the field.

My argument is based on the contention that people, including scientists, have a poor appreciation of the fundamentals which I shall later propose. Too often these are treated as *brute facts*, defined thus: "In contemporary philosophy, a brute fact is a fact that has no explanation. More narrowly, brute facts may instead be defined as those facts which cannot be explained. To reject the existence of brute facts is to think that everything can be explained."[2] I *do* accept the concept of brute facts, but will argue than in the context of this study, they lie much deeper than is generally appreciated. I will attempt to explain what can be explained below the level at which most scientific research occurs.

I have purchased samples of published works by specialists working in the neurosciences, including language, psychology, and other related disciplines, to which we will refer as the study proceeds. A question that I seek an answer to is this: Have any of these specialists asked themselves, in their published works at least, how physical matter can self-organise into conceptual patterns? For example, in her study, *"Language in Our Brain"*, the author unashamedly admits that "I start from the assumption that language is a biological system that evolved through phylogeny"[3] [Phylogeny is the study of relationships among different groups of organisms and their evolutionary development]. Why does she start with that assumption? Has it not occurred to this scientist that language, whilst undoubtedly assisted by biology, may not have biological origins, and quite likely, could not have arisen through biological evolution alone? As I study each of these works, this question is foremost in my mind. I apologise in advance if my continual revisiting this issue causes irritation, but it is fundamental to the question of whether the mind can be an *emergent property of the brain*, or whether the mind must be a separate, non-material entity, which interacts with, and controls the non-autonomous functions of the brain, much as software directs the operations of a computer.

My favourite sage of old, Maimonides, commented:

"It could be said that we're each essentially an orchestra of many instruments led by a single conductor, i.e., the self. But our analogy breaks down at a certain point, because there are some 'instruments' that seem to function on their own without direction, and others that will only play when directed. That is to say, there are parts of our beings that seem to have a mind or will of their own, and to thus defy direction, for example, our autonomic nervous system. In contrast, other parts do accommodate conscious direction."[4]

On the subject of psychology, I am bemused by psychologists holding to philosophical materialism when in truth they have no scientific explanation for their craft. Their goal is to achieve a reorganisation of the synaptic connections in the brain, to disconnect some that are causing the unwanted behaviour, and make new connections which will lead to more socially acceptable behaviour. Their advice to a patient is a verbal communication of the conceptual. Thus, they are assuming that at some point in the auditory sensory process, the conceptual is reliably converted into the physical as represented in the brain's neural network. The question becomes: what part of the cortex or brain is controlling that conversion, and how does it know how to do that? Is the brain capable of understanding the conceptual meaning of its own neuronal organisation such that it can autonomously reorganise itself? I have very strong doubts.

It could be said, with some justification, that I have an *obsession* with origins – I want to know how things got started. You see, part of the difficulty that I have with evolution theory is not just that absent of external direction, evolution could not have accomplished what is claimed for it, but even more, it could not have even got some processes started. Of particular interest to me are knowledge and language, and an implementation of those, communications. My background in both Air Traffic Control and Information Technology piqued my interest: firstly, in the technology of communications, and secondly, in the intellectual aspects of these domains. In an earlier study, "*Information, Knowledge, Evolution and Self*"[5], I offered why I believed that *undirected* evolution could not have been responsible for the development of the human mind. To clarify my meaning of undirected evolution, it is not that I do not believe in evolution, it is just that I do not know what to believe about evolution.

This study continues that same theme, reiterating some earlier material, and expanding on it through additional avenues of research.

My primary focus is the mystery of the mind-brain complex, but to explain the issues, I need to dig a little deeper into the evolution hypothesis. The first part of the title, *Crossing a Chasm*, was chosen to highlight the vast gulf between non-life and life as we humans know it, and the second part, *in Small Steps,* offers the irony that such is not possible. Quite simply, you cannot get from one side to the other that way, unless someone builds a bridge, and it is that bridge building which is in question. This, to my mind, has long been the *Elephant in the Room of Evolution*, a sub-title that I wanted to use on the cover, but considered it too lengthy. It does, however, reveal the very obvious but difficult problem that even evolving evolution theory cannot solve. Some scientists have made the effort and published their solutions, which we will get to in their proper place. In contrast, I intend to demonstrate that their solutions fall far short of being plausible, because they fail to deal with the fundamentals that I describe in this study.

Additionally, I would offer that just as Charles Darwin thought that some processes were possible because he lacked an understanding of the complexity of human biology, the same paucity of understanding is evidenced in the thinking of evolutionists today. American biochemist, Michael Behe correctly noted: "In private, many scientists admit that ... Darwin never imagined the exquisitely profound complexity that exists even at the most basic levels of life". We can only wonder how he might have rethought his hypothesis had he so imagined. In a later chapter, we will review the complexity of the human nervous system, paying attention to the interconnectivity of neurons in the Central and Peripheral Nervous Systems, and in the neural network of the brain. Neuroscientists are getting closer to understanding the "how", but not the "why", of neurons connecting to some neurons but not others. The physiology and biochemistry are understood to some extent, but the biological mechanisms of individual neurons synapsing with some neurons, but not others, is not understood at all. So far as I can understand from my research, no one is able to reverse-engineer these processes to substantiate evolutionary claims.

Because evolution sceptics such as myself are often scorned as Young Earth Creationists or similar, let me assure the reader that I am utterly convinced that our planet is billions of years old, and that over past eons, various life forms have come and gone, mostly evidencing greater complexity with each iteration. Just why this is so is beyond my knowing, but my contention is that undirected evolution has not been scientifically proven, and if my understanding of various subjects is correct, likely never will be. Part of my rationale for publishing this study, is to encourage others to ask as I do. If more people did this, perhaps there could be a grass-roots groundswell (excuse the pun) to encourage scientists to seek truth beyond the prevailing paradigm of philosophical materialism, which itself has no scientific foundation. Let me repeat that: there is no scientific basis for philosophical materialism – it is just the preferred position of scientists, owing its popularity to the Enlightenment Era. Some scientists accept the existence of the immaterial, and conduct research as best they can, but they are almost invisible in the scientific community. I applaud their courage, especially as grants for such research are hard to come by, and they are unlikely to ever receive recognition for their work. However, I believe such research is necessary if we are ever to divine the truth of human existence.

I begin by offering *foundational propositions* around which this study is built. Whilst some areas have been researched by scientists at a technical level, much has been glossed over superficially, leading to terminology which is misleading. In some instances, this lack of precision in terminology has led even highly qualified scientists to unknowingly contradict themselves. Here is one example found in two consecutive sentences in a book, *"How The Mind Works"*[6]. The author, a Professor of Psychology, and at the time of writing, Director of the Centre for Cognitive Neuroscience at MIT, is clearly an intelligent and educated man, but perhaps he pays insufficient attention when writing outside his speciality. I cannot know, I can only judge by the evidence. Consider this:

"Information is a correlation between two things that is produced by a lawful process (as opposed to coming about by sheer chance). We say that the rings in a stump carry information about the age of a tree because their number correlates with the tree's age."

Superficially, the second sentence appears logical, but at a deeper level, it is not. Pinker is partially correct in stating that *information is correlation*, but correlation can only be performed by an intelligent agency, which a tree is not. The tree cannot perform the correlation to produce information - scientific research was required to determine that there is a correlation between a tree's rings, and its age, just as has been discovered between the thickness of a ring, and the local climate that year. By themselves, the tree rings may be considered as *facts*, but not as *information*. I hope to explain why tree rings are not even facts, but that will take a little longer.

Another topic to be dealt with is the philosophical aspects of our human existence. I believe these to be the necessary conditions separating us from every other life form on earth. In our everyday lives, we take them for granted, being largely unaware of just how far separated we are from, for example, dolphins or chimpanzees, despite the efforts of naturalists trying to bring us ever closer together. Without an adequate understanding of these conditions, people are easily fooled into accepting that the genome is THE differentiator, and thus undirected evolution is the best explanation for the differences.

With no little trepidation, I will also visit the subjects of genetics and microbiology. I am not qualified to comment at a technical level, but I can review the works of scientists who are, and appraise the logic of many conclusions and suppositions, especially where contradictory conclusions are offered by equally qualified scientists. I often read that chimps share 99% of our DNA, and gorillas 98%. I have read that even dogs and rats can be taught how to behave in response to human training, so obviously, we must be of the very same lineage, using the very same processes. Well, that is the evolutionists' story.

The goal of this book is to convince you otherwise: if it does not convince, at least it may cause you to ponder.

From the outset, let me confess that I subscribe to philosophical *dualism*, as opposed to philosophical *materialism*. There is "me", and there is the body that I inhabit. The two are not the same entities, nor of the same form – they co-exist, and whilst being separate entities, they are nevertheless interdependent – one cannot exist without the other. I can investigate the components of my body, because they are material, but not the "me", because it is immaterial. The evidence of their respective abilities has convinced me of their separateness: the body can only do what the prevailing rules of chemistry and physics allow biological entities to do, whereas the "me" can do so much more, even contrary to the rules by which biological entities are constrained. This book provides the evidence, and my analysis of it, which have led to this conclusion.

Prospective publishers always ask: "Who would want to read your book?" My answer is treasure hunters, and those who, like me, are curious about all manner of things, especially science and the nature of themselves. Socrates would advise that we should never stop questioning, and that is my sense of my own existence. I do not know the truth, and likely never will, but I share the excitement of archaeologists as they carefully scrape away centuries of debris, unsure of what they will find, nevertheless continuing tirelessly and enthusiastically in their search. The truth may not be where I am searching, but I believe that by uncovering layers of untruth, I am likely to get closer. That is the best that I can hope, and I am satisfied with that.

It is curious, is it not, that space has often been declared the last frontier, and even the deep oceans are little understood, offering the opportunity for even more exciting research and discovery. Yet, closer to home, there remains an even greater mystery – the nature of the human mind. Are we too afraid to venture therein? Is there a fear that we are more than just the material substance of our bodies, and have an as yet, unrealised potential that could transform our understanding of our existence? I have a sense of this, and although I am unable to penetrate the mystery myself, there remains an undercurrent of excitement of what yet may be discovered. For me, the human mind is the last great frontier to be explored.

Like-minded readers should enjoy this book.

Wayne Talbot

Kelso NSW Australia

August, 2021

References:

1. Kerkut, G.A., *Implications of Evolution*, Pergamon, Oxford, UK, 1960, p 157

2. https://en.wikipedia.org/wiki/Brute_fact

3. Friederici, Angela D., *Language In Our Brain: The Origins of a Uniquely Human Capacity*, The MIT Press, Cambridge, MA, 2017, p. 2

4. Feldman, Rabbi Yaakov, *The 8 Chapters of the Rambam: A Classic Work on the Fundamentals of Jewish Ethics and Character Development*, Targum Press, Southfield, MI, 2008, p. 41

5. Talbot, Wayne, *Information, Knowledge, Evolution, and Self: a question of origins*, Xlibris, Bloomington, IN, 2016

6. Pinker, Steven, *How the Mind Works*, Penguin Books, London, UK, 1998, p. 65

What Piqued My Interest?

"One of the tenets in Quaker meditation is that you 'go inside to greet the light'. I am interested in this light that is inside greeting the light that is outside."

~ James Turrell ~

Perhaps if you, the reader, succumbs to the same level of curiosity that I do, you may be interested to know why I am so intensely interested in this subject, despite not being a scientist of any description, and am in so many ways, scientifically inept. Generally, I have no interest in science, other than it results in technology which has enhanced my life enormously, perhaps more so than any lives in the history of the world.

In short, my interest in writing this book arose from a penchant, ingrained from life, for intellectual paths less travelled. The trains conveying my thoughts have almost always departed from unnamed stations, along lines that seemingly led nowhere. Certainly, there have been no departure boards notifying of times, routes, and/or destinations, just in small print: "adventure starts here". I have little to no resistance to such invitations, just as Harry Potter for his very first time, unhesitatingly walked into the solid brick wall of Platform 9¾ on 1 September at 11 AM sharp, to catch the scarlet steam engine named the Hogwarts Express to the Hogwarts School of Witchcraft and Wizardry. I doubt that he had any more idea of what lay before him than I do on my many detours from what others would suggest, should be the normal progress of life.

As best as I can remember from my earliest primary school days, I was somewhat of a dreamer, not in the sense of wandering off-topic, but in the sense of looking below and beyond what I was being taught. Why does my teacher believe that? Did he believe his teachers for the same reason that he expected me to believe – because he was taught by someone in acknowledged authority, and he expected me to perceive him in that same way? As I progressed from primary into high school, with admittedly quite good achievements up to that stage, my results undertook a downward trend. Even in my callow youth of the time, I suspected that something was awry. Whilst I had been very proficient in what was termed at the time, *mental arithmetic*, my progress in more advanced mathematics was unusually slow. Eventually I came to the realisation that formulae were the problem. Instead of just memorising them as examinations required, I was distracted by trying to understand them, and unsurprisingly, I was doing a very poor job of it.

Thus, mathematics, and those disciplines predicated on that competence such as physics, chemistry, geometry, and engineering, had my examination results being sub-par for a person of my assumed intelligence. On the other hand, I was demonstrating an unusual competence in storing, recalling, and correlating otherwise unrelated subjects, from poetry, history, religion, and culture, to language, song lyrics, and philosophy, not that the latter had any prominence in a strict Catholic education. Nonetheless, the subject intrigued me, because it pointed some way in the direction of my intellectual difficulties – impairment even. I am not suggesting that I have any competence in poetry, lyrics, or even literature, nor can I remember the dates of significant historical events, but I do capture the relevant timelines of themes and cultures, and fit them into a tapestry of societies as they have evolved. I have noted how ideologies have had a marked impact even on supposedly objective disciplines such as science.

I think that I was fifteen years of age at the time, or thereabouts, and was working during school holidays as a time clerk, back in the days when pay envelopes were made up by hand, and even a single farthing discrepancy was an irrefutable cause to redo the whole lot, as painstaking and time consuming as it undoubtedly was. So much so that in many cases, a shortfall was made up from the pockets of the unfortunates engaged in that activity. They wanted to go home for they had no incentive to remain, as they would not be paid overtime for their own perceived incompetence. Sadly, it was often the fault of the bank in determining the correct apportioning of coinage to make up the individual pays (there was a technical term for that process but it has departed memory). Part of my job was to tally hand-written time cards to ensure that employees were paid for their time worked, neither more nor less. The other time clerks used mechanical calculators, clacking away interminably, as keys were mis-struck and calculations restarted. I, on the other hand, simply scanned the cards and wrote down the answer. Our supervisor, noting my apparently improvised and casual approach to this, the most important of tasks, subjected me to censure for being so irresponsible. I was in no position to defend myself, nor my method. Fortunately, a senior clerk, of great experience and respect, came to my aid, pointing out that I never made a mistake and was the most accurate and competent clerk in the office. The supervisor wasn't buying it, and so subjected me to a battery of tests against whom he perceived to be, the best clerks there present. Despite my unerring accuracy, and demonstrated speed, he never came to respect me, and was glad to see the back of me when school resumed. There was a lesson there, but I had yet to learn its significance. Not surprisingly, I remember numbers far better than names, much to the embarrassment of my wife on social occasions.

My competence in language extends no further than the Romance languages, those derived from Latin in which I was required to be more than competent, as any Catholic student would know. These included English, French, Italian, and Spanish, largely I suspect because the script was similar. When it came to Greek, and other Asian and Eastern European languages, the symbols simply failed to compute, much in the same way as mathematical symbology had failed me. On another note, what I found particularly curious was my competence in comprehending poor hand writing. One of my jobs in my otherwise undistinguished military career, was to summarise hand-written personnel reports. The writing varied from barely comprehensible to truly appalling. Other officers had great difficulty with the task, sometimes I suspect, just guessing to get through the volume of work. However, my competence was soon recognised and I was often asked to interpret. This proficiency confused me: why was I so good at some symbology, but not in others?

During my time studying religions, and in particular Judaism, I was encouraged to learn Hebrew. I quickly learned word formation, grammar, and related disciplines, but had trouble recognising the alphabetic symbols, especially as so often, the script (font) varied significantly. Try as I may, it just wasn't working, and so I abandoned that avenue of education. There was another lesson to be learned, but what did it mean?

Throughout my life of what should be identified as "trade training" rather than formal academic education, I have demonstrated useful proficiency in some areas, and utter uselessness in others. As a military air traffic controller in the RAAF (Royal Australian Air Force), I was competent in the administrative aspects of Base Operations and Flight Planning; as good as any in Tower and Surface Movement Control; but struggled with Area and Approach/Departures which in those pre-radar days, required a high degree of spatial ability. This failure was one of the reasons for my resigning my commission, before my incompetence proved dangerous for all to see. Embarrassment before my peers is not something that I enjoy.

When I began studying computer programming, I quickly recognised two career options: technical and business. The former was clearly not a path that I could successfully pursue, whilst the latter proved to be entirely suited to the way my mind works. Over time, I became very competent understanding large, complex, integrated systems designed to run entire operations, mostly manufacturing, finance, and supply. Despite my lack of formal academic qualifications, so necessary today even to operate Stop/Go signs in traffic management, or being a typist/secretary, I had a very successful career including senior management roles. As an analyst in many business applications, my talent was recognised and appreciated, and even renumerated at a very satisfactory rate, thank you very much. Clearly, I was a man of my times, because today, I doubt that anyone would employ me in any related role, devoid as I am of recognised academic qualifications.

What I am getting to by narrating these experiences, which on the surface may appear unrelated, is the evidence that human minds work in a variety of strange and mysterious ways. The animal mind-brain complex develops in a predictable and reliable manner: horses walk and run; birds fly; fish swim; worms burrow; and so on. The human mind-brain complex, as it matures, is not entirely like that. Unlike all other creatures, so far as we can determine, we are the only species capable of acting contrary to natural instinct, or to put it in medical terms, we do as contra-indicated. I am especially given to that behaviour. Yes, instinct still plays a part as it does in lower-order organisms, if I can use that term without attracting adverse criticism, but humans have no natural instinct to swim rather than walk, or leap off tall cliffs expecting to fly. Ok, when very young, I did attempt such foolishness from the single-storey roof of our house, but that is an entirely different matter.

According to the scientists who have studied genetics and neurology, "a child's brain undergoes an amazing period of development from birth to three—producing more than a million neural connections each second."[1] Another quote: "From birth to age 5, a child's brain develops more than at any other time in life. And early brain development has a lasting impact on a child's ability to learn and succeed in school and life. The quality of a child's experiences in the first few years of life – positive or negative – helps shape how their brain develops."[2] I have no memory of those first few years of life, so cannot comment on why my brain developed as it did. Unsaid, in these and many other articles which are primarily focused on how to raise your baby, is that these very early neuronal connections are the basis of *instinct*, just as happens in other developing young creatures of all species.

At some stage, or stages, of neuronal pattern connections in humans, development proceeds beyond forming instinctual patterns in the biological *material brain*, to what I would term, the *immaterial* mental patterns of the *mind*. Scientists of all disciplines offer explanations as to why this is so, but what they fail to offer, with anything approaching substantiation or plausibility, is why siblings with equivalent genomes and DNA, and undergoing equivalent cultural development, often fail to develop along similar lines of character, cognitive abilities, interests, and thought patterns. Yes, health does indeed play a part in some abilities, but there is no evidence that health plays any part in the direction of ambitions or interests. My brother, sister, and I share commonality in our genomes, likely better than the 99.9% we share with all other humans, but that is where the similarities start and end.

Why is that so?

This, as the original question asks, is what has piqued my interest in the mind-brain conundrum. Certainly, some similarities are entirely predicated on biology, but so many dissimilarities cannot be so. DNA, genes, biology, etc., have no choice but to be determined and regulated by what we understand to be the laws of physics and chemistry. Brains, being entirely biological, cannot do other than what they must do. Despite the proclivity of scientists to anthropomorphise the mechanical, and forced by their commitment to philosophical materialism into inverting their thinking to mechanicalize the anthropomorphic, their logic is an utter failure of incomprehensible proportions.

Follow me if you so desire, especially if you accept the foundation of my reasonings. I am all too aware that many do not.

References:

1. https://www.zerotothree.org/espanol/brain-development

2. https://www.firstthingsfirst.org/early-childhood-matters/brain-development/

Prologue

I am not an *evolution denier*.

Rejecting contrary opinions with pejorative terms, whilst evidencing a lack of civility, is also an example of the logical fallacy of the excluded middle. In emotionally charged issues, we have the true believers at one end of the spectrum, and dogmatic deniers at the other: in between are many shades of gray. I do not reject all evolutionary processes, but I question whether the undirected *microbes to man* evolution hypothesis could be true.

Whilst scientific enquiry is generally based on deductive reasoning, the over-arching hypothesis of evolution seems more inductive, attempting to derive conclusions from a superficial interpretation of observations. Effectively this is reasoning within a paradigm which, as with religion, has its own dangers. I accept the reality of many evolutionary processes, such as genetic mutation, speciation, genetic drift, and even natural selection to some extent, but modern literature on the subject has convinced me that no-one truly knows how life could have arisen from non-life, nor how microbes could have evolved into man. I have my own reasons for doubting, which are explained in this study. To reiterate, I have well-substantiated doubts that absent of external direction, *that all life on earth arose from a single common ancestor which itself arose from an inorganic form* (GTE) Some of my reference sources may suggest to the reader that I am arguing from the stance of creationism, but such is not the case - there is no deistic, theological, or religious basis for any of my arguments. One of my goals in this study is to evaluate in the context of the evolution hypothesis, what is widely known from the sciences, and the thoughts and experiences of philosophers down through the ages.

The term "evolution" is so elastic that any discussion in this area must be preceded by a definition in context. For example, when Professor Futuyma states that *"evolution is the single most pervasive theme in biology, the unifying theme of the entire science,"*[1] is he referring to *descent by modification* in its broadest sense, or more simply genetic inheritance generation by generation? Can it really be said that biological research into common diseases cares a whit about the overarching narrative of evolution?

In his book, *"Science on Trial – The Case For Evolution"*, Professor Douglas Futuyma stated that Darwin "drew his evidence from comparative anatomy, embryology, behaviour, geographic variation, the geographic distribution of species, the study of rudimentary organs, atavistic variations (throwbacks), and the geological record to show how all of biology provides testimony that species have descended with modification from common ancestors."[2] In recent years, research into genetics and related fields has brought the basis of Darwin's conclusions into question, with some scientists now asking whether the essential mechanisms of evolution have been correctly identified, if at all. Published works by Jerry Fodor & Massimo Piatelli-Palmarini[3], Suzan Masur[4], Stephen Meyer[5], John Sanford[6], James Shapiro[7], Lee Spetner[8], and Robert Wesson[9] testify to the discussion. Putting that aside, but keeping in mind that these scientific fields of research were totally unknown in Darwin's days, there is to my mind a far more important field of research about which I have been unable to find any published works. That is not to suggest that such do not exist - simply that I have been unable to unearth them.

Of particular interest is a symposium held at Cornell University is the Spring of 2011 with the proceedings published in 2013: *Biological Information - New Perspectives*[10]. I have not studied this particular work, it being highly technical and initially very expensive to purchase, but I have reviewed *The Synopsis and Limited Commentary*[11] by Dr. Sanford, to determine the degree of overlap with the material being presented here in my own study. There is considerable overlap; for example, Dr. Oller's paper on Pragmatic Information refers to "simple mathematical fact that the number of possible strings at any given level in any natural language, or any language-like biological signalling system, grows exponentially as we progress up the hierarchy of information layers" - this supports my arguments regarding data ancestors and that data only becomes cognitive information when processed through a referential framework of concepts which provide context. Dr. Donald Johnson, with PhDs in both Computer Science and Biology, validates my comparison with digital information processing and storage, when he "demonstrates that the information networks found within living cells are remarkably similar to computer networks", although I disagree with *remarkably similar*, arguing that such is by analogy only. Dr. William Dembski, "widely known for his work in information theory and information search strategies", notes that "a search program cannot be designed to do any better than a random search, unless the designer has vital information on which to base the search". In a later chapter, I will discuss the concept of a "search space" which is framed by search arguments, and how in cognitive, as opposed to biological information processing, the wide gap between the conceptual and the physical cannot be bridged without an intelligence capable of cognitive abstraction. There is more, a great deal more, such as the limits of self-organisation, but we will come to these in their proper place.

The take-away, if I can use that term, is that most of the principles that I discuss in this study have been validated by recognised scientific experts in relation to *biological information*, and given that the mind-brain complex is asserted to be biological, thus there should be no rejection of them in relation to *cognitive information*. The important issue though, is that there is a substantive difference in the application of these two types of information, and it is this difference that I wish to illuminate in this book.

I have studied works by William Dembski[12], Daniel Dennett[13], John Eccles[14], Werner Gitt[15-16], Martin Heidegger[17], Thomas Nagel[18], Denis Noble[19], Karl Popper[20], Walter ReMine[21], Gilbert Ryle[22], and numerous essays by less well-known authors (by me, at least), but none treat the subject of information and knowledge in quite the way that I do here. I will leave it to the reader to adjudge whether I have added any new thoughts.

There are many variations, even in evolutionary terms, in the understanding of how life came to exist in its present forms. The material-monist asserts that the evolutionary processes were entirely undirected, whilst others like the former head of the US Human Genome Project, Francis Collins, perceive Divine guidance, offering a synthesis of science and theology[23]. Dr Denis Alexander, Director of the Faraday Institute for Science and Religion has written similarly[24]. More recently, Professor Edgar Andrews, an English physicist and engineer, and Emeritus Professor of Materials at Queen Mary, University of London, has added additional perspectives in his book, *"What is Man?"*[25]. This is not an arena that I choose to enter in this study, and I would like to remind readers that nothing in this study owes anything to belief in the supernatural.

Though I am not a scientist in the accepted sense, I will nevertheless attempt to tread the path of the scientist in pursuit of the evidence that is readily available. In an earlier work[26], I sought to expose the errors in Richard Dawkins' *"The Greatest Show on Earth"*[27], but in this book, I seek to expose an area of study that to my knowledge, has not been pursued with any vigour: that concerning the *origins* of cognitive, as distinct from biological, information and knowledge, and the implications for evolution theory. There continues to be a great deal of work in the neurosciences, and energetic debate over mind (conceptual) versus brain (matter) as we shall see, but such is primarily in the domain of what is happening now rather than how it came to be.

Though the evidence and reasoning in this book have contributed to my personal conclusions regarding the origin of life, I would not presume that it should so conclude for others, for each of us should carefully weigh the evidence for ourselves. This is not a science book in the accepted sense, for I am not a scientist. The target readership is people like myself, well educated in a number of fields, enthusiastic amateurs if you like, but willing and able to see through the fog of technical language and unsupported assertions to discern the truth for themselves. Of course, I would welcome readership amongst the scientific community, but such people should understand that some of the more rigorous norms of scientific publications are absent. Hopefully, so are the many unsubstantiated hypotheses that proliferate through the literature.

Finally, in the context of the mind-brain conundrum, I will seek to justify my conclusion that psychology is an incoherent pseudo-science suffering from a degree of cognitive dissonance. That is not to suggest that it lacks value, for it surely does not, but far too many conclusions are predicated on a sandy foundation. Psychologists almost invariably hold to undirected evolution, yet cannot refrain from describing the mind in the teleological terms of purpose and design. I hope that if you do not already understand this, you soon will. Similarly, those proposing the *computational theory of mind* seem not to understand some fundamentals of digital computing, ignoring a significant issue related to the architecture of the brain, and what makes digital computing possible. Either that, or in realising the significance, choose not to mention it because they cannot resolve it. We will come to that in its proper place also.

References:

1. Futuyma, Douglas J., *Science on Trial – The Case for Evolution*, Pantheon Books, New York, 1982, p. 5

2. *Ibid*, p. 36

3. Fodor, Jerry and Piatelli-Palmarini, Massimo, *What Darwin Got Wrong*, Picador, New York, 2011

4. Mazur, Suzan, *The Altenberg 16: An Expose of the Evolution Industry*, North Atlantic Books, Berkeley, CA, 2010

5. Meyer, Stephen C., *Signature in the Cell: DNA and the Evidence for Design*, HarperCollins, New York, 2009

6. Sanford, Dr John C., *Genetic Entropy & The Mystery of the Genome*, FMS Publications, Waterloo, New York, 2008

7. Shapiro, James A., *Evolution: A View from the 21st Century*, FT Press Science, Upper Saddle River, NJ, 2013

8. Spetner, Dr. Lee, *The Evolution Revolution - Why People are Rethinking the Theory of Evolution*, Judaica Press, Brooklyn, NY, 2014

9. Wesson, Robert, *Beyond Natural Selection*, A Bradford Book, The MIT Press, Cambridge Massachusetts, 1991

10. Biological Information - New Perspectives, Proceedings of the Symposium *Cornell University, USA, 31 May – 3 June 2011,* Edited by: Robert J Marks II (*Baylor University, USA*), Michael J Behe (*Lehigh University, USA*), William A Dembski (*Discovery Institute, USA*), Bruce L Gordon (*Houston Baptist University, USA*), John C Sanford (*Cornell*)

11. http://www.biologicalinformationnewperspectives.org/#!synopsis/c1294

12. Dembski, William, *Being As Communion - A Metaphysics of Information*, Ashgate Publishing Company, Burlington, VT, 2014

13. Dennett, Daniel C., *Consciousness Explained*, Penguin Press, London, England, 1991

14. Eccles, John C., *How the SELF Controls Its BRAIN*, Springer-Verlag, Berlin, Germany, 1994

15. Gitt, Dr. Werner, *In the Beginning was Information*, First Master

Books, Green Forest, AR, 2007

16. Gitt, Dr. Werner, *Without Excuse*, Creation Book Publishers, Atlanta, GA, 2011

17. Heidegger, Martin, *An Introduction to Metaphysics*, Anchor Books, New York, 1961

18. Nagel, Thomas, *Mind and Cosmos: Why the Materialist Neo-Darwinian Conception of Nature is Almost Certainly False*, Oxford University Press, New York, NY, 2012

19. Noble, Denis, *The Music of Life: Biology Beyond Genes*, Oxford University Press, Oxford, UK, 2006

20. Popper, Karl R., and Eccles, John C., *The Self and Its Brain: An Argument for Interactionism*, Routledge & Kegan Paul, London, England, 1983

21. ReMine, Walter James, *The Biotic Message: Evolution Versus Message Theory*, St. Paul Science, Inc., St. Paul, MN, 1993

22. Ryle, Gilbert, *The Concept of Mind*, Penguin Books Ltd, London, England, 1990

23. Collins, Francis S., *The Language of God*, Free Press, Simon & Schuster, New York, 2006

24. Alexander, Denis, *Creation or Evolution: Do We Have To Choose?* Monarch Books, Oxford, UK, 2008

25. Andrews, Professor E.H., *What is Man? Adam, Alien, or Ape?* Thomas Nelson Publishers, Nashville, TN, 2018

26. Talbot, Wayne, *The Dawkins Deficiency – Why Evolution is Not the Greatest Show on Earth*, Deep River Books, Sisters, OR, 2011

27. Dawkins, Richard, *The Greatest Show on Earth – the Evidence for Evolution*, Bantam Press, London, 2009

Introduction

"If you don't know where you are going, any road will get you there."

~ misquote from Lewis Carroll's, *Alice's Adventures in Wonderland*[1]
~

Lest anyone be misled, or suspect that I am attempting to mislead, I am not a credentialled scientist, in truth, not a credentialled anything in academic terms. I have nothing in the way of formal scientific training, beyond what may be termed "trade training" for the various occupations in which I have been engaged over the past half-century or so. I am an analyst, with many years of experience in commercial applications, but my level of competence I will leave to others to adjudge. In addition to the subject of this book, I have researched, analysed, and written a number of studies on religions and theology generally, such is my interest in these things as they pertain to *origins*. In my analyses of books on both theology and science, I look for unstated presuppositions which are foundational to the propositions of the authors. In seeking truth, I find it to be common that most studies start somewhere in the middle of the subject, innocently failing to substantiate their presuppositions. This is both natural and logical, for if we were all to start from the beginning, on the occasion of every study, nothing would ever be finished.

However, in my passion for understanding origins, I must always dig a little deeper than the substance of the texts before me. I accept that people genuinely believe their propositions, but are likely unaware of their foundational presuppositions, and even more often, are unwilling to substantiate them. For example, when discussing religion with Christians, I am told that they have faith in Jesus Christ as their Saviour. Now, understand that I am not being critical of Christians, nor do I intend any offence. I use examples from Christianity because that is the religion with which I am most familiar. Had I been raised in another faith such as Islam or Buddhism, I would tailor my arguments based on that experience. No Christian over the past many centuries has ever met Jesus, or the people who wrote about him. So, in the first instance, the faith of Christians is in the people who have taught them, and so on back through the ages. They have been taught that the bible before them is the inerrant Word of God, and so they must not question. They have been taught that Christians who, like me, have walked away, were never truly Christians in the first place, not having been born again through the Holy Spirit. As to a lack of faith, faith is the gift of God (so it is taught) – you either have been so gifted, or unfortunately, you have not. Studies such as Systematic Theology and the like are premised on beliefs which have been taught, and carried on through tradition to the extent of establishing a consensus.

Thus, for the vast majority, the foundation of their faith in Jesus, is their faith in the person who informed them of him, their faith in the truth of what was written about him, and subsequently interpreted to formulate the doctrine and theology of Christianity. These are their foundational presuppositions which, for many reasons, they fail to validate. Some, in their attempt to defend their faith, contend that their faith is validated by what they feel in their hearts - civility prevents me from pointing out that the same can said by people of all sects and denominations, of all religions, and thus cannot be indicative of truth. Emotion is the least reliable of human subjective behaviour. I would like to extend my apologies to my Christian friends, but you know me well enough to accept my blunt approach to such issues.

Scientists often fail similarly, having been taught in their formative years what to believe, thereafter using such "truths" as their foundational presuppositions on a subject. They seldom, if ever, revisit to substantiate the truth, even when later research openly questions what was earlier taught and believed. One notable exception is Dr. John Sanford, who has questioned and later came to reject what he was previously taught. In his book, "Genetic Entropy", he explains:

"Modern Darwinism is fundamentally built upon what I will be calling "The Primary Axiom". The Primary Axiom is that man is merely the product of *random mutations* plus *natural selection*. Within our society's academia, the Primary Axiom is universally taught, and almost universally accepted. It is the constantly-mouthed mantra, repeated endlessly on every college campus. It is very difficult to find any professor on any college campus who would even consider (or, should I say, dare) to question the Primary Axiom.

Late in my career, I did something that would seem unthinkable for a Cornell professor. I began to question the Primary Axiom. I did this with great fear and trepidation. I knew I would be at odds with the most "sacred cow" within modern academia. Among other things, it might even result in my *expulsion* from the academic world."[2] [italics in original]

I wonder how many other scientists privately reject this Primary Axiom, yet cannot do so publicly for fear of scorn, ridicule, and the threat to their careers?

Some scientific research has little or no scientific foundation whatsoever. Astronomers and cosmologist can be found searching for extra-terrestrial intelligence (SETI), in the belief that as life evolved on Planet Earth, and there exist billions of other planets, life *must* exist elsewhere. But as will be demonstrated in a later chapter, there is a mathematical challenge even to the evolution theory. As so many do, stating that given enough time, something *must* happen, is simply untrue. There is a wealth of published studies challenging Charles Darwin's ideas, often because Darwin's understanding of biology and his logical ignorance of the, as yet undiscovered, science of genetics, led him to debatable conclusions.

Thus, in this book, I will challenge conventional scientific wisdom, without necessarily offering better substantiated hypotheses. I often encounter the logical fallacy that one ought not reject a proposition without having a counter offer; I will let you ponder that for yourself. I am not discomfited by not knowing, but am reluctant to accept as truth, propositions which lack substantiation, are directly refuted by more reliable evidence, and/or more plausible propositions.

George Sarton (1884-1956) was a Belgian-born American chemist and historian, considered by many to have been the founder of science history as a discipline worthy of study. I entirely agree with him, as science history helps us to understand how science managed to achieve its largely unchallenged standing in modern society. This quotation exemplifies his thinking and that of many scientists, and why science has taken the power that it has:

"Truth can be determined only by the judgement of experts ... Everything is decided by very small groups of men, in fact, by single experts whose results are carefully checked, however, by a few others. The people have nothing to say but simply to accept the decisions handed out to them. Scientific activities are controlled by universities, academics and scientific societies, but such control is as far removed from popular control as it possibly could be."[3]

I would offer that Sarton paid insufficient attention to the corruptible nature of power when one asserts that *knowledge is power*, and how men very quickly find ways to use that power to their own advantage, most commonly to protect their own reputations. His idea of results being carefully checked by a few others, found its form in Peer Review, which whilst once being a reputable practice, has since sadly degenerated into *Pal Review* according to many dissenting scientists. He likely did not foresee the rapid dissemination of knowledge beyond the hallowed halls of academia, nor the watering down of academic standards as education became a commodity to be sold. From experience, I find this earlier observation by G.K. Chesterton to be more prescient:

"The Fabian argument of the expert, that the man who is trained should be the man who is trusted, would be absolutely unanswerable, if it were really true that a man who studied a thing and practiced it every day went on seeing more and more of its significance. But he does not. He goes on seeing less and less of its significance."[4]

One of my goals is to offer perspectives on the true significance of some scientific discoveries.

Logic and Reasoning

"Why, sometimes I've believed as many as six impossible things before breakfast."

~ The White Queen, *Alice's Adventures in Wonderland*[1]~

One of the things that I have learned from other enquiries, is that whilst scientific reasoning and logic are tremendously important, they are not equipped to deal with questions of the mind. As one wag quipped, *"How does the mind know that it is?"* Each component of our bodies has its genesis in the genes passed down from parents, but is the peripheral or parasympathetic nervous system sentient? The brain has the same material source, but is said to be sentient – how can that be? All scientific inquiry depends on accepted scientific laws which, if they are to be implemented in reliable technology, necessarily exclude the possibility of exceptions. Nature's constancy is what allows for scientific inquiry, and why the results of the latter can be so powerful. But the mind deals with, and in truth originates, what we might call "revelations"; these demonstrate an existence beyond the demonstrated order of things until they are instantiated, e.g., music and poetry.

This is also true of logic. Logic is a difficult concept to pin down, for one must ask: What is logical about logic? In most cases, it refers to reasoning which is conducted or assessed by strict principles of validity. But how do we determine "validity"? One could offer that validity is determined or affirmed by logic, but now we have a circular argument, leaving the question open to interpretation. Let me confess that the kind of revelation that I am attempting to explain, *conceptual thinking*, may not in fact be logical: I am using my mind to explain my mind, which contradicts my first axiom[5] as we will discuss.

It is important to be aware that we are taught to think in terms of categories and sameness, a train of thought inherited from the Greeks. Hebraic thought patterns can be disarmingly different, especially as regards anything extraordinary, as we find in the Hebrew Scriptures. In Western thought, the truth of these mystical experiences is automatically excluded and considered impossible. What this really means is that the possibility of a non-material existence, as I claim is the realm of the mind, is not so much rejected because of any proper *a-priori* reason, but rather because we have been indoctrinated to believe that whatever cannot be demonstrated via the scientific method, must not be taken seriously. *Philosophical materialism* dominates Western science, with the scientific method the only path to truth.

But to my mind, this is clearly a fallacy.

A common issue that I find in debating contentious subjects is peoples' unwillingness to deal with ambiguity, leading them to logical fallacies such as false alternatives and straw man arguments. When I deny the Christian view of Jesus, I am told that I am arguing for agnosticism. The Christian difficulty is that in believing Jesus to be God, arguing against Jesus is seen as arguing against God – thus the accusation of agnosticism. When debating the mind-brain conundrum, one response was similar in offering a form of a straw man argument, to wit – "I could posit a more Tolkienesque version of the 'minds are beyond materialism' line of thinking, and who could gainsay me?" In the context of the discussion, this revealed that my opponent was not objectively considering the conundrum as I had proposed it, that simply, the mind was likely not material, but was contending with a deeper subject reflected in earlier parts of the conversation: whether or not God exists. Evolutionists do something similar when they reject the notion of Intelligent Design, whilst at the same time, other scientists use that very proposition in SETI. Intelligent design is the foundation of their thinking in the search – the real issue is acceptance of the identity of the designer. Being comfortable with ambiguity is an essential skill for any enquirer; objectively acknowledging a proposition without requiring the comfort of an answer, and more particularly, not being afraid of what the answer might be. The latter is why people rush to offer an alternative, usually in the form of a straw man which they can ridicule. The scientific method is what I attempt to follow. I observe a phenomenon, test hypothetical solutions against what is known, and reject those solutions that fail to explain the phenomenon. I do not require a solution, although I would like one.

I accept that in many cases, the best I can do is to understand what the phenomenon is **not**.

I was recently asked whether I knew a lot about climate change, but as that requires a subjective answer, I responded that I had no idea, because I did not know how much was known across the scientific community to be able to give either a quantitative or qualitative answer. However, I said that I was likely in a better position to understand scientists than most people, and even perhaps some scientists. The issue is that scientists use common terms in uncommon ways, understood as meta-language. For example, when Stephen Hawking said (as reported), that in the presence of the laws of gravity, the Universe can and will create itself out of nothing, an acceptance of his credentials indicated that he did not mean what he appeared to mean. Laws do nothing, and gravity relates to masses, so by mentioning gravity, he implied masses. His "nothing" is a quantum vacuum, a space in the lowest possible energy state with generally no particles, but nevertheless particles jump in an out of existence and when in existence, gravity gets them together and builds a universe. Now, I know nothing about quantum anything, but that is not the point.

In a similar vein, well-respected evolutionary biologist, Professor Douglas Futuyma stated that *"evolution is the single most pervasive theme in biology, the unifying theme of the entire science."*[6] He was both right and wrong: right in a specific context, and wrong in others. In medical research, microbiology and genetics are the primary fields of research, including genetic inheritance, but only over a limited number of generations. That other animals share a significant percentage of our genome is also useful for experimentation. However, medical researchers care not a whit about the *"microbes to man*" theory, and in fact, many researchers believe in Divine Creation just a few thousand years ago.

My point is that unless one is aware that a meta-language underlies and explicates the language of scientists, one is likely to be an unwitting disciple of The White Queen in Alice's Adventures in Wonderland, unaware that they *sometimes believe as many as six impossible things before breakfast.*

The Logic of Criminal Prosecution

I introduce this issue to explain why I think as I do, and why I conclude, or fail to conclude, and why I can be entirely comfortable with not knowing. When someone is suspected of a crime, the authorities may choose to not proceed to trial because of there being insufficient evidence to be confident of a conviction. Pragmatically, the guardians of the law choose to put such cases behind them, and move on to more pressing matters. In my studies of the New Testament, for example, I chose to ignore the claims of miracles, because only presuppositions could lead to a conclusion – there is no evidence, or logic, to conclude of their truth either way. Thus, I ignored them. I take the same approach with some aspects of evolutionary theory, where there is insufficient evidence to even proceed to a detached analysis.

Where I do find evidence or reason, I then proceed to analysis, or trial in this analogy. I either start with my own hypothesis (accusation), or that of others. I use the same logic that ought to be followed by members of a jury, although I suspect from personal experience of such matters that few jurors either comprehend, are willing, or are even capable, of such intellectual activity. My point concerns reaching a conclusion. In a jury trial, where members of the jury are not convinced beyond a reasonable doubt that the prosecution has proven its case, then a verdict of "Not Guilty" must be returned. This is not equivalent to a verdict of "Innocent", although it is often perceived as such; not guilty means *not proven*.

On the subject of the overarching theory of evolution, *microbes to man* as it were, the charge is claimed to have been proven: evolution is guilty of creating humankind in its current form. This may be a majority opinion by scientists, but not unanimous, and becoming less so over time. I would contend that the trial of evolution's guilt more resembles a *kangaroo court*, hastily assembled before there was sufficient evidence on which to base the accusation. Charles Darwin was unaware of the existence of so much contrary evidence. The authorities should not have proceeded to trial, but as is so often the case, politics and personal ambition won the day. In my opinion, and that of truly well-credentialled scientists, a retrial is overdue.

Thus, in this study, I suspend belief on the truth of the *microbes to man* theory, and offer the evidence and reasoning why I cannot, as yet, find undirected evolution guilty as charged. I cannot dogmatically assert that evolution is innocent; on the contrary, I do believe that evolution is complicit to some degree, but to what degree continues to be beyond my comprehension. That I offer no other suspect as the cause of the human condition is not sufficient reason to accept evolution as the culprit. We may never know the true culprit, but it is better that the guilty go free, rather than the innocent by sentenced as a last resort.

Correlation vs Causation

"One of the first things taught in introductory statistics textbooks is that correlation is not causation. It is also one of the first things forgotten."

~ Thomas Sowell, The Vision of the Anointed: Self-Congratulation as a Basis for Social Policy ~

My point here is that we need always be watchful for confusing or conflating the two. That we see activity in the brain associated with thoughts or volitional activities is correlation. To assert causation, there needs to be scientific proof. If scientists arbitrarily exclude possible causes, simply because they prefer to not accept the possibility of such a cause, then they succumb to confirmation bias.

I chanced upon this narrative, but omitted to secure the source, a failure which occurs far too often as I get lost in ever-increasing levels of detail. For those with IT experience, you may recall the concept of a *push-down pop-up stack* – that is where I spend a lot of my time. You may choose to accept or reject the truth of this example.

"Back when I was a medical student (in the Cretaceous Period) we were taught that someone once did a study comparing folic acid levels in the blood of cancer patients compared to the blood of healthy patients. The cancer patients had, on average, significantly lower folic acid levels. And the ones with the largest, fastest growing tumors tended to have the lowest folic acid levels. "Aha," they thought. "Something about folic acid deficiency predisposes them to cancer. We should give folic acid to cancer patients." Bad idea. A randomized trial showed that cancer patients given folic acid died sooner than those given placebo.

What happened? **Low folic acid levels are a consequence, not a cause, of cancer.** Folic acid is needed to synthesize DNA, and DNA synthesis is necessary for one cell to divide into two cells. So folic acid gets used up by rapidly dividing cells – like cancer cells. Giving cancer patient folic acid just gives their tumor a helping hand. (This roundabout insight led to medications that block folic acid metabolism which are used as chemotherapy to this day.)

The main lesson here is that correlation tells us almost nothing about causation." [emphasis in original]

I have experienced similar situations in my work environments over decades, so I believe that this is a common human predilection. Here is another example, with potentially serious ramifications – climate change. Associated with this is the claim of a 97% consensus amongst scientists, but when examined closely, it is difficult to find the substance of to what the scientists (and many were not) were said to have agreed. This online review[7] is worth your time, should you be so interested.

Research by Negation and Elimination

"An argument is seldom about what it is about"

~ Jewish Wisdom ~

I sense, but cannot know, that some opposition to considering the mind as immaterial, evidences the logical fallacy of the false alternative. The opposite to atheism is deism, but the opposite to material is not deism, it is immaterial. Immaterial does not imply the existence of God, although it would seem that some seem to fear that is does. It has been pointed out that science is ill-equipped to investigate God, because God is spiritual, but science only has the tools to investigate the material. One cannot argue with that. However, whilst science cannot directly investigate the immaterial, it can do so indirectly, by negation, as it so often does in other phenomena.

Scientific research is predicated on *causation*; observing an effect and seeking to understand the cause. The cause is always predicated on known properties. When observing a pattern on a rock which appears to closely represent, for example, an animal, water and wind erosion are eliminated as causes because they are not known to behave like that. Similarly, a sand-castle on a beach, or a pile of stones arranged in an intricate pattern – these evidence intelligent design, rather than being chance arrangements caused by wind or wave action.

This logic of negation and elimination should be used to investigate the mind-brain conundrum. When an effect is observed and a cause proposed, the properties of the cause need to be investigated as to whether such properties could cause that outcome. If no scientific link can be established, then either the capabilities of the properties need to be re-evaluated, or an alternate explanation should be sought. The properties and capabilities of biological entities are reasonably well understood. As research in chemistry and microbiology proceeds apace, more is understood concerning why cells work as they do, and the nature of signal transmissions in neuronal networks. In researching the relevant sciences, I have found nothing that even suggests that these behaviours owe their cause to other than chemistry and physics, and no suggestion of how the material can offer a bridge between the *physical* activities of the brain, and the *conceptual* activities of the mind.

I contend that if scientists are to claim intellectual integrity, they must endeavour to eliminate improbable causes of mental activity, by evaluating the properties of biological entities in terms of their capabilities. I am convinced, by the published scientific knowledge of today, that the biological is incapable of conceptualising, as happens in the mind.

Summary

Adopting the logic hopefully used in Courts of Law, I will structure my defence against the charge that undirected evolution is solely guilty for the human condition, as follows:

1. Foundational Propositions and Axioms.

2. The Exceptional Human Condition.

3. Foundations of Knowledge.

4. Principles of Information Processing.
5. Science and Evolution; and
6. The Mind-Brain Complex Conundrum.

To begin at the beginning, so to speak, in a following chapter I will offer the foundational propositions upon which my thinking is based.

A notable failure in many evolutionary text books and apologetics, is the focus on individual components and functions, to the exclusion of discussing how these must work in a system. Whilst it may be argued that individual components are not *irreducibly complex*, the same is seldom if ever true of systems. By way of example, my first installation of solar panels had a five-year warranty on individual panels, and a ten-year warranty on the entire system. A panel failed after six years, but the supplier refused to replace it under the individual panel warranty. However, the output of the remaining panels was insufficient to kick-start the inverter, so effectively, the system was down. I successfully argued that the failed panel should be replaced under the ten-year system warranty. Each panel was complex, but the system was irreducibly so.

Many of my ten primary axioms, to be later discussed, are systems oriented, and for good reason: the human body is composed not just of organs, but of complex communication systems, most especially as used in our sensory organs. Evolution theory must deal with this issue, but in this study, I will explain why I believe that it cannot.

I can hear your minds whirring already, but that is good, for these are complex subjects with many subtleties, and objections which need to be answered. I shall do my best to cover all bases, and to do this, I will make use of computer analogies where appropriate, which may be foreign to many readers, most especially if one's appreciation of computing is recent where much of what is happening underneath is black-boxed. This term refers to processes which are hidden from the observer, who only sees what goes in and what comes out – what happens in between being a mystery for most, particularly if they have no need to know. Our brain is, perhaps, the most frustrating example of a black box, as we shall see. As one who started in computing at a time when it was *essential* to know, because *nothing* was black-boxed, I have chosen to attempt to understand the brain in that same way. Of course, I cannot know whether my analogies are apposite for everybody or not. That notwithstanding, I will take the reader on a journey through the entrails of information processing. Let me start with a temptress, to motivate you to read further.

In the simplest definition: *information is that which informs*. Clear enough, but in truth, it is an illusion. Take, for example, a traffic light with green, amber, and red lenses. One can state that when driving along a road and you observe that the red light is lit, that the traffic light is conveying information that you must stop. Logically, it can be argued, it must have communicated information because you have been informed. However, book a passage with Dr. Who in the TARDIS, carrying the traffic light with you, and travel back in time to the days of the Roman Empire. Anywhere will do, as long as you can find a road. Erect the traffic light, wait for evening, and light a candle behind the red lens. Will chariots, soldiers, and passers-by stop? Perhaps, but probably only out of curiosity. The traffic light does not communicate traffic information to an ancient Roman, because said chap does not have pre-knowledge of the purpose of a traffic light. It might convey something to him, just what I can only guess, perhaps he might think it a message from the gods. But my point is that he has no context, and without it, he will not get the message. Keep this in mind, as I shall expound on it further.

Previewing my arguments: I will contend that the *mind* and the *brain* are not one and the same thing, and that the scientific contention that *"the mind is an emergent property of the brain"* is as one scientist confessed, "nothing but a label for our ignorance!" This will lead to a discussion on why AI (Artificial Intelligence) will forever remain as artificial as war time coffee, which, as we know, was not a satisfactory substitute for the real thing. When we turn to biology and genetics, I will offer amongst other facts, evidence that a single cell contains more "information" than found in DNA, a quite recent and significant discovery.

Along the way, we will review the works of numerous experts on the subject of information, how the mind works, and how language developed. Not being an expert myself, I am limited in this review to examining whether their studies acknowledge my primary axioms. It is common for scientists to begin with foundational presuppositions without offering evidence to substantiate them. Often, they are little more than wishful thinking, because without foundations, their hypotheses lack wings. As Samuel Coleridge put it, *"He that would fly without wings must fly in his dreams"*. Evolution theorists are experts at this game, and to be fair, the religious are not exempt. On the latter, I have published my own theological studies, ever conscious of the fault that I find in the writings of others. Hopefully, I will not make the same errors here, becoming a victim of my own theories and overlooking many issues.

Revisiting the maxim at the beginning: *if you don't know where you are going, any road will get you there*, it has application in two senses. Firstly, undirected evolution has never known where it was going, it had no purpose, no goal, and no chosen path. That it arrived somewhere, very special, by chance, must be considered highly improbable, yet here we are. Secondly, I offer the maxim as an explanation for the parlous state of the evolution hypothesis. Not knowing where they were going, in terms of a comprehensive understanding of the human condition, evolutionists have wandered off into numerous dead-ends, yet despite the evidence before them, remain reluctant to admit that the path they have chosen was never going to get them there. Charles Darwin misunderstood the complexity of cells and the genome, and as I have contended in other works, had he known then, what we know now, I doubt that he would have proposed the mechanisms of evolution as he did. History may have had little cause to remind us of him beyond his voyages and his work as a naturalist. In a later chapter, we will discuss the evidence that has led many scientists to proclaim that *Darwinism is broken*!

I trust that this has stimulated your interest, sufficient to have you continue turning the pages. If you like a good mystery story, hopefully you will enjoy this as a welcome contribution to that genre.

References:

1. Carroll, Lewis, *Alice's Adventures in Wonderland*, Macmillan Classics Edition, New York, NY, 2014

2. Sanford, Dr John C., Genetic Entropy & The Mystery of the Genome, FMS Publications, Waterloo, New York, 2008, pp. v-vi

3. Sarton, George A.L., Introductory essay, in J. Needham, ed., *Science, Religion and Reality*, Braziller, New York, NY, 1955, p. 12

4. Chesterton, G.K., *The Twelve Men*, Tremendous Trifles, 1909

5. Axiom: a statement that is taken to be true, to serve as a premise or starting point for further reasoning and arguments.

6. Futuyma, Douglas J., *Science on Trial - The Case for Evolution*, Pantheon Books, New York, 1982, p. 5

7. https://www.youtube.com/watch?v=ewJ6TI8ccAw&feature=youtu.be&fbclid=IwAR0AUnmRHd-4vE8j154-yolFHHXE--BdnsCZPolhmMkr56H6AA4rsAXCyy8&app=desktop

In Defence of Philosophy

"Man is the only creature who refuses to be what he is."

~ Albert Camus ~

And there, in a nutshell, is my defence of philosophy, and why I contend that the mind has an existence independent of the brain, at least whilst our bodies remain alive. A recurrent irony is materialists claiming to not believe in philosophy, being unaware of making a philosophical statement in doing so. Belief *is* philosophical. Philosophy is defined as "the study of the fundamental nature of knowledge, reality, and existence, especially when considered as an academic discipline, or a theory or attitude that acts as a guiding principle for behaviour." Materialism is a philosophy, because it acts *as a guiding principle for behaviour*, especially in the pursuit of science.

All belief systems are philosophical in nature. A belief can also be defined as an attitude regarding whether a proposition is true or false. Quoting from this Wikipedia entry[1]:

"Mainstream psychology and related disciplines have traditionally treated belief as if it were the simplest form of mental representation and therefore one of the building blocks of conscious thought. Philosophers have tended to be more abstract in their analysis, and much of the work examining the viability of the belief concept stems from philosophical analysis.

The concept of belief presumes a subject (the believer) and an object of belief (the proposition). So, like other propositional attitudes, belief implies the existence of mental states and intentionality, both of which are hotly debated topics in the philosophy of mind, whose foundations and relation to brain states are still controversial.

Beliefs are sometimes divided into core beliefs (that are actively thought about) and dispositional beliefs (that may be ascribed to someone who has not thought about the issue). For example, if asked "do you believe tigers wear pink pyjamas?" a person might answer that they do not, despite the fact they may never have thought about this situation before.

This has important implications for understanding the neuropsychology and neuroscience of belief. If the concept of belief is incoherent, then any attempt to find the underlying neural processes that support it will fail."

It is incontrovertibly true that *the existence of mental states and intentionality* are hotly debated topics, especially in relation to brain states. But propositions, by their very nature, express purpose and intentionality – they are expressions of concepts that have no physical reality until instantiated. In each and every chapter in this book, we cannot help but bump up against philosophical propositions, as we encounter opinions on various aspects of reality, most especially concerning the interrelationship of the mind and brain. I would like to impress upon the reader that none of us are free from philosophical thoughts, for we all have beliefs.

A question to be pursued is: Do we have substantiation for our beliefs?

As an aside, if you are interested in philosophy but have been discouraged by academic texts on the subject, I can highly recommend the two books listed below. They are easy to read, and are both entertaining and informative.

References:

1. https://en.wikipedia.org/wiki/Belief

2. Phillips, Christopher, *Socrates Café: A Fresh Taste of Philosophy*, Norton Paperback, New York, NY, 2001

3. Phillips, Christopher, *Six Questions of Socrates*, W.W. Norton & Company, New York, NY, 2004

Foundational Propositions

"You can't build a great building on a weak foundation. You must have a solid foundation if you're going to have a strong superstructure."

~ Gordon B. Hinckley ~

I apologise for the number and length of these opening chapters, but I am attempting to avoid some deficiencies that I find in the works of others; namely, that assertions and conclusions are not substantiated in the study itself. I frequently find myself asking why expressed beliefs are held and assertions made, and what evidence has been used to substantiate them, if any. Could it be that as with religions, a belief system establishes the paradigm within which all such thinking occurs? Can we draw an analogy with computer logic being "black boxed", such that all dependent logic is based on a paradigm which is hidden from the observer? The evolution hypothesis has been so well established as a paradigm for practically every aspect of the human condition, that no contrary argument can succeed without challenging the foundations of the paradigm itself. In a sense, that is not as difficult as it may at first appear, for in most cases, the evolution paradigm has no foundations other than the paradigm itself. So, here, I will set about establishing foundations that will undermine those of evolution.

The Nature of Things

In researching scientific studies on the human mind, I struggle to accept the coherence of any of the explanations I have encountered, because they all fail to identify what it is that they are referring to: what is the mind? In Substance Dualism, it is separate from the brain and of a difference substance (form); another offering is that it is an emergent property of the brain; and another, the mind is not the brain but what the brain does. Interestingly, way back in the 12th century, philosopher Maimonides offered an explanation which somewhat parallels that approach, but not entirely:

"Man, before comprehending a thing, comprehends it *in potentia*. When, however, he comprehends a thing, e.g., the form of a certain tree which is pointed out to him, when he abstracts its form from its substance, and reproduces the abstract form, an act performed by the intellect, he comprehends *in reality*, and the intellect which he has acquired in actuality is the abstract form of the tree in man's mind. For in such a case the intellect is not a thing distinct from the thing comprehended. It is therefore clear to you that the thing comprehended is the abstract form of the tree, and at the same time it is the intellect in action; and that the intellect and the abstract form of the tree are not two different things, for the intellect in action is nothing but the thing comprehended, and that agent by which the form of the tree has been turned into an intellectual and abstract object, namely, that which comprehends, is undoubtedly the intellect in action. All intellect is identical with its action: the intellect in action is not a thing different from its action, for the true nature and essence of the intellect is comprehension, and you must not think that the intellect in action is a thing existing by itself, separate from comprehension, and that comprehension is a different thing connected with it; for the essence of the intellect is comprehension."[1]

Here, Maimonides is focusing on the abstraction of realities as represented in the mind, and how comprehension is a function of intellect, not the physical brain as he explained further in his works (not quoted above). In a later discussion on our faculty of sight, I contend from the evidence that whilst the eye "sees", the brain can only "visualise", for it has no interaction with light. Notice how this parallels Maimonides' understanding from centuries ago: "It is therefore clear to you that the thing comprehended is the abstract form of the tree, and at the same time it is the intellect in action".

Philosophically, and perhaps more pragmatically, when we encounter inexplicable contradictions, or questions that are unanswerable, we realize that we have come to the end of the road. We then have no choice but to resort to what is intellectually possible, and thus, such is the path that I have sought to pursue.

Definitions

To apply rigour to what follows, I offer the following definitions:

Entity – something that exists in physical reality, at whatever level of simplicity or complexity. Entities are identified by their:

a. Properties.

b. Activities; and

c. State.

At one time, it was thought that the molecule was the simplest form of matter, but scientific research has taken us ever deeper into the complexity of the material. As this knowledge has increased, so too has knowledge of the properties of these lower-order entities, and what activities are both permitted and regulated by these inherent properties. It is these activities, in conjunction with the activities and states of associated entities, that determine the state of an entity in time and space. Fortunately for the reliability of technology, materials only do what they can, and *must*, do. Some evolutionists understand the implications of this when they discuss the concept of free will. For example, William Provine, a Biology Professor at Cornell University asserted: "There is no way that the evolutionary process ... can produce a being that is truly free to make choices."[2] Provine's thinking was at a higher level of causation, stating as he did: "Modern science directly implies that the world is organised strictly in accordance with mechanistic principles", but his thinking clearly paralleled mine, in this instance at least. "*Mechanistic principles*" implies that materials, alone or in conjunction with other materials, can only do what they *must* do as determined by their properties and their then current environment. This being so, there is no opportunity for choice, although ironically, scientists in their endeavours to explain biological behaviour, will often resort to purpose and choice in their explanations. In the context of the mind-brain conundrum, scientists cannot allow the mind to be a separate, immaterial entity, capable of choice – to do so would undermine philosophical materialism, and that would not do. This is why there is so much confusion in scientific literature concerning the mind: the mind evidences choice, amongst other amazing qualities, none of which can be attributed to the behaviour of the biological brain, limited as it must be, to mechanistic principles. Scientists understand that the material can only do what it must do, even if they are not prepared to express the issue in those terms. In this investigation of the mind-brain conundrum, I will often resort to this axiom.

Conceptually, entities are identified by names assigned to them by intelligent agencies – the entity cares not a whit whether it has a name or not. Whilst "a name" is a physical entity instantiated symbolically, "name of" is not a physical entity that exists in reality – it is just a conceptual label that we apply to an entity. This can be demonstrated by the fact that people can assign different names to the same entity, but the nature of the entity does not change. You cannot walk into store and ask to buy a "name" (well, maybe you can), just as you cannot walk into a store and buy a "one", because these are just arbitrary descriptors which intelligent agencies apply to realities. As we will discuss a little later, entities are not data – entity names are data, but data is conceptual, not physical.

Another aspect of entities which needs to be understood for later discussions, is that all entities have a relationship to some level of physical activity. A stone or rock is described as inert, or in a state of inertia, because it has a passive relationship with its surroundings: Newton's first law states that every object will remain at rest or in uniform motion in a straight line unless compelled to change its state by the action of an external force. Whilst this is true of rocks themselves, there is nevertheless a great deal of sub-atomic activity occurring within – little particles whizzing about as identified by the atomic number of its constituent elements. Thus, at some level, all material entities have a relationship with physical activity – anything which does not is not a real physical entity, and thus must be immaterial in nature. It stands to reason that unless science can identify the intrinsic physical activities of anything, such as the *mind* or *time*, then they cannot be real entities in accordance with philosophical materialism.

In discussions on the mind, it is treated as any, or all of, *entity*, *property*, or *activity*, depending on the point the author is attempting to make: logically it can only be one, not simultaneously all three. I would suggest that the cause of such confusion is the attempt to explain the mind within the paradigm of philosophical materialism, refusing to acknowledge that it simply does not fit. As an aside, the same confusion can be found in religious discussions on God. Maimonides, a renown Jewish philosopher of the Middle Ages, asserted that God being infinite, and us finite, God is unknowable, and thus we can only speak about God in negation; i.e., what God is not, rather than what God is. The author of Psalm 19, believing in God as Creator, wrote "the heavens declare the glory of God", referring not to God as an *entity*, but to the *activities* of God which might lead one to a conclusion concerning the *properties* of God. Curiously, evolutionists seem not to admit to such thinking even in the material world.

Evolutionists will often assert that "the mind is an emergent property of the brain", whilst in truth having no science to explain this phenomenon. In later chapters, we will discuss the concepts of *self-organisation* and *emergence*, and I will explain why it is entirely invalid to apply such concepts to the mind-brain conundrum. One observation that can be offered, however, is that as many scientists do distinguish the *mind* from the *brain*, it is thus incumbent upon the scientific community to identify which properties or characteristics are separately discernible. I believe it to be useful here to invoke Leibniz's law of the *indiscernibility of identicals*. J.P. Moreland puts it this way:

"In general, if 'two' things are identical, then whatever is true of the one is true of the other, since in reality only one thing is being discussed. However, if something is true of the one which is not true of the other, then they are two things and not one. This is sometimes called the indiscernibility of identicals and is expressed as follows:

$$(x)\,(y)\,[(x=y) \rightarrow (P)\,(Px \leftrightarrow Py)]$$

For any entities x and y, if x and y are really the same thing, then for any property P, P is true of x if and only if P is true of y."[3]

To apply these principles to the mind versus brain conundrum, we need to approach the problem from a different perspective, lest we simply affirm the consequent. Rather than start with the presupposition of separateness, we need to identify specific properties and ask of each: can this be a property of a physical organ such as we know the brain to be, or are these characteristics of a phenomenon incapable of explanation by the physical sciences? If all properties or phenomena can be explained by reference to the material brain itself, then no discernible separateness of the mind and brain can be asserted.

Taken from the perspective of Maimonides on *essential* and *accidental* properties, drawing a chart comparing these for the mind and brain reveals the essence of the mind, versus the essence of the brain, and highlights the disparities. On the basis of Leibniz's law, the mind and brain cannot be of the same substance for they do not have the same properties.

We must approach discussions on the mind-brain complex with this in mind (excuse the pun). We accept the reality of the activities of the mind, but know not the *how* or *why*. We can to some degree fathom the workings of the mind, and the products of the mind, but not much beyond that. We will discuss this further in later chapters, but the point to note is that whilst we can identify the physical properties of the instantiations of the mind, such as music, poetry, etc., we cannot identify the physicality of either the properties or activities of the entity that we believe to exist – the mind. It is the same conundrum that we find in both cosmology and evolution theory. We observe the phenomenon that we and the universe exist, and seek to reverse engineer back to the origin of the process, without having substantive evidence of that process. We observe the out-workings of the mind, and seek to explain them, but constrained by the paradigm of philosophical materialism, we can but speculate without substantiation, *flying without wings* as it were.

Observation, Aspects, and Appearance

This returns us to the subjects of correlation, causation, and especially *identity*. In a rather mysterious fashion, what we "see" in the mind's eye is a conceptualisation based on a stream of electro-chemical signals targeted to the visual cortex. If, for example, the object seen is a brass door-knocker which exists as a real entity, you will not find that object existing beyond itself. In the entire communications sequence: from light reflecting from the object onto our eyes; the light energy transduced to a complex, multiplexed, specifically encoded, neural signal stream; with further signal processing aligning multiple neurons in the brain; which the "mind" somehow decodes back into a mental image of the object; everything is only an encoded, symbolic, *representation* of the real thing. As we will discuss later, even the location and arrangement of the relevant neurons is more of a mystery than originally considered.

As Maimonides commented, the representation of the object in our neural network is not the same as the object itself, and thus is not of the same *identity*. There is definitely a *correlation* between the object and the neural arrangement, we can accept that. But can we ascribe *causation*? Well, we can accept that there is a *relational* cause, but not a *direct* cause. Throughout multiple signal transitions in the neural system, various biological protocols and encode/decode systems come into play, which determine the final neuronal arrangements. "Seeing" is the first cause of the sequence, but the object itself cannot be said to be the cause of the neuronal arrangement – that is a function of biology. This all seems logical, but the reason I mention it is to refute those who argue that experience and neural activity are one and the same. It should be apparent that practically any shared experience will not result in the same neural activity – there is far more going on which distances the actuality of the experienced event from whatever neural activity results in the brains of each individual experiencing that event. Thus, the event, whilst initiating a response, cannot be said to be the determining cause of the specific neural activity.

Which brings us to the "double aspect" theory, as espoused by American philosopher, John Searle[4] amongst others. The proposition is that the experience itself, and the neural activity in the visual cortex, are but two *aspects* of the same thing. In our example above, the suggestion is that the neuronal representation is just another aspect of the brass door-knocker. This is plainly silly. "Aspect" refers to a particular feature of something, or a position of something in a stated direction. In both cases, they are properties of the entity itself. Neural activity associated with an object or experience, is no more an aspect of that object or experience than a reflection in a mirror is an aspect of an object – they are two entirely separate entities with their own properties and behaviours. The reason for this absurd proposition was to argue that observing brain activity through neural imaging was to observe consciousness. *Consciousness* is one of those terms which has been subject to redefinition over the years, and seems to take on whatever meaning one wants it to have in a particular context. In one form, it is "awareness or sentience of internal or external existence". Extended, it can be "the state of being aware of, and responsive to, one's surroundings", or "a person's awareness or perception of something". So, in observing neural activity, of what is the brain said to be conscious? I strongly doubt that any *f*MRI (functional MRI) scan could make such a determination. It can be argued that a level of neural activity is indicative of a level of consciousness, such as asleep or awake, but little more than that. The specific activity of consciousness cannot be determined by mapping neural activity: consciousness is a state.

There is a relationship between what we observe, hear, or otherwise experience through *exteroceptors*, those receptors responding to a stimulus in the external environment, and neuronal arrangements in the brain. However, we know neither the process whereby those neuronal structures are determined, nor the intricacies of the structures themselves. An observed entity has its own intrinsic nature, but the encoded symbolic *representation* in the brain is an entity in itself, related but physically independent of the reality represented.

Again, we know that in many cases, a shared observation or experience will result in different neural patterns in each person's brain, because they report the observation or experience differently. *Appearance* is a cognitive conceptualisation determined by many factors beyond the reality of the experience itself.

Neural Activity versus Consciousness

Some working in the neurosciences contend that not only is neural activity *necessary* for consciousness, but it is also *sufficient*. It is expressed by others that nerve impulses actually cause consciousness. In the material that I have researched, there is a lack of definition as to the type and source of nerve impulses. From the nature of the discussions, I gained the impression that the theme concerned neural activity stimulated by *exteroceptors*, those receptors that respond to our external environment, such as the somatosensory receptors that are located in the skin. However, I remember reading of experiments concerning sensory deprivation by suspending a person in water in a darkened, sound-proof room, so that there could be no stimuli of the five (six) senses. The results of these experiments showed that short-term sessions of sensory deprivation can actually be beneficial, in that they were relaxing and conducive to meditation. In such cases, one has to ask: What is the causation of the meditation process absent of any external stimulus? On the other hand, extended or forced sensory deprivation can result in extreme anxiety, hallucinations, bizarre thoughts, temporary senselessness, and depression. Not surprisingly, sensory deprivation has been used as a method of torture.

What does that tell us? Firstly, the subjects were still conscious without external sensory stimuli, refuting any argument that such stimuli are the necessary and sufficient cause for consciousness. However, neural activity was continuing in the brain - there can be no disputing that such activity is *necessary* for consciousness, for without it a person would likely be pronounced as being dead. However, I suspect that is not the issue being raised as expressed in the opening sentence. Raymond Tallis addresses this subject in his book, "Aping Mankind"[5], based on his medical experience, and we will review some issues where appropriate, but here I wanted to convey one primary contention: consciousness is *conceptual*, not *physical*, even though there are necessary physical manifestations of consciousness.

Let me repeat: without brain activity, there is no consciousness – the two are indisputably related, but they are not the same thing.

Consciousness relates to the mind, and as evidenced in sensory deprivation experiments, the relationship between the brain and the mind is extremely fuzzy, to put it mildly. We know that sensory experiences lead to cognitive awareness, but curiously, a shared sensory experience does not necessarily result in the same cognitive awareness, evidence that other factors come into play. We also know that volitional cognitive activity leads to physical activity, further evidencing the two-way relationship of the mind, which originates the volition, and brain, which conveys the necessary instructions via motor neurons to the relevant parts of the body.

An Aside

I have found it interesting that the fundamentals of the natural world have been a point of curiosity, and indeed study, as far back as extant literature can take us. In this case, I would offer these comments by the aforementioned medieval Sephardic Jewish philosopher, Moses ben Maimon, aka Maimonides or Rambam (1135-1204).

Maimonides states that "Every description of an object by an affirmative attribute, which includes the assertion that an object is of a certain kind, must be made in one of the following ways."[6] These he gives as (1) by definition; (2) by part of its definition; (3) by something different from its true essence; (4) by its relation to another thing; and (5) by its actions. In Maimonides' way of thinking, a thing is what it is, and everything that we say about it is an addition. In describing a man, we could say that *"man is man"*, but apart from being a tautology, it adds nothing to our knowledge. Thus, we need to define man, which we do in biological, physical, moral, intellectual, occupational, and other terms.

I am given to taking that thinking to biology, and all aspects of the human condition. A thing is what it is, i.e., an entity. What differentiates one entity from another are properties and activities, or as Maimonides puts it, by its description, definition, and actions. Aristotle thought along similar lines, but using different terminology, and gave us his philosophical theory of *hylomorphism*, which conceived *being* as a compound of *matter* and *form*. Matter, which corresponds to my term, *entity*, is defined by its composition, such as a molecule is formed from atoms, and bricks are formed from clay. I do not fully understand where he was going with his use of the term, *form*, in that he speaks of substantial and accidental form, the former being essential to essence of a thing, and accidental being non-essential. I believe that he was on a different path to the one I am treading here. He applied his theory to living things, defining a soul as that which makes a living thing alive. There has been considerable debate over just what Aristotle meant when applying hylomorphism to living bodies, most especially humans, and if you are interested in following the debate, this online entry[7] is useful.

Revisiting my definition of an entity, its properties and activities are *essential*, whilst the names we give to entities are *accidental*, as they are not intrinsic to the entity. The point that I wish to emphasise is that our investigations into the mind-brain conundrum must consistently apprehend these truths.

The Information Misconception

Firstly, we must differentiate between *genetic* information and *cognitive* information – they are not at all similar. With regard to the former, I prefer the term, genetic *instructions*. There is a common misconception that brains and computers store information, and that communications convey information. In a philosophical sense they do, but in the technical sense, they do **not**. If we are to properly understand the mind-brain conundrum, we need to define our terminology more precisely. Communications and storage, whether inorganic, organic, or biological, are *physical*, whereas data and information are *conceptual*. The material world consists of entities, whose activities are allowed, constrained, and otherwise regulated by their intrinsic properties, the intrinsic properties of their components, and those of other material entities with which they interreact in their then current environment. This all happens whether we humans are aware of these things or not. The history of science is the history of increasing awareness of what is happening in the physical world, at ever increasing levels of detail and complexity. I believe it a truism that the more we learn, the more we reveal our ignorance. In the biological sciences, the more that is learned, the more an explanation for the origin of life recedes from our grasp.

A defect that I have found in my research is the lack of attempt to differentiate between *cognitive* information, and *biological* information. Quoting Stephen Meyer, "DNA contains data in a digital form, with its four chemical subunits (called nucleotide bases) functioning like letters in a written language or symbols in a computer code."[8] I am unsure of why Meyer used the term, *digital form*, but of one truth I can assure you: the biological constructs of DNA are not digital, and any such analogy is inappropriate. Nucleotide molecules (base pairs) are the building blocks of DNA and RNA. The analogy of letters is acceptable, but only because scientists have assigned the letters A, C, G, T, representing the chemical compounds Adenine, Cytosine, Guanine, and Thymine, with the codon sequences (of three) written using those letters. The point to note is these are symbols conceptually devised to represent real entities. If you would like a brief overview of nucleotides, this online source[9] is useful, but take note of the chemical complexity of each compound. By analogy, we can say that they function like language, but we must be very wary of allowing the *analogy* to replace our understanding of the *reality*.

A Data & Information Primer

No doubt you are somewhat familiar with MP3 and MP4 files, but perhaps not consciously aware that MP3 is short for MPEG Audio Layer-3. MPEG is the acronym for the *Moving Picture Experts Group* that developed the international standard for encoding and compressing video images, so that movies can be squeezed into ever smaller storage systems. Similarly, large music files can now be made more portable: a 30 MB music file on CD can be compressed into a 3 MB MP3 file. Notice the word, "encoding". Everything stored on magnetic or similar media is **EN**-coded, and must be **DE**-coded before it is human comprehensible. It should be obvious that such storage devices do not store pictures, music, or any sounds, but only encoded representations of those realities. The same applies to what people loosely refer to as *data* or *information* in computers. Computers do not, and cannot, store data or information, but only coded representations of data, but not information. The relevance of this to the mind-brain conundrum is that just like computers or any digital storage device, the brain is material, and thus only capable of storing symbolic representations of data. Whilst the encoded symbols can themselves be considered as data, they are not the same as the data they are intended to represent.

Human awareness is what establishes data – essentially the names we give to identified entities, properties, activities, and states. They all exist whether or not we give them names, and hence names are accidental properties of the realities to which they refer. Information is at the next level of cognitive awareness, where we correlate data into conceptual patterns, at times being entirely independent of, and at other times, entirely contradictory to, any physical reality. Knowledge is a form of information, being a retained or retrievable pattern, but as it has its origin in cognitive correlations, it can be true or false. People are known to have false memories (I suspect that many of my childhood memories are of that nature). In communications, of any form, through any medium, neither data nor information are conveyed. Communications, whether via biological or other mediums, deal only with physical entities such as electric currents, sound waves, or electro-chemical interactions. At best, they are symbolic representations of data, as I will later demonstrate. Whilst data can be encoded into symbols, information can only be represented by correlations, such as we find in many data storage strategies including networks.

These facts lead to the following definitions:

1. *Symbol* – physical entities used in communications, and in the storage of encoded representations of data.

2. *Data* – conceptual understanding of physical entities.

3. *Information* – conceptual derivations, often transient, from the correlations of data.

Two authors, whose works I have studied, misuse the term, *information*, as I understand it and have defined. In his book, "*In the beginning was information*", Dr. Werner Gitt asserts that "information ... is a fundamental entity on equal footing with matter and energy."[10] I entirely disagree and will later explain in detail my disagreement. In a subsequent work, "*Without Excuse*"[11], Gitt introduced the term, *universal information*, and again I disagree. There are universal entities, with properties and activities, but cognitive information only exists when there are intelligent entities capable of comprehending such entities. Information is conceptual, not physical. William Dembski takes a different approach in his book, "*Being as Communion*"[12], where he deals with the meta-physics of information. I find myself having greater agreement with Dembski, for as he says, "intelligent agents convey information to other intelligent agents by making meaningful statements within a system of language" (p. 17). I contend that he still does not have it quite right, but he is closer. Information theorist, Douglas Robertson, opined that the defining characteristic of intelligent agents (i.e., teleological causes that act for an end or purpose) is their ability to create and communicate information[13]. He is right in a philosophical sense, in that only intelligent agents can conceptualise information, but technically, what is communicated between intelligent agents is symbolic representations of conceptual data. I suspect that the primary cause of misunderstanding is a failure to differentiate "information" which can be said to be present in material substances, and that derived cognitively. Cognitive information is derived from patterns of conceptual data, and is always subjective, being subject to the pre-knowledge, beliefs, and worldview of the one informed. For all people, including me from time to time, these are the contents which are "black boxed" in our mind which we seldom attempt to access, even if we are able to.

I do not expect the reader to immediately accept these definitions, nor to immediately grasp their significance in the context of the mind-brain conundrum, but I am confident that later all will become clear. If we are to properly comprehend the difficulties with evolution theory, we must precede our discussions with such precise definitions.

Approaching the subject of information from a philosophical perspective, information is not simply to *have*. It is there to be converted into something much larger than itself; it is there to produce ideas that make sense of all other data gathered in order to move it forward to higher latitudes. Information is not there to be possessed, but to be *comprehended*. Somebody commented, I cannot remember who, that in educational institutions of today, we are producing a generation that believes its task is to tend potted plants, rather than plant forests. We offer our young people prepared experiences in which we tell them *what* to think instead of teaching them *how* to think. We rob them of the capacity to learn what thinking is really all about. Young people have fallen victim to the disease of information for the sake of information itself. The ideas of great thinkers are presented to them as information, instead of as challenges to their own thinking, or as prompts to the development of their own creativity. Far too often, information is reduced to an indisputable truth statement, such as we hear from evolutionists. It is too brief, unsupported by proper arguments, and yet still presented as "the answer." From my philosophical perspective: *questions are the true answers*, but I will leave you to ponder that for yourself. Alternatively, you can wait for the sequel to this book to be published.

In the context of the exceptional human condition, to be discussed in a later section, *the quest for certainty paralyzes the search for meaning*. It is *uncertainty* that is the very stimulus impelling man to unfold and exercise his intellectual capacity. If, as many claim, we have no free will, then any quest of our own would make Don Quixote's tilting at windmills a far superior endeavour. In passing, this mythical chap is a hero of mine.

The religious, for all their failings, often expound with profundity, wisdom applicable to both the religious and secular alike. Put aside, if you can, the religious context, and consider the message in these two offerings:

- "We desperately need to heed what Kierkegaard said about Christianity: The greatest proof of Christianity's decay is the prodigiously large number of [like-minded] Christians."[14]

- "Insight has been replaced with clichés, flexibility with obstinacy, and spontaneity with habit. What was once one of the great pillars of Judaism—the esteemed value of spiritual, intellectual and moral dissent—has become anathema. Instead of teaching the art of audacity, we are now educating a generation of kowtowers."[15]

Whilst leaders in both Christianity and Judaism express alarm at the intellectual decay of their respective flocks, we have no reason to believe that this is other than a human trait, in both the religious and secular – atheists are as prone as theists. From another perspective, evolutionism and scientism are in their own ways, practised with the same mindlessness as religions.

Primary Axioms

I herein contend that the following axioms are irrefutable:

1. Nothing can explain itself.

2. All knowledge is built upon prior knowledge.

3. Before you can know what someone is talking about, first you must know *what* they are talking about.

4. No coding system or communications protocol can devise or maintain itself.

5. No coding system or communications protocol can be implemented other than by an external agent at two or more nodes simultaneously.

6. No measurement device, nor sensory system, can calibrate

itself.

7. All systems must undergo a process of verification and validation to be implemented reliably.

8. A system, of whatever simplicity, is irreducibly complex.

9. Nothing can be done without prior knowledge that it can be done (knowing that you can); and

10. In a mechanistic world, as defined under philosophic materialism, entities can only do what they *must* do as determined by their properties, and those of other entities with which they interreact.

At this stage, you, the reader, are likely inclined to challenge the truth of these axioms, especially #9. I welcome such critical thinking, for it has long been my belief that we should own truth for ourselves, rather than simply accept what is truth to others, living vicariously through their beliefs. If you cannot accept these axioms, you are unlikely to accept my arguments based on them. I can only encourage you to read my later explanations, consider them deeply, and apply them to what you believe about evolution.

Finally, a further word about psychology, as many of my study references are by psychologists, or by scientists writing in that context. I contend that psychology is *a product of the mind about the mind*. If you accept my primary axiom that *nothing can explain itself*, then clearly, there is a conundrum which, whilst we are unlikely to ever solve, we should at least be aware of the implications. Psychologists, in general, sidestep this issue by conflating the concept of the mind, with the physicality of the brain, as they must do when they believe that evolution is the determinant of the whole human condition. They could be right, but they cannot know that they are right, just as I cannot know that they are wrong. However, if we are to seek the truth, we ought to at least evaluate the evidence in as detached a manner as possible, rather than within a paradigm, lest we succumb to confirmation bias.

Forewarned is to be forearmed, and thus I endeavour to avoid that trap.

References:

1. Maimonides, Moses, *The Guide for the Perplexed*, Digireads Publishing, 2018, translation from the original text with annotations by M. Friedlander, Trubner & Co., London, UK, 1881, pp. 179-180

2. Johnson, P., *Darwin on Trial*, 2nd ed., Illinois, USA, 1993, p. 127, quoting William Provine

3. Moreland, J.P., *Scaling the Secular City*, Baker Books, Grand Rapids, MI, 1987, p.83

4. Searle, J., *Intentionality: An Essay in the Philosophy of Mind*, Cambridge University Press, Cambridge, UK, 1983

5. Tallis, Raymond, *Aping Mankind*, Routledge Classics, New York, NY, 2016, Chapter 3

6. Maimonides, *Ibid*, p. 143

7. https://en.wikipedia.org/wiki/Hylomorphism

8. Meyer, Stephen C., *Darwin's Doubts: The Explosive Origin of Animal Life and The Case for Intelligent Design*, HarperCollins, New York, 2013, pp. v-vi

9. https://en.wikipedia.org/wiki/Nucleotide

10. Gitt, Dr. Werner, *In the Beginning was Information*, First Master Books, Green Forest, AR, 2007, p. 11

11. Gitt, Dr. Werner, *Without Excuse*, Creation Book Publishers, Atlanta, GA, 2011

12. Dembski, William, *Being as Communion - A Metaphysics of Information*, Ashgate Publishing Company, Burlington, VT, 2014

13. Douglas Robertson, "Algorithmic Information Theory, Free Will, and the Turing Test", *Complexity* 4(3) (1999), 25-34, cited in Dembski, p. 48

14. Thulstrup, M.M., "Kierkegaard's Dialectic of Imitation," in *A Kierkegaard Critique*, ed. H.A. Johnson and N. Thulstrup, (New York: Harper, New York, NY, 1962, p. 277

15. Cardozo, Nathan Lopes, *Jewish Law as Rebellion: A Plea for Religious Authenticity and Halachic Courage*, Urim Publications, Jerusalem, Israel, 2018

Part 1: Primary Axioms

"Definitions are the foundation of reason. You can't reason without them."

~ Robert M. Pirsig, *Zen and the Art of Motorcycle Maintenance*: An Inquiry Into Values ~

The subjects which I attempt to explain in this book are of such complexity, that from the outset, I must confess that I have but an incomplete understanding. My ideas have been developing over time, as earlier published in *"Information, Knowledge, Evolution and Self"*[1]. I have since undertaken a great deal more research and puzzling to produce the publication of this study, but I am still far from finished with it. One of my biggest difficulties is translating thoughts, which appear to be coherent in my mind, into words on paper. Many of the metaphors and analogies I choose simply fail to convey my thoughts to others. However, for me, if for no-one else, it provides a foundation upon which I can continue to work.

In the previous chapter, I previewed some foundational propositions, which I shall begin to explain in more detail, and substantiate them as best I can.

References:

1. Talbot, Wayne, *Information, Knowledge, Evolution and Self: A question of origins*, Xlibris, Bloomington, IN, 2016

Chapter 1-1: Primary Axiom #1
Nothing Can Explain Itself

"Progress occurs where truths are questioned."

~ Aniekee Tochukwu Ezekiel ~

A principal belief underlying all scientific research is *causation*: that each and every observed or hypothesised phenomenon is an *effect*, has a *cause*, and that such cause is not explained by the phenomenon (effect) itself. In intellectual endeavours, such as reading, no word can explain itself, and no sentence or text can explain itself. Language, alphabet, words, semantic range, phraseology, concepts, grammar, colloquialisms, and the like, must all be learned and applied before sense can be made of the written word. Similarly, with our senses: whilst we experience sights and sounds, and even touch, taste, and smell, we cannot identify these realities until an external agency explains them to us. This is of significance for evolution, as we shall see.

It is often the case that imprecise language is the direct cause of misunderstanding, leading to over-ambitious scientific pronouncements. A case in point, in the context of this study, is the word, *information*. It is almost universally believed that information can be communicated via speech, written words, pictures, or other visual signals. In truth, it *cannot*, and the following chapters will attempt to substantiate two axioms that underpin that truth, beginning with:

1.   Nothing can explain itself.

It is believed that to *communicate* is to share or exchange information, news, or ideas. This is technically incorrect. I explain this more fully under later Primary Axioms, but what passes between us in any exchange are encoded, symbolic representations of conceptual data. Mentally, we deconstruct ideas into their component data elements, encode them into language, transmit them symbolically using a method appropriate to the communications medium, whereupon the recipient decodes them into their corresponding data elements, and the mind correlates them with pre-existing data to cognitively construct the concept that the sender attempted to communicate. As we all know, often the recipient "fails to get the message", and I shall attempt to explain why.

Circa 1830, Samuel Morse and others developed the telegraph, the first technology enabling long-distance communication. It was primitive by modern standards, but effective nonetheless. A person would convey their message requirement via voice, or a paper form, to a telegraphist, who would mentally encode the message into Morse, and then tap on a key, transmitting electronic signals to a nominated location, where another telegraphist would decipher the message. Included in the prerequisites for this process was a shared language between the originator and telegraphist, and the telegraphist's knowledge of Morse Code. Morse Code is a system of short and long signals, often referred to as dots and dashes, or dits and dahs. Thus, dit (pause) dit-dit-dit (pause) dit-dah-dit-dit, or (. ... ._..) decodes to ESL. The telegrapher activates the telegraphic key in this pattern, sending electronic signals down a wire to a receiving key, which audibly repeats the input pattern. In later versions, the received message was encrypted onto paper tape.

The question becomes: Has information been conveyed between sender and receiver?

Well, not really: all that was communicated were symbols in the form of electronic signals. The same happens when someone speaks into a microphone, and the corresponding sounds (hopefully) emanate from loud speakers. Most would have experienced sounds from a loudspeaker that are muffled or cannot be reconciled with the voice of the orator. The technology failed at some point, and we should keep this in mind when we consider biological equivalents. Again, the speakers do not convey information, just sound waves characterised by wavelength, frequency, amplitude, time-period, and velocity. Each of these characteristics need to be decoded by the listener. What you are seeing on these pages are coded symbols collectively known as an alphabet. Arranged, or coded, into certain patterns, these arrangements are known as words. Words are then arranged in strings known as phrases and sentences, further arranged into paragraphs, chapters, and books. This is still not information: such arrangements are encoded, symbolic representations that will prompt conceptualisations in our minds. The point to note is that as with voice or electronic signals, these arrangements of symbols cannot explain themselves. They are coded messages: to decode them, the receiver has to pre-know a great number of things - not without reason is English one of the most difficult languages to learn. Even more significantly, and we will get to this under Primary Axiom #4, what we term our five senses: sight, sound, touch, taste, and smell, have sensors which detect particulars of our external environment, but from the beginning, these sensors could not have explained their purpose, nor what it was that they were attempting to convey. At the other end, the brain, could not have known either.

Bear in mind this observation: "Even a banknote would just be a *blank* note, a worthless piece of paper, if humans didn't have a mental conception of money."[1] The bank note, as with all other objects, cannot explain itself.

You may be familiar with the *Cosmological Argument*: the argument that claims that all things in nature depend on something else for their existence (i.e., are contingent), and that the whole cosmos must therefore itself depend on a being which exists independently, and is itself *uncaused*. Most commonly, this is often used to evidence the existence of God. Either the cosmos had a beginning, or it did not, in which case it is eternal – infinite in time, but this creates more problems than it solves. If the cosmos did have a beginning, did it create itself out of nothing, and is that even logical: to create itself, it must first exist. Off topic, I confess, as that is not the subject of this study, but I introduced it here to get your *little grey cells* working – very little is what it appears. My purpose with the side-track is to get you thinking about beginnings, the very origins of everything and anything, especially as this relates to the source and usage of knowledge, the topic of Primary Axiom #2.

For now, understand that nothing can explain itself – the explanation of anything and everything must lie outside itself. I once had cause to berate, quite gently of course, a Technical Writer who in composing the User Instructions for a system we were developing, provided the following explanation for the term, *part number*: "The part number of the item". True!

When it comes to chemistry and biology, I am attempting to only wade in very deep waters, a process unlikely to yield useful results, but I shall try. In a sense, chemicals can explain themselves to other chemicals, insofar as their intrinsic properties determine their interactions – each "knows" about themselves and each other. But chemicals are not sentient - they are not self-aware. As chemicals combine to form amino acids which, under very particular conditions, combine to become proteins, and eventually form cells, nothing has been added beyond chemical complexity. From the perspective of sentience, the same limitations that apply to individual chemicals apply to any combination of chemicals. If an individual chemical cannot explain itself, then logically, no combination of chemicals can explain itself. If biology is a combination of chemicals, and the brain is constructed of biological materials, then the brain cannot explain itself. I acknowledge that in certain circumstances, the whole is greater than the sum of its constituents, but that cannot be the case here, as similar physical constraints that apply to individual chemicals also apply to chemical compositions and derivatives thereof.

As I read somewhere, perhaps even in one of my own earlier publications: You can't prove a system of mathematics from within the system, and you can't derive an information-rich pattern from within the pattern. The information in a book, for instance, cannot be derived from the paper and ink used to print it. It is impossible to bootstrap a book from the bare ingredients. Kurt Gödel, in his Incompleteness Theorem, expressed something similar:

"In any axiomatic system, if the system is consistent, it cannot be complete, nor can the consistency of the axioms be proven with the system, since the system itself is 'part of the problem'. We can only know something to be true when regarded by an observer outside the system."

Thus, this great intellectual agrees with my Primary Axion #1: nothing can explain itself.

Keep this in mind as we attempt to understand the mind.

References:

1. Verschuuren, Gerard, *What Makes You Tick? A New Paradigm for Neuroscience*, SOLAS Press, Antioch, CA, 2012, p. 33

Chapter 1-2: Primary Axiom #2
All Knowledge is built upon Prior Knowledge

> "No man can reveal to you nothing but that which already lies half-asleep in the dawning of your knowledge"
>
> ~ Khalil Gibran ~

That quotation does not completely address what I am attempting to explain, but it does serve as a useful introduction.

There is a common term, *a meeting of the minds*, referring to the situation where there is a common understanding between parties. Whilst mostly used in a legal context, it is applicable to all communications. Paradoxically, for you to know what the other person is talking about, first you must know what he/she is talking about (Primary Axiom #3). You must have shared knowledge of terminology, concepts, and context, and we need to ponder how this could have come about via evolution where sensory systems were the only source. As it is unlikely that a solution will ever be found to that conundrum, we can at least review the process in our own lives, which will partly explain why evolution cannot explain it.

The branch of philosophy concerned with the theory of knowledge, is called *epistemology*, "from the Greek *epistēmē*, meaning 'knowledge', and λόγος, *logos*, meaning 'logical discourse'. There are four areas: (1) the philosophical analysis of the nature of knowledge and how it relates to such concepts as truth, belief, and justification, (2) various problems of skepticism, (3) the sources and scope of knowledge and justified belief, and (4) the criteria for knowledge and justification."[1] I touch on each of these in this study, but of initial interest is the *source of knowledge*, and later, the criteria for *justified belief* (authentication). A good analyst will attempt to follow an answer with "why", endeavoring to arrive at a foundation. In the context of understanding the source of knowledge, an analyst must always ask: How do you know that? The goal of such enquiry is to have our beliefs match objective reality. In philosophy we find the term, *virtue epistemology*, which, according to The Cambridge Dictionary of

Philosophy, is a "subfield of epistemology that takes epistemic virtues to be central to understanding justification or knowledge or both. An epistemic virtue is a personal quality conducive to the discovery of truth, the avoidance of error, or some other intellectually valuable goal." With no undue modesty, epistemic virtue is a trait I seek to cultivate, having no preference for what the truth should be, wanting only to know what it is, as uncomfortable and disconcerting as such discoveries may be. If you are of a mind to cultivate this trait for yourself, set aside a great deal of time, for it is far more complex than you might imagine.

In passing, I would comment that during the content evaluation of this manuscript by an earlier publisher (not the one who published this version), those making decisions regarding the validity of quotations repeatedly stated that "It is solely and wholly the responsibility of an author to ensure that the content of their works meets all applicable laws and regulations in areas including, but not limited to, libel, pornography, hate speech, and fair use / copyright." In brief, the rule for *fair use* is that no quotation should exceed 10% of the original article, or 150 words, depending on the length of the article. Now, I must confess that in a few instances, I failed to comply with these rules and corrected my errors when pointed out to me. On the other hand, the content evaluators stubbornly refused to accept that there could be no breach of the fair use / copyright rules where the authors and publishers explicitly stated that there were no restrictions on the use of their material published online. I referred them to this website[2] concerning OER (Open Educational Resources) which I would have thought self-explanatory, but apparently not. Based on my extensive experience in computer software and business systems analysis, I had my suspicions as to why we reached an impasse for several months: I challenged them on my suspicions but they never responded. My point in mentioning this unfortunate and aggravating experience is that in my opinion, those responsible for content evaluation had no commitment to *virtue epistemology*, or even perhaps, intellectual integrity. Where you find errors in these pages, or even just suspected errors, as you may well do, understand that none are intentional or even just due to intellectual laziness: I have done my best to be true to my own intellectual values. On the other hand, perhaps, *methinks I protest too much*.

Continuing as before this distraction, the logical corollary to Primary Axiom #2 is that knowledge had an origin, which raises the question: what was the cause of this first unit of knowledge, and how was it evaluated without complementary knowledge? Put another way, what was the first sensory experience that caused the brain to store the encoded symbolic representation in a neural network, why did it do so, and how did it conceptually understand the external reality that it represented? When further sensations arrived, what caused the correlation in a conceptual sense, if they happened, with the previous or any prior sensations? A difficulty that evolutionists have in dealing with arguments concerning mind versus brain, is that they do not know what they know, or how they came to know it, when dealing with the first knowledge ever acquired by an organism, and most especially humans.

Our education system has a number of predefined levels, building from pre-school to university. The reason is obvious; for example, one cannot use a medical textbook as a reading primer for children in kindergarten. This second axiom is foundational to all education systems, although I suspect that few are aware of the deeper implications, which is where we are heading in this study. Returning to our example of telegraphy, what pre-knowledge was required of the receiving telegraphist? Firstly, he (they were always men in the early days) had to understand that a chattering key meant that someone was attempting to communicate. To you, this may seem to be blindingly obvious, but I hope to convince you otherwise. For example, if a five-year old boy, with no pre-knowledge of telegraphs, wandered into an unoccupied office and noticed the chattering device, would he understand the purpose, or would he, as I was accustomed to do at that age, attempt to dismantle it to discover what made it tick? (Our house was littered with dismantled mechanical devices) The issue here is knowledge of *context*, which we will delve into later.

Let me give an example to which we will later return from a different perspective. My dictionary defines an *elephant* as a: "large pachyderm with proboscis and long ivory tusks". Without pre-knowledge of the terms, pachyderm, proboscis, ivory, and tusks, this description adds nothing to our knowledge. It would be unusual to learn about these terms, or an elephant, without visual aids, although the term *pachyderm*, meaning a very large mammal with thick skin, would likely be avoided. My point is that knowledge of an elephant *could* be learned by first gaining knowledge of its composite parts, although that would be unusual. Other subjects, however, must be learned from the ground up: mathematics, engineering, medicine, and biology being prime examples. My belief is that anyone interested in reading this book would already understand that latter point, thus obviating the need for clumsy explanations on my part.

Wisdom is defined as the quality of having experience, knowledge, and good judgement. It need not be stated that it would be unusual for a child or adolescent to be generally wise, simply because he/she has not had the opportunity to gain sufficient foundational knowledge and experience. Even many adults never exhibit that trait. It is said that one learns from experience, but that can only be true if a person has particular knowledge to apply to that experience: some will learn much, some not much, and some will learn entirely the wrong lesson altogether. The reasons are many, but at the heart of the process is not just knowledge, but the differentiation between *true* and *false* knowledge. Not everything that we think we know or believe is true. The issue for evolution is this: how could a biological process validate any "knowledge" acquired through sensory inputs? The obvious answer, offered by "experts" and others, is trial and error, but that is faulty and even wishful thinking. Firstly, "trial and error" is a teleological process, and one is forbidden to speak of teleology in the context of evolution. Secondly, the organism would have to know, or suspect, that its knowledge was false, before it would have reason to question or reject it. I trust that you can see the difficulties with that. Questioning generally only arises from doubt, a feeling of uncertainty or lack of conviction, based on other knowledge or convictions. If an organism has none to begin with, it is essentially gullible and will believe anything, true or false. Philosophers, on the other hand, question everything as a matter of course, whether they believe the truth of a proposition or not – they seek to know how or why they know of or believe it, which is a teleological process which is antithetical to the evolution hypothesis.

Now, one may argue that this is all explained by natural selection. Those organisms which correctly interpreted sensory inputs would survive, and those that did not, did not. That sounds plausible, until one considers how an organism *could* interpret correctly. With no pre-knowledge whatsoever, a sensation, which is nothing more or less than an electro-chemical impulse, or series of impulses, will forever remain just that – a physical sensation representing an external reality. To *interpret* is to explain the meaning of something, which cannot be done without knowledge relevant to the subject. *Meaning* is a conceptual term, and as should be obvious, physical phenomena have no meaning in and of themselves, and are incapable of conveying the conceptual to other physical entities. The embryonic brain of evolution theory was nothing but a physical entity, incapable of understanding itself, let alone making sense of its external reality through sensory inputs.

Truly, this is the *chicken-and-egg* conundrum of evolution – how could a mind, which deals only with the conceptual, arise from a brain which could only deal in the physical? This may not be scientific proof against evolutionary theory, but it is certainly a logical path that must be trodden in the search for truth.

How Computers Compute, and Sometimes Do Not

"Computers use logic, but not all programmed logic is logical."

(Me ... from experience)

To begin, programmers assign names to the real entities which they wish to represent in their computer program. For example, part number, customer number, etc. The software compiler encodes these terms in binary: ones and zeroes. In binary logic, these are interpreted as – "on" or "off", with nothing in between. Technically inexact, but endeavouring to keep this simple, these data names are recorded on magnetic media by magnetising very small regions where each region represents a "bit", set to *on* or *off*, and logically treated as *one* or *zero* for computational purposes. Printed, an eight-bit group, known as a byte, might look like this "01110011". This is meaningless for most people, but for those of us who started our computing career using binary, this could be meaningful, but only if we knew which standard was being used at the next level of encoding. Debugging programs involved a core (memory) dump onto paper, of varying heights depending on the size of memory. Things got a little easier and the dumps smaller, when they were at the next higher level of translation, EBCDIC (Extended Binary Coded Decimal Interchange Code) used on IBM, or ASCII (American Standard Code for Information Interchange). Programmers were now faced with character sets of either letters and numbers (EBCDIC), or just numbers (ASCII). Curiously, we learned to read these almost as easily as you are reading these words, but it required education (pre-knowledge) and practice. These second-level codings were further decoded into letters of the alphabet, and then words depending on the language required. I have no idea of just how languages such Chinese and Japanese work, but there must be an equivalent process.

The point that I am emphasising is that from a physical perspective, computers store neither data nor information. They store encoded symbolic representations of names assigned by the programmer to represent real entities and states. Different programmers, especially of different local languages, will assign varying names to the very same real entities and states. What should be evident from this is that as computers work with arbitrarily assigned data names, they are in truth dealing with conceptual entities, not real entities. The correlation with real entities occurs at the human interpretation level where the observer has access to the definitions of the data names. Individually, these names only have meaning in context, which the software application provides, but which the computer software itself does not understand – all meaning and logic is provided by the programmer, which is where the trouble starts - *not all programmed logic is logical*. Debugging programs written by other programmers could be one of the most frustrating tasks imaginable.

Computers of any construction, including the biological brain, work only at the physical level, using encoded symbolic representations of conceptual data. There are numerous interpretive steps to get from there to cognitive information or knowledge, all of which are external to the physical storage. The same is true of any form of communication.

The lesson here is that in any communication originating from a sensory device, or from a storage device, and that includes our brains, there are one or more levels of encoding at one end, and decoding at the other. Any and all communications occur at the symbol level, depending on the type of transmission medium; in short, they are encoded symbolic representations of conceptual data. The receiver must be of a type appropriate to the sending device, and most importantly, there must be a shared communication protocol. To decode any protocol, both sender and receiver must have common pre-knowledge to advance the communication to the next level. This must be taught, or learned, from another source, either from an already educated agency, or from the intelligent agency that devised the protocol. This again, is where the evolutionists struggle, as explained in the next chapter on Primary Axiom #3.

Computers can only do what they are programmed to do: even heuristic learning requires the programming of parameters and their weighting or relevance. Despite what you may have been told, computers cannot autonomously learn anything. When scientists attempt to compare the human mind-brain complex with computers, and develop such concepts as the "computational theory of mind", they ignore a foundational truth: a computer requires intelligently designed software and architecture to be capable of doing anything. It is this truth which also tells us why sometimes, computers fail to compute as intended. Programmers, and I was one, are fallible. Evolution, I contend, is even more fallible in this context.

In the early days of commercial computing, before the advent of database technology, files were organised sequentially or indexed-sequentially, depending on the method of storage. There were three categories of files: master, application, and transaction files. A transaction file, as the name suggests, contained data on activity such as payroll, inventory, etc. The master (reference) file provided additional data concerning the entities in the transaction, and the application file was used for computation of results and the storage of those results. During one period, IBM found that the most common source of error in programs was incorrect file matching. As the inhouse developer of software tools in the days when there was no third-party software, one of my tasks was to optimise coding standards to minimise errors, and so I developed standardised code segments which reduced file matching errors to zero. This experience was invaluable in developing my understanding of information processing, in a way not evident to modern practitioners. As needs must, they often use black-boxed routines, much as modern technology utilises boxes labelled "No user serviceable parts inside".

My point is that information is derived from intelligent correlation of conceptual data in context. Context is also conceptual, derived from the correlation of other data that is already acquired. Data is the conceptual representation of real entities. To arrive at a correct understanding of anything, the mind must perform a series of conceptual file matchings irrespective of how the symbolic representations of data are stored. Because even the storage of such symbolic representations is subject to the same cognitive processes, it is inevitable that some correlations will be false. In short, all cognitive processes are subjective, whereas all biological processes are objective in accordance with the constituent chemical and physical properties.

The best programmers not only understand logic, programming languages, and factors related to the hardware and software, but also the commercial or technical subject details to which their programming relates. Sometimes, the most difficult task was not just demonstrating that your program did what it was supposed to do, but that it did *not* do what it *ought not do*. The process here was to develop scenarios in which the software would be used, including attempts to replicate how it would be misused. Far too often, intelligent technocrats failed to appreciate that their products would be used by people with a greater, equal, or lesser understanding, often pushing the software beyond its design boundaries. We will return to this issue later in the context of the claimed evolution of the mind, the point being that without scenario testing, computational processes lack validation. The paradox is this: how could biological evolution develop an intelligence which is greater than that demonstrated by those exercising that intelligence? In an article he wrote in 2006, renowned anti-theist Sam Harris once asked a similar question without, it would appear, appreciating what he was saying:

"As many critics of religion have pointed out, the notion of a Creator poses an immediate problem of an infinite regress. If God created the universe, what created God? To insert an inscrutable God at the origin of the universe explains absolutely nothing. And to say that God, by definition, is uncreated, simply begs the question. (Why can't I say that the universe, by definition, is uncreated?) Any being capable of creating our world promises to be very complex himself."[3]

Here, I am not arguing for the existence of God. I am pointing out that it is illogical to argue against the possible existence of a being, who is sufficiently complex to be responsible for creation, by acknowledging that such a being *promises to be very complex himself*. The complexity of God, far greater than anything in the Universe, is used by deists as an argument in support of their position, but it is illogical for atheists to use the same argument for the contrary position. As for asking: Why can't I say that the universe, by definition, is uncreated? The simple answer is that one can if one so chooses, but that is a philosophical statement, not a scientific one. Sam Harris does have a tendency to shoot himself in the foot. Of course, the irony here is that the biological processes of evolution are said to have serendipitously managed to achieve a level of intelligent design beyond the capabilities of what was created. Does that not sound similar to the creationist God scenario?

For now, hopefully, I have explained sufficiently why these first two axioms are fundamental to our understanding of knowledge, and how it is derived. Keep these two in mind: nothing can explain itself, and all knowledge is built on prior knowledge.

Simple as it sounds, the implications are profound.

References:

1. https://en.wikipedia.org/wiki/Epistemology

2. https://rmit.libguides.com/openeducationalresources

3. Harris, Sam, *The Language of Ignorance*, 15th August 2006, http://www.samharris.org/site/full_text/the-language-of-ignorance

Chapter 1-3: Primary Axiom #3
Before you can know what someone is talking about, first you must know what they are talking about

"The trouble with her is that she lacks the power of conversation but not the power of speech."

~ George Bernard Shaw ~

As a part of this study, I purchased a book on the Human Nervous System[1]. I was expecting to encounter many unfamiliar medical terms, but I knew not the half of it. In the early chapters, I spent as much time with a medical dictionary as with the text itself. Paradoxically, I had to understand what the authors were talking about, before I could understand what they were talking about. This experience validated Primary Axiom #2, that all knowledge is built upon prior knowledge; in this case, terminology. Yes, this should be obvious, and it is, but in the context of origins and evolution, it is a critical point.

Before I could make sense of the explanations concerning the nervous system, I had to comprehend the terminology, and to do that, I had to access an external intelligent agency, or the works thereof, because as expressed in Primary Axiom #1, nothing can explain itself. We need to take this thinking to our five senses: sight, hearing, smell, taste, and touch, and even to a sense which rarely gets a mention, balance. I have been privileged in my working life, and even more so in my holiday travels, to encounter new sensations associated with all five senses. Of course, on each occasion, I asked: What's that? If someone had given me the wrong answer, I would not have known that it was wrong, and would forever have had *false knowledge* until my ignorance was exposed, and likely ridiculed. The issue here is that until corrected, we cannot know whether our knowledge is right or wrong. As for my sixth sense, balance, the less said about that the better.

If we take that thinking to the evolution of senses, in whatever stage of development, in whatever organism, we have a problem. Certainly, the perception of some sensations could be invalidated very early;

for example, attempting to walk on water, or *crossing a chasm in small steps*. The land creature would soon realise that such was a practice that was not in the best interests of survival, but without the capacity for imagination, it would have neither reason nor method to conceptualise how travel across such spaces could be achieved. Letting my own imagination run loose, I wonder how early creatures learned that walking off a cliff was not a good idea, when as a child wearing a cape, jumping off a roof was. Think about it.

There is another issue regarding sensory knowledge that I will touch on just briefly at this point. Inputs from sensory receptors have both quantitative and qualitative characteristics. The *quantitative* is the physical activity transmitted through the nervous system, and somehow instantiated in the brain's neural network so that it can be "remembered", and appropriate responses developed for future occurrences. I have no idea of why or how that occurs, but experience suggests that it does. The *qualitative* dimension is conceptual, not physical, and is the assessment of good or bad, favourable or unfavourable, pleasant or not so. Is this qualitative assessment also instantiated in our neural networks, the quantitative tagged with the qualitative, or is that a function of the mind, being entirely conceptual? I suspect that latter. Either way, we know that both dimensions are stored. The curious aspect is that whilst the quantitative is generally the same for all humans, the qualitative is not. My wife likes avocado, I do not. I like liquorice, my wife does not. Others can ponder why this is so, but I have no interest in this subject at this point, beyond noting it to be true.

But back to the axiom as stated. When we converse with one another on any particular subject, the words of each of us must be understood by the other parties if the conversation is to progress in a coherent manner. I have a friend who was a jet fighter pilot, and I delight in his stories irrespective of whether they are embellished or not. The point is that not having aviator experience, I cannot fully comprehend what he is telling me, just as he cannot fully comprehend my motorcycling experiences. The words are common to us both, and we can both visualise the entities being spoken of, but not the essence of the experiences, which brings me to my next point. Words will evoke memories of sights, sounds, fragrances and odours, and even physical experiences and their effects. The converse is also true: a reoccurrence of any sensory experience can evoke memories. Trusting that my wife is not reading this, certain flowers and fragrances remind me of past romances even to this day. The linkages in the mind-brain complex must truly be beyond counting.

When we wander back in time to the supposed beginning of intelligent organisms, we have to question how these complexities of the sensory systems could have evolved, especially the qualitative dimensions. In the context of this Primary Axiom, "you" and "someone" are generic for the parties involved in any communications exchange - sense organs and the brain can be thought of in that way. On the very first occurrence of a signal being sent from a primitive sensory receptor to the brain, I would contend that the brain could not have comprehended what the sense organ was trying to tell it. If the brain was unaware of the source of the signal, repetitive signals of different strengths, from the same source, could not register as being related in any way – they were just discrete, unrelated, sensations. You may argue for similar "patterns", but pattern is a concept which must be learned – biological organs may recognise other organs of similar chemistry, but not physical, external arrangements. The issue of recognition of source we will later discuss in the context of the human nervous system.

Without belabouring the point, I hope that I have shown that in any communications exchange, whether between people, or between organs of the body, a degree of pre-knowledge must exist for the communication to be effective. We understand how that is achieved for human communications, but so far as I have researched, there is no scientific explanation for how such pre-knowledge could have arisen through the processes of evolution. We do know that messaging does arise based on physical and chemical properties, and this process can go some way to substantiating evolution theory. However, this can only explain the *quantitative* (physical) dimensions, not the *qualitative* (conceptual), which is the primary issue for understanding the mind-brain conundrum. Appropriating the observation of George Bernard Shaw quoted at the beginning of this chapter, in the context of evolution: biology has given the human nervous system the *power of speech*, but not the *power of conversation,* which is beyond its capabilities. Thus, how did it arise?

I have no answer, and have been unable to find anyone who does.

References:

1. Barr, Murray L., and Kiernan, John A., *Barr's the Human Nervous System: An Anatomical Viewpoint* (Periodicals), Lippincot-Raven, Philadelphia, PA, 1998

Chapter 1-4: Primary Axiom #4
No coding system or communications protocol can devise or maintain itself

"Truth can only be found in one place: the code."

~ Robert C. Martin, Clean Code: A Handbook of Agile Software Craftsmanship ~

Think about that for a moment, in the context of the coding systems which exist in biology, and especially in the brain. If you understand the difficulties and complexities of coding systems, of any form, it should cause you to wonder how such could arise through random genetic mutation and natural selection. Even more, how evolution could have devised the coding system in the genome itself.

Coding Systems and Communications Protocols

The significance here relates to how messages are conveyed through the central and peripheral nervous systems, most especially those originating in our six senses; how they are initiated and encoded at one end; how they are decoded at the other; and how the intent of the message is understood as to whether, and/or what action is required. I cannot comprehend how this all works in the human body, but contend that the principles later explained do hold in all forms of messaging, including biological. My purpose is to expose the difficulties that evolution would face, highlighting that as far as my research has taken me, evolutionists have not only *not* explained these issues, but have, in the main, chosen to ignore them entirely.

We must pause for a moment to further reflect on terminology. Scientific descriptions often use the words *information*, *messages*, and *signals* interchangeably, as if they were synonyms. Perhaps they are in general conversation, but here I wish to be pedantic and differentiate, even if such differentiation cannot be found in dictionaries. Information we have already discussed, so next is messaging. A *message*, irrespective of the medium through which it is conveyed, has a *semantic* layer – a meaning. Conversations can be thought of as messaging. On the other hand, signalling is at a less complex level of communication. Wikipedia defines a signal as "a gesture, action, or sound that is used to convey information or instructions, typically by prearrangement between the parties concerned; an electrical impulse or radio wave transmitted or received." A multi-word communication would be a *message*; a single word communication would be more in the realm of a *signal*. Signals do not generally have a semantic layer – the meaning must be prearranged between the parties. The communication function of a signal is to prompt an understanding in the mind of the receiver. Signals may or may not be explicit, depending on the context where used: for example, in an argument, running your hand across your throat may be taken as a threat, or it may suggest ending the dispute. In aviation, it means to cut your engine(s). Thus, meaning can be derived in multiple ways, but it can generally be said that a signal conveys "information" only in the sense that an action is to be initiated. As will be explained, this is primarily how the human nervous system works.

We must acknowledge that the mind-brain complex participates in two modes of communication: *autonomous* and *volitional*. The autonomous system is what prevents us from falling down dead; the volitional is what can prevent us from falling down at all (certain drunken episodes in my youth testify to that truth). Both systems use the same protocols, the primary difference being in the cause of the initiation of the communication, whether or not a response is required, and what that response should be. For example, the endocrine (hormone) system is autonomous, being where glands secrete hormones into the blood that travel to the target organs to address a particular state. I can't imagine how that system could have evolved, firstly the brain understanding the message sent to it by the organ needing attention, and the brain knowing how to respond to another organ, which in turn needing to know how to respond to that particular situation. That is a very complex system.

Biologists understand the autonomous system in response to internal activities; I do not, and have no intention of venturing into that field. However, there are two mysteries that I contend, biologists cannot solve:

1. A biological organ, the brain, cannot conceptualise an external reality and structure a neuronal network accordingly; and

2. No-one truly understands the volitional at a biological level, although many hypotheses have been offered.

We will get to one well-documented attempt which I contend was a failure. In my view, the primary reason why all such propositions are false, is that this axiom is either not properly appreciated, or not recognised at all. As the technical details behind this axiom are complex and require extensive explanation, we will leave this here for now and come back to it in a more appropriate place.

Chapter 1-5: Primary Axiom #5
No coding system or communications protocol can be implemented other than by an external agent at two or more nodes simultaneously

"A relationship is based on communication."

~ John Cena ~

I venture that the quote above may be more profound than the author intended, or even knew. His context was likely human relationships, but here, I shall purloin it to apply to the relationships between various organs of the body – they too are based on communication. The relevance in this study is how a biological nervous system could evolve in small steps. The endocrine (hormone) system is even more complex in some ways, but at least it is restricted to the purely physical, and has no association with the conceptual, so far as I understand it, although it does have a relationship with feelings and other mental states.

Communications can only occur when the appropriate connections are made. If you are out of earshot of somebody, the speaker is speaking only to him/herself, as no communication is occurring between you and him/her. I will often speak to myself, audibly and otherwise, but I choose to ignore that specific complexity in this chapter. Now, let us consider the communications channels and connections within the human body. Anatomy and physiology are not amongst my library of knowledge, but I appreciate this definition: the human nervous system conducts stimuli from sensory receptors to the brain and spinal cord, and conducts impulses back to other parts of the body. In the context of sight and sound especially, note *stimuli* and *impulses*, not *data* or *information*. I am grateful when authors get this right, as it seldom occurs. In later chapters, we will discuss this issue more fully, but here I just want to cover enough detail to substantiate my primary axiom in the context of evolution.

The peripheral nervous system is comprised of communications channels: nerves made up of multiple nerve fibres constructed of neurons (nerve-cell bodies, axons, and dendrites) and synapses. Synapses are the junctions across which neurons pass signals to one other, and to individual target cells. My point is that a nerve is not at all like a length of wire or waterpipe – it is a highly organised system of components where communication is achieved by each component (neuron) within a nerve fibre passing a signal to the next one. Evolution is said to occur in small steps, over a long period of time, with genetic mutations fixed in populations where such mutations result in either survival or reproduction benefit. Some mutations are said to be neutral in their effects, and can persist in populations, or may benefit another component or system before being finally utilised in the way that we currently understand. The question arises: what mutation of what type of cell caused neurons to talk to one another, and what type of mutation(s) devised the "language" of communication?

Quoting from an online source: "A typical neuron has a cell body containing a nucleus and two or more long fibres. Impulses are carried along one or more of these fibres, called dendrites, to the cell body; in higher nervous systems, only one fibre, the axon, carries the impulse away from the cell body. Bundles of fibres from neurons are held together by connective tissue and form nerves. Some nerves in large vertebrates are several feet long. A sensory neuron transmits impulses from a receptor, such as those in the eye or ear, to a more central location in the nervous system, such as the spinal cord or brain. A motor neuron transmits impulses from a central area of the nervous system to an effector, such as a muscle."[1]

That is as much as I wanted to say about the anatomy of a neuron at this point, my purpose being to highlight the complexity of not just the individual components, but of the system itself. I have no idea of the sequence in which these components could have evolved, or began to interact, but the evolution narrative does sound problematic. The next issue is bridging the gap between a sensory receptor at one end, and for example, the brain at the other: why, and in what sequence, did that happen? A communications channel has no function until connections are made at both ends. Are we to imagine that in the case of nerves, the first connections consisted of just one neuron, which evolved offering no advantage, and then progressively, more neurons joined in as the end nodes got further apart? Consider this description of neuronal development during the growth of an embryo:

"The remarkable events of this early development involve an orderly migration of billions of neurons, the growth of their axons (many of which extend widely throughout the brain), and the formation of thousands of synapses between individual axons and their target neurons. The migration and growth of neurons are dependent, at least in part, on chemical and physical influences. The growing tips of axons (called growth cones) apparently recognize and respond to various molecular signals, which guide axons and nerve branches to their appropriate targets and eliminate those that try to synapse with inappropriate targets. Once a synaptic connection has been established, a target cell releases a trophic factor (e.g., nerve growth factor) that is essential for the survival of the neuron synapsing with it. Physical guidance cues are involved in contact guidance, or the migration of immature neurons along a scaffold of glial fibres."[2]

I cannot begin to imagine how that complexity arose through evolution. Richard Dawkins likes to scoff at people for being incredulous about such things, but in the absence of any explanation, let alone a plausible one, our incredulity is justified. I am fascinated by the above description, most particularly because it suggests that within the brain, not all neurons are alike. There must be something in our genome that results in specific neurons connecting to only predetermined other neurons (appropriate targets). This results in the patterns of neural connections and pathways that a baby starts with, but note that such connections are based on molecular signals. That makes sense at a biological level, but I wonder how that process fits with the creation of memories which are *conceptual*, but are instigated in *physical* patterns.

In the nervous system where messaging is said to occur, the "coding system" is a function of chemical and physical properties at each end of a nerve fibre, and those at the connecting synapses. Reviewing the description above, note the terminology: *apparently recognize, appropriate* and *inappropriate* targets – these are hardly scientific terms. I mean no offence to scientists who have done such prolific research – I simply wanted to illustrate how far we are from understanding. In any neuronal system, recognition of appropriateness or inappropriateness can only be at the physical level, not the conceptual, the latter being a function of the mind, not the brain. What is the origin of these "molecular signals", and can their development be traced from the genome? As prose, the quotation is quite beautiful in its simplicity; as a scientific explanation, it falls far short.

At issue is that a single neuron has no useful function, and thus from an evolutionary perspective, no reason to exist or develop until connections can be made by multiple neurons to form a signalling pathway. Before connections could be made, the electro-chemical coding system implicit in the *molecular signalling* had to be devised, and be instantiated in each copy of the neuronal cell. Additionally, there needed to have been an electro-chemical definition of appropriate and inappropriate connections, meaning that neuronal cells must be differentiated by function. This is an important point which deserves our attention. Whilst cells of a given type generally replicate closely, neuronal cells have another level of complexity: they must replicate with differentiation of connectivity. I have been unable to locate a source which describes where, when, and how that happens, but if neurons self-organise to form nerve fibres, which later merge to form larger pathways, which in turn feed the self-organising neural networks of the brain, where is the orchestration written in the genome?

Irrespective of from which end the nerve grew, it had no function until a connection was made at the other end. The growth and organisation of the structure of the cell, as determined by the genome and whatever factors, had no function and no reason to exist by itself. Even accepting that this did happen, that a nerve fibre managed to evolve and grow between a sensory receptor and the brain, we still have the issue of how the correct response developed. When a touch sensory receptor experienced a sensation, how did the brain "know" to transmit a motor neuron impulse to an effector, such as a muscle? There is clearly a chemical relationship established by the genome via the cellular structure, but the development of that relationship must have taken a very long time. In the meantime, was the embryonic sensory system not yet in use? Reviewing the earlier quotation on the development of neural networks within the brain, what evolutionary process was involved in determining which neurons should connect with which? Each branch of the network is a communications channel, and each connection (synapse) is implemented via chemical gradients which determine which axons connect with specific dendrites. In human development as we understand it today, it all goes according to plan, but how did that plan arise through genetic or epigenetic mutation? The complexity of the brain's neural network is staggering and dare I say, incomprehensible, yet we are to believe that this evolved through genetic mutation, and natural selection in response to a then-current environment.

I know insufficient about the central and peripheral nervous system to say much at all, other than to quote the public literature on the subject. It is of such enormous complexity, still beyond a full understanding by the most experienced scientists, that to suggest that it evolved in small steps via genetic mutation and natural selection, or whatever is the preferred paradigm, is drawing a very long bow indeed. I cannot be dogmatic and assert that it did not happen that way, but any explanation must take into account what I believe to be axiomatic:

No coding system or communications protocol can be implemented other than by an external agent at two or more nodes simultaneously (irreducible complexity).

References:

1. https://www.britannica.com/science/neuron

2. https://www.britannica.com/science/human-nervous-system

Chapter 1-6: Primary Axiom #6
No measurement device, nor sensory system, can calibrate itself.

"I have spent more than a decade developing the calibration techniques we needed to obtain these results."

~ Mark B. Reid, MD, self-styled academic hospitalist ~

You have, perhaps, not thought of yourself from this perspective, but our bodies contain an amazing telemetry system which monitors the workings of our internal organs, and provides guidance for our interaction with the external environment. Sensors, especially *exteroceptors*, those located near a stimulus in the external environment, can fail or deteriorate, leading to deafness, blindness, or as in cases such as leprosy which affects the Peripheral Nervous system, even our sense of touch. Not often mentioned with the five senses is *balance*, which attempts to keep us the right way up.

We know from experience that the precision of such sensors varies with individuals, for any number of reasons. I have always done quite well in this area, apart from balance, whereupon activities such as skiing, skating, and even dancing, have not figured on my list of favourite pastimes. Ah, the despair of an empty dance card, for those who remember such things. Medical specialists have set "standards" by which the performance of our sensors can be compared with a "norm", however determined. This has extended even into the legal world where disability is determined by a prescribed level of impairment. If we are to believe in evolution, then at some stage, an adequate level of sensory performance was achieved, and as far as I can determine, not much progress has continued since, although different species exhibit different levels of performance in sight, hearing, smell, etc.

Technically speaking, a sensor is a *transducer*, which transforms energy from one form into another. To be effective, such transduction must be performed reliably. Transduction involves the interpretation of the characteristics of the input energy, such as light or sound, in such a way as to "understand" how to faithfully reproduce the semantic layer in one energy form into another. For example, a light sensitive sensor must react appropriately to that range of the electromagnetic spectrum which it is designed to detect. The cone shaped cells in our eyes are capable within the narrow range 380 – 700 nanometres; other portions of the spectrum have wavelengths too large or too small and energetic for the biological limitations of our perception. If evolution of the supposed, primitive light sensitive cell did not manage just the right biological properties of the cell, then we would not have the visual acuity that we do. Perhaps we would only be able to see in the infra-red or ultra-violet. But that was just the first step.

The functionality of sensory neurons and the types of connectivity vary with the type of sensor, so I will avoid any discussion of the technicalities in this chapter. Other than understanding visible light as a narrow range within the wider electromagnetic spectrum, the semantic layer of light is entirely beyond my understanding, although I will attempt to guess based on my experience with digital photography and software such as Adobe Photoshop®. As displayed, a digital photograph is composed of pixels of varying colours and intensity. The differentiation in colour is what allows us to determine shape, light, and shade. In a picture of meadows, hills, and forests, the vertical dimension is interpreted as depth and distance: at the bottom are objects that are close, and at the top are objects further away. However, in a photograph of a person standing against a neutral background, the vertical dimension is understood as height, not depth. The photograph is an encoded symbolic representation of reality, yet somehow, we comprehend it as if we were seeing the three-dimensional reality itself. There is a great deal of signal processing going on, which cannot be a function of the neuronal signalling process itself.

Encoded within a light stream are details of intensity, shape, pattern, texture, and distance. To be effective in providing sight, any light sensitive cell must be capable of interpreting the embedded encoding, and transforming it into differentiated neural signals for transmission to the visual cortex. At that end, the cortex must "understand" the external reality of each differentiated signal and recombine them into an "image". The difficulty for evolution is how both the sensor, and the visual cortex, knew how to get that right. At a technical level, scientists are still struggling to understand the embedded coding structure of light, such that it conveys symbolic representations of the details of the object from which it is reflected. Evolutionists would have us believe that genetic mutation and other biological processes somehow solved this difficult technical problem, without having any purpose, and most particularly, no method of verification that its internal communications system achieved the necessary results.

Let me stress that without a method of verifying the communications process, through calibration and validation, the chance of a reliable signalling system arising through undirected biological evolution must approach zero.

Imagine that you had to design a system, such that as light strikes a sensor, the scene is faithfully represented on an LED or plasma screen. If you have followed the development of television sets over the past half-century or so, you would have noticed the emphasis on picture quality, i.e., how faithfully the appearance follows reality. In truth, the technology goes beyond that, but we can ignore that for now. The issue is the design and calibration of the intermediate communications system components. When your local service person tuned your TV set, as they did in earlier days, they were adjusting various elements to their calibrated norm. You see, calibration is the key to effective sensory perception, whether in biological systems or modern technology applications. The question becomes: How could evolution have calibrated the sensory systems of the human body, when there was no design goal? An *interoceptor*, one that interprets stimuli from internal organs and tissues, is possibly capable of self-calibration in that the system would fail to work effectively until a sufficient level of calibration was achieved, but that would have left a multitude of dead and dying organisms. However, *exteroceptors*, those that tell us about our external environment, are different in that the only method of verification was via the very same sensory systems. The brain did not know that the ear was *hearing*, nor did it know that the eye was *seeing*. Though both systems used physical signalling, the intent of that signalling is conceptual and could not be learned from the signalling itself.

The system could not know its own level of performance, for it had no frame of reference for the verification of results. In some instances, the organism could understand that its perception was faulty; for example, if it kept bumping into immovable objects or falling into holes, but it could not know why, or how to go about recalibrating the faulty sensor. The only argument that evolutionists can offer is that those species which got it right, purely by chance, survived, and those that did not, did not. In later chapters, I will describe sensory systems in far greater detail so that you can appreciate the likelihood of that ever happening.

Chapter 1-7: Primary Axiom #7
All systems must undergo a process of verification and validation to be implemented reliably

"Uncontrolled variation is the enemy of quality."

~ Edward Deming, Basic Statistical Tools for Improving Quality ~

If we apply Deming's quote to evolution, it should be apparent that random genetic mutation is the enemy of developing a robust and reliable biological system. Quality does not happen by chance, and ignoring the thermodynamic context, the theory of entropy tells us that systems tend from order to disorder. The theory of evolution is contrary to that of entropy – a *disordered system* brought about by genetic mutation is *ordered* via natural selection, which also acts as the QA (Quality Assurance) mechanism. Most of my experience in QA, involving validation of data and verification of systems, has been in Data Processing and Management Information Systems environments, but I have also attended courses and implemented QA systems in manufacturing. This is the background from which I will explain this axiom.

The two terms *verification* and *validation*, are often treated as synonyms, but in a technical sense, they are not, although even in some technical domains, such usage is often inconsistent. It is necessary that we differentiate these two important processes, which we need to consider in the context of the integrity of the human nervous system. Here I offer my own (non-authoritative) definitions for the purposes of this study:

• Verification – proving the integrity of a communications system hardware componentry, irrespective of message content.

• Validation - proving that a message has been faithfully conveyed in the sense that it was intended.

In our discussion here, verification relates to system integrity, validation to data integrity.

Verification

Not wanting to bore the reader with technical specifications, you can study those for yourselves, I will give a non-technical description of a multi-layered communications verification process. Initiating a communication, the sender identifies the receiver, and asks the receiver whether it is ready to receive. The receiver responds and the sender transmits the first packet which contains a method for the receiver to validate at a summary level, that what it receives is what was sent. The protocol ensures that packets arrive in sequence and without error, by swapping acknowledgments of data reception, and retransmitting lost packets. Eventually, the sender transmits an "end message" notifying the receiver to stop listening. All very neat and tidy, and in some protocols, such as simple file transfers within a network, a hash total is included as a mathematical verification. It should be noted that mathematical definitions of information, as espoused by Claude Shannon[1], are useful only in *verification*, for they completely ignore meaning – sense and nonsense can have the same mathematical equivalence. Note also that a protocol is a way of communicating data, but says nothing about the data itself; that is the subject of *validation*.

Before we get to that, there is one, out of many other technical issues, which I would like to consider. *Multiplexing* is a method by which multiple analogue or digital signals are transmitted simultaneously over a shared medium. For example, several telephone calls may be carried using one wire, or multiple video streams through the same optical fibre. We must consider the implications of multiplexing in the central and peripheral nervous systems, for many nerve fibres in the PNS appear to merge into single pathways in the CNS, but I am open to being corrected on that issue – the literature is not clear on that point. As we will discuss in more detail later, the commonly observed physiological phenomenon of "referred pain" is the result of signal processing errors in the Central Nervous System. Evolution did not get it quite right, in the sense that it failed to correctly verify the source of a sensation. The question therefore arises: what is the biological mechanism that allows the CNS to get it right most of the time?

Two other issues, which may or may not be relevant, are *bi-directional* communications, and the requirement for some form of *synchronisation* to prevent everyone talking at once and no-one listening. I know insufficient about the physical structure of the human nervous system to know whether these are of interest, so for now I will just float them for your consideration – you may well know far more than I do. Consider the difficulties if telegraphists at both ends attempted to transmit simultaneously before multiplexing became available. You can see that whilst multiplexing allows multiple simultaneous communications on a single channel, there needs to be a method of controlling the sequence of messages backwards and forwards in the one conversation. Simplifying, imagine a hand being burned on a hot plate: a nerve ending sends a continuous stream of message to the cerebral cortex and the brain responds: "Get your hand off!" Unfortunately, the nerve ending is so busy talking that it is not listening. I know that it is more complex than that, but that is the point – it is complex. If we attempt to reverse engineer the evolutionary process, we encounter innumerable, and perhaps, unanswerable questions.

We know that synchronisation is not a problem in the human nervous system, in that *sensor* neuron signalling (input) uses different pathways than *motor* neuron signalling (output). However, this evidences another level of complexity that evolution would have needed to have solved. I have no idea of where to begin with that one.

All this evidences the technical conundrum of the body's internal biological communications system. Whilst my knowledge of how that works is limited, I contend that the technical specifications must largely parallel what has been developed in modern communications technology. There are functional requirements that must be met, and as our bodies work wonderfully well, mysteriously even, I must assume that those requirements *are* met. The question remains: how was that accomplished?

The Nerve Network

Our human nervous system comprises two primary arrangements: the Central Nervous System (CNS) and the Peripheral Nervous System (PNS). The CNS comprises the brain subtended by the spinal cord, and represents the body's primary control system. The PNS branches from there, and is the body's connection with the outside world. The PNS has two primary divisions: the Somatic and the Autonomous, which is further divided into the Sympathetic and Parasympathetic. The Somatic division transmits stimuli from outside and within the body to the CNS, and transmits motor neurons from the CNS to organs of the body. There are nerves with specific functions, such as olfactory, oculomotor, and facial nerves, but they are of no specific interest just yet. There are thirty-one pairs of spinal nerves in the PNS, which have both sensory and motor functions. There are several hundred peripheral nerves throughout the body. The many sensory nerves that bring sensation from the skin and internal organs merge together to form the sensory branches of the cranial and spinal nerves. The motor portions of the cranial nerves and spinal nerves divide into smaller nerves that divide into even smaller nerves. So, one spinal or cranial nerve may divide into anywhere from two to thirty peripheral nerves. It should be obvious that this presents challenges in the system to differentiate specific messages to and from the body where nerve fibres merge and diverge. The question to be asked is this: what is the content relayed from one neuron to another that contains the necessary addressing and purpose? It is accepted that since "the 'language' of the nervous system is electrical signals (impulses), each of the many types of receptor cells must convert, or *transduce*, its sensory input into an electric signal."[2]

Neuron signalling works as follows:

1. When neurons signal another neuron, an electrical impulse (action potential) is sent down the length of the axon.

2. At the end of the axon, the electrical signal is converted into a chemical signal. This leads to the release of molecules called neurotransmitters.

3. The neurotransmitters bridge the gap, called a synapse, between the axon and the dendrites of the next neuron.

4. When the neurotransmitters bind to the dendrites of the next neuron, the chemical signal is again converted into an electrical signal and travels the length of the neuron, and the electrical potential of the sending neuron returns to its rest state.

This is relatively simple (?) to understand, except that from one neuron to the next, only an electrical pulse is communicated (ignoring the chemical neurotransmitters which stimulate an action potential). Unfortunately, the terminology of many scientific descriptions is far from accurate, when they state that a neural pathway carries information or messages, when at the intermediate neuron level, the truth is simply a signal (action potential). At each end of a nerve fibre, there is a chemical action which represents the transduction of the input sense, but in between, just the electrical pulse. Any degradation in the nerve fibre should result in a degraded signal, but from the beginning of the evolutionary process, one would expect that the process was far from the necessary level of quality. Now, hypothesising without direct application to the true structure of the nervous system, which I do not completely understand, consider the architecture of single spinal nerve dividing into multiple peripheral nerves. In telecommunications, a router is used at the junction of a single input cable and multiple output cables. The incoming signal contains data that is interpreted by the router to determine the correct output channel. How does the PNS manage this? What differentiates motor neurons to activate one finger or another? If a single spinal nerve can carry messages for multiple recipients, what differentiates the signals passed between neurons? If evolution was responsible for this development, what was the verification process, and how did that evolve?

I have limited understanding, and fewer answers, my point being that I have been unable to find literature which attempts an explanation of this complexity. Every narrative that I have found deals only with one-way, single-channel, single message, single recipient, communications. One might also introduce the issue of inverse or demultiplexing, but I suspect that you get the point. I do not claim that scientists have not dealt with these issues, only that as yet, I have been unable to find any such studies. To be honest, I do not expect to be able to find any.

Validation

Validation is even harder to explain within the evolutionary paradigm, and here I am talking not of the conceptual, but of physical representations. The issue is simply this: how can a biological organism validate that those sensations received via a receptor are reliably transmitted, and then stored in a neuronal network that is a faithful representation of the external reality? Not only that, but could all variations of organisms accurately and consistently replicate that process? Variations in biology result from random genetic mutations, and as Deming asserted, *"Uncontrolled variation is the enemy of quality"*.

By way of example, the sense of smell of my wife and myself differ. This could have resulted from our genetic inheritance and/or our upbringing, I cannot know, but that would not argue the evolutionist's case, for the same would apply to organisms of whatever generation. For example (not real examples), what I may detect as wood burning, my wife may sense as rubber burning; whereas my wife smells dust, I smell nothing; where my wife can smell rain in the air, I sense nothing. Where I detect ozone during lightning strikes, my wife cannot. Therefore, the representation of an external reality is conveyed and stored differently in our respective nervous systems. Either that, or our conceptualisation of the sensation has been varied by previous environmental conditions or teachings. If as children one was taught that a particular odour was of rubber, and the other was told that the same odour was of wood, this could account for it. The general case would be that of direct experience. Our internal biological systems cannot validate a shared experience as the true representation of an external reality: so, who is right? Now, it matters not what biological explanations are offered for this variation in behaviour – the fact remains that our nervous systems are incapable of reliable validation as experience shows. In truth, our interpretation of sensory inputs is subjective, not objective, but on what basis?

At the next level, that of cognition and validation of the conceptual, the issue becomes even more mysterious, but we will leave that for now. Getting a little ahead of ourselves, consider the above requirements in relation to the autonomous functions of the human body. The protocol is implemented in biology, such that when the brain detects the need for whatever, the message it sends must know where to go, how to get there, and must be understood by the receiving organ to respond appropriately. Modern medicine is progressively understanding how this works, but the question must be: how did it become so?

Similar, but more complex issues arise in our six senses (remember balance), especially sight and hearing. Human body autonomous functions operate within a closed system, but the senses communicate with external systems, the circumstances of which must be learned. We will come back to this in more detail as the complexity forms the basis for my argument – it is the elephant in the room of evolution.

References:

1. Shannon, C.E., and Weaver, W., *The Mathematical Theory of Communication*, University of Illinois Press, Urbana, IL, 1949

2. https://www.ncbi.nlm.nih.gov/books/NBK21661/

Chapter 1-8: Primary Axiom #8

< A system, of whatever simplicity, is irreducibly complex >

"It is always easier to destroy a complex system than to selectively alter it."

~ Roby James, Commitment, Starfire Saga #2 ~

It can generally be said that if you remove any part of a system, the system will fail to function as intended, very much like removing a link in a chain. If any part of a system is compromised in any way, then again, it will fail to function as intended. This is why in critical applications, such as aviation, multiple redundancy is built-in to avoid catastrophe, not that it always does as experience has shown. Industrial espionage has the intention of stealing a competitor's secrets, often in the form of acquiring a product and then seeking to reverse engineer it. If a manufacturer attempts to build a product similar to that of a competitor, with inadequate understanding of the detail, an inferior product will result. Sometimes, the product is so inferior that it hardly works at all.

Development of entirely new types of systems involve a degree of trial and error. In my IT career, I had the good fortune of working with new technologies to develop such systems, experiencing varying levels of frustration and delight, depending on achievements up to that point. Sometimes, despite the most diligent efforts of many people over a period of time, it just did not work! Some new system projects were eventually abandoned altogether, because a solution could not be found within the existing technologies to achieve the intended results. My IT career came to an abrupt and inglorious end over one such dramatic, and very expensive, abandoned project. In my defence, it was not my fault, as I had no say in which technologies were being utilised.

When I consider the evolution of biological systems from that perspective, I wonder how that process could have achieved systems of greater complexity than I, or anyone, has ever attempted. The false starts must have been beyond counting. There is an issue in systems development which can be illustrated by the question: could a typing pool of monkeys ever write a Shakespearean play?

You have no doubt heard of this contention that given enough time, typewriters, and monkeys, one will eventually write one of Shakespeare's plays. Reject the contention, for it is fallacious. If you grew up, as I did, in the age of typewriters, typing for even the semi-proficient was a chore because of the ease of making mistakes. With important documents, a mistake required one to start again at the beginning of the page, but not necessarily at the beginning of the document, unless of course you did not find the error until much later, and if you missed a word, sentence or phrase on page one, well, your were going to be very late for dinner. This is an important point: progress requires a method of storage of correct results so far, but if storage is not available, and checkpoints not taken, each mistake will require you to start again from the beginning. To type a single word of eight letters correctly, by chance, is $(1/26)^8$. Now assuming 52 keys on a typewriter and a play of 100,000 words (ignoring spaces and punctuation) of an average length of 5 letters, the chance of correctly typing the play is $(1/52)^{500,000}$. Note that the monkey would need to be extremely dextrous and not hit two keys simultaneously, jamming the typewriter (I have done that). One could make further calculations and introduce additional complications, but I think that you can see the point: something may be possible mathematically, but impossible in a practical sense.

But that point is not the one I wish to emphasise here: it is checkpointing.

If you begin building a system, at even the lowest level of complexity, but it fails, you need to understand the cause of failure and replace it, whilst still retaining the functional parts. In my youth and that of earlier generations, we had fun building primitive communication systems, comprising two jam tins and a length of string. For whatever reason, sometimes the type of string did not give the desired results, and at other times, we found that the type of tin was the problem. Through intelligent (?) trial and error, we retained what worked and discarded what did not, effectively checkpointing progress. Now, consider that in relation to the development of the human nervous system, which provides the communications channel between our sensory receptors and the brain – how did it checkpoint progress?

We later continue our discussion on the complexity of neuronal pathways, but for now, we just need to accept the complexity, and ponder how it became so. At some stage in the evolutionary process, a neuron of some level of complexity and capability, came into existence. The function of a neuron is to connect with other neurons to complete a pathway – a single neuron, as best as I understand, could have no other funtion. It is analogous to a chain link. One link by itself has no function. If one of more links in a chain fails, the chain ceases to have a function. If neurons were replicated, and that is a big "if", but their early capability was limited, they could not have completed a functional neuronal pathway. The question becomes: why would the mutation that gave rise to that cellular arrangement have persisted when it offered no advantage to survival? In passing, and we will come back to this additional complexity, not all nerve fibres are constructed the same, and for neurons to connect in a network, there needs to be differences in their chemical composition. Clearly, replication of neurons must follow some form of pre-programmed variation.

Revisiting the quotation at the beginning of the chapter, *it is always easier to destroy a complex system than to selectively alter it*. The same is even more true of building a complex system from scratch through *random* alteration, relying on *selection* only when the primitive but fully functioning system provided a survival or reproductive benefit. One could speculate endlessly of possible scenarios, but I doubt whether any could deny the truth that systems are irreducibly complex, and that a biological system which was little better than a work in progress, would more likely be deleterious than beneficial. Referring back to Primary Axiom #4, it is far easier to destroy a complex coding system than to selectively alter it.

Fritjof Capra, an Austrian-born American physicist, systems theorist and ecologist, offered: "With the subsequent strong support from cybernetics, the concepts of systems thinking and systems theory became integral parts of the established scientific language, and led to numerous new methodologies and applications -- systems engineering, systems analysis, systems dynamics, and so on." It is my contention that if the *microbes to man* hypothesis of evolution is ever to be proven, evolutionists must attempt to explain their science in terms of *systems theory*, most especially in terms of systems development and maintenance. If they cannot, or if they are unwilling to try, then we have no reason to accept their hypothesis.

As Deming opined, "If you can't describe what you are doing as a process, you don't know what you're doing." Coming from a systems background, I must agree: If evolutionists cannot explain their theory as a detailed process, they have no idea of what they are talking about. Comparing bones with similarities simply will not do.

Chapter 1-9: Primary Axiom #9
< Nothing can be done without prior knowledge that it can be done >

"The first step is you have to say that you can."

~ Will Smith, Hollywood actor ~

You might need to pause and grab a mug of coffee, or a glass of your favourite tipple, before beginning to read this one. As I apologised at the beginning of Part 1, many of the metaphors and analogies I choose simply fail to convey my thoughts to others. I suspect that this is one such subject, as I have attempted to reword it numerous times, but am still not satisfied with my explanations. Hopefully, you will get at least a sense of what I am attempting to convey.

I am not usually given to quoting celebrities, but I recently came across this observation on a calendar, and realised how apt it was for this discussion. A point to note, however, is that only sentient creatures with intellect, imagination, and free will can make that sort of decision – the first step for others is to *know* that they can. But even for humans, the first step is, in truth, that they must know what it is that they are saying that they can do. Can any reader contemplate performing an *entrollapation*? (I just made that up, in case you hadn't guessed.)

In this chapter, I am extending my use of the term, *knowledge*, to include *instinct*. Instinct in all animals occurs at two levels; the first is simply a behaviour that occurs within organs of the body: in a very broad sense, we can say that the organs know what they are supposed to do, and carry on doing it as determined by the cellular construction. This is not a cognitive process, but philosophically, we can say that chemicals *know what they can, and must do*. Barbara McClintock, an American cytogeneticist and Nobel Prize winner in Physiology, would seem to agree with me in a sense when she wrote, "A goal for the future would be to determine the extent of knowledge a cell has of itself and how it utilises this knowledge in a 'thoughtful' manner when challenged."[1] It is interesting how scientists cannot escape this analogy of the cognitive when discussing the purely chemical, even when they have no scientific basis for doing so. As I ponder my navel, I wonder what thoughts it is having.

The next level of instinct occurs, for example, in the manipulation of limbs, such as when a foal struggles to get to its feet and walk soon after birth. We can say, again in a broad sense, that animals and humans know how to do what comes naturally. The question becomes: how and why? A great deal of research continues into motor psychophysics and electromyographic activity in muscles to enable treatment of people with disabilities; the complexity of these systems is truly amazing. Tracing the chain of cause-and-effect from the mind to the primary motor cortex, responding for example to a visual stimulus and being guided by that sensory input, through nerves, muscles and tendons, and being further controlled by sensory nerves which tell the system when an object is finally touched, has resulted in an enormous body of research and literature, much to the advancement of medical treatments.

In humans, the complexity increases exponentially when we examine what I might call, *sensory systems management*, as in the control of playing a piano in response to a music sheet, accessing each of the keys in particular arrangements and touching them *just-so*: light, heavy, short, long, and so on. The ability of any a fish, animal, or human to move in any manner appropriate to the appendage (arm, leg, tail, fin, etc.) is a function of the cellular arrangements, which in the evolutionary context, must have occurred before the movement was possible. It is claimed by scientists that this incredible complexity and astounding accuracy has all occurred purely by whatever evolutionary mechanisms are now in vogue. But as one sceptic suggested, we are witnessing a professional performance by an accomplished orchestra, but who is the conductor, who wrote the music, and who taught the musicians how to play?

Knowledge always precedes actions. Endeavouring to explain this to people, a common response is that this does not apply to all actions, especially those of animals: they simply do things instinctively. At a superficial level this is true, but *instinct* must at some level be a form of *biological knowledge*, just as Barbara McClinton suggested. A dictionary definition of *instinct* is: an innate, typically fixed pattern of behaviour in animals in response to certain stimuli. We can ask: what caused the animal to behave that way, or how did the animal learn that behaviour in the first instance - paraphrased, what biological knowledge lies behind that behaviour? A more important question in the context of the mind-brain conundrum, is how we humans manage to behave in other than fixed patterns, and often contrary to instinct. How do humans choose to take heed of, or ignore, certain stimuli?

If we see an object hurtling toward our head, we instinctively duck. Why? Because we "know" that if struck, it will hurt. How do we know that it will hurt even before experiencing such an event, and why is being hurt something to avoid? We can *will* to overcome instinct, which is evidence to my mind of the existence of the "free will" in humans. If you are a follower of the noble game of cricket, you no doubt marvel at the bravery and skill of batsmen, oops, sorry, I should use the gender-neutral term *batter* (not what you put on fish), when seeing a ball in flight to their head, they can stand their ground and swat the ball away to the boundary. If they miss, they will be much the worse for wear. We shall come back to that as well but for now, I just want to introduce the relationship between knowledge and actions. For a bird to flap its wings, its mind-brain complex must be preloaded with the instructions that allow it to correctly activate the motor functions that cause the muscles and tendons to move in a defined manner, in response to the type of flight intended. Those instructions are a form of knowledge, and once again we must ask: where did that come from? If you answer: "trial and error", you accept the principle of teleology (purpose), and refute a fundamental plank in the theory of undirected evolution.

Examining Examples

"I think I can," puffed the little locomotive, and put itself in front of the great heavy train. As it went on the little engine kept bravely puffing faster and faster, "I think I can, I think I can, I think I can."

~ The Little Engine That Could, Watty Piper, 1930 - ~

Except for humans, every activity of every organism is underpinned by the "knowledge", conscious or otherwise, that such activity is possible. I cannot know whether animals ever attempt activities that may be beyond them: do dogs watch birds and wonder if they too could fly, and how they would achieve flight? Do dogs look at a stream and wonder whether breast-stroke may be a more efficient method than paddling? Do birds ever get tired of the cold at high altitudes, and decide that it would be more comfortable to walk south that winter, or maybe just secretly winter over in a cave? Do animals even have that remarkable trait of humans: *trying* when failure is a distinct possibility? Watching our recently acquired kitten, they most certainly do. But, is it only a human trait to strive to get better at some activity, to compete and beat others? What particularly bemuses me about those who hold to the evolutionary concept that our free will is but an illusion, nevertheless live as if free will is real and a powerful force for human advancement. Even those scholarly types who advocate the absence of free will are at a loss to explain why they chose the profession they did, and why they tried so hard to succeed. If evolutionary processes are indeed absent of purpose or teleological drive, how can this be so?

The issue of self-awareness has long mystified scientists of numerous disciplines, leading to questions regarding the differentiation of the mind and brain, as we are questioning here. Evolutionists, and I suspect, most neuroscientists, subscribe to the philosophy of *material monism*, the ontological position that all is material and that all phenomena can (and **must**) be explained in physical or material terms. John C. Eccles, an eminent neurophysiologist and Nobel Prize winner in that discipline, warned:

"But one must not underestimate the materialist-monists. There has been developed by Feigl (1967) and others the strange belief in an identity theory – that mental events such as consciousness are in some manner 'identical' with brain events. So, with this enigmatic identity, mental states are just brain events! This dominant materialism gives the brain complete superiority over the mind, even in the experience of consciousness ... We each of us have, though perhaps not fully realized, the primal certainty of the conscious self at the center of our being ... The great problem confronting each of us is the manner in which our experienced self relates to our brain."[2]

The belief that mental events are identical to brain events led to the "double aspect" theory, which we earlier discussed.

The opposing position is that the mind controls the brain, or even that the *Self* controls the brain as proposed by John Eccles[3], but whilst I agree to an extent, this hardly advances our knowledge: what is the fundamental nature of the mind, or self? As I have sought to explain in the preceding chapters, an agent external to the material brain is required to manage the multiple levels of abstraction from the symbolic representation of conceptual data to cognitive information and knowledge. In other words, the faculties of cognition have yet to be explained in material terms and in my opinion, never can be. The nature of consciousness, or self-awareness, continues to be the subject of scientific contention, the complexity being such that perhaps it is best left to the experts. In this chapter, however, I would like to consider awareness at a more fundamental level: the interaction of the brain with muscles, tendons, and the like from an evolutionary perspective.

The term "muscle memory", synonymous with motor learning, can be defined as a form of procedural memory that involves consolidating a specific motor task into memory through repetition. All very well, but before a creature can perform such tasks, it must know that it can, and it must have a purpose in mind. We know that physiologically, the brain controls movement of our limbs, apart from involuntary spasms of course. That means that the brain is aware of the limb, its functions, and the possible range of movements. The issue is how to explain the evolution of awareness of an evolving feature such as a fin, arm, leg, wing, or whatever, and the speculation or awareness of its potential function. More importantly, what is the impetus to "forget" one muscle memory and substitute it with another? I acknowledge theories of self-organization processes in cell-differentiation and body-plan morphogenesis, as proposed by the Santa Fe Institute, and as described in the book, *"The Origins of Order: Self-Organization and Selection in Evolution"*[4] by biochemist Stuart Kauffman, but we deal with that in a later chapter. For now, I wish to focus of the relationship between the brain and evolving appendages.

From here, I shall let my imagination wander free – join me if you so wish, *flying without wings*, as it were.

Stumpy the Fish

Imagine if you will, some primitive fishy creature, more like an eel being bereft of fins, swimming in the waters of the early Earth. A genetic mutation occurs which leads to the growth of a protuberance, either within or upon the external surface of the body. That mutation is subsequently inherited by successive generations and continues to grow until it is quite prominent. By chance, attendant muscles or tendons evolve along with the ever-lengthening protuberance even though at that stage no survival advantage is offered. From here, the protuberance could evolve into a fin, a leg, or even evolve no further and just be an ugly lump.

Let us suppose it will become a fin.

At some point Stumpy has to come to the realisation, consciously or otherwise, that this embryonic feature is capable of function: how does that happen? Let us assume that for reasons yet to be uncovered, the muscles or tendons begin to expand and contract. Now either the muscle movement was autonomous and informed the brain, or the muscle moved in response to signals from the brain, although just what the brain intended for the movement is likely unknown to even the brain at that point; after all, it has never had one of these stumps before. Of course, it could have been just an involuntary spasm but that would hardly serve the progress of evolution. One further problem relates to the positioning and symmetry of the stump. If we are speaking of chance mutations, then the stump could have arisen anywhere on the creature, in any number, and if more than one, could have been symmetrical or asymmetrical, or of unequal growth and capability. Perhaps stumps grew all over various progeny until the right combination or positioning favoured selection. As with many other phenomena, evolution must explain symmetry and the optimum placement of appendages.

On the subject of symmetry, a brief note on *chirality*. A chiral molecule has a mirror image that cannot line up with it perfectly - the mirror images are non-superimposable. This pair of non-superimposable mirror image molecules are called enantiomers ... It turns out that many of the biological molecules such as our DNA, amino acids and sugars, are chiral molecules. Another example is our hands and other appendages – they are symmetrical but non-superimposable mirror images. Just how evolution could manage that design is questionable.

Evolution would have us believe that for whatever reason, the brain commanded the stump(s) to move in various ways until eventually the movement actually became useful. One could imagine two stumps moving in opposition to one another with Stumpy simply revolving in place, or perhaps the movement was vertical and Stumpy either crashed to the bottom, or broached the surface, struggling to breathe. Perhaps, by chance, the movement of one or more stumps became coordinated, but in any event, what was the mechanism that allowed Stumpy to come to the realisation that such movement was or would become useful? Had Stumpy been quite content to glide gracefully through its watery element for whatever number of generations, should not the addition of protuberances for which there was at that time, no useful function, be seen as degrading rather than enhancing survival fitness? Would that not be like bolting a bucket to each side of an F-16?

Having achieved movement, Stumpy had to be aware that he had done so. Controlled movement requires a feedback mechanism, whereby the motions commanded by motor neurons are confirmed as having occurred as requested, by the corresponding sensor neurons. We discuss this complexity in a later chapter.

Up from the Sea

Evolution would also have us believe that over time, similar protuberances evolved to become legs which enabled fish to leave their natural watery environment and venture onto land in search of whatever. Perhaps as their ponds dried up, something prompted them to follow a type of sensory perception that informed them of the presence of a larger body of water, but all sensory perception needs a method of verification, lest what is perceived is at best a mirage or at worst an unrecognised danger. In all likelihood, this could only be accomplished through the creature venturing into the unknown, with significantly greater chance of standing into danger. After all, there were likely predators which it had never before encountered. This raises the question: what is the probability of survival over the long-term when so many chance occurrences have to work in combination in the right way? There is also the issue of the probability of the embryonic organism that inherited the mutation, surviving to the adult stage to be able to continue to develop and pass on the experience, given that cognition of experience is believed to be heritable. Given what is now known of genetics and reproduction, the biological mechanisms for heritable traits are still being investigated.

In offering climatic conditions as a selection pressure that guided evolution, author Richard Dawkins mentions the development of amphibians with the words "Fishes that made their living in water could benefit from a temporary ability to survive on land while they dragged themselves from a shallow lake or pond that was threatened with immanent desiccation to a deeper one in which they could survive until the next wet season"[5]. At first, reading this seems eminently sensible, and one could not dispute the logic that such temporary ability would indeed provide survival benefit, but if you try to imagine the small, gradual, incremental steps needed to achieve that capability, it becomes a somewhat different problem. How would a fish discern immanent desiccation anyway? If it had occurred before, the fish would have died and thus unable to pass on its experience to its progeny. Richard Dawkins has a far greater imagination than I do.

I can only deal with this briefly here, but let me cast some bones that might magically re-arrange themselves in a way that would achieve amphibian evolution. Dawkins acknowledges these issues, in a way, but offers no explanations. Firstly, we have the fins to legs issue and how to overcome that handicapped stage (to be later explained). Then we have the gills to lungs, or whatever mechanism gets added to extract oxygen from air rather than water, and how that manages to develop: Did it develop to maturity before the exposure to air (i.e., before it offered any selection advantage), or did it continue to develop in parallel with periodic exposure, still with no selection advantage until it actually worked?

How about the skin covering? Fish skin/scales offer no ultra violet or other protection from the sun. How about a mechanism for regulating body temperature, not a problem for fish in the same way that it is for land dwellers; eye covering or mechanism to prevent the eyes drying; bowel activity in the absence of water? Maybe none was required.

There is the question of all of these, and perhaps more, mutations happening or accumulating over time in the same animal family; they may have happened independently in different populations over a period, but for the transition to be successful, there has to be a minimum set of capabilities that will allow the fish to successfully navigate dry land. Of course, we could posit that many, if not all, occurred in a semi-wet environment that amphibians first evolved in mud and then gained the capability to tackle the desert, but that alters the problem only in degree.

Next, consider bio-feedback mechanisms: the ability of the brain to take impulses from a sensory receptor, understand what it means in context, process that data, formulate an appropriate response, and communicate the required action back to the appropriate organs, with a complementary process, or perhaps both. Genetic mutation of organs and limbs needs to be supported by corresponding mutations in the brain for the animal to know what to do with these new features, and the bio-feedback mechanisms need corresponding, simultaneous mutations to offer any survival benefit.

Finally, volition, the mental ability to formulate the thought to crawl from a shallow pond across dry land to another pond: Where did that come from? Putting legs on a fish will not enable it to walk; it has to conceptualise that it can walk, it needs to want to walk, and all of this has to be enabled by genetic mutation, presumably in the brain. Evolution of fish to amphibian is far more complex than evolutionists admit in the generally available literature, and my scepticism does not arise from ignorance or incredulity; it arises from knowledge. Continuing research has revealed that the species known as "mudskippers" are even more complex than I knew. They have numerous unique attributes that allow them to survive on exposed mudflats such as stereoscopic vision; liquid-filled skin 'cups' to keep their eyes moist when out of the water; a special retina that allows them to focus through the upper and lower sections of their eyes depending on whether they are in or out of the water; the ability to breathe through their mouth and skin (rather than gills); and a lubricant slime covering with antimicrobial properties[6]. The number of genetic mutations required to achieve that additional functionality must attract serious questions about the ability of a variation-selection mechanism to produce such an animal, capable of surviving whilst all these improvements were being added.

The most problematic issue is this: the evolution of a fin, tail, leg, arm, or whatever had to continue in the right direction, checkpointing what eventually would become useful, for generations with no survival advantage, and before the creature's brain became aware of a possible future function. This is another significant issue which evolution has yet to explain: *complementarity*. When we examine the complexity of an appendage such as a leg, it is valid to ask whether each capability evolved in sequence, and the right sequence at that, or in parallel? The evolutionists posit that each characteristic evolved for a different function, and only later combined, thus contending with the argument of *irreducibility*. That is a fine proposition, and can be validated in some cases, but as an overarching explanation, it fails dismally as so many have shown.

If indeed the brain controls the muscles and tendons as modern physiology teaches, what possible mechanism could there have been for the brain to eventually devise the right command sequence to have the feature perform in a useful way, and why would it have done so anyway? Whilst there may be valid arguments against the irreducible complexity of specific components, there can be none against the *irreducible complexity of systems.*

The Incomplete Bird

There is a variation of the albatross called a "gooney bird" which has titillated observers for centuries with its ungainly antics, particularly when attempting to land. As a former military air traffic controller working with ab-initio cadet pilots, I recall how we were simultaneously entertained, amused, horrified, and even frightened at times as the students sought to imitate the much-derided gooney bird, albeit unintentionally. Clearly, graceful landing is an acquired skill, with humans at least having the knowledge that such is attainable, even if the gooney bird seems not to have such hope. Just as importantly, humans understand the difference between a good landing and a bad one, other than just being able to walk away from it.

Evolutionists tell us that birds evolved from some form of dinosaurs or lizards, though some tell it the other way around. No matter, the problem remains the same, so let us consider the dinosaur-to-bird hypothesis. The anatomical and physiological differences between dinosaurs and birds have been well documented elsewhere, so we have no need to revisit them here, save to note that the differences are significant and the mutations required to achieve the transition must have numbered in the multi-millions at least. If evolution is a gradual process, then the transition to flying must have taken some considerable time. At whatever mutation rate science would posit, this timescale is troubling.

In a previous study[7], I coined the term "handicapped fossils" to highlight the duplicity of evolutionists like Richard Dawkins when they assert that there are no "missing links" in the fossil record: that there are sufficient fossils of transitional stages (intermediates) to support the overarching narrative. I called these fossils *handicapped*, because there must have been animals where some characteristic was in transition, but was no longer fully functional in its original role, and not yet fully functional in its new role. Given the vagaries of evolution and the extended periods over which transitions would have occurred, handicapped fossils should outnumber fully functional fossils by several orders of magnitude, yet they are entirely absent. It is logical that whatever dinosaur-to-bird-to–dinosaur evolution scenario one prefers, it must be the case that for an extended period of time in even geological timescales, there were lots of intermediates running around that were neither fully functional birds, nor normal functioning dinosaurs.

The evolutionary steps between the two have not been calculated to my knowledge, but it must have involved hundreds of millions of mutations over perhaps thousands of generations, and even these numbers may be understated. It would be entirely optimistic to think that the evolutionary path proceeded in just one direction; that there were no branches in the intermediate stages that led to unsuccessful development; that somehow the genetic development knew where it wanted to go and just went there in incremental steps, one building on the other, each proceeding in the direction that eventually resulted in a bird, or dinosaur, depending on which direction you believe it proceeded. It is difficult to imagine that both evolution scenarios occurred; that would require a great deal of explaining. If we accept that evolution proceeds via random mutation and natural selection, there must have been an enormous number of animals—neither fully bird nor fully dinosaur, some leading to evolutionary dead ends, some on the path to success—during a very extended period. So, where are the fossils of these innumerable partly functional animals? It was this conundrum that prompted the late Stephen J. Gould and others to advance the concept of *punctuated equilibrium*, and ponder the causes of the Cambrian explosion.

Natural selection would have de-selected innumerable evolutionary false starts, and given that intermediates between dinosaurs and birds would have diminished survival capacity, I find it difficult to understand why they were not all de-selected, let alone just some, but apparently, they were not. Otherwise, dinosaurs could not have evolved into birds (or birds into dinosaurs). Before putting aside those particular difficulties, we should note that in some respects, they are trivial in comparison with the primary issue raised in this chapter: that of knowing that you can, and especially in the case of proto-birds, knowing that they could fly.

Stumpy must have had considerable difficulty in coming to grips with the potential of his/her evolving protuberances, but at least it was all happening within the one watery environment: not so for the unfortunate birds seeking to leave the safety of Mother Earth. If forelegs evolved into wings, there would have been a long period when the dinosaur's brain still considered them as legs, and would have continued to try to use them as before, leading no doubt to great frustration. Attempting to use embryonic wings as legs could have offered no survival advantage whatsoever, quite the opposite in fact, and would argue for deselection. The issue becomes: What caused the brain of the dinosaur to give up on leg function, and seek another usage for whatever form the appendage that used to be legs now took on? Evolution of a physical appendage is one thing, but along with that process, the brain needs to be informed of the new function. The answer is, I suspect: no-one knows. More to the point, nothing in the research I have undertaken gives the slightest hint of how meaning can be imparted to the brain through the claimed evolutionary processes of growing new features such as fins, legs, or wings.

Of interest is this extract from the EAA Sport Aviation magazine. The "E" stands for Experimental, the magazine reporting the efforts and results of entrepreneurs in the sports aviation field. A great deal of experimentation continues in exploring the intricacies of flight, which makes these comments authoritative. In a discussion on a new airframe design:

"The Mk II airframe is mechanically unstable in all axes and requires complex control algorithms to fly. The EFRC is exploring the "learn to fly" technology that NASA researchers are also developing.

The fledgling Mk II could not be analyzed using computational fluid dynamics (CFD), Digital DATCOM (a computer program that calculates static stability control, and dynamic characteristics of aircraft), or standard aerodynamic textbooks. Instead, researchers chained it to the ground, built a cocoon structure around it to protect it from crashing, and began experimenting with parameter identification. "That's a fancy way of saying we tell it to do something, but we didn't know what it would do", Pat explained. "You start learning that when you tell it to do this, **this** is what happens."

Pat likens the process to how the human nervous system works. "It's the equivalent of isolating a single nerve ending and sending an electrical signal down your arm watching with your eyes what your arm does. You don't know what the muscle is going to do when the signal gets there. You eventually learn what sending the signal does."

"What has that to do with evolution?" you may ask.

Well, everything, but I suspect that you already know that. You see, if limbs and wings evolved, and nerves found their way from somewhere to there, how would the controller (brain) know what would happen when a signal was sent? A human scientific researcher can intelligently watch, study, and experiment, but such is beyond the remit of evolutionary processes."[8]

Maybe if evolutionists spent more time studying aviation, they would be less confident of their assertions that the capability of flight arose through evolutionary processes. Modern technology tells us otherwise. How was it that the proto-bird was mechanically and aerodynamically stable from the outset? If it wasn't, how did it become so without becoming extinct?

Perhaps the most important issue for proto-birds is why they would attempt to fly in the first place. Flying squirrels are said to have developed that capability over time, but I wonder. At a first attempt at flight, any animal type would crash to the ground, much as I did as a five-year old thinking that with a Superman cape, I could leap off the roof and fly. That didn't end so well, and I quickly learned to not do that again (at least, not whilst mother was watching). The early lessons of attempted flight would be pain and disillusionment, so why would an animal or proto-bird continue to do the same thing? What survival value existed in repeated crashes? What is natural about selecting for failure?

Summary

We could speculate endlessly on the sequence of the evolution of legs to wings, or even the evolution of wings from an entirely different anatomical source, and we could wonder about how long it took for the proto-bird to find the right muscle movement before successfully achieving controlled flight, but it would still be nothing other than speculation. There is a more fundamental problem to be solved: What informed the brain to firstly abandon natural usage and experiment with new muscle movements when it had no idea of what such movements might achieve? If new movements were entirely random, given that evolution is undirected, the likelihood of attaining the right combination of complex movements, by chance alone, is something that only wishful thinking could contemplate.

The phenomenon of *acquired characteristics* remains a subject of contention, largely because it is needed for evolution to be true, but the mechanism is (as yet) unknown. It is suspected that epigenetic inheritance through generations may be controlled by long noncoding RNAs, but this is yet to be proven. Curiously, though, Lamarck may have been onto something even though he has long been refuted. Even if true, the difficulty for the evolution narrative is that undesirable traits have the same likelihood of inheritance as desirable traits until deselected in later generations. Given that in the case of the proto-bird with their unintentional actions, unsuccessful experiments with proto-wing movement would far outnumber successful outcomes. In my mind's eye, I see generations of clumsy proto-birds trying to get airborne but crashing back to earth like the unfortunate gooney bird without achieving much at all. The Wright Brothers first flight was monumental by comparison!

Humour aside, evolution proponents must come to grips with this important issue if their narrative is to be in any way plausible. As best we know today, the brain controls muscle movement and for it to do that effectively, it must "know" that it has functional appendages; must "know" that it can move them; must "know" that such movement will be functionally useful; and must "know" that having commanded movement, that the movement occurred successfully. From a technical perspective, a fully-functional telemetry system must be in place before successful controlled movement can occur. We also have learned from amputees that the brain retains such "knowledge" long after it has been proven to be false. The brain does not automatically purge false knowledge of muscle function. If fins, legs, arms, and wings did indeed evolve, there would need to have been a process or mechanism to inform the brain of their existence, and an impetus within the brain to cause their movement. According to the evolution narrative, there is neither an ontological nor teleological basis for such activity - it is all simply a matter of chemical interactions, which raises the question as to what chemical reactions would arise in an undirected and autonomous manner to cause muscles to move in specific ways. Most importantly, if an appendage changed in function, how did the brain "forget" the original function? What erased the muscle memory of before?

I have said little in this chapter regarding *knowing that you can* beyond what we know as *instinct*. In the next part, we will investigate further.

References:

1. Bray, Dennis, Wetware: *A Computer in Every Living Cell*, Yale University Press, New Haven, CT, 2009, quoted in introduction

2. Eccles, John C., *How the SELF Controls Its BRAIN*, Springer-Verlag, Berlin, Germany, 1994, p. x

3. *Ibid*

4. Kaufmann, Stuart A., *The Origins of Order: Self-Organization and Selection in Evolution*, Oxford University Press, Oxford, UK, 1992

5. Dawkins, Richard, *The Greatest Show on Earth – the Evidence for Evolution*, Bantam Press, London, 2009, p. 165

6. Bell, P., *Mudskippers - marvels of the mudflats*, Creation **34**(2), 2012, pp. 48-50

7. Talbot, Wayne, *The Dawkins Deficiency – Why Evolution is Not the Greatest Show on Earth*, Deep River Books, Sisters, OR, 2011

8. EAA Sport Aviation, Volume 67 No. 1 / January 2018, *Seed of Technology – Heurobotics Mk II*, Beth E. Stanton, pp. 16-17

Chapter 1-10: Primary Axiom #10
< Material Entities Can ONLY Do what they MUST Do >

"Don't mistake activity for achievement"

~ John Wooden, American basketball player and coach ~

Obviously, these words above were delivered in an entirely different context, but I sense that such advice could be given to scientists, especially those committed to the neurosciences. In the context of cognitive functions, activity in the brain, in a particular region, should not automatically lead to a conclusion as to what is achieved in that region. For example, consider this statement:

"Ganglion cells transmit distinct visual characteristics in multiple parallel pathways; such parallel processing allows the separation of different aspects of the visual stimulus, such as color, form, and motion. All of these signals are processed and interpreted by the part of the brain called the *visual cortex.*"[1]

In what way does the visual cortex *interpret* electrical signals? To interpret is defined as to explain the *meaning* of a communication or action. The more interesting question is: to whom or to what are the signals explained? The recipient of messages from the visual cortex are some other parts of the brain, which no doubt further "interpret" what has been communicated to them. No doubt, a great deal of signal processing is performed in the visual cortex, and whilst experiments with animals have allowed the identification of other parts of the brain involved with sight, the process of visualisation remains a mystery. One source opines: "The brain extracts biologically relevant information at each stage and associates firing patterns of neuronal populations with past experience"[2]; and another, with a trifle more modestly, "Neurons in the visual system do not form a picture in the brain, but somehow we can interpret all of these multiple, parallel signals generated by "seeing" an object"[1].

Correlation of psychometric and neurometric testing results has successfully identified areas of the brain implicated with specific cognitive activities. Whilst the domain of the cognitive activity is known, e.g., associated with reasoning, spatial ability, memory, processing speed, and vocabulary as done in IQ tests, the specifics of the mental activity cannot be interpreted from that neural activity. What cannot be known, at least as yet and as I believe, never, is whether the brain activity represents the totality of the mental activity, or whether, as I contend, there is a contribution from the immaterial mind.

A theme to which I will regularly return is the use of "we" as in "somehow **we** can interpret all" (emphasis mine). Who or what is this *we*? Mentioned earlier, Feigl (1967) and others believed in identity theory – that cognitive events are identical with brain events. If the brain, a biological lump of exquisite complexity, is the conductor of events, then there is no valid concept of a personal identity: I or we. Human identity is therefore no different in explanation than the identity of a chimpanzee or elephant.

Scientific research has identified areas of the brain associated with specific sensory inputs, but use of the term "interpret" is inappropriate. Interpret is a cognitive function dealing with conceptual entities, but signalling activities from sensory receptors to the cortex and other parts of the brain are entirely physical. Physical (material) entities are incapable of conceptualisation, because their properties do not allow it.

As earlier noted, William Provine asserted "Modern science directly implies that the world is organized strictly in accordance with mechanistic principles. There are no purposive principles whatsoever in nature."[3] The foundations of this opinion are evolution and philosophical materialism. "Mechanistic" relates to theories which explain phenomena in purely physical or deterministic terms, or determined by physical processes alone. It should be obvious that such principles ought to lead to the belief that material entities can only do what they *must* do, in whatever circumstances they are subject to, based on their properties and those of other entities with which they interact. This is foundational to technology and its applications. I have a passing acquaintance with quantum theory and the Uncertainty Principle, but for all practical purposes, I would offer that what I have stated is true, and accepted by the scientific world as true. Well, not quite. I have been intrigued by explanations that we will come to, that biological entities evaluate and make decisions.

Again, quoting Johnson quoting Provine, ""There is no way that the evolutionary process ... can produce a being that is truly free to make choices", this opinion is predicated on the assertion that *there are no purposive principles whatsoever in nature*. If nature has no purpose and no goals, then the behaviour of material entities cannot evidence these attributes – every effect (behaviour) is subject to a predetermined mechanistic cause. Broadly speaking, this philosophy is to our benefit, and is supported by the orderly nature of our Universe – if it were not so, our existence would be one of chaos. It would seem that only humankind is capable of upsetting the orderly nature of things, and indeed, the modern Greens movement argues that Planet Earth would do far better without us.

In the context of the mind-brain conundrum, we should use this axiom to assist in identifying the behaviours of which the brain in capable, and those of which it is incapable. We have sufficient knowledge of the anatomy and physiology of the brain to make these determinations. When scientists offer that the mind is *an emergent property* of the brain, or that that the mind is *the activity* of the brain, we have sufficient knowledge to categorise such properties and activities. Interpretation in a cognitive sense is not one of those activities, despite scientific narratives which presume that it is. Similarly with the mind, but starting from the observed behaviours of the mind, we can determine whether such activities exhibit compliance with this axiom, leading to a conclusion as to whether or not the mind is material.

This we shall do in later chapters.

References:

1. https://www.ncbi.nlm.nih.gov/books/NBK21661/

2. https://www.brainfacts.org/thinking-sensing-and-behaving/vision/2012/

3. Johnson, P., *Darwin on Trial*, 2nd ed., Illinois, USA, 1993, p. 126, quoting William Provine

Part 2: The Exceptional Human Condition

"Perhaps our greatest distinction as a species is our capacity, unique among animals, to make counter-evolutionary choices."

~ Jared Diamond ~

If humans are more exceptional than other life forms on Planet Earth, why is that so? C.S. Lewis, in a theological context, but apposite in this discussion, offered the following advice: "As long as you are looking down, you cannot see something that is above you."[1] The point, as Gerard Verschuuren[2] notes for us, is that if we keep looking down at biological and physical levels of existence, such as DNA, genes, neurons, brains, computers, and such, we will lose focus on what makes us "tick" – the mind. Certainly, the brain operates autonomously for primary functions of the body, so long as it is supplied with the necessary resources, but there are volitional and creative aspects to our existence which from an intellectual perspective, should interest us far more. If we do not ask: Why is it so? We fail to exercise the attributes which make us *unique among animals*.

Clinical neurologist, Raymond Tallis wrote: "Human exceptionalism, so it is argued, is a hangover from religious doctrines about the special relationship between Man and the Creator and the former's consequent special place in Creation. Hence the paradox that much humanist thought is motivated by the desire to put humanity in its place, and to emphasise that we are, above all, animals. If we differ from our nearest animal kin, it is argued, this is a matter of degree rather than kind."[3] I am not interested here in discussing creationism or God, but would suggest that much of this *desire to put humanity in its place* arose from revolutions in recent centuries against the monarchy and its collaborative partner, religion. If the monarchy were to go, so too religions, and even more, the gods they worshipped. Sadly, I contend that the baby has been thrown out with the bathwater, arguing against the reality of the so-called *Age of Reason*. Explaining his own motivation, Tallis commented: "*Aping Mankind* was written in the belief that we can accept a less restricted sense of human possibility than naturalism permits, and acknowledge what is in front of our eyes – namely our unique mode of being – without risking a regression to religious dogma or embracing irrationalism and throwing away the freedoms we have gained from the creeds that proved such a mixed blessing for humanity. While we are not supernatural entities, we are not entirely parts of nature, either; we are extra-natural creatures, best understood as human beings rather than smart chimps."

Whether we recognise our own, or even accept the concept, everybody has a *worldview*. We all perceive the world through a lens coloured by race, religion, culture, nurture, tradition, education, the circumstances of our upbringing, our later life, and our preferences. I was raised in the Catholic tradition, and though I have since discarded much of the theology in which I was indoctrinated, some does still remain, even a sense of guilt for abandoning the faith of my forebears. I still enjoy listening to, and even singing along with, some of my favourites from the *Missa Cantata* (Latin for "sung Mass"), the Latin words more in my mind than rolling off my tongue as they did in the days of my youth. There is comfort in these memories, memories which likely will never leave me, and perhaps never should. We see examples of this in the Western Christmas tradition, where public performances are held of Christian hymns being sung by people who adhere to various, or no, religious faith, yet delight in the evocative sounds that such hymns offer. I especially enjoy musical performances at Christmas, even though the occasion no longer has any spiritual significance for me.

On such is Western civilization built.

In a very real sense, this and previous studies represent my worldview, some of it informed and rational, some of it quite likely not. But in interrogating my own worldview, I have come to better understand the worldview of others. For me this is important, for I am curious to know why my thought patterns are so different to those of other people, and why what interests me is of so little interest to others. Of course, this is all for the good, for if most were like me, there would be no music, no art, no medicine, no athletic ambition, no engineering, and lots of other pursuits which have never been part of my playbook or bucket list. That said, I am curious as to why, even when we have shared interests, our worldviews are so different.

I often encounter literature quoting Jonathan Swift: "*You cannot reason a person out of a position he did not reason himself into in the first place.*" I am unsure of the original context, but can state unequivocally that this is not an absolute truth. Most of us were indoctrinated: some in religion, and more recently, in the overarching narrative of evolution, but many, such as myself, have reasoned our way out of positions which we *never* reasoned our way into, and probably would never have done so.

From the extremes of science at one end, and fundamentalist religion on the other, comes dogmatism, as if any of us could ever know the truth of our existence. Science and religion seem ever in conflict, in the minds of many, but for an objective theist (if there ever could be such a person), they could never be truly in conflict, because they both have the same source. Ever since mankind began to think and form opinions, truth became a victim. This is true of both science and religion, as I shall later contend, offering the evidence that has led me to this conclusion. At the same time, I acknowledge that, I too, may be part of the problem, rather than the solution.

But a thirst for, and acquisition of, true knowledge, is the only path that I can see which will lead to greater truths, and so it is the path that I have chosen to follow. Mind you, I have led walks in the Australian bush, and motorcycle rides elsewhere, and have been known to lead people astray.

Follow at your own peril.

References:

1. Lewis, C.S., *Mere Christianity*, Harper, San Francisco, 2001, p. 74

2. Verschuuren, Gerard, *What Makes You Tick? A New Paradigm for Neuroscience*, SOLAS Press, Antioch, CA, 2012, p. 77

3. Tallis, Raymond, *Aping Mankind*, Routledge Classics, New York, NY, 2016, pp. xi-xii

Chapter 2-1: Sex

"An intellectual is a person who's found one thing that's more interesting than sex."

~ Aldous Huxley ~

Perhaps I have started my discussion on the *Exceptional Human Condition* on this subject, because I suspect that Aldous Huxley was speaking of me. Mind you, approaching my 77th birthday, I do recall the term, but my mind says that it has something to do with ancient, primitive, religious customs - such is the infirmity of age, in both mind and body. Yes, I confess that the title of this chapter may be seen as disingenuous, but let us see how we go.

I blame Sigmund Freud for the modern obsession with sex, not that humans have not always had a fascination with, and a natural desire for, sex, but let us put that into a modern perspective. Psychiatric professionals have much to say on this subject, and it is worth pondering how importantly they treat it in relation to the human condition, and how their views would seem to refute the notion that our minds are nothing more than the workings of the biological lump in our heads, for our minds can overcome our natural biological predilections. More to the point, science cannot explain psychology. Consider this quote from the American Psychiatric Association:

"In our current oversexualized culture, sex has become a commodity, immaturity is often idealized, and sexual conquests have been valorized as sport. These pervasive exhibitionistic displays undermine the psychological value of intimate long-term personal attachments. While aspects of Sigmund Freud's theories have undergone revision, the central place the founder of psychoanalysis gives to sexuality and intimate personal connections remains valid. Freud's ideas teach us the value of intimate personal attachment and its key place in mature sexual fulfillment."[1]

Whilst intimacy leading to sexual congress is often the start of long-term relationships, relationships do mature in the absence of sexual intimacy. Psychologists attempt to explain this in *non-materialist* terms relating to the *material*, which is where their explanations lack both coherence and credibility. There is clearly something else going on which neither genetics, nor biology, seem able to explain. Intimacy is more than sexual attraction, for as evidenced so often, the former outlasts the latter. In relation to the sexual urge, which no-one doubts is biological in origin, humans evidence abilities and behaviours not found in lower order species, primarily *restraint*. Restraint is a function of will power, or free will, an attribute that evolutionists deny exists because in their better moments, they admit that evolution is incapable of developing such an attribute. More on that later. Here, I am content to let the reader consider their own experience related to sexual urges: how they do, or do not, respond to them; and whether they believe that "they" are in control, or whether they are at the mercy of the biological organ known as the brain. Who is the master here?

In this context, I now want to turn to another aspect of the mind-brain conundrum: indoctrination and other forms of mind control including subliminal suggestion. There can be no doubt that our minds are conditioned by culture and other factors, and if the evolutionists are correct, such conditioning is instantiated in the neural networks of our brains. In evolution terms, this is, *adaptation*, but unlike the morphogenic phenomena discussed by Charles Darwin, such as the changes in the beaks of finches, adaptation in neural networks is fluid, responding to changes in fashion, protest movements, and the like. Again, if the evolutionists are right, the brain is the most adaptive organ of the human body, totally unlike any other claimed biological predecessor such as chimpanzees. No other creature on the planet responds in any similar way.

If you were brought up with the adventurous literature of my youth, you would have enjoyed the works of authors such as Hammond Innes, Neville Shute, Ion Idriess, Nicholas Monsarrat, Alistair McLean, *et al*. In these books, explicit sex featured not at all – romance, yes, but the explicit descriptions, page after monotonous page, scene after forgettable scene, of today's authors and movie producers? One could excuse it on the prudery of earlier generations, but that would be far too easy. Perhaps, instead, it was morality, and a refusal to accept the evolutionists' desire to *devolve* us to nothing more than animalistic instincts – evolution in reverse as it were. Are attractions between men and women nothing more than a sexual desire to leap into bed, or as Hollywood would have it, coitus as they press the bell for the elevator, stripping clothes from one another, as if they cannot wait for the elevator doors to open? Why would they care? They might as well have it off in the foyer of the hotel, for all that they would appear to care, or notice the reactions of other people.

Are the representations that we see on movie or television screens, or as we read in modern books, truly typical of normal human behaviour? If not, why does society at large accept such representations without protest? Why do so many succumb to the depravity of publishers and movie producers? Have you ever been to Indonesia, India, or other parts of Asia, where monkeys eagerly leap upon the lower orifices of other monkeys, male or female? Do you accept this animalistic behaviour as representative of where you came from, and in fact, still are? Is that truly whom you believe yourselves to be, nothing more than an evolved version of the primitive animals that you see around you, with no plausible explanation for how evolution made you so different, so unbelievably different, in such short a time? How far different are chimpanzees from lions or tigers, other than their method of locomotion? How far different, by comparison, are humans from chimpanzees? What was the posited timeline of such evolution?

My point here is to emphasise the volatility of human culture in contrast to the fixed cultures of lower order species. There is something truly mysterious about the mind of humans in that they can both succumb to indoctrination, and fight against it, and at times struggle to come to terms with where they stand on a particular issue. From the outset, I took a firm stand against what I perceived to be the lowering of moral standards over the past half-century or so, no doubt due to my strict Catholic upbringing, but others of my time, with a similar upbringing, took a different path.

One more point, perhaps irrelevant to some, but of interest to me. Whilst reading authors such as noted above, or listening to stories on the radio, our minds were actively involved – our imaginations adding more to the stories than the narrators could convey. In modern times, where entertainment is graphically visual, the mind goes into recess. I recall a quote from Albert Einstein where he stated (not his exact words), that reading is the most intellectually intensive activity we can pursue. On the other hand, he also said: "Reading, after a certain age, diverts the mind too much from its creative pursuits. Any man who reads too much and uses his own brain too little falls into lazy habits of thinking." The irony is that evolutionists would say that Einstein had it backwards – it is the brain that uses man.

Scientific research into all domains, including genetics and biology, relies entirely on the premise of "cause and effect", based on the current understanding of the laws of chemistry and physics, and in some areas, quantum mechanics. The question becomes: how can scientists, with any level of confidence or authority, explain the variety of behaviours of the human mind, when the brain of humans has the same biological construction as that of the claimed last common ancestor, the chimpanzee?

References:

1. https://psychnews.psychiatryonline.org/doi/full/10.1176/pn.40.15.00400018

Chapter 2-2: An Unfortunate Experience

"But like so many unfortunate events in life, just because you don't understand it doesn't mean it isn't so."

~ Lemony Snicket, *The Bad Beginning* ~

In my seventy-fifth year, during a trip to Norway, Finland, and Russia, whilst the snow was still deep on the ground and inland overnight temperatures dropping to -30°C, I contracted a serious double-lung infection, causing me to wonder why this was ever on my bucket list. What was wrong with the sunshine in my native Australia? The illness was the worst I had ever experienced in my life (up to that point). I couldn't sleep due to continually coughing up phlegm; I had no appetite and even less energy, and lost all interest in life. I thought that I was lucid and rational, but on later reflection, I realised that I had been delirious. There was no suggestion of suicide, but life no longer had any meaning, and there was nothing remotely resembling hope for the future. Had someone wanted to take my life, I had neither the energy nor interest to resist. I wanted more care, attention, sympathy, and compassion than I was receiving, and I later realised that my wimpish side, normally easily suppressed, had come to the fore. I am a hungry reader, but I could barely achieve two lines. I am a prolific writer, but no words came to mind. I have been an avid motorcyclist for over fifty years, but my motorcycle sat in the garage unridden and unwanted – it could sit there forever, for all I cared.

The initial recovery period was about seven weeks, attracting three courses of antibiotics and the same number of x-rays, and numerous months thereafter before I considered myself back to where I should have been. Slowly, my normal character and personality re-emerged, and sitting out in the comforting Australian sun, cold beer in hand, I thought to myself: What was that all about? The illness had temporarily changed me, not only physically, but mentally and emotionally as well. Why was that? No doubt my brain had suffered a level of oxygen deprivation as my breathing had been shallow and laboured for weeks. My body had suffered a lack of proper nutrition, and I had experienced some of the normal side-effects of a prolonged course of antibiotics. My brain had ceased to work as normal, but what had that to do with motivation?

There are numerous studies on the power of positive thinking, and how thoughts and emotions can affect health and physical well-being, but here I was experiencing the apparent reverse, or perhaps a closed-loop system with physiological feed-back. Health can affect the mind which can affect one's health – no doubt, there is a relationship. When René Descartes opined, "I think, therefore I am", was he also aware that perhaps *what* he thought had implications for *what* he was?

That said, I remain convinced that the mind and brain are not one and the same entity, and do not believe that the mind could be an emergent property of the brain. However, here was irrefutable evidence that physiology had ramifications for both mental and emotional states. What was the connection with the mind, if there was one, and could this experience help me to define the mind as a separate entity? I have no training in neurology nor any related discipline, but I do have access to the writings of acknowledged experts, and am well versed in the fundamentals of epistemology and knowledge itself. My extensive career in Information Technology has also given me insights into the requisite processes for the creation, transmission, and storage of what is loosely termed, information.

It is these disciplines which give substance to this study, without in any way explaining the conundrum which this unwelcome experience had highlighted. Nonetheless, I seemed to have learned something without really understanding it.

Chapter 2-3: Official Theories of Being

"I am, therefore I think."

~ my own take on René Descartes' more famous saying ~

In his book of the same name, Steven Pinker notes that *"BLANK SLATE is a loose translation of the medieval Latin term tabula rasa – literally, 'scraped tablet'. It is commonly attributed to the philosopher John Locke (1632-1704), though in fact he used a different metaphor."*[1] Let us understand what this refers to, but before I begin, I wish to acknowledge that I have sourced much of the substance of this chapter from Steven Pinker's book – I could not have researched this on my own.

The context here is what some describe as the "official" theory of the human condition. When the Christian religion dominated Western thought, man was made in the image of God; had fallen from grace and in need of redemption; had an innate conscience of what represented good and evil but nevertheless was inclined to evil; but also had guidance in the form of the bible – the Sacred Scriptures. This was largely predicated upon Augustine's doctrine of original sin, proclaiming that all people were born broken and selfish, saved only through the power of divine intervention. I believe Augustine to have been wrong on many issues, largely due to his commitment to Christian apologetics, whereby the human condition had to be seen in terms of Christian dogma, especially the doctrine of *The Fall*. I contend that he failed to differentiate between *processing* and *being*, as discussed in a later chapter. We are not our dispositions, because we can overcome them. We do have dispositional properties, arising from the culture of our development, and perhaps even from genetics, but these do not define who we are. In a similar vein, Viktor Frankl noted: *"between stimulus and response there is a space. In that space is our power to choose our response. In our response lies our growth and our freedom."*

Christians say that we sin because we are evil, whereas the Jew says that we are evil when we sin. The Christian makes the category mistake of equating a dispositional property of man with man himself, an unavoidable mistake if they are to be true to their religion. We are not our dispositions, but how we respond to those dispositions. To possess a dispositional property is not to be in a particular state; it is to be possible or liable to be in a particular state when a particular condition is realised. A China cup has the dispositional property of being brittle, but that is not to say that it will break or shatter without cause, nor that it will necessarily break when dropped. Dispositions (propensities) do not have uniform outcomes.

In recent years, some of the work of psychologists has been taken over by people described as *Life Coaches*. That this is recognised as a formal discipline, becoming ever more popular, is evidence that many believe that we can overcome our dispositions and change our responses and behaviour, even later in life. Kimberly Giles advised regarding overcoming fear:

"First, understand you are not choosing this behaviour consciously. Your subconscious programming (that you got from your family and your past experiences) is causing you to react to certain people with fear. The first step to changing this is just recognizing it consciously when it happens, so you have a choice. Becoming more mindful and paying attention to your reactions, thoughts and behaviour is something that will require practice and patience, but you can do it and you can change your programming. I help people do it every day, but it takes commitment and work. I'll give you some principles and tips to help you do this."[2]

During the period of Western colonial expansion, new theories arose, one being that of the *noble savage*. Jean-Jacques Rousseau (1712-1778), though he did not coin the term, wrote:

"So many authors have hastily concluded that man is naturally cruel, and requires a regular system of police to be reclaimed; whereas nothing can be more gentle than him in his primitive state, when placed by nature at an equal distance from the stupidity of brutes and the pernicious good sense of civilized man."[3]

Curiously, this concept remains evident in the minds of many moderns who, condemning colonialism, assert that indigenous tribes lived in peaceful cooperation. Sadly, history tells us otherwise. It is suggested that Rousseau wrote as he did to refute the earlier opinions of people such as Thomas Hobbes (1588-1679), who perhaps was closer to the experience of the early colonialists. He wrote:

"Hereby it is manifest, that during the time men live without a common power to keep them all in awe, they are in that condition which is called war; and such a war as is of every man against every man ... In such condition there is no place for industry, because the fruit thereof is uncertain: and consequently, no culture on earth."[4]

One can sense a foundational Christian view in this opinion, and although Hobbes' father was a vicar, there is uncertainty concerning his own religious views. At one stage he was accused of atheism, responding: "atheism, impiety, and the like are words of the greatest defamation possible"[5]. Renowned French philosopher, René Descartes (1596-1650), opened the door to the concept of the mind versus body, or as some are wont to ridicule his idea, the *Ghost in the Machine*. Descartes opined:

"There is a great deal of difference between mind and body, inasmuch as body is by nature always divisible, and the mind is entirely indivisible ... When I consider the mind, that is to say, myself inasmuch as I am only a thinking being, I cannot distinguish in myself any parts, but apprehend myself to be clearly one and entire, and though the whole mind seems to be united to the whole body, yet if a foot, or an arm, or some other part, is separated from the body, I am aware that nothing has been taken from my mind."[6]

The ridicule was from the pen of philosopher, Gilbert Ryle (1900-1976), whose book, *The Concept of Mind*[7] is one to which we will also refer in later chapters. In brief he wrote:

"There is a doctrine about the nature and place of minds which is so prevalent among theorists and even among laymen that it deserves to be described as the official theory ... The official doctrine, which hails chiefly from Descartes, is something like this. With the doubtful exception of idiots and infants in arms, every human being has both a body and a mind. Some would prefer to say that every human is both a body and a mind. His body and his mind are ordinarily harnessed together, but after death of his body his mind may continue to exist and function. Human bodies are in space and are subject to mechanical laws which govern all other bodies in space ... But minds are not in space, nor are their operations subject to mechanical laws ...

... Such in outline is the official theory. I shall speak often of it, with deliberate abusiveness, as the dogma of the 'Ghost in the Machine'."[8]

Ryle describes this a *Descartes Myth*, but I suspect that his self-proclaimed abusiveness clouded his mind to some demonstrated truths, most especially that the mind is not *subject to mechanical laws*, at least, not as so far understood. Had I had the opportunity, I would have challenged him on the evidence or logic that he used to propose to the contrary. Here is a perspective on Ryle, for it is always worth understanding the man, before appraising his work.

"Although Gilbert Ryle published on a wide range of topics in philosophy (notably in the history of philosophy and in philosophy of language), including a series of lectures centred on philosophical dilemmas, a series of articles on the concept of thinking, and a book on Plato, *The Concept of Mind* remains his best known and most important work. Through this work, Ryle is thought to have accomplished two major tasks. First, he was seen to have put the final nail in the coffin of Cartesian dualism. Second, as he himself anticipated, he is thought to have argued on behalf of, and suggested as dualism's replacement, the doctrine known as *philosophical* (and sometimes *analytical*) *behaviourism*. Sometimes known as an "ordinary language", sometimes as an "analytic" philosopher, Ryle— even when mentioned in the same breath as Wittgenstein and his followers—is considered to be on a different, somewhat idiosyncratic (and difficult to characterise), philosophical track.

Philosophical behaviourism has long been rejected; what was worth keeping has been appropriated by the philosophical doctrine of functionalism, which is the most widely accepted view in philosophy of mind today. It is a view that is thought to have saved the "reality" of the mental from the "eliminativist" or "fictionalist" tendencies of behaviourism while acknowledging the insight (often attributed to Ryle) that the mental is importantly related to behavioural output or response (as well as to stimulus or input). According to a reasonably charitable assessment, the best of Ryle's lessons has long been assimilated while the problematic has been discarded. If there are considerations still brewing from the 1930s and 40s that would threaten the orthodoxy in contemporary philosophy of mind, these lie somewhere in work of Wittgenstein and his followers—not in Ryle.

But the view just outlined, though widespread, represents a fundamental misapprehension of Ryle's work. First, Cartesianism is dead in only one of its ontological aspects: substance dualism may well have been repudiated but property dualism still claims a number of contemporary defenders. The problem of finding a place for the mental in the physical world, of accommodating the causal power of the mental, and of accounting for the phenomenal aspects of consciousness are all live problems in the philosophy of mind today, because they share some of the doctrine's ontological, epistemological, and semantic assumptions."[9]

As dualism is fundamental to the ongoing debate over the reality of the mind-brain complex, we need to revisit the issue. *Dualism* is the philosophical position that the mind exists as something separate from the brain, or body. *Cartesian Dualism* is simply a term for the dualism expressed by Descartes, to differentiate from other philosophical positions. All forms of mind-body dualism claim that the mind is not physical or material in the common understanding of those terms. *Substance Dualism*, as mentioned above, is a variation on that position, stating that two sorts of substances exist: the physical and some other. I am unclear of the meaning of the word, *substance*, in the context of the "other", and the only clarification that I can find on that issue is in Christian theology. As advised by my good friend, Edgar Andrews, who knows far more about these issues than I do, he offers that this is an ancient usage meaning "essence" or "nature", as used in the Nicene Creed, "being of one substance with the Father". This theological overtone would be cause for rejection by those not believing in God. Another suggestion is that physical things exist in our space-time continuum but do not possess any thought, whereas mental things have thought as their very essence, but have no physical properties. If I accept that this "other" substance is nonphysical, but without any connotation of God, then I can concur with *Substance Dualism*.

And then we have *Property Dualism*, as expressed by "the brain is an emergent property of the brain", arguing that the mind and body exist as ontologically distinct *properties* of a single, physical substance. It is asserted that the brain exists as a physical substance with both mental properties, such as thoughts and feelings, and physical properties, such as size, shape and chemistry. Whilst property dualism places an ontological distinction, it denies Substance Dualism that they are separate, independent substances. I entirely reject Property Dualism, for nothing physical can give rise to the conceptual – the material only does what it can and must do.

As can be seen, nobody truly knows, and I suspect that those who take varying positions do so based more on their philosophical positions than on scientific or logical evidence. *Philosophical Materialism*, strictly speaking, is not synonymous with *Material Monism*, which is a pre-Socratic belief that objects of the world (entities) are all composed of a single element. However, given that modern physics is burrowing ever deeper into the composition of physical matter, expecting to find the smallest identifiable particle, it could be said that the ancients were correct – it was just their understanding of "element" which was at odds with a modern definition.

J.B.S Haldane (1892-1964) was renowned for his work in the studies of physiology, genetics, evolutionary biology, and mathematics. He wrote on the subject of *abiogenesis*, attempting to explain the chemical origin of life. He made useful contributions in the fields of statistics and biostatistics, and is credited with being the first to suggest the possibility of *in vitro* fertilisation. He was a religious sceptic, and vigorously promoted Darwinian evolution. Clearly a scientist of credibility, but on the subject of the mind he wrote:

"It seems to me immensely unlikely that mind is a mere by-product of matter. For if my mental processes are determined wholly by the motions of atoms in my brain, I have no reason to suppose that my beliefs are true. They may be sound chemically, but that does not make them sound logically. And hence I have no reason for supposing my brain to be composed of atoms. In order to escape from this necessity of sawing away the branch on which I am sitting, so to speak, I am compelled to believe that mind is not wholly conditioned by matter."[10]

Here we have a scientist, well versed in the sciences associated with biology and evolution, who has concluded that the mind cannot be "conditioned by matter", to use his words. If it is not matter (material), then it can only be conditioned by the immaterial, whatever that may be. Haldane was emphasising that chemicals are reliable in what they do, but that they are incapable of logic relating to conceptual matters. This is similar to my own contention that the material can only do what it can and must do, and no other, whilst the mind is capable of conceptualising the most improbable and impossible things, just as the White Queen did.

My own position leans toward the *Ghost in the Machine*, but as this study will explain, not for any philosophical reason. Perhaps knowing a lot more about information processing than the philosophers of earlier ages, and having access to the results of recent neurological research, I am able to shed a different light on the subject. In essence, I concur with René Descartes, but with additional arguments. I can appreciate why those accepting evolutionary theory, and those agreeing with Carl Sagan that "the Cosmos is all that is or was or ever will be", must hold to either Property or Substance Dualism, for to do otherwise would undermine their foundational beliefs. However, the evidence leads me to conclude that neither of these philosophical positions can be valid. I cannot offer a coherent philosophy of my own – the mind-brain complex remains a mystery to me, but I must reject philosophies which are demonstrably contradictory to the evidence, and that is what I have set out to do in this study.

From the sources I have researched, it does seem that *Property Dualism* is the favoured position of modern psychology and related neurological sciences. However, there remains a great deal of debate about just what that means in physical terms. Another more recent philosophy is *Predicate Dualism*. This view, which is espoused by non-reductive physicalists, maintains that while there is only one ontological category of substances and properties of substances (usually physical), the predicates that we use to describe mental events cannot be redescribed in terms of (or reduced to) physical predicates of natural languages. Whilst I am yet to reach a conclusion of my own, I find myself somewhat aligned with non-reductionist philosophers who recognise this conundrum with regard to mind–body relations: if physicalism is true, then mental states must be physical states; on the other hand, mental states cannot be reduced to behaviour, brain states or functional states, using the standard terminology of physicalism.

Those holding to *Substance Dualism* seem unable to articulate or define what they mean by *mental substance*. Those holding to *Property Dualism* seem unable to coherently articulate or define what they mean by *mental property*. Those holding to *Predicate Dualism* understand why the foregoing difficulties exist. In the next part of this study, we will address some foundational issues which many scientists ignore in their attempted explanations of the mind. Whether that is because they do not understand, they do not wish to reveal unexplainable difficulties, or simply a matter of choice of words, I am not in a position to conclude. However, what is evident from their writings is that they fail to understand the difference between entities, properties, and activities (see earlier discussion on *Foundational Propositions*). I offer that nothing can at the same time, be all of these things. If we are to properly understand the mind, or accept the writings of those who claim to do so, any explanation must begin with a definition of the thing that is being discussed. It is not uncommon for scientific explanations to fail in this way, as I have found when researching the nature of space, time, energy, and matter: it may surprise the reader that I have been unable to find a coherent definition to which scientific texts faithfully adhere. Just like the mind, these entities are whatever the context requires them to be.

Raymond Tallis alerted me to "Avrum Stroll's superb historical essay on epistemology in the 1993 editions of the *Encyclopedia Britannica*"[11]. I have included it here because of its relevance to dualism:

"There have been explosive advances in neuroscience, psychology, cognitive science, neurobiology, artificial intelligence, and computer studies. These have resulted in a new understanding of how seeing works, how the mind forms representations of the external world, how information is stored and retrieved, and the ways in which calculations, decision procedures and intellectual processes resemble and differ from the operations of sophisticated computers, especially those capable of parallel processing.

The implications for epistemology of these developments are equally exciting. They promise to give philosophers new understandings of the relationship between common sense and theorizing, that is, whether some form of materialism which eliminates reference to mental phenomena is true or whether the mental-physical dualism which common sense assumes is irreducible, and they also open new avenues for dealing with the classical problem of other minds, (Stroll, 1993)"

In this study, I attempt to demonstrate that *the mental-physical dualism which common sense assumes is* indeed *irreducible*. Examined closely, the terminology used by those who reject dualism, in explaining the workings of the mind-brain, glosses over the physiology of the brain to such an extent that I cannot recognise the entity they are discussing. I can highly recommend Raymond Tallis' book referenced here, for it explores issues which, with a sense of envy, I had not considered. I pick up on some of those ideas, but with nowhere near the clarity of his own writings.

Those holding to *Cartesian Dualism*, and I am one, have no idea how to define the mind other than by the *activities* of which we know it to be capable. Its inherent nature is beyond our comprehension. I have a suspicion, from reading the arguments against Descartes' position, that such rejection may well be predicated on an antipathy toward any suggestion of something continuing to exist after the death of the body. The obvious questions arise: If the mind is immaterial, in what way is its existence dependent on the material (body)? Can the mind exist prior to the body, and persist after the death of the body? Even worse, does this segue into a discussion on the existence of God and the immortal soul?

It should be obvious why some people do not want to start down that track, just as they reject the concept of Intelligent Design – they are concerned about the identity of the designer. I have no such fears; on the contrary, I research with a degree of excitement, wondering what I will next discover. I have been an inquisitive wanderer all my life, travelling through numerous countries, and experiencing numerous cultures. What I have long done physically but no longer, I continue to do intellectually, knowing that I will never fully know, but content to at least unearth false knowledge in the quest to find true knowledge.

That is my philosophy of being (for the time being at least).

References:

1. Pinker, Steven, *The Blank Slate: The Modern Denial of Human Nature*, Penguin Books, London, UK, 2002, p. 5

2. Giles, Kimberly, *Choosing Clarity: The Path to Fearlessness*, Thomas Noble Books, Wilmington, DE, 2014

3. Rousseau, Jean-Jacques, *Discourse upon the origin and foundation of inequality among mankind* (1755), Oxford University Press, New York, NY, 1994, pp. 61-62

4. Hobbes, Thomas, *Leviathan* (1651), Oxford University Press, New York, NY, 1957, pp. 185-186

5. Hobbes, Thomas, *Opera Philosophica Quae Latine Scripsit Omnia* (1669), Molesworth edition, p. 282

6. Descartes, René, *Meditations on first philosophy* (1641), in R. Popkin (Ed.), *The philosophy of the 16th and 17th centuries*, Free Press, New York NY, 1967, p. 177

7. Ryle, Gilbert, *The Concept of Mind*, Penguin Books Ltd, London, England, 1990

8. *Ibid*, p. 13-17

9. https://plato.stanford.edu/entries/ryle/ (First published Tue Dec 18, 2007; substantive revision Wed Feb 4, 2015)

10. Haldane, J.B.S., *When I am dead*, in *Possible Worlds and Other Essays* (1927), Chatto and Windus, London, UK, 1932 reprint, p. 209

11. Tallis, Raymond, *Why the Mind is Not a Computer*, Imprint Academic, Exeter, UK, 2004, p. 12

Chapter 2-4: Philosophy

"The study of the fundamental nature of knowledge, reality, and existence, especially when considered as an academic discipline; a theory or attitude that acts as a guiding principle for behaviour."

In my (mostly futile) debates with atheists, especially those whose lives are spent in scientific pursuits, I find that they generally decry philosophy as having any real existence, let alone significance. They are adamant that they have no philosophy, however defined, and deny its legitimacy. Just as often, when I question some statements by scientists, I am told that I am insulting the intelligence of these dedicated people – how could they believe in something which was not right? I then point to their own argument that "science is self-correcting", and thus from time to time, is in *need* of correction. That being so, there must be times when scientific beliefs are false, or to put it bluntly, *just plain wrong*. The natural deduction is that throughout the ages, scientists have often believed what is not right. So often, in so many ways, philosophy triumphs science.

There was a time when scientists were polymaths: people whose expertise spanned a significant number of different intellectual disciplines, enabling them to solve problems by viewing them from numerous perspectives.

"In the Middle Ages, one's education began with the trivium, which consisted of grammar, rhetoric, and dialectic. Once these were mastered, the student was ready to begin the more advanced topics of the quadrivium: music, arithmetic, geometry, and astronomy. Only when he had mastered all seven liberal arts of the trivium and quadrivium was the student considered competent to proceed to theology."[1]

In the nineteenth century, American educator and philosopher, William T. Harris had a similar view of the curriculum that should be taught in public schools:

"To Harris, the nature of course study in the public school was largely reducible to what he saw as the five great divisions in the life of civilization, which he labelled "the five windows of the soul." Two of the windows (or areas of inquiry), mathematics and geography, were committed to humanity's conquest and comprehension of nature. The other three, literature, grammar, and history, were more connected to human life: literature speaking to literary works of art; grammar, to the study and the use of language; and history, to a multifaceted understanding of the nation's institutions."[2]

Notwithstanding the religious context, these fundamentals of education should have remained the same, but they have not. It is perhaps, inevitable, that as knowledge in specialised domains expands, the ability of people to be truly conversant with multiple domains diminishes accordingly. Not only are some scientists unable to see the woods for the trees, they are unable to see the trees for the twigs and leaves. I mean no criticism: I believe it to be inevitable. One consequence is that when people become experts in a domain, especially the technical, they fail to appreciate the significance of their own human frailty when placing faith in what they believe to be true. Let me offer that as education, from even the earliest age, focuses more on technology than the trivium, the ability of societies to exercise discernment will diminish quite dramatically. I believe that the lowering of academic standards for numeracy and literacy is a significant omen, a harbinger of the dumbing down of societies, despite an increase in technocracy, or perhaps because of it. In the Australian education system, at this time of writing, English is not a required subject. How younger generations can learn from literature without being literate is beyond my imagining, but such is the stupidity pandemic of modern times.

Science history is replete with misunderstandings, academic arrogance, protection of reputations, and even fraud. A well-respected professor of evolutionary biology had cause to remark:

"In fact, scientists are just as human as anyone else. They believe that one or another hypothesis is most likely to be true, and they engage in sometimes bitter battles to defend their ideas. Scientists' beliefs are also shaped by their political, social, and religious environment."[3]

Concluding a discussion on how people can be misled, he went on to say:

"Thus, the common image of scientists as abstracted, unbiased, detached intellects has no foundation in reality. Scientists are often highly opinionated, even in the face of contrary evidence; and they are often not particularly intelligent either. The spectrum of scientists, as of any other group of people, runs from the brilliant to the fairly stupid."

That is not to decry scientists in general, but to accept that, as with any other demographic, we cannot ignore the human side of people – all people. As we find from the very definition of science in our modern world, scientists' beliefs are shaped by *philosophical materialism*, or as Carl Sagan opined: "The Cosmos is all that is or was or ever will be". However, as Robert Wesson, evolution proponent, political scientist, and Senior Research Fellow at the Hoover Institute put it:

"The contention that reality consists of only material particles and their modes of interaction is not even a clear-cut theory. It implies a narrow definition of reality, making the thesis true by definition: if only material substance is real, then material substance contains the whole of reality. But are the laws of nature not real? Are mathematical theorems real? Are patterns real? Are thought and consciousness? It is paradoxical to deny their essentiality, for science could not exist without them."[4]

Later, amongst other things, we will review the modern theory that the laws of nature, as we observe them, are the expression of just our reality, which is only one of an infinite number of realities, each with their own laws, likely different to ours.

Whether scientists are willing to accept it or not, their behaviour and endeavours are underpinned by their worldview, and thus philosophy. It is my contention that you cannot understand scientific endeavour, unless first you understand human endeavour. Whilst many refuse to accept that they are bound by philosophy, the truth lies elsewhere. As discussed in the previous chapter, research into the mind-brain complex is predicated on a philosophy of dualism, and in most cases, *Property Dualism*, largely because other perspectives are unacceptable under the principles of philosophical materialism, aka, material monism.

One may claim a belief in science, but that is philosophy not science, for beliefs themselves are conceptual, whereas science by its own modern definition, deals only in the physical. One should ask such people: What are the physical and chemical properties of a belief? Albert Einstein, that doyen of science, was one of those who understood the truth and reality of how philosophy guides the human mind. Robert Thorton, Physics Professor at University of Puerto Rico, had written to Einstein on persuading colleagues of the importance of philosophy of science to scientists (empiricists) and science. Einstein replied:

"I fully agree with you about the significance and educational value of methodology as well as history and philosophy of science. So many people today — and even professional scientists — seem to me like someone who has seen thousands of trees but has never seen a forest. A knowledge of the historic and philosophical background gives that kind of independence from prejudices of his generation from which most scientists are suffering. This independence created by philosophical insight is — in my opinion — the mark of distinction between a mere artisan or specialist and a real seeker after truth."[5]

I am comforted that such an intellect agrees with my own conclusions. Suffering from prejudice, so many choose to not see the wood for the trees, and thus cannot claim to be genuine seekers after truth.

--- ∫ --- ∫ ---

"We only hear what we want to hear. We only see what we want to see."

This statement about human nature, pessimistic though it may be, rings true to many of us. It helps to explain all sorts of strange human behaviours. The following comments are by Rabbi Dr. Tzvi Hersh Weinreb, published here[6]:

"We are all familiar with the experience of listening to a speaker and discovering that we heard a very different message than did our companion ... beside us. We hear and see what we want to, and fail to hear and see the proverbial "writing on the wall," perhaps because it is so unpleasant to us that it simply does not register ... we place our own "constructs" upon everything that we see or hear so that you may hear one message, and I may hear an entirely different one."

I have quoted this to affirm my belief that philosophy is a real science worthy of study, as do many genuine intellectuals, and that the workings of the mind cannot be limited by nothing more that neural activity entirely dependent upon the laws of chemistry and physics. And finally, other observations by Albert Einstein:

"Everyone who is seriously involved in the pursuit of science becomes convinced that some spirit is manifest in the laws of the universe, one that is vastly superior to that of man."[7]

"A new idea comes suddenly and in a rather intuitive way. But intuition is nothing but the outcome of earlier intellectual experience."[8]

Dare I be so bold as to contend that far too many scientists are guilty of shallow thinking when it comes to understanding the human condition, ignoring what they prefer to not see despite the evidence that has convinced others. I am given to wonder: How is it, that the human who has evolved through a process without purpose, seeks purpose and meaning in life? Even material-monists live with, and by, purpose, yet purpose can never be a feature of biology, especially as espoused by the evolution hypothesis.

References:

1. Buchanan, Thomas S., *Disciplined Science*, Touchstone, Sept/Oct 2017, p. 25

2. https://education.stateuniversity.com/pages/2030/Harris-William-T-1835-1909.html

3. Futuyma, Douglas J., *Science on Trial – The Case for Evolution*, Pantheon Books, New York, 1982, p. 164

4. Wesson, Robert, *Beyond Natural Selection*, A Bradford Book, The MIT Press, Cambridge Massachusetts, 1991, p. 4

5. Letter to Robert A. Thorton, Physics Professor at University of Puerto Rico (7 December 1944) [EA-674, Einstein Archive, Hebrew University, Jerusalem]. Thorton had written to Einstein on persuading colleagues of the importance of philosophy of science to scientists (empiricists) and science.

6. http://www.jewishworldreview.com/0618/weinreb_balak_perspective

7. Letter to Phyllis Wright (January 24, 1936), published in *Dear Professor Einstein: Albert Einstein's Letters to and from Children* (Prometheus Books, 2002), p. 129

8. Letter to Dr. H. L. Gordon (May 3, 1949 - AEA 58-217) as quoted in *Einstein: His Life and Universe* (2007) by Walter Isaacson ISBN 9780743264730

Chapter 2-5: Curiosity & Imagination

"What is a scientist after all? It is a curious man looking through a keyhole, the keyhole of nature, trying to know what's going on."

~ Jacques Cousteau, 1910-1997, explorer, conservationist, scientist, photographer ~

Curiosity is a quality related to inquisitive thinking such as exploration, investigation, and learning, evident by observation in humans and other animals. It is an expression of the desire to learn or know about anything – i.e., an expression of inquisitiveness. Inquisitive, on the other hand, is defined as having or showing an interest in learning things – i.e., being curious. So, having come full circle, let is investigate its relationship to science. As one commentator advised:

"The important thing is not to stop questioning. Curiosity has its own reason for existence. One cannot help but be in awe when he contemplates the mysteries of eternity, of life, of the marvellous structure of reality. It is enough if one tries merely to comprehend a little of this mystery each day. Never lose a holy curiosity. ... Don't stop to marvel."[1]

This book that you are reading, and many others before it, has been instigated by my endless curiosity about all manner of things.

Let me first attempt to differentiate the curiosity of humans, and that of other living organisms, for I believe them to be essentially very different. As earlier discussed in another context, inherent in all life forms is what we can broadly refer to as *instinct*, commonly defined as: "an innate, typically fixed pattern of behaviour in animals in response to certain stimuli". To establish a fundamental of life, I would extend that to plants, their stimuli and responses being purely chemical. In the main, instinct even in animals is chemical, although I would not venture a scientific explanation other than that it is formed genetically during development of the embryo. However, if evolution theory is true, what else can instinct be, other than chemical in a life form constructed by biology? I cannot know the nature of, for example, the curiosity of a cat. Observing our domestic pet, now past her fourteenth birthday, I sometimes wonder whether her apparent curiosity is a function of declining eyesight. However, I have witnessed what appears to be curiosity in all manner of animals and birds, and whilst not knowing the essence of it, I would suggest that it is related to instinct, rather than being an intellectual pursuit.

Humans have a similar instinctual curiosity as animals, but we also have curiosity born of intellect and knowledge. It is my proposition that other life forms do not possess the ability to acquire knowledge, beyond that which is within the domain of their instincts, and is related to survival and/or reproduction. As my cat stares out the window on a rainy day, I doubt that she is pondering whether anyone has invented a raincoat for cats. As she watches the magpies digging in the garden, I doubt that she is wondering what it would be like to eat what they eat. In her lazy moments luxuriating in the sunshine, I doubt that she is composing a poem. I cannot know these things, but I am confident that she is incapable of the curiosity that I am demonstrating here.

Human curiosity extends to the esoteric, and to subjects which have no immediate benefit for either survival or reproduction. As Raymond Tallis noted: "If, for example, we humans are no different from all other animals, how did we dream up the idea that we are different? Come to that, how did we ever arrive at an idea of ourselves at all? As far as I know, centipedes do not have the concept 'centipede'; nor do they relate that concept to a higher-order concept such as 'insect'; they do not compare themselves to other centipedes or calibrate centipedes against other insects. Not even our biological next-door neighbours, the chimpanzees, do this."[2] Of course, we cannot know, but there is no evidence that creatures other than humans have that innate sense of self.

Perhaps the motivation behind the breadth and depth of human curiosity is best expressed by the answer to the question: Why do you seek? As in my case: Because I can, but that hardly answers the question. In truth, this raises even deeper questions: Why can I, and how do I know that I can? In an earlier chapter, we explored the implications of *knowing than you can*, or more especially, the implications for evolution of *not knowing* that you can. For now, however, I will just leave you to be curious about your own areas of curiosity, and ponder what it is that differentiates humans and animals in that regard. You might even ponder why your particular curiosities differ from mine and that of other people.

I am tempted to ask my cat, but imperious as ever, she would treat me with ignore.

As an aside, I remember reading that a stone is more intelligent than a cat, because a stone knows to get out of your way when you kick it (don't tell my wife).

Imagination

"The true sign of intelligence is not knowledge but imagination."

~ Albert Einstein ~

From that perspective, animals are not truly intelligent if they lack imagination, which is a form of conceptualization. Can one imagine, or explain, imagination, that faculty for forming new ideas, of visualising images of objects not subject to sensory stimulation at the time? Could it be true that imaginations arise only through activity of our neural networks over which we have no control? If knowledge is stored only in our neural networks, it must be limited, as Albert Einstein observed:

"Imagination is more important than knowledge. For knowledge is limited, whereas imagination embraces the entire world, stimulating progress, giving birth to evolution. It is, strictly speaking, a real factor in scientific research ... I believe in intuition and inspiration ... At times I feel certain I am right while not knowing the reason."[3]

Centuries earlier, Jewish philosopher, Maimonides, commented:

"The imagination encompasses the capacity to retain impressions of experiences when they have vanished from the senses involved, and to compound some and separate others. It is the capacity that enables a person to combine certain experiences he has never had, nor ever could grasp – for example, to imagine an iron ship sailing in the air, or an individual whose head is in the heavens while his feet are on the ground, or an animal with a thousand eyes, as well as many other such impossible things that are bred when the imagination combines things and produces a phantasm ... That is where the Mutakellium[4] made their great, absurd mistake upon which was built an erroneous basis about the difference between what is essential [i.e., what is necessarily so], what is possible [i.e., what may or may not be so], and what is impossible [i.e., what simply cannot be so]. They themselves believed or convinced others to believe that whatever can be imagined is possible. They did not realize that the imagination can concoct impossible things, as we have said."[5]

Of course, Rambam was describing something quite ordinary in our age and not at all impossible, the aeroplane. It is interesting to note that iron was not used for the hull of ships until the mid-19th century, and heavier-than-air machines did not appear for forty to fifty years after that! The question must be asked: What is the stimulus and process for combining concepts which have never been combined in the individual's experience?

The point can then be made that the imagination can quite brilliantly concoct *illusions* of the *impossible*, whilst in truth, nothing truly impossible can ever be constructed in reality, despite our imagining that it could. In the field of entertainment, we are satisfied with simulations of the impossible, to a degree accepting them as the real thing. At one level, we know that they are not real, but at another level, suspect that they could be, which is why we don't just get up and leave a theatre unable to accept nonsense. Do famous composers have illusions of the impossible, when they compose such complex symphonies, even to the extent of writing the music sheets for a variety of musical instruments? Are neural networks, composed of nothing but chemicals, capable of forming relational connections with so much beauty and complexity, all by themselves, and not "knowing" how such music would sound?

In a later chapter, we discuss Maimonides' view of existence, especially these definitions of attributes, what I earlier referred to as properties of an entity:

- Essential – an attribute with which, an entity cannot be what it is said to be.

- Accidental – an attribute that may or may not belong to an entity, without affecting its essence.

According to philosophical materialism, genetics and biology are solely responsible for the *essential* attributes of human existence. Is imagination, or humour, essential attributes of humanity, or simply *accidental*, given that so many people fail to demonstrate, or appreciate, those attributes? Imagination of the genuinely impossible, the ability to compose music or poetry, the ability to conceptualise new theories of physics, the variety of senses of humour from the politically correct to the absurd, are all accidental attributes whose lack does not affect the essence of any particular human. On the other hand, these attributes partially differentiate the human race from all other organisms on Planet Earth. In that sense, they could be said to be *essential* to what the entity, humanity, is said to be. Given that according to evolution and the underlying materialistic beliefs, humans are just some more evolved organisms, subject to the same genetic and biological imperatives, how is it that humanity displays attributes not found in lower order organisms, and cannot be explained in mechanistic terms?

Among the many disciplines which, to my mind, unambiguously evidence the imagination of humans, there being no mechanistic explanation for them, I would mention just two: philosophy, and Albert Einstein's theories of relativity.

Humour

I suspect that only humans have a sense of humour, although I cannot know that. What I do know is that cultural humour varies from country to country. Whilst volunteering at International Expo 1988 in Brisbane, I had occasion to visit the representatives of many countries, and made a point of asking them about their jokes. Not surprisingly, everybody from the West had Irish and Catholic jokes. In Eastern Europe, there were innumerable jokes about Russian and Polish tractors; some I understood, others not. Inscrutable Orientals seemed not to want to understand, although I apprehended from experience that such people were not entirely humourless. In Australia, Britain, and some parts of the USA, "black humour" is, or was, common, but political correctness is waging war against what it perceives to be uncouth or hurtful.

My point here is simply this: would any evolutionist like to propose a materialist explanation for humour? In the shower one morning, this little ditty composed itself:

There was a young lady from Woking,

Who hid in the toilet, smoking.

A cleaner on hire,

Smelled smoke and yelled fire,

And the lass got a bloody good soaking.

I have others that perhaps should not be shared in public.

References:

1. *Old Man's Advice to Youth: "Never Lose a Holy Curiosity,"* LIFE magazine (2 May 1955) statement to William Miller, p. 64.

2. Tallis, Raymond, *Aping Mankind*, Routledge Classics, New York, NY, 2016, p. 4

3. *Cosmic Religion: With Other Opinions and Aphorisms* (1931) by Albert Einstein, p. 97; also, in *Transformation: Arts, Communication, Environment* (1950) by Harry Holtzman, p. 138

4. Feldman, Rabbi Yaakov, *The 8 Chapters of the Rambam: A Classic Work on the Fundamentals of Jewish Ethics and Character Development*, Targum Press, Southfield, MI, 2008, p. 172, Notes: "The *Mutakallimun* were members of a sect of Arabic religious thinkers and dialecticians who tried to reconcile Islamic ideology with Aristotelian philosophy, and arrived at some very serious flawed notions."

5. Feldman, *Ibid*, pp. 32-33

Chapter 2-6: Knowledge

"I know nothing, because I know too much, and understand not nearly enough and never will."

~ Anne Rice, The Vampire Armand ~

I cannot be sure of when I first encountered the term, epistemology (*the branch of philosophy concerned with the theory of knowledge*) or why it held such fascination for me, but over recent years, I have become increasingly more interested in "knowledge", not so much from a philosophical perspective, but in relation to the mechanisms of data acquisition and processing. At some point, I came to the realisation that my brief career in Air Traffic Control, and a more extensive career in Management Information Services, had provided me with a sound knowledge of many technical aspects that I could apply to the theory of *undirected* evolution. Albert Einstein had this to say on the subject:

"The reciprocal relationship of epistemology and science is of noteworthy kind. They are dependent on each other. Epistemology without contact with science becomes an empty scheme. Science without epistemology is — insofar as it is thinkable at all — primitive and muddled. However, no sooner has the epistemologist, who is seeking a clear system, fought his way through to such a system, than he is inclined to interpret the thought-content of science in the sense of his system and to reject whatever does not fit into his system. The scientist, however, cannot afford to carry his striving for epistemological systematic that far. He accepts gratefully the epistemological conceptual analysis; but the external conditions, which are set for him by the facts of experience, do not permit him to let himself be too much restricted in the construction of his conceptual world by the adherence to an epistemological system. He therefore must appear to the systematic epistemologist as a type of unscrupulous opportunist: he appears as *realist* insofar as he seeks to describe a world independent of the acts of perception; as *idealist* insofar as he looks upon the concepts and theories as free inventions of the human spirit (not logically derivable from what is empirically given); as *positivist* insofar as he considers his concepts and theories justified *only* to the extent to which they furnish a logical representation of relations among sensory experiences. He may even appear as *Platonist* or *Pythagorean* insofar as he considers the viewpoint of logical simplicity as an indispensable and effective tool of his research."[1] (Italics in original)

Thus, I would offer, it is critical that we understand how we determine whether what we know is true, or false. False knowledge is as common as true knowledge, or perhaps even more prevalent; if we do not analyse "how we know", we cannot trust the veracity of "what we know". This is particularly important in evaluating the truth of evolution, for instinct and knowledge in all forms of living creatures had to have had an origin. There is a process behind developing knowledge, one that requires another process to validate what is believed to be true. I have doubts about whether knowledge can evolve through biological processes alone. The irony surrounding the acquisition of knowledge, as Anne Rice acknowledged, is that the more we learn, the more we realise how much there is to learn, and how unlikely it is that we will ever learn enough.

Whether implemented in what we know as hardware, software, and firmware, or implemented in *bioware* (my own term for biology related to this subject), the functions to be performed are the same. For example, a principle behind the development of sonar is that, knowing the speed of sound through water, one could use the time interval between sending and receiving the returned pulse to determine distance. It seems reasonable, to me at least, that the echo-location capabilities of bats and dolphins must depend on the same principle: the time interval between sending and receiving a pulse of sound. Some questions arise:

1. What was the impetus for the animals to first emit a sound pulse?

2. Why did they consider the returned sound to be of significance, or value, to them?

3. Why did they even understand that the sound that they heard was in fact, a return of the sound emitted by them?

4. What was the impetus for the animals to continue experimenting with this faculty?

5. How did these animals learn what that time interval meant, especially as the speed of sound varies with the temperature and

density of the medium through which it passes?

Could such a capability arise by undirected evolution, without a method of verification that what was interpreted, actually matched reality? If successive pulses resulted in a shortening or lengthening of the interval, how could undirected evolution "teach" that this meant that the distance was closing or opening? Trial and error? Trial and error suggest intelligence and purpose, together, the antithesis of undirected evolution. If, hypothetically speaking, the bat first understood that an increasing time interval meant that it was getting closer to the object, resulting in a sore nose, would it consider giving up on that method, or would it "learn" that the longer time interval meant the opposite? For a computer programmer, this is synonymous with "if then else" logic, but was a primitive bat capable of such logic? Before evolutionists posit that the bat, or dolphin, learned how to use that facility, they should offer an explanation of how such learning was possible. I would also like to know whether the echo location and ranging of dolphins, for example, can compensate for variations in water temperature, and thus density.

In my time as a military air traffic controller, I was trained in the operation of the AN/FPN-36 Quad Radar. This device was primitive in the sense that it lacked any computerisation, as more modern radars have. It had numerous controls for fine-tuning the emitted pulses. Thus, we learned about pulse length (duration), pulse width, and pulse repetition frequency, which could all be varied for better target discrimination. We also learned about anti-clutter circuits, such as circular polarization to eliminate returns from rain drops (and unfortunately, other circular objects such as the wooden nose of De Havilland Vampire jets, but that is another story). In short, I learned a great deal about echo-location and ranging, and the variety of factors that influence the accuracy of interpretation of returns. My mind thus turns to the biological equivalents, and how evolution alone could result in such useful faculties.

But back to my Primary Axiom #9, that to perform any physical action, an animal must first *know that it can*. Equally, or perhaps even more importantly, there must be an impetus for the animal to perform that action beyond a response to stimuli. From the perspective of learning, all knowledge is built upon prior knowledge *in context*, so we must ask: How was the first fact stored and recalled in context, such that new facts could be interpreted within the appropriate context?

The more I pondered the subject, the more avenues of exploration I found to pursue. In an earlier book, *The Dawkins Deficiency*[2], I touched on some of the issues of volition and information processing. In a later work, *Information, Knowledge, Evolution and Self*[3], I approached the subject from the perspective of digital information processing, and introduced the concept of *knowing that you can*. This current study examines undirected evolution from the perspective, amongst others, that all activities must be preceded by conceptualisation, before they can be instantiated in animal or human behaviour. To implement such activities, there is a long series of causally dependent events, both in the initial development of the faculty, and continued usage. The background controller of such behaviour, other than instinct, is knowledge derived from sensory inputs (data) and a series of relevant data ancestors. We come back to this later, but sensory inputs are *encoded symbolic representations* of external realities, not the realities themselves. Thus, a deeper question: what biological process could have been capable of developing reliable encode/decode processes (Primary Axiom #4)?

In passing, note that implied in my usage of the term, *undirected evolution*, is my acceptance of the possibility of *directed evolution*. If that brings to some minds, *Intelligent Design, Theistic Evolution*, or *Evolutionary Creationism*, then so be it. One should not shy away from research because one is afraid of to where it will lead: let us follow the evidence, not with trepidation, but with excitement. There may be few unexplored areas on the surface of this Earth, but there remains much that lies beneath, the exploration of which, requires intellectual, rather than physical courage.

I am encouraged by the following excerpt, which demonstrates that intellectual integrity does exist, in some minds at least:

"It is, of course, nothing but a truism, and not a scientific theory, to say that living systems do not survive if they are not fit to survive. [Ed. *I am entirely confident that this is not necessarily true, especially of humans.*]

To postulate, as the positivists of the end of the 19th century and their followers here have done, that the development and survival of the fittest is entirely a consequence of chance mutations, or even that nature carries out experiments by trial and error through mutations in order to create living systems better fitted to survive, seems to me a hypothesis based on no evidence and irreconcilable with the facts.

This hypothesis wilfully neglects the principle of teleological purpose, which stares the biologist in the face wherever he looks, whether he be engaged in the study of different organs in one organism, or even of different subcellular compartments in relation to each other in a single cell, or whether he studies the interrelation and interactions of various species. **These classical evolutionary theories are a gross oversimplification of an immensely complex and intricate mass of facts, and it amazes me that they are swallowed so uncritically and readily, and for such a long time, by so many scientists without a murmur of protest.**"[4] [emphasis in original]

The author, Ernst Boris Chain (1906-1979) was awarded the Nobel Prize in Physiology / Medicine, 1945, jointly with Sir Alexander Fleming and Sir Howard Florey: "for the discovery of penicillin and its curative effect in various infectious diseases".[5] Given his achievements and the recognition thereof, we should consider Chain's views as being those of a scientist of valuable knowledge and competence.

That said, so many scientists restrict themselves to the physical, not unexpectedly, as only the material surrenders to the scientific method, as science has defined it. My area of enquiry is additional to the material, being concerned with meta-physics and how the material is so obviously subject to immaterial forces. Those wedded to *philosophical materialism* refuse to acknowledge the existence of the immaterial, and eschew any research into that subject. Consequently, reference material is scarce, leaving me to struggle as best I can. If some of my explanations lack scientific support, I can but beg your understanding.

Revisiting "This hypothesis wilfully neglects the principle of teleological purpose", it is worthy of note that in so many explanations or descriptions by scientists, they cannot avoid using terms expressing design or purpose, contrary to their own expressed beliefs that there is none.

References:

1. Contribution in *Albert Einstein: Philosopher-Scientist*, p. A. Schilpp, ed. (The Library of Living Philosophers, Evanston, IL (1949), p. 684). Quoted in *Einstein's Philosophy of Science*

2. Talbot, Wayne, The Dawkins Deficiency: Why Evolution is Not the Greatest Show on Earth, Deep River Books, Sisters, OR, 2011

3. Talbot, Wayne, Information, Knowledge, Evolution, and Self: a question of origins, Xlibris, Bloomington, IN, 2016

4. Chain, 1971, "*Social Responsibility and the Scientist in Modern Western Society*". Perspectives in Biology and Medicine, Spring 1971, Vol. 14, No. 3, pp. 367

5. "*The Nobel Prize in Physiology or Medicine 1945*". Nobelprize.org. Nobel Media AB 2014. Web. 2 Jul 2017. http://www.nobelprize.org/nobel_prizes/medicine/laureates/1945/

Chapter 2-7: Processing versus Being

"The human brain works as a binary computer and can only analyze the exact information-based zeros and ones (or black and white). Our heart is more like a chemical computer that uses fuzzy logic to analyze information that can't be easily defined in zeros and ones."

~ Naveen Jain ~

The assertion that the human brain works as a binary computer is unsubstantiated – nobody really knows how the brain works – and examining the architecture of the brain's neural networks, I am inclined to the opinion that the workings of the brain more resemble a chemical computer. The human brain is nothing like a binary computer, and is incapable of analysis using binary logic. One might be excused for attempting to explain by analogy, but even then, such an analogy does not fit the biological facts. Curiously, if the brain can only analyze the exact information-based zeroes and ones, then it would be incapable of imagining beyond those binary elements. That we do so supports the proposition that the mind deals in other than physical elements.

Nevertheless, it is the computing analogy of the human brain that has convinced me that the mind and the brain must be separate entities. In the above quotation, the author appears to use the heart as his metaphor for the mind, perhaps in an attempt to avoid contention, but his assertion nevertheless evidences some truth, whilst at the same time adding even more confusion. We shall get to this critical point in more detail later, but here I shall put a stake in the ground to get your attention: Technically speaking, no computer ever analyses *information*. Computers deal in encoded, symbolic representations of conceptual data, with the analysis function provided by an external agent; i.e., whoever wrote the program code. It is critical that this distinction is understood for when we later discuss functioning of the human brain. In this chapter, for simplicity of prose, I will refer to encoded, symbolic representations of conceptual data as more simply, data. Now back to what computers compute (processing), and what they *do not* (being).

In a very broad sense, humans *are*, and *do*, and it is our state of being that does the doing. "Are" is our state of being – alive, sentient, aware, awake, asleep, happy, sad, and so forth. Whilst what *we are* may be a determinant of *what we do*, it is not necessarily so. This is one of the aspects that differentiate humans from all other life forms on Planet Earth, with the possible exception of some animal forms. A "state" can be represented computationally, but is conceptual and not the state itself – a state just "is", and relates to being. Computing relates to "doing": processing data, but whilst a computer is a real entity, having existence, it has no sense of self – it is not sentient, is not capable of cognition, and is not aware of the context of its processing. Computerised business systems process transactional data to derive the "state" of some aspect of the business, relating to manufacturing, finance, logistics, or other domain. For example: where is my order; what is my level of debt; how much do customers owe us; how much do we owe in wages; etc. A *representation* of a state can be reached via a computational process, but it is not necessarily true to the reality of the state that it attempts to represent. A humorous example is found in how inventory was processed in the early days of computing, but it is instructive nonetheless. The quantity on hand (QOH) can be computed by:

Previous QOH + Receipts – Issues = New QOH

Programmers unfamiliar with the realities of inventory processing, and more especially, transaction processing, decided that there could be no such reality as a negative quantity, and therefore they did not allow for such in their programs. This was, as later learned, a major cause of poor inventory accuracy. What the programmers failed to understand was that sometimes, the purchasing or inwards goods department would get behind in processing receipts, whilst the production inventory department would diligently process issues as they occurred, resulting in non-chronological transaction processing. There are many such examples in computing, especially in modelling, where what is represented fails to eventuate as predicted; meteorology and climate change spring to mind.

In the inventory processing example, each of those four quantities had an existence in reality, but their computational representation was not necessarily accurate. The point is that if the brain behaves like a binary computer, then its function is primarily *processing*, not *being*. This suggests that some other part of human existence does the being: what else can that be other than the mind? Keeping it simple, computers have four basic components: a central processing unit (CPU), input-output (I/O) devices, management software, and storage devices. Our brain is said to be the physical CPU; our sensory receptors are the I/O devices; the mind is the software; and according to conventional scientific theory, the brain is also the storage device. The management software (mind) knows where to store new data in the brain; how to retrieve it for processing; which other data sources are needed for correlation; how to process the data to create new data; and where to store it. But wait: whilst the mind represents the software of the brain, modern science contends that the mind and brain are one and the same physical construct (*identity theory*). So who wrote the software?

I omitted a critical component, which we might call the "bootstrap" - that which in a computer, is initiated when the power is turned on. This is the software that "wakes up" the processor, knowing the memory size of the CPU; the existence, location, and function of the I/O devices; and where to find the executable code for the required processing. It is said that once again, this is all part of the physical brain construct, which raises the question: What is the *bootstrap* process in the brain? Of course, the brain requires no bootstrap for its autonomous processes as it is always running for autonomous functions, but what causes it to run in a non-autonomous (volitional) mode to initiate creative activities? What is the bootstrap process for volitional activities? If philosophical materialism is true, then absent of internal and external stimuli, even volitional activities are autonomous. But given that the bran is material, just like a computer, the brain can never do anything which is not physical in nature, i.e., whatever the laws of physics and chemistry allow. Likewise, the brain can do nothing other than that for which it already has an internal representation.

But here is the really tricky part. As earlier discussed, digital computers do not store information, and they do not even store data: they store *encoded symbolic representations* of data. At the physical level, they store just two binary states representing 1 and 0, or *on* and *off*. Everything that we recognise as data or information is derived through multiple levels of decoding and computing. I cannot be certain of the precise method of encoded data in the human brain, as physically, it is a complex network of neuronal connections, where perhaps the pattern of individual paths in the network represents the encoded symbols corresponding to a physical reality. As a storage device for representing an external reality, the brain cannot represent the *state of being* of itself. Which brings us back to the question: where in the brain is the *human state of being*? It may well be derived from both internal and external stimuli, but that is processing: wherein the being?

Using the computer analogy, the part of the brain storing the instruction set (mind) cannot be the same part as does the processing and data storage. Overwriting the instructions with data is not recommended. The question becomes: Where in the brain is the mind stored and updated? If the mind is in the brain, being a part of the brain and physical, a set of mental modules as some would claim, then neuroscientists should be able to read thoughts, once they have learned how to decode ... what exactly? Scientists have been able to associate different parts of the brain with particular mental processes, but that is no different to an engineer knowing where in a network a program is running: that is processing, not being. As Egyptologists learned long ago, they could not decipher Egyptian hieroglyphs by studying the hieroglyphs themselves, confirming the axiom that nothing can explain itself (Primary Axiom #1). It was only after the discovery of the so-called Rosetta Stone, inscribed with the same story in three different languages, one of which, Greek, was well known, that the hieroglyphs could be decoded. What equivalent of the Rosetta Stone could there be for decoding mental processes? The brain is all we have, and is no more capable of knowing how and why it works than a steam engine, most especially as is claimed that it is formed from the same fundamental components as a steam engine, i.e., atoms and molecules.

Perhaps the most important human state of being is *self-awareness*. We are conscious of being ourselves, something that no digital computer can do, despite the creations of science fiction writers, and the aspirations of many modern scientists. Computers can process at unbelievable speeds, but cannot be self-aware, because self-awareness is a state of being which might well be derived, but cannot exist in the same entity that derives it. Let me explain.

Computers are run by software, but cannot be autonomously aware of the context of the software on which their processing is dependent. Computers can, and have been, programmed to recognise specific segments of the software being executed, and even alter instructions on the fly (I have done that), but they cannot be programmed to be cognisant of the nature of their own instructions, nor to alter the function of an instruction. A *move, add*, or *subtract* will always be so, because software is incapable of comprehending its own instructions. A computer will never autonomously decide that when instructed to add, it will instead subtract; fortunately, in that regard, computers can always be relied on to do the right thing. Gilbert Ryle, in his discussions on the mind, asserts similarly of us. In the chapter on *Distinction between Voluntary and Involuntary*, he notes that in the background processes of the mind, we always get an answer right insofar as we understand it; he opines: "it is incorrect to say that he could have avoided getting it right"[1]. I do agree with him, even though we may later choose to do otherwise, but his point is important to understand. There is an autonomous aspect to the processing of the mind, and also a volitional aspect. These two contradictory states cannot simultaneously exist in the one physical organ. If the mind and brain are of the same properties, then the mind is no more or less capable than the organic brain, whose capabilities are limited by its physical and chemical properties as regulated by its constituent cells. The brain, as a biological organ, is no more self-aware than a heart, lung, or kidney. For self-awareness to arise, the *being* must transcend the *doing*.

We investigate this further in later chapters, but first I must take a step back and ask: What does it mean for something to exist? This might seem like an odd question with perhaps an obvious answer, but it is not as simple as it seems. John Walton, a Professor of the Old Testament, provides a perceptive understanding of this issue as we dabble briefly in philosophy:

"For example, when we say that a chair exists, we are expressing a conclusion on the basis of an assumption that certain properties of the chair define it as existing. Without getting bogged down in philosophy, in our contemporary ways of thinking, a chair exists because it is material. We can detect it with our senses (particularly sight and touch). These physical qualities are what makes a chair real, and because of them we consider it to exist. But there are other ways to think about the question of existence. For example, we might consider what we mean when we talk about a company "existing". It would clearly not be the same as a chair existing … The question of existence and the previous examples introduce a concept that philosophers refer to as "ontology". Most people do not use the word *ontology* on a regular basis, and so it can be confusing, but the concept it expresses is relatively simple. The ontology of X is what it means for X to exist. If we speak of the ontology of evil, we discuss what it means for evil to exist in the world. The ontology of a chair or a company would likewise ask what it means when we say they exist. How would we understand their existence? What is the principal quality of its existence? The view represented in our discussion of the chair would be labelled a *material ontology* – the belief that something exists by virtue of its physical properties and its ability to be experienced by the senses. The example of the company might be labelled a *functional ontology*."[2]

Walton also provides the example of a new student's first visit to a university. The student is taken on a tour of the campus, visiting lecture rooms, gymnasium, sporting fields, canteen, dormitories, etc., and at the end of the tour is asked, "What do you think?" The student answers, "Very impressive, but where is the university?" From this one should ask: What does it mean for a university to exist? Similarly, we are led to ask the question: "What does it mean for the *mind* to exist?" Like, the company and university, it is a question of functional ontology – what is the purpose and functions of the mind? Again, the *being* of the mind must transcend the *doing* of the mind, and if we offer the proposition that there are no material components of the mind, that does not argue that the immaterial mind does not exist. Its *doings* are immaterial, which suggests that its *being* is immaterial. We know of no material entity capable of immaterial activities.

Now, as for the heart being *more like a chemical computer that uses fuzzy logic to analyze information that can't be easily defined in zeros and ones*, let me offer that such is just one way of describing it, but not a particularly useful one. The heartbeat is triggered by electrical impulses that travel down a special pathway through the heart. The impulse starts in a small bundle of specialized cells located in the right atrium, called the SA (sinoatrial) node, which is effectively the heart's natural pacemaker. The electrical activity spreads through the walls of the atria and causes them to contract. A more important question than *how* the heart works, is *why* it works as it does. The most curious aspect of the heart is that provided it is supplied with the correct mixture of oxygen and nutrients, a heart can beat on its own without other external control. We know that the brain continues to work under the same conditions, so using the logic of *cause and effect* that underpins all scientific enquiry, it is reasonable to assume that in a sense, the Sinoatrial Node is like a small, but specialised, form of brain, that performs autonomous functions. If that be true, then the SA works in much the same way as the brain, and if we use the analogy of *ones and zeroes* for the brain, it should be appropriate for the heart as well. Now, if we reverse engineer from the small heart to the bigger brain, we begin to see that at a fundamental level, the brain by itself can do nothing other than what it is: a collection of nerve bundles kept alive by a supply of blood.

The heart continues to pump autonomously until interrupted by an external agent. The biological brain continues to be active autonomously, just as a biological entity like the heart, doing whatever it does naturally, until interrupted by an external agent to do something different or additional. The question becomes: what is that external agent? It is illogical to assert that the brain itself awakens itself from its reverie, where it was just idling along, and decides autonomously to engage in some additional intellectual activity, or do nothing at all. The brain cannot be both itself and an external agent. Newton's first law relating to inertia states that every object will remain at rest or in uniform motion in a straight line unless compelled to change its state by the action of an external force. I contend that this definition is apt in the case of voluntary cognitive activity: the brain must remain at *intellectual* rest unless compelled by an external force, in this case the mind.

Let me offer that a computer can never be self-aware in a state of consciousness, for it can never decide for itself to not execute its initiating program: "Nope, it's Monday. I don't feel like working today, I am going to sleep in!" Whilst a computer can present a representation of a state of being, as in a hologram, the representation is its own being, and not the being that is being represented. Computers have their own being (reality), and their processing has its own being, but the processing cannot be the being of the entity, computer.

The same holds true, I would offer, for the brain. The brain exists as a real, biological entity, which performs activities consistent with its properties. But its *processing* cannot also be its *being*. It is often said that *we are what we do*, which is why when introducing themselves to others, people often ask, "What do you do?" Athletes are athletes because they indulge in athletic pursuits; boxers in boxing; cricketers in cricket; bankers in banking; and even intellectuals in activities of the intellect. But this is more of a metaphor, than a reality. Evolutionists in particular use simplistic evolutionary arguments without substantiation. In refuting an argument concerning behaviour in Patricia Churchland's book, *"Touching a Nerve"*[3], Professor Andrews, an English physicist and engineer, whose specialities included large molecules and polymers, contended:

"Churchland is surely hiding behind a form of 'psychological behaviourism' – which is defined as the belief that: 'Behaviour can be described and explained without making ultimate reference to mental events or to internal psychological processes. The sources of behaviour are external (in the environment), not internal (in the mind, in the head).'[4]

Or to put it more simply, Churchland sidesteps any attempt to explain *how* brain states influence our actions and merely asserts that they do. That's fair enough, but she then smuggles in the claim that this eliminates the *need* for a conscious agency to be involved – a claim that reduces all living organisms to pre-programmed robots."[5]

I have mentioned Professor Andrew's credentials because, whilst he is not a biologist, he has demonstrated understanding of material properties and behaviours. He has consulted on such issues during his long career. Both animate and inanimate matter are comprised of the same raw materials, albeit organised in very different ways, and subject to varying influences both internal and external. I have yet to read any scientific argument, or been presented with any scientific evidence, that convinces me that different arrangements of chemicals can give rise to sentience and cognitive abilities in one instance, but not in another. Processing is not being, and whilst processing can give rise to being, such processing cannot be its own being. It is rather like the illogical claim that something can create itself out of nothing – the being which resulted from the processing was the same being that pre-existed the processing to create itself – sheer nonsense.

Another useful insight has been provided by David Gelernter, a Professor of Computer Science, who like many other computer professionals (including me), cannot accept the computational theory of mind as understood in other professions, such as psychology. In his book, *"The Tides of Mind"*[6], he proposes that the mind changes in a spectrum, as its activities ebb and flow; from low focus, where we become mere experiencers of our existence, to high focus, where we use memory and thought processes in a disciplined and focused way. I suspect that we all are familiar with this phenomenon. At either end of the spectrum, we are not easily distracted, because we are totally absorbed in our own thoughts or daydreams. Whilst it is reasonable to posit that at the low end, the mind has succumbed to both our internal and external environment, where feelings and/or sensations dominate, the same cannot be said of the high end, which is intentional and purposeful.

Curiously, no-one seems to be able to avoid the term "we", from which I would deduce that subconsciously, every person believes themselves to exist. Ontologically, what does it mean for "I/me" to exist? Is it plausible, given our experience of life, that our *being* is sufficiently defined by the *doings* of the material brain, as constrained by its material construction?

References:

1. Ryle, Gilbert, *The Concept of Mind*, Penguin Books Ltd, London, England, 1990, pp. 67-68

2. Walton, John H., *The Lost World of Genesis One: Ancient Cosmology and the Origins Debate*, IVP Academic, Downers Grove, Illinois, 2017, pp. 21-22

3. Churchland, Patricia S., *Touching A Nerve: Our Brains, Our Selves*, W.W. Norton & Company, New York, NY, 2013

4. https://plato.stanford.edu/entries/behaviourism/

5. Andrews, Edgar, *What is Man: Adam, Alien or Ape?* Elm Hill Books, Nashville, TN, 2018, pp. 188-189

6. Gelernter, David, *The Tides of Mind: Uncovering the Spectrum of Consciousness*, Liveright Publishing Corporation, New York, NY, 2016

Chapter 2-8: Sentience versus Consciousness

"Why, if it weren't for this 'internal illumination' [i.e., sentience] the world would be nothing but a pile of dirt!"

~ Albert Einstein ~

If we consider a pile of dirt, it is composed of physical particles of a particular type. If we consider human organs, they are not that different to a pile of dirt, other than their arrangements of atoms and molecules into cellular structures, which when fed by forms of nourishment, give them life. If the expression, *from dust to dust* is true, then we as humans are in the state of potentiality of becoming a pile of dirt! The question becomes: can a particular arrangement of atoms and molecules achieve sentience? Dennis Bray, emeritus professor, University of Cambridge, with a particular interest in Integrative Systems Biology and computational modelling of cell signalling, is adamant:

"I say repeatedly in the book as clearly as English words will allow that in my opinion single cells are not sentient or aware in the same way as we are. To me, consciousness implies intelligent awareness of self and the ability to experience introspectively accessible mental states. No single-celled organism or individual cell from a plant or animal has these properties. An individual cell, in my view, is a system that possesses the basic ingredients of life but lacks sentience. It is a robot made of biological materials."[1]

My question is this: if a single cell, a system made of biological materials which by itself cannot be sentient, is thus only a robot, why is a larger system comprised of these very same cells, made of the same biological materials, not also a robot? What is it about a particular arrangement of these cells that would endow it with sentience? Why would a brain, composed of the same biological materials but in different arrangements, be sentient, when a heart, lung, or kidney is not? Bray continued, "Cells are undeniably the 'stuff' from which consciousness is made", but how can he know that other than his subscribing to philosophical materialism and evolution? His book does not explain his assertion, significantly I would offer, because he cannot. Other scientists agree, for example Colin McGinn, who states "that science as it stands lacks the ideas and intellectual framework to explain subjectivity and consciousness", continuing "not only do we not understand consciousness; there are no grounds for believing we ever will"[2].

Rupert Sheldrake observes:

"The central doctrine of materialism is that matter is the only reality. Therefore, consciousness ought not to exist. Materialism's biggest problem is that consciousness does exist. You are conscious now. The main opposing theory, dualism, accepts the reality of consciousness, but has no convincing explanation for its interaction with the body and the brain."[3]

I do not expect that a convincing explanation will ever arise, because we do not have the tools to investigate the immaterial, other than the human mind, which I contend, cannot explain itself or any other immaterial phenomenon (Primary Axiom #1). Richard Sheldrake has offered an explanation: *morphic resonance*, but his foundational arguments do not resonate with my mind (excuse the pun), and thus I have not researched that issue any further. His proposition has its points of interest, but they are for another time.

Dangerous Precedents based on Ignorance

In 2019, legislators in the ACT (Australian Capital Territory) decided to consider legislation that would enshrine animal "sentience" in law, recognising for the first time in an Australian jurisdiction that animals have feelings, sensations, and emotional states - what scientists call "affective states". I suspect that anybody who has worked with animals, in any capacity, would have no doubts on that score. However, no matter how one defines sentience, animals are not sentient in the same way as humans. Sentience is a minimalistic way of defining *consciousness*, which otherwise commonly and collectively describes characteristics of the mind, such as creativity, intelligence, sapience, self-awareness, and intentionality (the ability to have thoughts about something). Sentience, in the context of the proposed legislation, is limited to the capacity to feel, perceive or experience subjectively. I believe that they are heading into dangerous territory.

The point here is that *sentience*, from the perspective of having a *subjective experience*, is far removed from human *consciousness*, which includes abilities not present in lower order organisms.

In 2015, a revolutionary decision was made in a U.S. court: chimpanzees were acknowledged to have rights of their own. This was the first occasion legal rights of any kind have ever been accorded to anything other than a human. The story started in 2013, when an organisation called the Nonhuman Rights Project filed a lawsuit in the New York Supreme Court on behalf of four chimps kept for research by Stony Brook University. The eventual conclusion of Justice Barbara Jaffe was that they were not to be treated as property, but as legal persons. Not as persons with full human rights, but as persons with a right not to be held in captivity and a right not to be owned. I understand the emotion of such a decision, but it ignores the next question: Who has the right to decide the future of the chimpanzee, since the chimpanzee is neither able to comprehend, nor exercise, this newly granted right? What if the chimpanzee prefers to stay in captivity, even in a zoo, rather than be returned to the forest it may never have previously experienced? What is the chimpanzee's choice, and how would you know? Before granting rights, there must be an evaluation of the feasibility of such rights being exercised by the recipient of those rights. If it is not possible for rights to be exercised, it is entirely illogical, and dare I say, irresponsible, to grant them. Whilst Justice Jaffe ruled those chimpanzees were not to be treated as property, she failed to understand that without the ability to decided their own fate, chimpanzees would still be treated as property, insofar as their future disposition was concerned. I wonder whether that implication ever crossed her mind?

That is the first point of ignorance; the second arises from modern genetic studies revealing the truth of the nearly 99% commonality of human and chimpanzee genetic material, yet we are so different. The conclusion of many is that chimpanzees are more like cousins of humans than a separate species. Progressions in genetics and biology have revealed that commonality of an earlier understanding of DNA has less to do with commonality of species than previously thought. The concept of *genetic determinism* has been proven to be true in the context of the different species, but for a different reason; there is much more going on, and here we need to differentiate between protein coding DNA (about 1%) and what was previously considered as "junk DNA" (about 99%). In a sense, protein coding DNA can be said to be a *passive partner* in genetic expression, not the driving force. Studying these so-called junk DNA sequences, researchers have identified large genomic gaps in areas adjacent to genes which have the effect of turning these genes are on or off. It would appear that the cause is likely the activity of retroviral-like transposable element sequences, which would explain why human and chimpanzees are so different.

"Our findings are generally consistent with the notion that the morphological and behavioral differences between humans and chimpanzees are predominately due to differences in the regulation of genes rather than to differences in the sequence of the genes themselves," said McDonald."[4]

Sentience is the capacity to feel, perceive, or experience subjectively. Computers are not sentient because everything that they process is objective, even when programmed to give the appearance of subjectivity - physical entities are incapable of sentience, and one has to wonder how organic entities, constructed of the same basic materials, could be sentient either. That we are sentient suggests that the faculty is not of a material nature.

The Continuum of Consciousness

Fortunately, I am not the only one who struggles to understand what the term, *consciousness*, actually means, as this extract testifies:

"Consciousness at its simplest refers to sentience or awareness of internal or external existence. Despite centuries of analyses, definitions, explanations and debates by philosophers and scientists, consciousness remains puzzling and controversial, being at once the most familiar and most mysterious aspect of our lives. Perhaps the only widely agreed notion about the topic is the intuition that it exists. Opinions differ about what exactly needs to be studied and explained as consciousness. Sometimes it is synonymous with 'the mind', other times just an aspect of mind."[5]

Yes, I know, this excerpt comes from Wikipedia, but if you read the online original, you will note that it references credible sources. We will attempt to avoid the mind-brain conundrum here, as we deal with it extensively elsewhere, but we cannot ignore it entirely, because of the question: Where does consciousness reside? Everything material resides somewhere, uncertainties acknowledged, so if consciousness is material, scientists should be able to locate it. Continuing, consciousness can be defined as the awareness of our own thoughts, feelings, and perceptions of both our internal and external realities. As I explained in an earlier chapter, *An Unfortunate Experience*, our perception of our internal reality can often be false. Similarly, our perception of our external reality may also be false, because our biological sensors do not always discern the truth. In an earlier life, when I was more intrepid, I enjoyed night orienteering. I noted that my ability to correctly discern objects in low light was better than many of the other competitors. I have some idea of why, but nothing conclusive. In an earlier life, I was also taught how to "see" rather than just "look", this being related to the art of camouflage. In short, the perception of both our internal and external reality can be both optimised, and led astray. As we will later discuss, "seeing" is as much a cognitive activity as it is a perception of visible light. Only the eye *sees* because that is the organ that interacts with light; at the back end, our mind-brain complex can only *visualise* based on the electro-chemical messaging from the eye via the visual cortex. From this, the mind-brain constructs a *representation* of what is seen.

The question arises: can chemical interactions, as occur in biological activities, be led astray? Well, yes, in one sense, but not in another. Disease can cause disruption to what we perceive as normal biological workings, but the biological entities causing such disruptions are still working as normal, for them at least. At a fundamental level, the chemical constituents continue to faithfully do what their intrinsic properties determine they should do, as influenced by their external environment. What has that to do with consciousness?

The continuum of consciousness refers to the psychological construct of consciousness, during such states as normal waking consciousness, and altered states of consciousness including tiredness, partial alertness, partial awareness, and sleep. From my research, consciousness is described as a psychological construct because according to those who dabble in such things, it is believed to exist, but we are unable to physically measure it, so descriptions are "constructed" to explain it. I am unsure of why it is just *believed* to exist: I would have thought that it is *known* to exist, but perhaps philosophical materialism prevents psychologists from accepting the obvious. Consciousness is associated with awareness – the more conscious we are, the more aware we are of our thoughts, feelings, perceptions, and surroundings, the latter being termed in the military as situational awareness. Both consciousness and awareness, assuming that they are not just different names for the same phenomena, are subject to our circadian rhythm, the 24-hour internal clock that runs in the background of our brains managing our sleep/wake cycle. Basically, a part of the hypothalamus controls the circadian rhythm, but it can be influenced by external factors such as light levels, illness, and that most dreaded of all experiences, jet lag.

The point that I am getting to is that the above descriptions, or explanations, all relate to physics, chemistry, and biology, but we know from experience that no matter where we are in our natural state of consciousness, awareness, or sleep cycle, we can still choose to behave contrary to what is otherwise natural. We can be drowsy, absent-minded, or daydreaming, but can metaphorically jerk ourselves awake and pay more attention.

How is it possible, if the mind is but a material aspect of the biological brain, which has been going about as usual doing what it does, that the mind can act as an external agent to interrupt its own reverie? What can that external agent be if not a non-material entity, not subject to the processes which took us into the previous state? If the mind is running as a background processor, not subject to the then concurrent internal biological and external physical influences, is that not evidence that the mind is an entity entirely separate from the brain?

Consciousness and Sensations

> "What he needs is a *clarification* of the concept of consciousness, instead of an *explanation of it along scientific lines*."[6]

I am in the same boat: what *is* consciousness? I am back to Primary Axiom #1, that nothing can explain itself. As it is I who is conscious, how can I possibly explain my consciousness, especially when it is claimed by some in scientific circles that the mind, and thus consciousness, is the child of the brain, whose workings we have yet to understand?[7]

For context, let me repeat these definitions of *consciousness*: "the state of being aware of, and responsive to, one's surroundings", or "a person's awareness or perception of something". A little later, we discuss the argument that without the neural activity associated with sensory experiences (sensations), there is no consciousness. Here, I wanted to delve into the relationship between sensations, our awareness of them, and most importantly, our cognitive interpretation of them. Neuroscientists have coined a term for the latter, *qualia*. Raymond Tallis explains:

"Qualia are the very fabric of consciousness: the material of experience, of the 'what-it-is-like' feel of mental states. Although experience is gathered up into various kinds of wholes – objects, fields, situations, worlds – it is possible to pick out individual qualia to exemplify the notion. And so, somewhat at random, I pluck out from my rich sensory field the sound of a violin playing, the blackness of letters growing across the screen, the feeling of pressure on my buttocks, the redness of a hat next to my computer, the sensations associated with a present anxiety. If *these* components cannot be identified with nerve impulses then no aspect of human consciousness can. So let me set out some of the problems that arise when one tries to identify qualia with nerve impulses."[8]

Tallis proceeds with a longish explanation which I shall attempt to summarise. Firstly, there is no physical resemblance between a nerve impulse and a sensation. Consider "beauty is in the eye of the beholder". Well, actually, it is not in the eye, but in the mind, but if you were to study the neural activity of the beholder's brain, you will not find anything that you can identify as "beauty", especially as different beholders have different sensations of beauty, just as the quotation tells us. It is reasonable to expect that within the limits of the biology, the neural impulses reaching the brain of each beholder will be the same, yet the sensation (interpretation, experience) is different. Briefly, a neural impulse is a spike of electrical activity that travels down the length of a nervous system, neuron to neuron, via synapses. Each neuron has a membrane where the outside contains positively charged sodium ions, and the inside negatively charged potassium ions, resulting in an electrochemical differential. When a nerve impulse is generated, there is a change in the permeability of the cell membrane, and a series of electro-chemical reactions result. We need understand no further to appreciate that there is nothing there remotely resembling beauty, colour, shape, or texture, even though there must be a relationship.

The obvious conclusion should be that in the context of awareness of our external environment, neural impulses are encoded symbolic representations of that external reality which, if we are to have cognitive awareness, must then be decoded or otherwise interpreted by the mind. If it be true that *nothing can explain itself* (Primary Axiom #1), then the neural activity of the brain cannot explain itself. What, then, is doing the explaining?

To what do we attribute the *qualia*?

Legitimately, Raymond Tallis asks:

"If you dismiss the neural theory of consciousness because it is baffling then, to be consistent, you ought to reject quantum mechanics, which demands that you set aside many more of your common-sense intuitions, even such fundamental ones as that things have a definite location."[9]

Likewise, why would you believe in a singularity which was *nowhere* and in no *time* because according to the Big Bang model, its failing to be so created both space and time?

Another observation, and this is where we are heading in later chapters:

"When ... we want to assess the claim that consciousness boils down to neural activity, we need to have a clear idea about what that activity actually consists of. Likewise, if we are to resist the claim that it is the structure and complexity of the brain that creates consciousness, it is a good idea to know a little of that structure and the nature of its complexity."[10]

The Hard Problem

Fortunately, this term was chosen by scientists, not by me, so I am encouraged that scientists are struggling as much as I am on the subject of consciousness.

"David Chalmers presents the hard problem as follows:

It is undeniable that some organisms are subjects of experience. But the question of how it is that these systems are subjects of experience is perplexing. Why is it that when our cognitive systems engage in visual and auditory information-processing, we have visual or auditory experience: the quality of deep blue, the sensation of middle C? How can we explain why there is something that is likely to entertain a mental image, or to experience an emotion? It is widely agreed that experience arises from a physical basis, but we have no good explanation of why and how it so arises. Why should physical processing give rise to a rich inner life at all? It seems objectively unreasonable that it should, and yet it does. If any problem qualifies as *the* problem of consciousness, it is this one. (Chalmers 1995: 212)[11]

The Hard Problem can be specified in terms of generic and specific consciousness (Chalmers 1996). In both cases, Chalmers argues that there is an inherent limitation to empirical explanations of phenomenal consciousness in that empirical explanations will be fundamentally either structural or functional, yet phenomenal consciousness is not reducible to either. This means that there will be something that is left out in empirical explanations of consciousness, a missing ingredient."[12]

Responding to his own question, "why should physical processing give rise to a rich inner life at all", Chalmers answers "It seems objectively unreasonable that it should, and yet it does". The same could be said of our experience of free will. Evolutionists argue that it is *objectively unreasonable* that we should have free will, and thus they conclude that we don't have it. Chalmers would answer that despite the evidence that we ought not, we do.

References:

1. Bray, Dennis, Wetware: *A Computer in Every Living Cell*, Yale University Press, New Haven, CT, 2009, p. ix

2. McGinn, Colin, *The Mysterious Flame: Conscious Minds in a Material World*, Basic Books, New York, NY, 1999, as quoted in Gelernter, David, *The Tides of Mind: Uncovering the Spectrum of Consciousness*, Liveright Publishing Corporation, New York, NY, 2016, p. xxiii

3. Sheldrake, Rupert, *The Science Delusion*, Coronet, London, UK, 2013, p. 109

4. https://www.sciencedaily.com/releases/2011/10/111025122615.htm

5. https://en.wikipedia.org/wiki/Consciousness

6. Dilman, Ilham, *Philosophy as Criticism: Essays on Dennett, Searle, Foot, Davidson, Nozick*, Continuum, New York, NY, 2011, p. 2

7. https://academic.oup.com/brain/article/130/11/3050/333014

8. Tallis, Raymond, *Aping Mankind*, Routledge Classics, New York, NY, 2016, p. 95

9. *Ibid*, p. 103

10. *Ibid*, p. 28

11. Chalmers, David J, 1995, "Facing up to the Problem of Consciousness", *Journal of Consciousness Studies*, 2(3): 200–219

12. https://plato.stanford.edu/entries/consciousness-neuroscience/#MapBrain

Chapter 2-9: Objective vs Subjective

"Memory is not particularly linear - it is associative, repetitive, subjective and porous."

~ Dana Spiotta ~

The body, including the brain, consisting as it does of physical matter, can only experience *objectively*; the mind, on the other hand, experiences *subjectively*. If the material could experience subjectively, we could not do science and modern technology would fail. The proof of the pudding, if I may, is that for the most part, modern technology is reliable, and scientists continue in their work on that basis.

A *subjective experience* refers to the emotional and cognitive impact of an experience, as opposed to objective experiences which are the actual events of the experience. While something objective is tangible and can be experienced by others, subjective experiences are products of the individual mind, albeit, they can occur simultaneously, and in varied ways, in multiple minds sharing the same objective experience. Subjective experiences themselves are subject to multiple factors. For example, when I look in the mirror and notice my ever-increasing waistline, I feel a sense of shame for my lack of personal discipline. When my now deceased, fat cat looked into that same mirror (objective experience), I doubt that it ever shared my subjective experience – mostly it just purred in apparent self-satisfaction. No matter how many times I told our fat cat to go on a diet, it ignored me entirely, whereas were I to offer the same advice to my wife, well ... I am far too wise to go there. Over time, as I age, my perspective on my waistline has changed to the point where my subjective experience is edging ever closer to that of my cat.

My point is that whilst a human can have both an emotional and cognitive response to an objective experience, an animal is limited to an emotional response at best, depending on its ability to comprehend that experience. It is also true that just as with humans, animals can have a subjective experience (emotional response) which is not justified by the objective experience. Subjective responses can be managed by knowledge and experience. My cat reacts to loud thunder, often accompanied by a shock wave, much as primitive man is said to have done. Depending on the prevailing climate and my intentions for the day, my reaction can be anywhere from joy to frustration. Biological organisms are subject to external, objective experiences. Bones break, skin burns, and our bodies have an internal temperature-regulating system that responds to changes in climatic conditions. Curiously, I have found that my cognitive ability varies with temperature; it seems to have an optimal temperature range of 15°C to 25°C. This confirms to my mind at least, the strong relationship between mind and brain.

Biological responses to objective experiences are always objective, in the sense that the response is in accordance with chemical and physical properties of the organism. But then, in the human experience, subjectivity can be imposed on certain events, such as when you stub your toe and stifle your vocal response, or when you attempt to stifle a yawn or a sneeze. Does the brain do that, and if so, how? Without question, we can train ourselves to respond in what might be termed, a non-natural manner, and that training likely results in an imprint in the brain, just as happens with the muscle memory of a pianist, dart-thrower, or chef. I will not even attempt to slice vegetables the way they do, lest I no longer have fingers with which to do the slicing. But muscle memory can be defeated by other biological conditions such as arthritis, which imposes an objective limitation. The objectivity of chemical and physical processes is what allows us to do science, and implement proven science in reliable technology.

Scientists prefer to consider themselves objective in their evaluation of observed phenomena, but they are fooling themselves, for it is processes of the mind that are involved in such evaluations. We can have a degree of control over the activities of the mind, as we do when we think consciously, and some people seem to have even total control, lapsing themselves into a deep trance. But then, there must be a process operating in the background which enables them to awaken from that state, suggesting that there is a part of the mind over which they have no control. But even when people are in a trance, the manifestation is primarily physical. Because the mind cannot explain itself, we cannot know consciously everything that is influencing our thoughts. Analysed "objectively", we find that our mental processes are influenced, mostly unconsciously, by knowledge, experience, beliefs, worldview, etc. Scientists have also made us aware of subliminal influences, which we may never recognise even in our best moments of introspection.

Analysts have tools such as logical and physical data flow diagrams, which bring order to their thoughts, which otherwise would be floating around in their minds in a disordered state. I am particularly prone to that, which is why I write – it enables me to review and reorganise. Computer programmers may often flowchart their logic before committing it to code, because for them visualisation is the key to comprehension. If you have never tried it, think of a complex subject and then attempt to document it. As one quipster remarked, "Sometimes it comes out unedited!" As a writer, this is the challenge I continually face, attempting to bring order and cohesiveness to the strings of words which bubble forth from my mind, to my fingers on the keyboard. Left to their own devices, many such strings are incomprehensible even to me. It took me several attempts to structure the chapters in this book, and even now, I am not convinced that I have it right.

This evidences our normal chaotic state of mind, or as David Gelernter entitled his book, "Tides of the Mind"[1] which we have already briefly discussed. There is nothing orderly about our minds: thoughts ebb and flow, currents clash, whirlpools form, and fantasy crashes against breakwaters of reason. It takes an effort of will to calm these waters. Imagine the chaos if a robot had a human mind! You would be rushing to find the "OFF" switch. Imagine if our brains were truly like that, a far easier task, for we are faced with the evidence of brain disease in conditions such as Alzheimer's, Dementia, Epilepsy, Parkinson's, and stroke. In these conditions, cognitive function is sometimes impaired in direct relationship to the locus of the physical condition, but at other times not.

When faced with a mystery, it is often easier to determine what a thing is not, rather than what it is. When I had the opportunity to visit the plains of Africa, I saw many four-legged animals wandering about. I knew not what they were, but I did know some of what they were not: elephants, giraffes, sheep, or llamas, because I knew what they looked like. When attempting to understand the nature of the mind, I know not what it is, but I do know what it is not: biological, because my mind does not behave in the objective ways associated with the biological. If my brain behaved as my mind does, I would be taken away in restraints to a padded cell. The biological operates objectively, in accordance with its properties and how it interacts with other physical entities.

The mind is not like that.

References:

1. Gelernter, David, *The Tides of Mind: Uncovering the Spectrum of Consciousness*, Liveright Publishing Corporation, New York, NY, 2016

Chapter 2-10: Rationality & Reason

"It has been said that man is a rational animal. All my life I have been searching for evidence which could support this."

~ Bertrand Russell, British polymath, philosopher, logician, and Nobel laureate ~

Quoting from a commentary on the writings of the 12th Century Jewish philosopher, Maimonides (yes, I am impressed and influenced by him), Rabbi Feldman expounded:

"*Matter* is the material substance from which everything is made. *Form* is what gives definition (or spirit) to that particular matter. So, for example, before it can be said to 'take on life', a painting is basically a splash of colours on a canvas. What makes it a 'real' painting is the *form* that blob of paint takes, which then defines it and functions as its spirit. Likewise, a person is at bottom a splash of flesh and bones who only becomes a 'real' person by virtue of his or her *form* or *defining spirit*.

According to Rambam himself. Reason is the intellectual capacity to 'break things down into their parts', i.e., to analyze, systematize, differentiate, and correlate components; and to 'speculate about them abstractly and to attempt to determine their essences and what motivated their coming into being', i.e., to isolate a thing's essence from its contents and surmise things about its roots and sources. It also entails the ability to 'generalize and universalize things, and to separate the essential from the circumstantial', i.e., to think abstractly about a thing and to refer to it in general rather than specific terms."[1]

Now, compare that to the thought processes of Charles Darwin. In his book, *Science on Trial – The Case For Evolution*, Professor Douglas Futuyma stated that:

"Darwin drew his evidence from comparative anatomy, embryology, behaviour, geographic variation, the geographic distribution of species, the study of rudimentary organs, atavistic variations (throwbacks), and the geological record to show how all of biology provides testimony that species have descended with modification from common ancestors."[2]

On the surface, Darwin's conclusions were rational, i.e., he reasoned through the evidence that was before him to draw his conclusions. In recent years, research into genetics and related fields has brought the basis of Darwin's conclusions into question, with some scientists now asking whether the essential mechanisms of evolution have been correctly identified. What seemed reasonable to Darwin, based on his rationalisation of the evidence as he understood it, has now been shown to be unreasonable, admittedly with some exceptions. Darwin misunderstood embryology, as did Ernst Haeckel, who helped to spread Darwinism throughout Germany in the 19th century with his catch phrase, "ontology recapitulates phylogeny". This has since been proven to be false. Haeckel's drawings have been tainted with scandal, and although initially accepted as truth, have later been the rock on which his biogenic law has foundered. What seemed reasonable and rational to some, has now been pronounced as constituting scientific fraud.

Darwin sought "to 'generalize and universalize things, and to separate the essential from the circumstantial', i.e., to think abstractly about a thing and to refer to it in general rather than specific terms". Unfortunately for him, and later generations, through excusable fault on his part, he failed to differentiate the essential from the circumstantial, which led him to false generalisations. What Darwin unintentionally demonstrated was that being reasonable and/ or rational does not necessarily lead to truth. One needs to be in possession of sufficient truth before drawing conclusions – to do otherwise is not necessarily irrational or unreasonable, but it can lead to the spread of untruth.

In passing, Noam Chomsky observed, "Rational discussion is useful only when there is a significant base of shared assumptions", whilst Abhijit Naskar weighs in with, ""Talking reason with a fundamentalist is like talking balanced diet with a hungry tiger - you can't win."

This subject is substantially more extensive than I intend to deal with here. I have included this chapter primarily to give the reader food for thought on the mind-brain conundrum. What is it about the human brain, that biological network of neurons, that gives it the ability to be reasonable and rational? Do *animals analyze, systematize, differentiate, and correlate components*? Is there any life form, other than humans, that *speculate abstractly and attempt to determine their essences and what motivated their coming into being*? Why do humans dabble in the mystical, philosophical, and religious?

From an evolutionary perspective, how did the brain learn what these concepts mean? Does the brain really have the ability to conceptualise? Such thinking is said to occur in the prefrontal lobe of the cerebrum, but this organ deals only in neuronal signals, messages passing from one neuron to one or many other neurons. Such activity can be monitored by instruments, but only the *fact* of activity: there is no comprehension by an external observer of the content or context of that activity.

References:

1. Feldman, Rabbi Yaakov, *The 8 Chapters of the Rambam: A Classic Work on the Fundamentals of Jewish Ethics and Character Development*, Targum Press, Southfield, MI, 2008, p. 37

2. Futuyma, Douglas J., *Science on Trial – The Case for Evolution*, Pantheon Books, New York, 1982, p. 36

Chapter 2-11: Intentionality & Causation

"Intentionality is a philosophical concept defined as *the power of minds to be about, to represent, or to stand for, things, properties and states of affairs.*"

~ Stanford Encyclopedia of Philosophy[1] ~

Franz Brentano, a German philosopher and psychologist, explained:

"Every mental phenomenon is characterized by what the Scholastics of the Middle Ages called the intentional (or mental) inexistence of an object, and what we might call, though not wholly unambiguously, reference to a content, direction towards an object (which is not to be understood here as meaning a thing), or immanent objectivity. Every mental phenomenon includes something as an object within itself, although they do not all do so in the same way. In presentation something is presented, in judgement something is affirmed or denied, in love loved, in hate hated, in desire desired and so on. This intentional in-existence is characteristic exclusively of mental phenomena. No physical phenomenon exhibits anything like it. We could, therefore, define mental phenomena by saying that they are those phenomena which contain an object intentionally within themselves."[2]

Note that intentionality deals with the *"inexistence of an object"*, i.e., it is a mental phenomenon, not physical, and therefore could not be a function of a biological brain. It is nonsensical, to my mind, to claim that any philosophical concept could be attributed to a biological construct such as the human brain. As David Gelernter reminds us,

"All states of mind and nothing *but* states of mind are intentional, said Brentano. No mere physical state, on the other hand – no state of a tree, planet, proton, tomato – can be intrinsically *about* anything. *Only* states of mind can be intentional, but it is widely agreed that some states of mind are not."[3] [italics in original]

Such latter states exist, as Gelernter expresses them, "... being depressed is *not* an intentional state. Being depressed is about nothing. It is just *a way to be*. You are depressed in roughly the same way that a petunia is purple. Purple is how that petunia is. Depressed is how you are. (That depression is not intentional makes it no less important)." (p. 29) On the other hand, we can be *intentionally* happy or unhappy; some people can put misfortune behind them, but others seem to prefer to be miserable. Answering the question: What makes a state of mind different from all other states? The author believes that it lies in subjectivism. He avers: "All and only states of mind are two-faced, two-sided, double states. (The eminent philosopher Thomas Nagel has said so, in a different way.) Your mental states have an outer, objective side that anyone can see, and an inner, subjective face that is visible to you alone. Everything else in the universe, so far, is one-sided, with objective properties only. Everything else has only outer; minds alone have an outer *and* inner." (p. 30)

People, when inscrutable, hide their feelings, but animals lack that capacity, at least, not so far as we can understand. Animals are not introspective as we humans are – they have no ability to look inward. Animals are spared the human concept of an inner self, about which we often worry. They have no concept of "keeping up with the Joneses". People, especially children, can conduct themselves without a care in the world, living in the moment, but this is only a temporary state. Often, a wonderful day at the beach is later followed by introspection: why don't I have body like some I saw today? I doubt that animals ever have such cares.

These *tides of mind*, as David Gelernter accurately describes them, we will briefly review in a later chapter, but back to the subject at hand, *intentionality*. According to Rayment Tallis, and I agree, "There is nothing elsewhere in nature comparable to intentionality. It will prove, as we shall see, to be the key to our human differences; our subjectivity; our sustained self-consciousness; our sense of others as selves like us; first and second being; our ability to form intentions; our freedom; and our collective creation of the human world offset from nature."[4] It could be argued that animals exhibit intentionality as they walk toward a stream to drink, or a cat purrs to be picked up and cuddled, but these actions are part of instinct. I doubt that animals would walk toward a stream just to admire the waterfall, or a cat purrs with intentionality besides its comfort (attempting to compose a new song?) What distinguishes intentionality in humans from that in animals, is that we move toward goals which have nothing to do with survival or reproduction. As an aside, I often wonder why some indigenous tribes across the world have never ventured into agriculture, or have shown signs of developing technologies beyond the primitive. But I digress.

Raymond Tallis, whose referenced work is rightly worthy of study, observes:

"Nothing in physical science can even seem to provide an adequate explanation of why, or how, some (although not most) neural activity would reach causally upstream to events that led up to themselves; why or how a burst of impulses in my visual cortex should refer itself back to the interaction of the light with the hat (Ed. referring to an earlier context) and, out of this, construct a hat-in-the-light 'out there'."[5]

Scientific research is predicated on the principle of cause-and-effect, and it is the intelligence of scientists which has them reaching back (or upstream) from the effect to the cause. Nothing in nature moves upstream (well, ok, salmon do), but so far as we know, there is no effect in physics or chemistry that reaches causally upstream. For there to be so, we would have circular cause-and-effect which logically would amount to the much desired, but never achieved, *perpetual motion*. Intentionality is closely related, or perhaps even just a function of, consciousness. Continuing from Tallis as above,

"It is not merely a case of "registering: those events, as a photoelectric cell might register light from any source. For we not only register events, but also register them *as* belonging to something other than ourself: we are *aware* of them and aware of them as "over there". It is a *revelation*: of an object to a subject in which object and subject are kept separate and distinct, with the subject (me) being here and the object (the hat that I am looking at) being over there.

This difference between physical registration (if one can truly speak of something being "registered" by an entity, such as a photometer, that is not conscious of that which is registered) and perception is absolutely fundamental but quite elusive. It is easy to lose sight of it, particularly if one is a neuromaniac and has a vested interest in concealing it. It is even easier to conceal it if one treats the brain both as material object and as a quasi-person."

In a later chapter on our visual sensory system, I offer that the eye "sees", but the brain "visualises" a representation of what the eye sees. The reason is that light reaches the eye in the same sense as light reaches a photometer (or other physical object), but there is no cognitive registration of the event at that point. Sensory signals are sent via multiple nerve fibres in the optic nerve to the visual cortex, where further signal processing occurs. It is only when signal reach upper areas of the brain that visualisation occurs. As Tallis rightly questions, why should we believe that a physical object, in this case, a biological blob of complex neural networks but merely physical for that, can do more than other physical objects? His term, *neuromaniac*, is one he coined, to refer to those who have succumbed to neuromania – the appeal to the brain, as revealed through the latest science, to explain our behaviour.

To finish this chapter:

"The assimilation of consciousness to the causal net in which the organism is located has been the central pillar of materialist theories of mind, in particular of a highly popular theory called "functionalism". Functionalists argue that mind is not importantly about the phenomenal aspects of consciousness: actual awareness. No, its job – and consciousness, according to them simply *is* its job – is to refine the connection between inputs and outputs in such a way as to optimize the survival of the organism or the group to which the organism belongs. Any particular element of consciousness is constituted entirely by its functional role: its causal relations to sensory inputs, to other mental states, and to behavioural outputs."[6]

I will leave the reader to ponder this deeply: Are all your activities and thoughts optimising the survival of yourself, or the groups to which you belong? Is selfishness or self-centredness only about survival of self? Is altruism only about survival of the group?

References:

1. https://plato.stanford.edu/entries/intentionality/

2. Brentano, Franz, *Psychology from an Empirical Standpoint,* edited by Linda L. McAlister, Routledge Publishing, London, UK, 1995, pp. 88–89

3. Gelernter, David, *The Tides of Mind: Uncovering the Spectrum of Consciousness*, Liveright Publishing Corporation, New York, NY, 2016, pp. 29-30

4. Tallis, Raymond, *Aping Mankind*, Routledge Classics, New York, NY, 2016, p. 105

5. *Ibid*, p. 106

6. *Ibid*, p. 107

Chapter 2-12: Volition & Free Will

"We must believe in free will, we have no choice"

~ Isaac Bashevis Singer ~

When people say that we do not have free will, my immediate response is to ask them what part of their material existence forced them to say that, and what is different about my material existence that forces to me to say precisely the opposite?

Anthony Cashmore, a Biology Professor at the University of Pennsylvania, assures us that consciousness and free will are simply illusions. He wrote: "The reality is, not only do we have no more free will than a fly or a bacterium, in actuality we have no more free will than a bowl of sugar."[1] Charles Darwin, Thomas Huxley, Francis Crick, and numerous other evolutionists are of the same view, Susan Blackmore observing, "I think nature has played this enormous joke on us."[2] Some scientists and philosophers have taken this a step further, positing that as we do not really have free will, we are thus not responsible for our own actions: our behaviour is simply the result of complex chemical reactions over which we have no control. William Provine, a Biology Professor at Cornell University commented: "There is no way that the evolutionary process … can produce a being that is truly free to make choices."[3]

Taking Professor Provine at his word, if it can be demonstrated that *we are able* to make free choices, as he apparently did when making his assertion, would he then agree that evolution theory must be false? Does Professor Provine truly believe that his choice of profession, his dedication to hours of study and research, and the conclusions that he has reached, are all the results of complex chemical reactions over which he has no control? Should any of these academics be lauded for their achievements when according to them, they had no personal influence over the outcomes? Should the Nobel Prize ever be awarded to anyone when they are little better than chemical automatons?

As these scientists are talking about themselves as well as everybody else, we cannot ignore the obvious: their own theory strips evolutionists of any authority to make such assertions – we have no reason to believe them. One would hope that such scientific luminaries are joking when they deny the existence of volition, but I suspect not. Fortunately, other scholars recognise the obvious. Gilbert Ryle, former Waynflete Professor of Metaphysical Philosophy at the University of Oxford noted way back in 1949:

"Teachers and examiners, magistrates and critics, historians and novelists, confessors and non-commissioned officers, employers, employees and partners, parents, lovers, friends and enemies, all know well enough how to settle their daily questions about the qualities of character and intellect of the individual with whom they have to do. They can appraise his performances, assess his progress, understand his words and actions, discern his motives and see his jokes. If they go wrong, they know how to correct their mistakes. More, they can deliberately influence the minds of those with whom they deal by criticism, example, teaching, punishment, bribery, mockery and persuasion, and then modify their treatments in the light of results produced.

Both in describing the minds of others and in prescribing for them, they are wielding with greater or less efficiency concepts of mental powers and operations. They have learned how to apply in concrete situations such mental-conduct epithets as 'careful', 'stupid', 'logical', 'unobservant', 'ingenious', 'vain', 'methodical', 'credulous', 'witty', 'self-controlled' and a thousand others."[4]

If you consider these observations in the light of humans having no free will, you should be able to see that there is something seriously wrong with the perspective of the evolutionists. Could any of these phenomena occur in the absence of free will? However, Gilbert Ryle goes on to argue: "the doctrine of volition is a causal hypothesis, adopted because it wrongly supposed that the question, 'What makes a bodily movement voluntary?' was a causal question. This supposition is, in fact, only a special twist of the general supposition that the question. 'How are mental-conduct concepts applicable to human behaviour?' is a question about the causation of that behaviour."[5] It would appear that Ryle supports the notion of free will, except that it is not "free" in the sense that it is not our personal choice. He then offers what I contend is an illogical argument: "Champions of the doctrine should have noticed the simple fact that they and all other sensible persons knew how to decide questions about the voluntariness and involuntariness of actions and about the resoluteness and irresoluteness of agents before they had ever heard of the hypothesis of the occult inner thrusts and actions."[6] Philosophy is subsequent, not antecedent to, any human activity. Here, Ryle seems to be suggesting the expressed philosophy precedes understanding, which may at times be true, but not generally true. As one author put it: *"The author who benefits you most is not the one who tells you something you did not know before, but the one who gives expression to the truth that has been dumbly struggling in you for utterance."*[7] Ryle then proceeds to confuse me even further when he asserts, "Yet this correlation could, on the one hand, never be scientifically established, since the thrusts postulated were screened from scientific observation"[8]. Is he admitting that the mind is *not* material? In a later chapter, we will attempt to discern just what it is that Ryle proposes concerning the mind. Before getting there, however, let us examine some other of his claims. He wrote:

"Before we bid farewell to the doctrine of volitions, it is expedient to consider certain quite familiar and authentic processes with which volitions are sometimes wrongly identified.

People are frequently in doubt what to do; having considered alternative courses of action, they then, sometimes, select or choose one of these courses. This process of opting for one of a set of alternative courses of actions is sometimes said to be what is signified by 'volition'. But this identification will not do, for most voluntary actions do not issue out of conditions of indecision and are not therefore results of settlements of indecisions. Moreover, it is notorious that a person may choose to do something but fail, from weakness of will, to do it; or he may fail to do it because some circumstance arises after the choice is made, preventing the execution of the act chosen. But the theory could not allow that volitions ever fail to result in action, else further executive operations would have to be postulated to account for the fact that sometimes voluntary actions are performed. And finally, the process of deliberating between alternatives and opting for one of them is itself subject to appraisal-predicates. But if, for example, an act of choosing is describable as voluntary, then, on this suggested showing, it would have in its turn to be the result of a prior choice to choose, and that from a choice to choose to choose ..."[9]

I am not familiar with the theory to which Ryle refers, but it is not the *Cartesian* theory of mind, which places no such restrictions on decision making as Ryle assumes. Perhaps Ryle has his own version of the theory from which he conjures up a strawman argument, but I have not encountered similarly in the writings of other philosophers or psychologists. Decision making is not necessarily a one step process, nor does the process need to occur at just one moment in time. When Ryle wrote as he did, he deliberately chose words to express his thoughts - I am confident that as with all writers, the final version differed from the original as he reviewed his work, and made improvements, all decisions of his own making. In my own case, I make decisions which when made, are tentative, subject to further additional evidence, and contrary to Ryle's assertion, this theory of volition *does allow that volitions sometimes fail to result in action*. Ryle seems to be suggesting that opting based on appraisal-predicates must result in an endless regression of choices. There is a term for this: *analysis paralysis*, but it is not inevitable because consciously or otherwise, we usually manage a way out of this jungle of thoughts. Some people seem unable, but this cannot argue that Ryle is right in his hypothesis. The voluntary aspect of resolving decision making conflicts is to assign a level of importance to conflicting criteria, which we sometimes do consciously, and other times unconsciously, based on our predilections and world view. Ryle himself seems to be struggling with conflict resolution: on the one hand, he accepts that the mind exists, but on the other, asserts that it is not open to empirical language such as is used for the body.

Ryle gives the appearance of a lack of appreciation of the real world, or at least, his writings do. Perhaps in pressing his point, he fails to think through an issue thoroughly. For example, in discussing the *Distinction between Voluntary and Involuntary* (pp. 67-68), he notes: "We do not laugh on purpose". Has he not heard of polite laughter, when people laugh not because they perceive humour, but to avoid embarrassing the narrator? Preceded by a number of examples, he opines: "it is incorrect to say that he could have avoided getting it right". Ryle asks: "Could you have helped drawing the proper conclusion?" Well of course, I and others, could, and sometimes (often?) do, but this depends on the meaning of "proper" for the individual. Juries do get their verdicts wrong, from time to time, because they draw the wrong conclusion from the evidence before them. A "proper" conclusion is not necessarily the "right" conclusion. In politics and the media, I often find people drawing the wrong conclusion, likely influenced by their unconscious prejudices and worldview, not necessarily cognisant that they did so. On the subject of *cognitive dissonance*, we know that some seek escape by denying the truth of a proposition, or by downplaying its significance. They are not always aware of doing so until it is pointed out to them. These are cases where people cannot help but draw the wrong conclusion, contrary to Ryle's proposition. However, to be fair, I believe that Ryle should have included what he meant by a "proper conclusion" in that context. I suspect that he meant "proper" in the sense of what was rational to the individual, not what was empirically proper, if there is such a thing. I believe it to be true that whenever we reach a conclusion, it is "right" in the sense that it is right according to us – I doubt that any of us could ever internally, deliberately reach the "wrong" conclusion, even though we may later express it in denial of our own thoughts.

I must confess that when I purchased Ryle's book entitled, *The Concept of Mind*, I did so in the expectation that he would explain what in his opinion, the mind is. I was misled by the title. Ryle makes no effort to define the mind or the concept thereof: his thesis primarily concerns the language that we use to describe the mind. We discuss this further in a later chapter of the same name.

There is, perhaps, no empirical evidence of volition and free will, depending on how you interpret "empirical", but there is certainly an abundance of circumstantial evidence within the lives of every one of us, and of every person recorded in human history. What physical properties, arrangements, or processes of the human brain result in choices, decisions, preferences, fancies, or predilections? Consider two boys, brothers of the same parentage: from an early age, one wants to study history and dreams of tenure in academia, whilst the other sees himself as a fighter pilot and a lifetime as an aviator; on the basis of an almost identical genome, what physical properties are responsible for the differences and almost opposite polarity in the choice of careers? When aspiring pilots fail to demonstrate adequate proficiency, as I and many others have done, why did some continue to pursue careers in aviation whilst others chose accountancy, engineering, or alternate careers not at all connected to aviation? Raised as a Catholic, I pursued independent bible studies and eventually came to the conclusion that I no longer accepted all of the doctrines of traditional Christianity, yet many of my peers remained as devout Catholics, whilst others experimented with Buddhism and other belief systems. What was it in the physical make-up of our brains, based on our genomes, that predetermined these separate outcomes?

Decision making contrary to preference is best evidenced in *compromise*. Husbands and wives in most cases, live a life of compromise in all of their choices and activities. Diplomacy and politics are said to be the art of compromise. In these cases, initial decisions may not result in actions based on those decisions, sometimes not even expressing them to others.

Why is it that some individuals from generations committed to one religion eventually chose to abandon that religion, and even their children and families, to follow another path as Carma Naylor[10], Tony Coffey[11], Jerry Rassamni[12], and Walid Shoebat[13] did? Why did some atheists choose to accept the existence of God contrary to their previous teachings and beliefs as Francis Collins[14], Anthony Flew[15], Peter Hitchens[16], C.S. Lewis[17] [18], John Sanford[19], Lee Strobel[20], and many others have publicly attested? Why did an Egyptian woman, whose father was assassinated by the IDF (Israeli Defence Forces), eventually establish an organisation supporting Israel, contrary to her culture and her tragic experience at the hands of the Israelis[21]? Were these people, and the multitudes of whom we have never heard, all compelled by complex chemical reactions to choose as they did?

Did Tchaikovsky compose the *1812 Overture*, did Rembrandt paint *The Night Watch*, did Michelangelo sculpt *David*, or were these famous artists unwitting accomplices to chemical reactions over which they had no control? Luck seems to have played a part, according to the evolutionists.

The very same people who are proponents of evolution, also as parents and teachers, exhort us to develop character? How can we develop character if we have no control over who we are and how we react to circumstance? Why are some people honest and others dishonest; some gentle and others brutal; some altruistic and others selfish? Why do courts and secular authorities believe that criminals can be rehabilitated if the criminals have no control over their future behaviour? Why did despots such as Lenin, Trotsky, Stalin, Hitler, Himmler, Pol Pot, and Mao Zedong all believe that indoctrination was an effective tool in bending people to their will? Why do commercial enterprises spend exorbitant funds trying to convince people to buy their products if nobody truly has free will and the ability to choose? Why do universities have courses in political science, philosophy, psychology, psychiatry, psychotherapy, ethics, and related subjects if the imparted knowledge is incapable of changing behaviours?

I have read the works of psychologists, beholden to philosophical materialism and evolution, where they attempt to describe the mechanisms of the mind whereby previous beliefs and attitudes can be altered. In the main, I have wondered how deeply they have studied neuroanatomy and physiology, and seriously contemplated how the known behaviour of neurons could accomplish the actions they posit. As with the title of this book, there is a vast chasm between what is known, and what they posit.

How do people live their lives - do their actions testify to the assertions of the evolutionists, or to some other belief system? How do the evolutionists live their lives - do they live as if they truly believe what they teach, or do they give the lie to their beliefs and live as intellectual hypocrites? Why would anyone attend university or pursue a career when if the evolutionists are right, they can have no confidence that the chance chemical reactions in their brains won't have them heading off in some other direction just moments later? Why would anyone take out a loan and invest tens of thousands of dollars in an education, when they can have no assurance that their dreams won't be shattered by a later chemical reaction? What is the point of following a hope, dream, or ambition if indeed, you have no personal relationship to it other than an evolutionary event caused it to intervene in your life at some chance moment? Ryle and others are suggesting that such people are deluded concerning their participation in decision making, but there is the rub, there is no "they", "I", or "we".

Summary

I have no doubt that chemical interactions are involved in the implementation of decisions - it must be so otherwise I could not be typing the words that I choose to type. I also accept that cognitive functions are in some way related to electro-chemical reactions in the brain, but the question remains, what is driving what, or as Lewis Carroll expressed in another context: *'The question is,'* said Humpty Dumpty, *'which is to be master — that's all."* Have I deliberately chosen to write what I have written, or has the choice been made for me, and I only write because I am compelled to write, what I write, when I write, due to complex chemical reactions over which I have no control?

Are all our actions, words, decisions, choices, and preferences merely the outcome of complex chemical reactions over which we have no control, or do we in truth have some level of control expressed as *free will* or *volition*? If we have no free will, should anyone be held accountable for anything: why should achievers be rewarded and offenders punished?

We are faced with a *contradictory reality*. On the one hand the evolutionists, not being able to explain free will and volition in evolutionary terms, have chosen to deny that such exist even as they practice those very same attributes when they profess their denials. On the other hand, we have ample evidence that free will and volition do exist, even though we are unable to explain them in physical terms. It seems prudent, to my mind, to accept the existence of free will and volition, and to ponder whether their nature is immaterial. If that be the case, the logical corollary would be that we have an *immaterial* Mind that controls the *material* Brain in all but autonomous functions.

The denial of humans having free will is so absurd, and so contrary to the evidence, that at best we should describe this assertion by evolutionists as an *argument from desperation*. In any event, if we have no control over our mental processes, then clearly evolutionists can have no confidence in what they believe, preach, or write, and nor can we. Revisiting the observation by William Provine that "There is no way that the evolutionary process ... can produce a being that is truly free to make choices", I find myself free to make the choice that the evolution narrative must be false.

I was more than a little bemused to read that according to some, we have *free won't*, but not *free will*. "Won't" is a choice, just as "will" is a choice, so by what convoluted logic does one decide that deciding *against* is not an example of free will? The background is as follows concerning Libet's experiments:

"The authors concluded that there is a network of high-level control areas "that begin to prepare an upcoming decision long before it enters awareness." It looks as if we don't know what we are doing until we have found that we have done it.

This has certainly brought many philosophers up short. Among them is Alfred Mele, who in 2010 was awarded a $4.4 million grant by the Templeton Foundation to look into the whole issue of human free will. Libet's original interpretation of his own experiments was that they demonstrated that we do not have free will: the brain "decides" to move; the brain "initiates" movement. As Libet put it in a more recent paper, "If the 'act now' process is initiated unconsciously, then the conscious free will is not doing it. We do, however, have "free won't": we can inhibit movements that are initiated by the brain. We don't quite initiate voluntary processes; rather, we "select and control them", either by permitting the movement that arises out of an unconsciously initiated process or "by vetoing the progression to actual motor activation"."[22]

We have more discussion on Libet's experiments in a later chapter, but what should be evident from the above is that the crux of the matter is not that we do not have free will to make choices. The proposition by Libet is that we never consciously decide to initiate some action – that is done unconsciously by the brain (which belongs to whom?). But if the brain is the determinant of consciousness, how is it that at the same time, one part of the brain is conscious and another not? For an extended discussion on consciousness, I would recommend this online resource[23].

As a segue into the next chapter, let us consider the intellect, with its capacity to reason, imagine, acquire knowledge, and most importantly, based on such knowledge, differentiate between good and evil. Maimonides observes, in connection with the intellect:

"Some of its functions are practical, and others are speculative. Practical functions are either mechanical or conceptual. The speculative ones touch upon our awareness of the nature of the immutables (which are called 'the sciences'). Mechanical functions include the capacity to acquire skills like carpentry, agriculture, medicine, or navigation, and the speculative ones to touch upon the capacity to think about doing something one would like to do, when he wants to do it, whether it is feasible to do it or not, and if so how to do it."[24]

As an aside, the *speculative* function of our intellect comprises part of our *free will*, which material monists deny that we have. I am indebted to Maimonides' insights which significantly contribute to my thinking and understanding in this area.

Finally, putting aside the compelling evidence that we do in fact have free will, let us approach the *"free will is an illusion"* proposition from the perspective of logic. An "illusion" defined is a deceptive appearance or impression, or a false idea or belief. According to philosophical materialism, all appearances, impressions, ideas, and beliefs are functions of the biological brain. For the brain to be deceived or to hold a false idea or belief, it is itself which is doing it, and is thus delusional. Illusions cannot recognise themselves (Primary Axiom #1), and cannot be recognised as an illusion by the entity experiencing it. As there is no other entity involved in the experience, then it can only be the brain which is having the illusion, and thus the assertion that free will is an illusion can only be the outcome of a delusional brain. It "thinks" it has what it hasn't, or hasn't what it does have. Logically then, we do have free will.

So, enough of the nonsensical imperatives of those beholden to philosophical materialism, and let us open our minds to what is possible, and even probable.

References:

1. Cashmore, A., *The Lucretian swerve: The biological basis of human behaviour and the criminal justice system*, *Proceedings of the National Academy of Sciences* **107**(10), 2010, p. 4499-4504

2. *Ibid*, p. 4502

3. Johnson, P., *Darwin on Trial*, 2nd ed., Illinois, USA, 1993, p. 127, quoting William Provine

4. Ryle, Gilbert, *The Concept of Mind*, Penguin Books Ltd, London, England, 1990, p. 9

5. *Ibid*, p. 66

6. *Ibid*

7. Chambers, Oswald, *My Utmost for His Highest*, Barbour Publishing Inc., Uhrichsville, OH, 1963, p. December 15th

8. Ryle, *Ibid*, p. 66

9. *Ibid*, pp. 66-67

10. Naylor, Carma, *A Mormon's Unexpected Journey* - Volumes I & 2, WinePress Publishing, Enumclaw, WA, 2006

11. Coffey, Tony, *Once a Catholic: What You Need to Know about*

Roman Catholicism, Harvest House Publishers, Eugene, OR, 1993

12. Rassamni, Jerry, *From Jihad to Jesus – An Ex-Militant's Journey of Faith*, Living Ink Books, Chattanooga, Tennessee, 2006

13. Shoebat, Walid, *Why I Left Jihad*, Top Executive Media, USA, 2005

14. Collins, Francis S., *The Language of God*, Free Press, Simon & Schuster, New York, 2006

15. Flew, Anthony, *There is a God – how the world's most notorious atheist changed his mind*, HarperOne, New York, 2007

16. Hitchens, Peter, *The Rage Against God – How Atheism Led Me to Faith*, Zondervan, Grand Rapids, Michigan, 2010

17. Lewis, C.S., *Surprised by Joy*, Harper Collins, London, 1955

18. Lewis, C.S., *The Case for Christianity*, Collier Books, New York, NY, 1989

19. Sanford, Dr John C., *Genetic Entropy & The Mystery of the Genome*, FMS Publications, Waterloo, New York, 2008

20. Strobel, Lee, *The Case for a Creator*, Zondervan Press, Grand Rapids, MI, 2004

21. Darwish, Nonie, *Now They Call Me Infidel – Why I Renounced Jihad*, Penguin Group (USA), 2007

22. Tallis, Raymond, *Aping Mankind*, Routledge Classics, New York, NY, 2016, pp. 55-56

23. https://plato.stanford.edu/entries/consciousness-neuroscience/#MapBrain

24. Feldman, Rabbi Yaakov, The 8 Chapters of the Rambam: *A Classic Work on the Fundamentals of Jewish Ethics and Character Development*, Targum Press, Southfield, MI, 2008, p. 36

Chapter 2-13: Intelligence & Intellect

"The test of a first-rate intelligence is the ability to hold two opposed ideas in mind at the same time and still retain the ability to function."

~ F. Scott Fitzgerald, American fiction writer ~

This is similar to the thinking of the ancient Greek philosopher, Aristotle, when he opined, "It is the mark of an educated mind to be able to entertain a thought without accepting it." This evidences what largely differentiates the mind from the brain – *ambiguity*, i.e., the quality of being open to more than one interpretation. Chemicals, biological cells composed of chemicals, and biological organs composed of biological cells, are incapable of ambiguity – they can only do what they can, and must do (Primary Axiom #10) – their experiences and reactions are always objective. Whilst psychologists and biologists may speak of cells evaluating the environment and making decisions, the simple truth is that in any given environment, only *one* outcome is possible. When doctors or chemists are unsure of an outcome, it is not the reality that more than one outcome is possible; rather, the reality is that they are insufficiently apprised of all factors affecting the outcome. This is the same logical fallacy associated with randomness – we cannot know that a process or outcome is truly random, unless we have a comprehensive understanding of the process itself, and both the existence and influence of all factors affecting it.

So, what *is* intelligence?

Let me start with one expression of intelligence: *subjectivity*. No behaviour or activity in the natural world, as described under philosophical materialism, can ever be subjective, only ever objective, for nothing material can act based on, or be influenced by, personal feelings, cognitive preference, or opinions. Materials can only do what they must do, which leaves no room for subjective decisions.

Another expression of intelligence is the ability to think in abstract terms. Antonyms for *abstract* include *material* and *physical*, which are adjectives appropriate for the brain. From a social perspective, intelligence is a curious thing. Except in academia and certain sciences, society appears to scorn any demonstration of intelligence with epithets such as, "You think you're smart". Society accepts, applauds, and even adores demonstrations of prowess and excellence in football, tennis, athletics, swimming, music, singing, and acting for example, but woe betide those who choose intellectual activities (as I have found all too often). It is only in recent years that medical experts have started to recommend intellectual exercise to ward off the worst effects of aging. I wonder whether deep in the psyche of humans holding to philosophical materialism, there is the realisation that it is the quality of human intelligence that differentiates us from all other living organisms. That is not to suggest that other creatures do not evidence intelligence, for they surely do, but just as surely, their intelligence is severely limited when it is compared to that of humans. Pursued intellectually, our intelligence should cause one to question the *descent with modification* paradigm. Firstly, one must identify and describe in physical terms, just what is being modified. Such identification continues to elude scientists.

Defined in psychology, intelligence includes the ability to benefit from past experience, act purposefully, solve problems, and adapt to new situations. Another definition is: the ability to learn or understand, to deal with new or trying situations, or to reason, Alternatively, the ability to apply knowledge to manipulate one's environment, or to think abstractly as measured by objective criteria, such as IQ tests. I would offer that intelligence is the *potential for cognition*, and all that it implies; put in terms of my Foundational Propositions, intelligence is a *property* of the *entity* we call, the "mind". Revisiting Maimonides' 18th Proposition, "Everything that passes over from a state of potentiality to that of actuality, is caused to do so by some external agent". The *actuality* of intelligence is what we do with it, and I would contend, the external agent is the mind, not the biological brain. Continuing from Maimonides, "because if that agent existed in the thing itself, and no obstacle prevented the transition, the thing would never be in a state of potentiality, but always in that state of actuality." We can be confident from both observation and personal experience, that intelligence is not always in a *state of actuality*. At times, in some sentient beings, we can detect no evidence of intelligence whatsoever.

Now, for a little bit of political incorrectness – intelligence as a function of race. Can it be true that some races are more or less intelligent than others? If philosophical materialism is true, then I am persuaded to accept that possibility - why not? We know that races can be of different skin colour and other features, particularly the face, such as the shape of the periocular area (skin surrounding the eyes), nostrils and noses, etc. Other variations in physical attributes are evident; for example, height: compare African pygmies with the tall, slender, African Zulus. Irrespective of how these differences arose, they are subject to genetic inheritance. The brain is composed of physical features in the form of neurons and neuronal connections, also subject to genetic inheritance. It would be absurd to argue that whilst the genotype plays a significant role in determining the phenotype, it plays no part in determining the structure of the brain's neural network. The idea that we are all equal arose from the religious belief that we are all made *"in the image of God"*. If you are an atheist, or otherwise subscribe to philosophical materialism and evolution, then you have no reason to agree that we are all equal, because quite obviously, evolution does not result in equality, except in a very general sense of the phenotype. If variations do exist due to genetics, adaptation, and environmental circumstance, there can be no contention that none of these have implications for the development of neuronal networks and their efficiencies. If intelligence is purely a function of the biochemistry of the brain, wherein the substantiation that all races are of equal intelligence?

George Gilder wrote an intriguing book, *"The Israel Test"*[1], wherein he discusses as one reviewer put it, "the extraordinary virtues of the beleaguered Jewish State" (back cover). The relevance here is his recounting of the achievement of Jews, and studies comparing the IQ of Jews with other races. As regards achievements, he wrote:

"From the day Heinrich Hertz, whose father was a Jew, first demonstrated electromagnetic waves and Albert Michelson conducted key experiments underlying Einstein's theory of relativity, the achievements of modern science are largely the expression of Jewish genius and ingenuity. If 26 percent of Nobel Prizes do not suffice to make the case, it is confirmed by 51 percent of the Wolf Foundation Prizes in Physics, 28 percent of the Max Planck Medailles and 38 percent of the Dirac Medals for Theoretical Physics, 37 percent of the Heineman Prizes for Mathematical Physics, and 53 percent of the Enrico Fermi Awards."[2]

In recounting this, I am not making any case for Jews, but simply offering evidence supporting the possibility that some races, genetically, may be more or less intelligent than others. How else do we explain that one race, as of 2010 comprising just 0.2% of the world's population, is so over-represented in scientific achievements? As for intelligence measured by IQ tests, Gilder references a study[3] postulating that a "cognitive elite" are at the top of a new class structure in America, and that low intelligence levels are at the root of many social problems, challenging the fundamental belief that everyone is created equal:

"In *The Bell Curve*, Charles Murray and Richard Herrnstein pointed to the massive superiority in IQs of Ashkenazi (Eastern European) Jews over all other genetically identifiable groups. The upside tail of the curve is massively Jewish. As Murray later distilled the evidence in *Commentary*, the Jewish mean intelligence is 110, ten points over the norm."[4]

Again, I am doing nothing more than recounting the observations of others, which if true, point to the possibility of race being a genetic determinant of intelligence, as controversial and politically incorrect as this concept clearly is.

Returning to the definitions of intelligence as given in psychology texts, there is another more important issue as identified by Albert Einstein: "The true sign of intelligence is not knowledge but imagination". If one holds to philosophical materialism, then it would be natural to always think in terms of cause and effect, because that is how the material world works. However, imagination is a function of the mind, no doubt in part drawing on actual things and experiences, but adding to those in sometimes impossible ways. I have long been a devotee of science fiction. Some of what was considered impossible when the author penned as he did, has since become science fact. Other phenomena remain in the realm of truly science fiction. But that is the workings of imagination, as wonderful as it is.

Imagination *transcends* the material, which is further evidence that the mind cannot be material.

Intellect

From another perspective, *intellect* is our capacity to think *pragmatically*, the antithesis of what we commonly see on the streets during protests and violent demonstrations, where evidence of intellect is entirely absent. Intellect is our capacity to extrapolate, even irrationally, derive, and study the nature of things, and act on that knowledge, whether it be true or false. It is an active faculty, which transcends instinct, which is a passive faculty in humans and lower order organisms. It embodies the uniquely human capacity to take one's destiny into one's own hands by acting on one's own, and making circumstantial, career, and moral life choices. But there is one human capacity that is yet greater than intellect – reason, which we discussed earlier.

Animals and lower order organisms lack intellect – they are unable to act counter to instinct, yet their brains are constructed of the same materials, and in roughly the same neural patterns, as those of humans. It is reasonable to conclude on the evidence, that intellect is not a function of the brain, although there is a relationship. We must ever be wary of confusing correlation with causation.

References:

1. Gilder, George, *The Israel Test*, Richard Vigilante Books, USA, 2009

2. *Ibid*, p. 35

3. Murray, Charles, and Herrnstein, Richard J., *The Bell Curve: Intelligence and Class Structure in American Life*, Free Press, New York, NY, 1994

4. Gilder, *Ibid*, pp. 32-33

Chapter 2-14: Morality

"The assumption that animals are without rights and the illusion that our treatment of them has no moral significance is a positively outrageous example of Western crudity and barbarity. Universal compassion is the only guarantee of morality."

~ Arthur Schopenhauer, The Basis of Morality ~

A significant difference between animals and humans, is that the latter argue about what constitutes morality, whilst as far as we can discern, the former have no concept of it. Whatever they do, they do out of instinct, and never stop to wonder, as humans do, whether they are "doing the right thing". What constitutes *the right thing*, and who decides? Is Schopenhauer correct, which would suggest that it is immoral to kill animals for food? Why then is it not immoral for animals to kill animals, or even humans for food? We never expect morality from lower order creatures - morality is a human concept, which varies by culture and time.

Whilst in this study, I would prefer to not discuss religious views, when it comes to morality, such discussion is impossible to avoid. The question becomes: Which is morally superior, or even practical – the religious or the secular? The following observations by Jewish rabbi, Nathan Lopes Cardozo, are worth reviewing, especially in the context of which are the more restrictive of human activity. Secularists would argue the religious, but are they correct? Here I shall quote from his book[1], the chapter entitled "Halacha Means Full Liberty – To be Secular would be Hell: Everything would be Forbidden". To justify my usage of this religious text, remember that *religion is philosophy*, and philosophy is a very significant aspect of the *exceptional human condition*, as is religion. So far as we know, *Homo sapiens* is the only species that indulges in such fantasies of the mind.

Introducing the subject, Cardozo observes:

"I need space to breathe freely. I can't live in a cage where everything is forbidden and I am surrounded by prohibitions that will take all the joy out of my life and make it unbearable.

So, I can't be secular."[2]

The secular must be wondering whether the rabbi has been inhaling far too much incense smoke. Everybody knows that the religious, and especially those who practise Judaism, lead the most restrictive lives of all. So, what led him to that conclusion? Let us follow his thinking:

"How do we know that animals experience less pain than we do? How can we measure this? What is it that makes animals different from human beings and their lives of lesser value? It has been argued that animas do not possess the sophisticated level of consciousness that humans have. Humans are aware of their very being, of their thinking. They are much more intelligent, far more creative. They can think in abstract terms. Animals are unable to do so, at least not on the same level of humans. Surely there is an ontological difference between animals and humans. Human beings live life on a level that animals do not share, but we don't even know what this consists of.

But even when we acknowledge that these differences are real, what criteria determine that the lives of animals are less valuable than those of human beings? How do we know that ontological inferiority also means a lesser sanctity of life? Who says that less *gestalt* [an organized whole that is perceived as more than the sum of its parts] means less significance and meaning? What gives us the right to kill animals for consumption ... what moral criterion do we rely on ... we must also ask what gives us the right to pluck a flower and stop its life flow, or to kill an insect, even if it is dangerous? The astonishing conclusion is that there *are* no objective criteria to *follow*.

Let us take this matter even further. What gives a husband the right to impregnate his wife, knowing she will no doubt undergo serious pain and discomfort when giving birth? Is this entirely dependent on his wife's consent? Who says we are allowed to put her in the slightest danger even if she has fully agreed? What gives us the right to have sexual relations with a human being but forbids us to do so with another creature? Once again, we lack absolute moral criteria."[3]

More to the point, what if the wife has not agreed? Consider the many cultures where sexual relations in marriage, and even procreation, are considered the absolute, inalienable right of the husband, with the wife having no say in the matter. In the earlier chapter on *Sentience versus Consciousness*, we saw how various well-intentioned groups, and even legal authorities, are rather confused on the subject of animal rights, leading to some very dangerous precedents. These could be used to substantiate further absurd, and more significantly, *restrictive legal decisions,* as was the case in New Zealand when, believe it or not, the Whanganui River was granted *person status*, opening the way for rights to be given to the river itself. Cardozo continues,

"If there are no absolute criteria by which we determine how to deal with these questions ... This brings us to a most amazing conclusion: A strictly secular approach to major moral issues may have to be *much more restrictive* than that which any religion would ever demand. In fact, a secular moral attitude may make life extremely difficult and even impossible. I therefore declare that I cannot be secular. It's too difficult. I'd have to live with so many restraints that I would collapse.

It is religion, and not a consistent secular attitude, that has the more liberal approach to moral issues."

As I write, mid 2021, protests and riots have broken out across not just the United States of America, but in many countries across the world. The reason: the illegal killing of a black man, with a long criminal record, who was caught under the influence of illegal substances, with even more in his pocket, allegedly attempting to pass a counterfeit $20 note, but stupidly, resisted arrest. Consequently, he was feted as a wonderful son, incapable of harming anyone, even though he had spent time in prison for holding a gun to the stomach of a pregnant woman whilst his cohorts robbed her house. He was declared a hero by the protesters, a martyr, in the cause of protesting police brutality. As a result, a statue has been erected in his honour. Consequently, riots have erupted whenever a person of colour has been perceived as having been killed illegally by the police. So much for the presumption of innocence. Even worse, in New York, the mayor has decided that those who ran riot, destroying and stealing, will not be prosecuted for their crimes. Apparently, so it would seem, their criminal actions were excusable in the circumstances.

In law, and hopefully in the moral perception of rational people, if a person commits an act knowing, or should have known, that the act could lead to criminal actions, that person is considered complicit in any subsequent criminal actions that might arise. When the first protest erupted in Minneapolis, the organisers should have known that violence would accompany the protest, given that previous such protests had resulted in violence, destruction of property, assaults on police and other people, and even deaths. The organisers of the initial protest may possibly be excused, but only just, but most certainly, the organisers of subsequent protests in so-called, solidarity, were complicit in the violence that followed, and their actions considered equally criminal.

But here is the rub: lawful authority failed to see it that way, likely wanting to not exacerbate the violence even further, which one can grant as demonstrating restraint and wisdom. But what of the morality of the participants in the protests, even though their intent was peaceful, knowing that violence would surely result? Wherein the moral compass of such people, and on what basis are such secular moral decisions made? How do these people justify their actions to themselves?

Thus, the conundrum: secular morality is open ended, depending on circumstance. Animals, and even rivers, can be granted rights, even though they have no capacity to exercise them, and humans can be granted, or deprived of, rights, out of convenience or expediency. In truth, then, secular morality is based on shifting sands, instead of any firm foundation. We can have no idea of where future legal determinations on rights will take us. If we consider the history of people of many cultures on the imposition of laws, and the deprivation of rights, up to the present day, especially those which are *much more restrictive* than any religion would impose, it is difficult to not sympathise with the views of Rabbi Cardozo.

Again, on the subject of rights, if they can be granted to entities incapable of exercising them, why should not rights be granted for chairs to not be abused, or to cars to not be driven in a reckless or dangerous manner? It is absurd to my way of thinking, and a corruption of the true sense of morality, to grant rights to non-human entities rather than impose duties on human entities.

But back to the primary theme of this section: *The Exceptional Human Condition*.

It should be obvious that the flexibility, shifting positions, and open-endedness of the human perceptions of morality cannot be solely attributed to biochemistry. On the other hand, the randomness of expression would suggest that indeed, such expressions could only be attributed to uncontrolled, random shifts in the neuronal patterns of the brain, over which the "self" has no influence. If that be true, then no human can claim to be rational, on any subject. On the contrary, David Gelernter considered: "The essence of *rational thought* is building your case step by step so that each step is justified by the previous ones. The variety of techniques or rules to choose from is huge, but there is always some kind of rule."[4] But what if he is wrong, and rational thought is in truth, simply an illusion, modern technical innovation being like morality, simply the result of our biological minds going about their business as their chemical and physical properties dictate?

Thus, you get to choose whom you perceive yourself to be ... or do you?

References:

1. Cardozo, Nathan Lopes, *Jewish Law as Rebellion: A Plea for Religious Authenticity and Halachic Courage*, Urim Publications, Jerusalem, Israel, 2018

2. *Ibid*, p. 254

3. *Ibid*, pp. 254-256

4. Gelernter, David, *The Tides of Mind: Uncovering the Spectrum of Consciousness*, Liveright Publishing Corporation, New York, NY, 2016, p. xvii

Chapter 2-15: The Human Spirit

"My religion consists of a humble admiration of the illimitable superior spirit who reveals himself in the slight details we are able to perceive with our frail and feeble mind."

~ Albert Einstein ~

I would offer that Einstein had no doubts concerning the existence of the spirit as a function of the mind, rather than just the out-workings of neural networks in the brain, uncontrolled by anything but themselves. But, no doubt, others interpret his words differently.

Gabriel Garcia Marquez, in his book, *"Love in the Time of Cholera"*, wrote admiringly:

"It had to be a mad dream, one that would give her the courage she would need to discard the prejudices of a class that had not always been hers but had become hers more than anyone's. It had to teach her to think of love as a state of grace: not the means to anything but the alpha and omega, an end in itself."[1]

If one ponders how people can rise above adversity; can "discard the prejudices of a class"; can act contrary to their otherwise conditioned culture; can walk away from family as they reject religious practices that they find to be absurd or unwholesome, as narrated by Carma Naylor[2] concerning Mormonism; can turn from fanatical Islamic jihad as attested by Walid Shoebat[3]; or raised as an atheist, the brother of one of the most vocal antitheists of modern times, Christopher Hitchens, Peter Hitchens chose a different path[4].

As a younger man with pretensions, I threw myself wholeheartedly into games of rugby football, often coming off the worse for my foolhardiness, but absent of competition for food and sexual encounters, what other creatures have the ego, desire to win competitions, to be the best they can (or even can't) at the risk of injury? What is this innate desire of humans to strive and excel, when the activities have nothing to do with fitness for survival or reproductive success? It can be argued, I guess, that success in numerous fields does lead to betterment in life from riches, and correspondingly success in the pursuit of the more desirable members of the opposite sex, but in only a handful of cases, does success in the latter relate to reproductive success. There is something quite peculiar about the many facets of the human spirit, as I believe any of us can attest, which is why we celebrate heroic achievements. From ANZAC Day in Australia and New Zealand, Memorial Day in the USA, to Commemoration (Martyrs') Day in the UAE and the National Day of Commemoration in Ireland, practically all countries have designated a day to commemorate the sacrifice of their people in wartime. Why is that?

The evidence of the human spirit, transcending genetics and biology, can be said to be circumstantial and beyond empirical proof, but it is overwhelming nonetheless. Despite what they otherwise profess, I doubt that any scientist or evolutionist would say to him/herself: *there is no "me"* – I am nothing but what has been determined by my genes in mind and body. I have no mind of my own: what I perceive as my mind, my will, my choices, is nothing but an illusion – the neural networks of my brain make all such decisions.

The irony is that this so-called *autonomous* biological blob is professed to be the cause of such illusions, effectively deluding itself. Now that is truly strange!

References:

1. Marquez, Gabriel Garcia, *Love in the Time of Cholera*, Penguin Books, London, UK, 1989

2. Naylor, Carma, *A Mormon's Unexpected Journey - Volumes I & II*, WinePress Publishing, Enumclaw, WA, 2006

3. Shoebat, Walid, *Why I Left Jihad*, Top Executive Media, USA, 2005

4. Hitchens, Peter, *The Rage Against God – How Atheism Led Me to Faith*, Zondervan, Grand Rapids, Michigan, 2010

Chapter 2-16: Memory

"What you end up remembering isn't always the same as what you have witnessed"

~ Julian Barnes, The Sense of an Ending ~

Trolling through numerous online articles on the brain, the mind, etc., I came across this little gem, another of the "just so" narratives which are offered with no substantiation:

"Memory is a complex process that includes three phases: encoding (**deciding what information is important**)"[1] [emphasis mine]

To be fair, the audience for this website is not neurology specialists, but ordinary folk such as ourselves (I presume). Intended as an overview of the anatomy of the brain, it is excellent, and I highly recommend it for those wanting to know a little more (as I did). However, I wince every time I read explanations as above, for they tend to reinforce narratives which I contend, are the domain of the non-material mind, rather than the material brain. *Deciding what information is important* is a complex, subjective, cognitive task, one which is well beyond the capabilities of material structures. Most especially is the issue of the *subjective*. Most people accessing this website would be unlikely to remember the emphasised words above, likely because they are of no particular interest, but they stand out to people on quests such as mine. A daunting task for neuroscientists is to explain how the objective (material) can operate in a subjective (non-material) mode. Moving on ...

Alzheimer's, the most common form of dementia, is a progressive disease of the brain that slowly causes impairment in memory and cognitive function. Researchers have identified an enzyme and protein that they believe hold the key to curing this disease. Not only can cognitive function be restored, they claim, but lost memory can be retrieved. The obvious question here is that if "lost" memory can be found, where was it whilst lost? If memory is a function of the brain's neural network, then memories which were previously lost but later rediscovered, must be segments of that network which continued to exist, but for some reason or other, they were misplaced or otherwise inaccessible during periods of disfunction. We will investigate this phenomenon further in later chapters, after we delve into the anatomical structure of the brain.

With my experience in digital computing, an aspect of memory that I find interesting is its demonstrated lack of reliability. We remember sometimes accurately, sometimes not, as Julian Barnes noted in the opening quotation. In the context of the *computational theory of mind* advocated by so many scientists today, a number of possible causes of faults come to mind. A biological "programming" fault might have caused a misinterpretation of the context, leading to a false connection in the network whilst storing the memory. The correct connections may have been instantiated from the beginning, but a later fault, logical or biological, may have corrupted the authenticity of the original connections. A biological disorder may have corrupted the connections, such as is suggested for Alzheimer's. From the outset, the original occurrence or circumstance may have been misinterpreted due to the observer's worldview. Or finally, for whatever reason, our memories both fade and become muddled over time.

Due to its paramount significance, an issue which I cannot avoid questioning is the process whereby electro-chemical signalling through the nervous system is somehow correlated at the conceptual level so as to be recalled in context. The brain, a biological organism subject to the laws of physics and chemistry, is the only known instance of this occurring. No construct of scientists is able to transcend its intrinsic properties such as is claimed for the brain. All physical computing devices receive input, process, store, and deliver output in material mediums, which require human intelligence to decode and interpret in non-material, conceptual terms. Yet, the brain, composed of the very same atomic and sub-atomic particles, but arranged in different compositions, is seemingly capable of behaviour that no other physical compositions can attain. Even within the human body, the very same neuronal components only achieve conceptualisation when arranged in specific ways, and only in the mind-brain complex. Apart from perhaps the arrangement such as found in the visual cortex, no part of the central or peripheral nervous system has that peculiar capability of the brain, an organ of no physical difference other than the arrangement of otherwise similar components. It is as if a particular arrangement of common components can achieve a capability which the components themselves cannot achieve in any other construct, nor can they display any intrinsic capability suggestive of that displayed by this one, unique composition.

Scientists opine that memory is instantiated in the brain in the form of neural network patterns. At the physical level, that is likely true. The question to be asked, however, is what interprets those physical patterns? If all the brain is structured that way, then the suggestion is that one network pattern interprets another network pattern. Such a method leads to infinite regress. Patterns cannot explain themselves (Primary Axiom #1), nor can patterns explain the concepts behind other physical patterns. It should be obvious that this "scientific" explanation lacks explanatory power.

Neither logic nor science can explain this phenomenon, which suggests that what is believed to be happening in the brain is actually occurring elsewhere – a place whose existence is denied by those beholden to philosophical materialism. As this conundrum is at the crux of this study, we shall delve further into the mystery when we examine neuroanatomy and related disciplines.

Most fascinating is our ability to visualise what we remember - a common term for this phenomenon is the *"mind's eye"*. Our memory can be jogged by circumstances, or sensory phenomena associated with the original event or experience, such as sights, smells, fragrances, sounds, and even touch. I shall avoid embarrassing myself by recounting some of the more unusual associations from my own experience, but I am confident that we all have our own experiences, embarrassing or otherwise. One particular fragrance reminds me of a girlfriend of my youth, and in my mind's eye, I see her sparkling blue eyes and tender countenance. An occurrence of a certain tune brings to mind the first time I heard it, and whom I was with. The smells associated with the cockpits of early jet fighter aircraft remind me of my first flight in a De Haviland Vampire, and even now, some half-century later, I can still "see" the details of the flight controls, such is the impact of that experience.

All very romantic, to be sure, but from a scientific perspective, how can we explain this? A critical component of the process is missing. The memory was formed from the physical inputs from one or more of our sensory systems, as stored in our physical brains. Sometimes, at will, prompted by whatever, we can *"visually"* recall the event or experience. At other times, a serendipitous event will prompt the recall, lulling us into a reflective state, our mind's eye reliving the memory in real time, our emotions rising to the occasion as they originally did. Of course, one has to be careful as to who is watching at the time.

Thus, the conundrum of how a biological organ, subject to the same laws of physics and chemistry as the other organs of our human bodies, can inexplicably exhibit a unique behaviour that transcends those laws. What is it about the organisation of the brain's neural network that transforms physical sensations into immaterial conceptual phenomena, or do the philosophical materialists in the scientific world have it completely wrong? No doubt you have guessed by now that I believe immaterial phenomena to be in the immaterial domain which we have yet to understand, and which many refuse to accept as existing.

Nevertheless, neuroscientists continue to research the physiology of the brain, attempting to find the residence of memories. Quoting from this online source[2]:

"The idea that synapses store memories has dominated neuroscience for more than a century, but a new study by scientists ... may fundamentally upend it: instead memories may reside *inside* brain cells."

The difficulty that I have with the "synapse store" hypothesis is one of persistence. The article refers to: "each time a memory is recalled"; what triggers the recall of memory when a sensory experience is not involved, or do we never recall memory other than when it is triggered by an external stimulus? A memory must involve numerous neuronal circuits: are these inactive other than when the memory is recalled? What is the entry point to this specific set of circuits, and how does the brain "remember" that entry point? We review these issues in a later chapter. Continuing from this article, "Administering propranolol ... during this window can enable scientists to block reconsolidation, wiping out the synapse on the spot."

I was curious about the process of "wiping out the synapse". I found other articles which describe brain development in babies and children, and associated with neural plasticity, noting "Once the nerve cells are formed and finish migrating, they rapidly extend axons and dendrites and begin to form connections with each other, called synapses, often over relatively long distances."[3] This raises even more questions such as: What guides the *migration*? I am trying to imagine the space within the brain occupied by some jelly-like substance, and neurons obediently heading for their reserved spaces. I can only assume that scientists know this to be true, and it is not just a scenario of their own making. However, "Beginning 20 years ago,

Huttenlocher ... first showed that there is a pattern to synaptogenesis in the human cerebral cortex characterized by the rapid proliferation and overproduction of synapses, followed by a phase of synapse elimination or pruning that eventually brings the overall number of synapses down to their adult levels."[2] And then, concerning the pruning of surplus synapses, "Those that fit the intended pattern were retained, and those that did not were eliminated." Intended pattern? If the genome knows of this intended pattern, why the oversupply in the first place? I don't mean to criticise the work of these dedicated researchers - I am just disappointed that the articles never seem to raise the same questions that I do.

Moving on, the question becomes: Where is memory located if not in the synapse? Research has shown that despite a synapse being erased, molecular and chemical changes could be found within the soma. This suggested that "The engram, or memory trace, could be preserved by these permanent changes. Alternatively, it could be encoded in modifications to the cell's DNA that alter how particular genes are expressed."

I relate these excerpts to demonstrate that research into the *physiology* of memory is largely limited to experimenting with animals. Yes, humans have been subjected to experiments using drugs and electric shocks, and whilst these have been claimed to have been beneficial in the treatment of disorders, they have taken us no closer to understanding how physical neuronal structures give rise to cognition or visualisation. This is the crux of the mind-brain conundrum. The brain is inarguably critical to memory function: it is necessary, but not sufficient, to explain the conceptual aspects of human memory.

Keep in mind, also, that whilst there may be biological similarities between the brains of mice and men, there is no comparison of cognitive abilities, especially the ability to conceptualise, imagine, invent, and create.

Reference:

1. https://mayfieldclinic.com/pe-anatbrain.htm

2. https://www.scientificamerican.com/article/new-estimate-boosts-the-human-brain-s-memory-capacity-10-fold/

3. https://www.ncbi.nlm.nih.gov/books/NBK225562/

Chapter 2-17: Music

" ... and all men are brothers, their souls joined in music."

~ heard during an André Rieu concert ~

Not to everyone's taste, I know, but I thoroughly enjoy the performances of André Rieu and his orchestra. I find his music uplifting, joyous, and at times, I even get goosebumps when I experience the sights and sounds as found in this rendition of *"You'll Never Walk Aone"*[1]. I recall when it was released by Gerry & The Pacemakers in 1963 – it was a song for its time. It has long been adopted by the Liverpool Football Club as its anthem, and I doubt whether any member of that club would shrug and say: nothing spiritual there, just my brain doing its thing.

Over the centuries, literally millions of songs, lyrics, tunes, and poems have been written by hundreds of thousands of authors. The variations are seemingly unlimited, and yet, we are asked to believe that all this has been accomplished by nothing other than an unknown mechanism in the brain, which arranges neurons in particular patterns giving rise to artistic accomplishments.

As I will continue to stress throughout this study, the genetics and biology of the human body and its internal organs, including the nervous system and the brain, are dependent upon, and constrained by, whatever the chemical and physical properties and arrangements will allow (Primary Axiom #10). Whilst the outcome of specific arrangements of chemicals may be beyond our understanding, and at times seem random, the truth is that the outcome is never random and always inevitable – chemicals only do what they must do. There is no subjectivity – only objectivity.

Human creativity is entirely subjective, never objective, and is the antithesis of biological behaviour. No arrangement of biological cells can ever result in a subjective outcome - only objective. When one considers the complete works of composers such as Amadeus, Beethoven, Brahms, Chopin, Debussy, Haydn, Strauss, Stravinsky, Tchaikovsky, and countless others, it defies belief that genetically influenced arrangements of neural cells, mere chemicals, are capable of autonomous rearrangement to cause these composers to compose as they did. If neural cells and the arrangements thereof, are capable of expressing free choice cognitive behaviour, then any outcome of autonomous rearrangements could not conform to the disciplines of music, not even by chance. Any outcome would be like a five-year-old randomly banging away of a piano keyboard.

Music is not like that (except some modern music *does* sound like that to *my* ears).

Wherever there are birds, we hear "singing", for which I will defer to naturalists for an explanation, but I suspect that their calls are limited in variation, and likely only resemble tunes. Many creatures of land and sea communicate via sounds, whilst male humpback whales have been described as "inveterate composers" of songs that are "'strikingly similar' to human musical traditions". No doubt they are, but again, this is more correlation that causation.

In passing, it is said that both plants and animals are responsive to music. Just why this is so is interesting, but I have seen no evidence that plants and animals can compose music, not even our last common ancestor in evolution's tree. I cannot know, but I strongly doubt that the human brain, a biological entity, could be capable of music composition. Genetic expression plays the major role in the variance in humans who have a 99.9% commonality of the genome. I would be intrigued to find the chemical behaviour that accounts for the amazing capability and variation in music composers, poets, painters, sculptors, theoretical physicists and mathematicians, and why my genes were so poorly expressed in that regard.

Has evolution cheated me?

Postscript

Whilst reviewing the manuscript pre-publication, a friend referred this article[3] to me. I don't have time at present to fully comprehend the implications, but considered that it may be of interest to readers. Quoting from that article:

"One morning in late 1964, a 22-year-old Paul McCartney awoke with a tune in his head. It was lovely, and, sitting at his piano sounding it out, he was pretty sure he was only remembering something written by somebody else. After playing it for the other Beatles and assorted friends, he eventually realized that the dream melody was his alone. After a little more work, and with lyrics added, it became "Yesterday," one of the most recorded songs in history."

I have had similar experiences, though not with music. Sometimes whilst pondering a problem, I reach the stage of analysis-paralysis, i.e., I get stuck and can proceed no further with intentional cognitive activity. From experience, I know that if I sleep on it, the answer may resolve itself, and so it has on more than one occasion. There must be some reorganisation of neural network patterns for the memory of that resolution to persist until I wake up, but I have two questions at least: (1) what was the initiating stimulus to continue "thinking" about the problem; and (2) what was guiding what I consider to be, my *unintentional* thoughts? Consistent with my debate concerning the material versus the non-material, the physical versus the conceptual, I struggle to accept that the neural networks of the brain were alone responsible for this phenomenon.

References:

1. https://www.facebook.com/hashtag/neverwalkalone

2. Payne Roger, quoted in: Author(s): Susan Milius. "Music without Borders", p. 253. Source: *Science News*, Vol. 157, No. 16, (15 April 2000), pp. 252-254. Published by: Society for Science & the Public.

3. https://www.magellantv.com/articles/music-and-the-brain-discoveries-reveal-the-depth-of-melody-in-our-lives?fbclid=IwAR0PqKsdcEIaajfId-rZbGIgI_AnTI-MS-I4v_AcEM-BX7x0OTcILm6pcNk

Chapter 2-18: To Reiterate

"Man is only man insofar as he is essentially more than a human animal."

~ Simon Glendinning ~

Simon Glendinning is entirely correct – humans are more than simply evolved animals. The previous chapters have evidenced the truth of this. When we ask: In what ways is man more than *just* a human animal, we can answer in innumerable ways which, if nothing else, must give one pause for thought, and reason to question those who argue from the stance of philosophical materialism.

For centuries, mankind has seen itself as somewhere between gods and animals. Although contentions that many animal and plant forms arose through some form of organic progression, date back as far as the twelfth century, the first broad idea of the evolution of humans from other species only arose in the early nineteenth century, courtesy of Jean-Baptiste Lamarck. Charles Darwin and Alfred Wallace built upon Lamarck's work, although many of the latter's ideas have since been discarded. On the other hand, his belief in acquired characteristics has found new favour. Ever since, evolution has been widely accepted as true, and I am not about to contend with that, other than that I have yet to be convinced that it could have occurred without some form of external direction.

The first published work on intelligent design was in 1802 by British theologian William Paley. Whilst one could always contend that his religious beliefs underpinned his approach, there have been many scientists since, both secular and religious, who have remarked on how many aspects of our humanity *appear* to have been designed. Ironically, as I mention elsewhere, many explanations by evolutionists themselves include the terminology of design and purpose, such being antithetical to their professed positions. It is curious, is it not, that even evolutionists cannot avoid teleological arguments in their explanations? Could it be that they secretly agree with design, but are afraid to admit to such as they are afraid of the identity of the designer? Any mention of god or God has people trotting out their own version of this entity, which may or may not be true, and as I have no clear idea of how to describe such an entity, I prefer to avoid the subject altogether (in this context at least). I can only assure the reader that I have worked forward through science and logic, not backward from religion or other deist ideas. I have used the evidence before me, with no preference for what *should* be true.

I have some experience with animals, from domestic pets to horses, sheep, cattle, emus, kangaroos, and other native wildlife. I have observed their behaviours, though not with the eyes of a naturalist. I have assisted with training, domesticating, shearing, rounding up, and other activities of animal husbandry. In all this experience, I have observed very little behaviour that mimics human behaviour, other than as identified in the *Biological & Physiological* layer of Maslow's *Hierarchy of Needs*[1]. At the next layer, *Safety*, animals certainly display some motivation, but mostly they lack the capabilities that humans would consider as a cognitive threat assessment. As we move up Maslow's five layers, the uniqueness of humankind is highlighted even more. Maslow was an American psychologist whose theory of the hierarchy of needs was predicated on fulfilling innate human needs in priority, culminating in self-actualization. At the third level, *Love & Belonging*, we can see the same in some animals, especially the tender care of their calves by elephant mothers, but I would offer that this is instinctive rather than being predicated on a desire for self-actualisation. The final two layers are clearly in the human domain only.

Do cheetahs run as fast as they can demonstrating a competitive spirit, or is their speed related to nothing but catching food? Do dachshunds wish that they had longer legs? Do sheep wonder what humans want with their wool? Do cows choose to go on a diet to avoid the slaughter house? Do chooks object to laying eggs for human consumption? Do fish race each other? Do dogs peer into a mirror, wishing that they were more handsome? Do male giraffes choose a mate based on the eyes, lips, hips, or other feature that they might consider attractive? Do any animals desire to be something that they are not? Are any animals dissatisfied with what they are?

Why is it that mankind alone is dissatisfied with its essential being, and strives to be so much more? What is it in genetics that makes us unique amongst living creatures? Whilst lower order species may exhibit a sense of self, in no way does it approach the human sense of self, wherein we attempt to better ourselves in any number of fields, and are capable of introspection to discover our "real selves" as some would put it. We philosophise and ponder the meaning of life, and our place in the world. We are the only creatures who study not just our essential being, but *what makes us tick*, to borrow from Gerald Verschuuren[2]. What is it about our biology that encourages and allows us to study our biology? What is it about our biological brain that allows it to study itself?

In fact, what are the properties of any biological or physical entity that allows it to study itself? Without a plausible explanation for that question, the overarching hypothesis of evolution remains unproven.

References:

1. Maslow, Abraham H., *A Theory of Human Motivation*, Psychological Review, 1943

2. Verschuuren, Gerard, *What Makes You Tick? A New Paradigm for Neuroscience*, SOLAS Press, Antioch, CA, 2012

Part 3: Establishing Foundations of Knowledge

"The loftier the building, the deeper must the foundation be laid."

~ Thomas à Kempis (1380-1471) ~

All knowledge is built upon prior knowledge (Primary Axiom #2), but realistically, not all prior knowledge is true. Charles Darwin misunderstood the cell as little more than a "blob", and given the knowledge of his day, I cannot fault him for that. However, he based his theory of evolution on a misunderstanding, and thus, the ideas he developed were based on *false knowledge*. Incidentally, reading Darwin's works, I found him more uncertain of his ideas than those who subsequently have based a "science" on them. I am sure that you are familiar with many of the false steps taken in the pursuit of science. That is not a criticism - simply an observation of what is irrefutably true.

The natural corollary to this is that from time to time, when new phenomena arise, or are newly understood, that cannot be explained by conventional thinking, science must revisit its foundational presuppositions. In my opinion, and that of many well-credentialled scientists, such rethinking is well overdue on the subject of evolution. I contend that those beholden to *philosophical materialism* are increasingly becoming an obstacle to scientific research, being part of the problem rather than the solution. Another related definition is the *Philosophy of Naturalism* – the metaphysical conviction that nature is a closed system of causes and effects that cannot be affected by any outside factor like God. I present no argument for God – God is irrelevant to my case study. The issue is the accepted meaning of naturalism, with the implicit understanding that only material entities composed of physical attributes are natural, whereas the immaterial cannot be a naturally occurring phenomenon. On the evidence, I disagree, and the following chapters expose what I believe are some scientific domains and definitions much in need of revision.

Many of my arguments relate to *ontology*: the philosophical study of being, or more broadly, the concepts that directly relate to being, e.g., becoming (origins), existence, and reality, as well as the *basic categories of being* and their relations. Ontology raises questions concerning what entities exist or may be said to exist, and how such entities may be grouped, related within a hierarchy, and subdivided according to similarities and differences. Under philosophical materialism, only entities comprised of physical properties are said to exist, whilst in religion and spiritualism, non-material entities also exist, e.g., gods, angels, demons, and spirits of all forms.

As for the *basic categories of being*, this has required a process of abstraction to discover the number and names of the categories. Much research has been undertaken by many philosophers since Aristotle (384-322 BCE), involving careful inspection of each concept to ensure that there is no higher category or categories under which that concept could be subsumed. I have an opinion on the relationship between scientific law and natural law, believing the former to be a sub-category under the latter, itself being a sub-category of a higher category under which it should be subsumed. My dissertation on that subject follows at the end of this study (*A Law Hypothesis*). Later scholars of the 12th and 13th centuries further developed Aristotle's ideas, for example, dividing Aristotle's ten categories into two sets, primary and secondary, depending on whether they inherently exist in the subject or not:

- Primary categories: Substance, Relation, Quantity and Quality.

- Secondary categories: Place, Time, Situation, Condition, Action, Passion.

Around the same time, Maimonides identified these sets as *Essential* and *Accidental*. In the 18th century, Immanuel Kant recognised that we can say nothing about an entity or category, such as Substance, other than through its relationship to other things. Of interest is how Albert Einstein comments on that subject, refuting the understanding that many have since attributed to his theories:

"The meaning of relativity has been widely misunderstood. Philosophers play with the word, like a child with a doll. Relativity, as I see it, merely denotes that certain physical and mechanical facts, which have been regarded as positive and permanent, are relative with regard to certain other facts in the sphere of physics and mechanics. It does not mean that everything in life is relative and that we have the right to turn the whole world mischievously topsy-turvy."[1]

No doubt I will repeat this quote in a later chapter, but that aside, without perceiving myself as in any way comparable to Einstein, this accords with my own independently determined axiom: *nothing can explain itself* (Primary Axiom #1) As noted earlier, Maimonides stated that "Every description of an object by an affirmative attribute, which includes the assertion that an object is of a certain kind, must be made in one of the following ways."[2] These he gives as (1) by definition; (2) by part of its definition; (3) by something different from its true essence; (4) by its relation to another thing; and (5) by its actions. In Maimonides' way of thinking, a thing is what it is, and everything that we say about it is an addition. In describing a man, we could say that "*man is man*", but apart from being a tautology, it adds nothing to our knowledge. Thus, we need to define man, which we do in biological, physical, moral, intellectual, occupational, and other terms. I do wonder whether Immanuel Kant, although born into a Lutheran Protestant family, was familiar with the writings of Maimonides, or more likely, the writings of Aristotle.

I am given to taking that thinking to biology, and all aspects of the human condition. A thing is what it is, i.e., an entity. What differentiates one entity from another, apart from its *accidental* properties such as its position in space-time, are *essential* properties and dependent activities, or as Maimonides puts it, by its description, definition, and actions. Where these fit into the formally devised primary and secondary categories of being are of no interest to me, and of no relevance in this study. However, I do want to stress that ontology must be foundational to our thinking about the mind-brain complex, lest we fail to accurately identify entities, and falsely ascribe properties and activities of which they are intrinsically incapable.

But back to the quotation at the beginning of this chapter. Putting aside his religious convictions, the writings of Thomas à Kempis reveal him to be a deep thinker, and an accomplished philosopher. Like Darwin, he held to certain presuppositions which cannot be substantiated. Nevertheless, he understood the relationship between lofty goals and the foundations needed to support them. I do wonder, it must be said, how a material brain could even conceive of lofty goals, let alone strive for them. In the context of this study, understanding the human condition, and most especially the human mind, is perhaps the loftiest scientific goal that can be imagined.

To that end, we must establish foundations much deeper than are currently assumed.

References:

1. Relativity Theory), a talk given on 5 May 1920 at the University of Leiden, and *"Geometrie und Erfahrung"* (Geometry and Experience), a lecture given at the Prussian Academy published in *Sitzungsberichte der Preussischen Akademie der Wissenschaften*, 1921 (pt. 1), pp. 123–130

2. Maimonides, Moses, *The Guide for the Perplexed*, Digireads Publishing, 2018, p. 143

Chapter 3-1: Thesis

"We do not know where to look, or what to look for, when something is memorized. We do not know what it means, or what change there is in the nervous system, when a fact is learned. This is a very important problem which has not been solved at all."

~ Richard P. Feynman, American theoretical physicist ~

What is the physical or biological architecture of memory? Nobody knows, but my contention is that memory is not physical at all. Yes, I do acknowledge the relationship to the physical brain, but we will come back to that.

Based on the evidence that I have been able to research, I contend that the undirected, organic mechanisms of evolution, so far offered by the proponents of its over-arching narrative, cannot account for the *cognitive abilities* of humans. To constrain the context of the discussion I would offer three working definitions:

Teleology, (from Greek telos, "end," and logos, "reason"):

The explanation of phenomena in terms of the purpose they serve rather than of the cause by which they arise. Alternatively, explanations by reference to some purpose, end, goal, or function. Traditionally, it was also described as final causality, in contrast with explanations solely in terms of efficient causes (the origin of a change or a state of rest in something). Human conduct, insofar as it is rational, is generally explained with reference to ends or goals pursued or alleged to be pursued, and humans have often understood the behaviour of other things in nature on the basis of that analogy, either as of themselves pursuing ends or goals, or as designed to fulfill a purpose devised by a mind that transcends nature.

Cognition: the set of all mental abilities and processes related to knowledge.

These include the interpretation of sensory inputs, commitment to memory, organisation of memory, recall from memory, determination of relevance, rejection of "noise", correlation of abstractions, comprehension, invention, judgment, evaluation, logical reasoning, abstract reasoning, computation, problem solving, decision making, development and use of language, etc.

Evolution: The theory of how complex organisms arose on Planet Earth.

As noted earlier, I prefer the definition as given by Professor G.A. Kerkut: *"the theory that all the living forms in the* world *have arisen from a single source which itself came from an inorganic form"*[1]; I refer to it as the "General Theory of Evolution", GTE, or the overarching narrative of evolution.

I acknowledge that this particular definition of evolution is not one used by evolutionary biologists and other specialised disciplines, but I consider it disingenuous, perhaps even intellectually dishonest, to speak of origins yet start somewhere in the middle. This study concerns the *origins* of information, knowledge, and the cognitive applications by the human mind. As I will attempt to demonstrate, the evolution narrative for all its twists and turns, inventions of hypothetical processes, just-so stories, and elasticity in definition, fails entirely in its attempts at explanation.

Teleology is perhaps the biggest obstacle to evolution theory. As we proceed, you will notice how scientists seem unable to remove teleology from their descriptions of evolutionary processes, using words such as design, purpose, or goals. Evolutionists, if they are to be true to their self-proclaimed beliefs, must only speak of efficient causes, not of intended effects. Modern theories speak of heritable traits leading to improved survival and reproduction success, but if you wind back the evolutionary timeline to some unspecified point, one must ask: what was it about the organisms that survival and reproduction success were themselves considered favourable? Certainly, cellular replication is the result of chemical processes, but by the time primitive animals were driven by instincts to reproduce and survive, which can be considered teleological, how did such instincts arise simply through chemical interactions?

The irony of Darwin's ideas on natural selection in pursuit of survival benefits and reproductive success, is that these are teleological - purpose-driven traits which have never been substantiated.

Abstraction

The word "abstraction" can be used in two senses, and it is well that I clarify my usage. To *abstract* can mean to *separate*, *extract*, or *distillate*, but that is not what I intend here. *Abstraction* as used in a study of the mind, refers to the intangibles that result from cognitive processes and the mental conceptions so formed. When viewed, works of art result in different conceptions (subjective experiences) in different people, as do music, poetry, literature, scenery, and other physical entities. As is often expressed, "beauty is in the eye of the beholder": affirming that the perception of beauty is subjective. As I contended from the outset, cognitive processes are *always* of that nature – *subjective*, whereas all physical or biological processes are unavoidably *objective*: physical entities can only do what they *must* do (Primary Axiom #10).

A primary issue in this study is to what do we ascribe the link between the physical (organic) and the conceptual (non-physical): can that be entirely organic or is there some as yet unrecognised immaterial process involved? A related question, and one which perhaps lies at the very heart of the problem, is whether the purely organic, which is objective, can give rise to the conceptual, which is subjective?

Data, Information, and Communication

An insightful and concise definition of information is: *that which informs*, but in a sense, it is inadequate because absent is a statement of who or what is being informed. Without a recipient, it is quite simply a broadcast rather than a communication, and thus there is no information flow as such. Let me provide a simplistic overview to establish what I believe you need to understand to follow the remainder of this study.

When I was first employed in this field, I joined what was known as the EDP (Electronic Data Processing) Department. That's what we thought we did – we processed data electronically. As I progressed in my career, I became Manager of an MIS (Management Information Services) Department, with a greatly expanded role, and a much deeper understanding. Equivalent organisations later became known as IT (Information Technology) Departments, and in my opinion, lost both understanding and focus. As a pioneer in the industry, Edsger

Dijkstra, observed: "Computer Science is no more about computers than astronomy is about telescopes". Technology is an enabler - it does not represent the primary task, other than for technicians. My point is that computer systems do not store or process *information*, and at a technical level, not even *data*. Computers of all types including biological (brain), store and process encoded symbolic representations of conceptual data, appropriate to the medium employed. All communications, whether received aurally, electronically, by visual signals, or on paper, convey symbolic representations of conceptual data, but never information or data itself. This is easy to demonstrate.

Consider a restaurant with a multi-language menu. Each language conveys the same conceptual data (eggs, meat, fish, etc.), but the symbolic representations are different. The symbols, whether pictograms, photographs, or letters of the English, Greek, Hebrew, or Latin alphabet, all represent the same thing, but in different ways. Whilst commonly referred to as data, these symbols are but symbolic representations which need to be interpreted (Primary Axion #1). When, for example, the menu lists *Hoki fillets* and *Mongolian lamb*, you have four symbolic representations, but what do they mean? If, as a child, you had been taught that Mongolia was a region in north-western Australia, and a lamb was a geriatric camel, you likely would not choose that menu item. Data, for you, is what you have learned or been taught it is, as it is linked to a symbolic representation. Similarly, information is a transient derivation of data correlated in context. Whilst data can be said to exist in a symbolic form, information has no permanent existence or representation.

Symbolic representations of data are arbitrary, as a comparison of ancient and modern languages will demonstrate. For communication to occur, there must be agreement between the parties on what those symbols represent. Obvious enough, and you might wonder why I am going to such pains to explain the obvious. The reason is that these principles apply to the mind-brain complex, and to the nervous system supporting our sensory receptors. Our peripheral and central nervous systems transmit electrical pulses which in themselves mean nothing other than perhaps, the biological equivalent of "you've got mail". Understanding of that mail comes from the mapping of the nerve fibre originating in a sensory receptor, to the chemical processes at the destination of the electrical pulse. Overall, the nervous system communicates via symbolic representations in the form of electro-chemical signals, of what our sensory receptors detect. The brain is part of the central nervous system and operates biologically as do other parts of the nervous systems.

As contended above, symbolic representations are not data, and are not information. Thus, the brain can neither store, nor process, either conceptual data or information, but needs a supervisory process to interpret such physical representations into cognitive concepts. The question becomes: how can an organic mechanism, which is capable of only dealing with physical phenomena appropriate to its own environment, derive a conceptual representation (e.g., an image) from purely physical symbolic representations? We know that is a function of the mind, but how can that be of the same substance as the brain, an emergent property of the brain, or the brain itself?

A (Not So) Simple Test

In researching an earlier study, I set myself a test to discover the depth and complexity of what I have defined as Primary Axiom #2: that *all knowledge is built upon prior knowledge*. I wanted to map everything that I had to pre-know to derive information from "the cow jumped over the moon", or similar simple sentences. I was not surprised, but I was fatigued, by the intensity of the intellectual effort required. I remember reading an observation by Albert Einstein that reading is the most intellectual activity of the human mind, and I have come to agree with him, almost. Reading with comprehension requires our intellect to immediately and unconsciously correlate an innumerable number of facts, but it takes a great deal more intellectual activity to identify those facts. Einstein's letters reveal his deep interest in *epistemology*, and I find some comfort in that I share his curiosity in that subject.

But now, back to the test.

Take a clean sheet of paper (or electronic equivalent) and on it write one single, isolated fact or quantum of "information" that, on the surface, requires no other facts or information to comprehend it. It is my hope that after whatever period of contemplation, you will have concluded that no such single fact exists. For example, you might write the letter "a" but in doing so, you have subconsciously understood this to be a letter (concept) of the alphabet (concept), the latter being essential to providing context. You might even have realised that in order to be able to carry out the task, you had to pre-know certain other facts, such as what is meant by writing, paper, pencil, and so forth. Just as importantly, *you had to know that you could* (Primary Axiom #9). This latter truth is significant in the context of evolution, as I will attempt to further demonstrate in later chapters. Here, I will simply reiterate my previous contention, that all knowledge is built upon prior knowledge, being a chain of data ancestors.

The Evolution Context

Let us try to understand these lessons in the context of the overarching narrative of evolution. British philosopher, Gilbert Ryle, coined the term *the ghost in the machine*, to refute the description of René Descartes' mind-body dualism (*Cartesian*). I wonder whether Ryle used this term in the derogatory sense that he did because he was afraid of what this *ghost* may allude to, unsettling his attachment to philosophical materialism. I have concluded for a ghost of some form, although just what that may be is beyond my comprehension.

Examining what are loosely termed, *communications*, we should understand that the letters on this page are symbols devised for a specific coding system. To decode them, the reader must have pre-loaded data relating to language, alphabet, vocabulary, grammar, syntax, punctuation, and knowledge of the subject matter (Primary Axiom #3). If the symbol set (alphabet) is familiar to you but the language is not, no useful communication occurs. If the symbol set is unfamiliar, such as the Hebrew or Cyrillic script, or even worse, Oriental pictograms, then again, no communication can occur. I appreciate that this is so basic that it hardly needs restating, but we need to keep these basics in mind when examining how the acquisition of knowledge could have occurred through evolutionary processes for the very first time.

What may not be immediately obvious is that every one of these terms used above are "concepts", absent of understanding of which, no useful communication can occur. A dictionary is useful for building one's vocabulary, but no dictionary can convey foundational concepts: these the reader must bring with them to the book. For example, my dictionary defines an *elephant* as a: "large pachyderm with proboscis and long ivory tusks". That hardly helps in understanding what an elephant is, because there are at least four concepts which must be preloaded: pachyderm, proboscis, ivory, and tusks.

My point, on which we will elaborate and continue to emphasise as we proceed through the study is this: data only becomes cognitive information when intelligently processed through preloaded, or pre-acquired, concepts which provide context.

Now let us have a closer look at "data". Material storage devices do not store data; as earlier proposed, they store encoded symbolic representations of conceptual data, of a type appropriate to the method of storage. The concept is in the mid of the programmer, as different labels may be applied to the same concept to be represented. The storage and retrieval of such symbols must be based upon a coding system, itself several levels removed from the data that is intended to be represented. Examine any material storage device in terms of physical properties and the truth of this should be self-evident. Examine a printed page and in terms of physical properties, all you will find are ink-blots in what appear to be regulated patterns, but keep in mind that for you to comprehend that, you had to pre-know concepts such as ink and patterns. Examine a computer hard disk – with what? The only properties that you can discover are variations in magnetic regions which in themselves are meaningless. Examine a music CD and you will not find music, or pictures on a DVD. Examine the human brain as an entity (organ), and all you will discover at the physical level is a complex network, composed of neurons, synapses, dendrites, axons, etc., all of which are regulated by the chemical properties and arrangements of their cells. All organic substances are subject to the laws of physics and chemistry, but nothing that you can comprehend externally as informational content. Activity can be sensed, but not what is being processed, nor the context of that activity, other than that neuroscientists have now learned that certain processes occur in specific regions of the brain.

As with reading the symbols on this page, the physical or chemical symbols need to be processed through a hierarchical framework, implemented in a network, of concepts and contexts before information can be presented in a human cognitive form. Earlier I opined that tree rings are not data: they are but symbols. They only become data when an intelligent observer understands that they are in fact, annual growth rings, not just some fancy pattern discovered in a cut tree trunk. The tree rings cannot tell you what they are – observation over time and comparison with other tree trunks were required to determine the nature of those symbols. The same is true of practically all naturally occurring phenomena.

Repeating the earlier observation by Dr. Oller regarding Pragmatic Information: it is a "simple mathematical fact that the number of possible strings at any given level in any natural language, or any language-like biological signalling system, grows exponentially as we progress up the hierarchy of information layers." Note the reference to "hierarchy of information **layers**" [emphasis mine]. We will come back to that concept a little later, or as I have described it, a chain of data ancestors. Dr. Oller was speaking of genomic information, but the same holds true for cognitive information: given the exponential growth of possible paths as we proceed from raw data to cognitive information, the probability that this occurred by undirected evolution clearly approximates zero.

In this study, I have begun with what I contended are *primary axioms* which must not be contravened in any proposed solutions to the mind-brain conundrum. We will be reviewing numerous scholarly studies with attention given to this approach. We will also cover in detail, data and information processing principles, as I contend that irrespective of the medium in which such processes occur, the principles do still hold. The human brain is often compared to a computer, but I will later argue that it is much more than any computer that humans can devise, the principal reason why I believe that artificial intelligence will never achieve the reality espoused by science fiction writers, and some researchers and prophets of science.

When scientists discuss information in the context of evolution, they invariably refer to the biological information in DNA, but such information is substantively different to the cognitively derived information in the mind-brain complex. Biological information is represented by chemicals stored in particular arrangements, whereby autonomous activity occurs in accordance with properties of those chemicals, and the organisation both within and across cellular components. Cells can only do what their chemical properties and arrangements allow them to do, and nothing else, unless of course, they are affected by external influences. This is all to the good, otherwise our lives would be most uncertain. If specific chemicals or arrangements are altered, different results will ensue, which is precisely what scientists rely on when attempting to treat, or find cures for, diseases.

The *brain* is considered to be the organ, or the agent of, the *mind*. The mind is responsible for cognitive abilities including the interpretation of sensory inputs, commitment to memory, organisation of memory, recall from memory, correlation of abstractions, comprehension, invention, judgment, evaluation, logical reasoning, abstract reasoning, computation, and so on. As a biological organ, its autonomous behaviour must be subject to the regulatory rules as other human organs. The genome is chemical in nature: the processing is based on the intrinsic properties of those chemicals when arranged in specific ways, and transacted in specific sequences. We might say that in a sense, the intrinsic properties of the genome, and the cells which utilise them, allow them to "know" what they are on about. They do what they do because their behaviour is intrinsic to their properties. As the chemical constituents are the same, then only the cellular arrangements are what differentiates it from other biological organs, and such organs from each other. Experts in stem cell biology tell us that such cells can be manipulated to build every cell type in the human body. This being true, the basic chemicals must be identical, with only the arrangements responsible for the different behaviours. Scientists have some understanding of why organs such as skin, hearts, lungs, kidneys, and livers behave in different ways, but no idea of what makes the brain so special such that it can be responsible for sentience and cognition. Simply claiming that the brain stores and processes information will not do, because genetic information is so totally different to cognitive information.

In cognitive processing, there is no natural chemical or cellular relationship between the properties and structure of the biological storage medium, and the abstract concepts that are said to be derived from the symbolic representations of the data stored therein. In that sense, we can say that in contrast to genetic information, the stored symbolic representations do not know what they mean.

In cell development, the genetic instructions that specify the required relationships, and the processing sequence for the desired end results, are encoded in the chemical medium itself: absent of mutation or other errors, the genome can only do what it must do. Cognitive processing is entirely different in that there is nothing inherent in the chemicals that can give rise to imagination, creativity, volition, abstract thought, artistry, and the other wonderful capabilities of the human mind. Thus, there must be another agency at work that is not present in our DNA, for unlike the genome, the mind can do much more than it is.

Conventional evolutionary theory is never going to be able to solve this conundrum, which is why I have such strong doubts that microbes could have evolved into man, purely by a process of undirected evolution. Something else must be afoot - I just do not know what.

Philosophical Materialism

There was a time in history when the term "science" covered a much broader spectrum of human intellectual endeavour, and many of the most notable scientific discoveries were by *polymaths* - people who had expertise in numerous subject areas allowing them to draw on complex bodies of knowledge including philosophy and yes, even theology. In more recent years, perhaps tracing its origins back to the *Enlightenment Era*, the term "science" is more commonly limited to the physical sciences as understood by the material-monists. However, as G.K. Chesterton noted way back in 1909, specialisation also has its downside:

"The Fabian argument of the expert, that the man who is trained should be the man who is trusted, would be absolutely unanswerable, if it were really true that a man who studied a thing and practiced it every day went on seeing more and more of its significance. But he does not. He goes on seeing less and less of its significance."[2]

Asking himself the question: "Does the Mind Exist", Professor Edgar Andrews, an English physicist and engineer, and Emeritus Professor of Materials at Queen Mary, University of London, answered as follows:

"The view that only material objects are real is called "materialism" or "physicalism". A material object is a publicly observable object, accessible to everyone in space-time. Materialism therefore denies the existence of genuine *non-material* entities such as mind, spirit and soul, while treating consciousness as an illusion arising from the operation of a material organ, the brain. For example, Francis Crick, co-discoverer of the DNA double-helix, begins his book *The Astonishing Hypothesis* with the following words:

The astonishing hypothesis is that "You", your joys and your sorrows, your memories and your ambitions, your sense of personal identity and free will, are in fact no more than the behaviour of a vast assembly of nerve cells and their associated molecules. As Lewis Carroll's Alice might have phrased [it]: 'You're nothing but a pack of neurons.' This hypothesis is so alien to the ideas of most people today that it can truly be called astonishing.

Perhaps so, but the belief that mind is nothing more than a by-product of brain activity is widely held and vigorously promoted by many neuroscientists and popular science writers today. They ignore the self-evident problem highlighted by Haldane [Ed. See Haldane's comment in Chapter 2-1 *Official Theories of Being*], among others, and claim that mind has no independent existence but is simply our awareness of the activity of our physical brains. Haldane was right. Neurons, neural-circuits, or any of the other physical hardware of the brain, have no capacity to make the decisions that guide and characterize our daily lives."[3]

Our awareness of the activity of our physical brains. Think about that. What faculty is experiencing this awareness? Is the heart *aware* of its own activity, or the lungs or kidneys? If other biological organs cannot be self-aware to the extent of comprehending their own activities, what cellular (chemical) properties allow the brain to be self-cognizant? I know but a little in the fields of chemistry and physics, but am unable to reconcile the beliefs of *Property Dualism* with that knowledge, as meagre as it is. Later, we will discover whether the proponents of that philosophy have the answers.

The fact that the General Theory of Evolution provides the best *materialistic* explanation of origins is not, in itself, reason to accept it unquestioningly, unless of course, one's personal philosophy compels them to accept only materialistic explanations; all others should keep an open mind. Richard Lewontin, described as an evolutionary biologist, geneticist, and social commentator, had this to say:

"We take the side of science in spite of the patent absurdity of some of its constructs, in spite of its failure to fulfil many of its extravagant promises of health and life, in spite of the tolerance of the scientific community for unsubstantiated just-so stories, because we have a prior commitment, a commitment to materialism. It is not that the methods and institutions of science somehow compel us to accept a material explanation of the phenomenal world, but, on the contrary, that we are forced by our *a priori* adherence to material causes to create an apparatus of investigation and a set of concepts that produce material explanations, no matter how counter-intuitive, no matter how mystifying to the uninitiated. Moreover, that materialism is an absolute, for we cannot allow a Divine Foot in the door."[4]

Just as political correctness ought to have no place in a free society, ideological correctness has no place in honest science. We need to keep calling out ideological correctness in science for what it is, and we need to keep defying it. "Sir John Eccles, as a young man studying physiology, was not satisfied with the scientific explanations of the mind-brain interaction, and this influenced his decision to become a neuroscientist."[5] One of his books that I have researched in this study is *"How the Self Controls its Brain"*[6]. In another work collaborating with Daniel Robinson, Eccles noted: "Promissory materialism is simply a religious belief held by dogmatic materialists ... who often confuse their religion with science."[7] I agree with those who assert that we need an entirely new paradigm if science is to come to grips with the mysteries of the mind. In his Foreword to Verschuuren's book, Paul Camarata, Chairman, Department of Neurosurgery, University of Kansas School of Medicine wrote,

"The new paradigm that Verschuuren proposes will require an innovative way of thinking among those who are not used to thinking outside the box. Rather than the reductionist materialistic approach of neuroscience today, Verschuuren asks that we consider a paradigm that does not rely exclusively on biological substrata and physics. Far from dismissing science and physics, he demonstrates how any cogent theory of mind must also include metaphysical constructs."[8]

It is clear from these and similar quotes that many scientists suspect that there may well be more to our existence than can be described in purely materialistic terms. Not being equipped by inclination or training however, they prefer to not venture outside of safe harbour. Those committed to philosophical materialism claim to be pursuing truth, but it is as if they are faced with two doors both labelled "TRUTH", with one having a codicil in fine print that they are unwilling to read out aloud. They nevertheless choose that particular door.

Correlation, Causation, Ideology

In both statistics and logic, the phrase "correlation does not imply causation" refers to the inability to legitimately deduce a cause-and-effect relationship between two variables solely on the basis of an observed association or correlation between them. Correlation is a conceptual notion with the same force as the oft-quoted "opportunity, means, and motive" in detective stories, and just as often misused. Both concepts are useful for identifying "suspects", but in an of themselves, they are not evidence. An example of an invalid correlation can be found in the use of syllogisms, a form of reasoning in which a conclusion is drawn from two given or assumed propositions. My mentor from ancient times, Maimonides, spoke of essential and accidental properties of an entity, which highlights the need to differentiate the two in syllogisms. If the correlation in a syllogism uses accidental, rather than essential properties, then the conclusion is likely invalid. As I understand it, there are six rules for deciding whether a syllogism is valid or invalid, but here I want to focus on correlation, or more formally, distribution. For example,

Major Premise: All dogs run on four legs.

Minor Premise: Some chairs have four legs.

Conclusion: Some chairs can run.

Simplistic I know. Nevertheless, the point is that having four legs is not what identifies the essentiality of dogs (or chairs). The point to understand, as nebulous as it may seem, is that when considering the mind-brain conundrum, analysis must be performed on properties as to their essentiality, and whether such properties enable the activities or phenomena thought to be attributed. I will retract next time I see a four-legged chair run by.

Scientific research generally begins with an observation of a phenomenon, and then the question: What is the *causation* of that? The scientist looks for related phenomena, termed *correlation*, and develops hypotheses which he/she then attempts to prove. Some hypotheses can be verified, or falsified, by experimentation, but others not. As earlier briefly discussed, climate change is one such phenomenon. Scientists begin with the observation that the climate is warming, evidenced by melting ice caps and glaciers, and other observations. The suspicion is that gases emitted from the planet are fouling the upper atmosphere, blocking the release of heat from the Earth, much as the covering on a greenhouse traps the heat, and then aided by convection, assists in plant growth. Which gases are at fault? Based on the premise that such warming is quite recent, then recent human activity must be the cause, casting suspicion on gases such as carbon dioxide, nitrous oxide, ozone, chlorofluorocarbons, and hydrofluorocarbons, now termed, unsurprisingly, *greenhouse* gases. Methane is also a culprit, human complicity being livestock farming, especially that of cows for milk and meat. The most abundant greenhouse gas is entirely natural – water vapour, which accounts for anywhere between 35 – 75%, depending on the day. Greenhouse gases are not entirely bad: it is estimated that the planet's average temperature without them would be around 0°F (-18°C). Thus, what we are experiencing seems to be too much of a good thing.

There is sound science behind the effects of greenhouse gases – I hope nobody disputes that. The issue is the contribution of each of these gases to the overall effect. As this cannot be proven via the scientific method, enter mathematical modelling. Mathematical modelling, as the name suggests, is subjective, not objective. The model contains variables, the values of which are initially unknown. These values are established by the scientists' understanding of the weight that should be given to each of the arguments, much as we do in logic. In the climate change model, these are termed *forcing factors*. The values of these factors are modified until the predictions of the model match historical reality. In other words, the goal of the project is to demonstrate that what some might think as mere *correlation* is in fact, *causation*. Once that occurs, predictions for our future reality can be made with some level of confidence.

As an aside, one can ask, with some validity, whether the IPCC actually practices "science" as is usually defined: the process of observation, hypothesis, and experimentation leading to validation or falsification. Understandably, there is no scientific experimentation, only computer modelling, and as can be demonstrated from neural network modelling, modelling is known to be subject to both bias and error, as constants and variables used in the model are generally arbitrary, and often selected to obtain the desired result. If one were to study the history of the modelling, and I have only had access to limited data on the forcing factors, a trend can be seen where the values of the contributory factors have been changed over time, with no scientific basis. Most often, it seems to be on the basis of best fit to desired results, which in a way is valid, for what is being attempted is to develop a model using historical data which accurately predicts other historical data. Yes, the climate is changing, but whether or not by significant anthropogenic causes is unknown. Everyone should note that the IPCC's charter is limited to human factors, which practically guarantees the result, because if the "scientists" found against anthropogenic causes, they would all be out of a job, including the IPCC itself. Never under-estimate the attraction of bureaucratic tenure and power - those attracted to such positions are the least likely to relinquish them.

At issue is that once headed down this path, science descends into ideology, and then further into religiosity. Whilst some scientists manage to maintain a degree of detachment, others do not, succumbing to the herd mentality of politicians and other zealots. Before long, the truth is lost in the raucous braying of the mob.

This latter phenomenon can be shown to have occurred in the evolution debate. Charles Darwin thought that all species of organisms arose and developed through the natural selection of small, inherited variations that increased the individual's ability to compete, survive, and reproduce. The sciences of genetics and microbiology being unknown in his day, Darwin had no conception of how complex the issue really was. The Neo-Darwinian Synthesis asserted that genetic mutation and natural selection were the driving forces of evolution. This perception has persisted to this day, despite many scientists coming to the conclusion that such phenomena lack the power ascribed to them. What intrigues me most is that there is no explanation for natural selection itself. We refer to concepts such as self-preservation and the survival instinct, but no-one can explain where these come from. Naturalists marvel at the struggle for survival in the animal kingdom, but I wonder why the animals bother. Would it not be easier to simply lie down and die? Where did the motivation to live come from? The earliest chemical combinations, said to be the originators of life, could have had no desire to improve or survive – what do these mean for a chemical? Survival and reproductive success are teleological arguments – arguments for purpose, the very antithesis of randomness that evolution theory postulates. Evolutionists need to explain how purpose entered the equation.

My research has found that some scientists now admit that they have no idea of how evolution could have occurred, especially life arising from non-life. Nevertheless, their belief in evolution persists. Given that in the minds of some scientists, no plausible, coherent scientific explanation exists, the concept of evolution has descended from a scientific theory, to an ideology, or in scientific terms, a paradigm. An outcome of this doubt over undirected evolution, like the doubts over anthropogenic climate change, is that rational *scepticism* is now labelled as irrational *denial*. Steeped in the religiosity of their ideology, the acolytes refuse to listen, let alone contemplate, any objections to their beliefs.

Sadly, the same occurs in religions. Based on my studies, I contend that Judaism suffers to some extent, but Christianity even more so. Both religions start with a premise concerning the nature of God: He is omnipotent, omniscient, immutable, and infinitely prescient. But then, they develop doctrines and theologies based on an entirely different God. They fail to countenance that if God is truly as they describe him, then their interpretation of the events in their scriptures cannot be true. In that sense, some (not all) adherents to Judaism and Christianity are little different to the devotees of evolution and anthropogenic climate change. That is not to assert that the understanding of these latter phenomena lacks scientific foundations, for a degree of scientific proof does exist. The issue is that the understandings are inadequate, and does not justify the ideology and religiosity that surrounds them.

Inference to Best Explanation

I will be so bold as to contend that there is not a single, substantive, scientific fact, nor substantive scientific evidence, for the General Theory of Evolution. Let me explain. Substantive evidence is that which supports just one proposition, and one proposition only. If the evidence can be used to explain more than one proposition, then it can only be circumstantial, and this is what we discover when we evaluate the scientific evidence used to support the over-arching narrative of evolution. Whether from *microbes to man*, or inorganic *molecules to microbes*, substantive evidence has yet to be uncovered. Let us review generally accepted definitions of evidence, both from a philosophical and scientific perspective.

Evidence in its broadest sense includes anything that is used to determine or demonstrate the truth of an assertion. Philosophically, evidence can include propositions which are presumed to be true used in support of other propositions that are presumed to be falsifiable. The term has specialized meanings when used with respect to specific fields, such as policy, scientific research, criminal investigations, and legal discourse. In scientific research evidence is accumulated through observations of phenomena that occur in the natural world, or which are created as experiments in a laboratory. Scientific evidence usually goes towards supporting or rejecting a hypothesis. When evidence is contradictory to predicted expectations, the evidence and the ways of making it are often closely scrutinized (see experimenter's regress) and only at the end of this process the hypothesis is rejected: in that case we call that falsification of the hypothesis. The rules for evidence used by science are collected systematically in an attempt to avoid the bias inherent to anecdotal evidence: nonetheless even anecdotal evidence is enough to reject a theory incompatible with that evidence, if there are sufficient repeated examples. The question becomes: Is anecdotal evidence, of any quantity, sufficient to prove a scientific hypothesis?

There is a wealth of circumstantial evidence evaluated within the accepted paradigm, and much hypothesising and reasoning leading to claimed *inference to best explanation*, but all of this is to the arbitrary exclusion of alternative explanations. There *is* a great deal of substantive scientific evidence for some evolutionary processes such genetic mutation, genetic drift, speciation, changes in allele frequencies, descent with modification, and even natural selection, but none of these can be said to offer a satisfactory explanation for the *origin* of the genome. Similarly, scientists have coined new terms for phenomena that are believed to have occurred, but for which the evidence is circumstantial at best, for such "evidence" resides only in the mind of the researcher trying to explain anomalies in earlier hypotheses. Terms such as *convergent evolution* and *emergent properties of the brain* are examples, and are in truth, conceptual, without any evidence of physical instantiation. Being light hearted by nature, I wonder whether these scientists, by their own paradigm, accept that perhaps these ideas are simply figments of their own imagination, derived from the genetically determined motions of atoms and molecules in their brains, over which they have no control? I have yet to have the opportunity to make such a suggestion to any scientist.

It should be understood that an inference to best explanation is always subjective, not objective, for any conclusion is subject to a person's worldview, personal philosophy, education, training, and the prevailing paradigm within which such enquiry is conducted. If a particular possibility is arbitrarily ignored, then our understanding of "best" must be similarly limited. This is the case with opponents to Intelligent Design proponents – they are concerned about the identity of the designer, and thus ID is not an inference to best explanation.

It is not my intention to disparage individual scientists or scientific endeavour in general, particularly as my entire life has been substantially enhanced by the advances in medicine and various technologies. My working life in aviation and later, Information Technology, was based on new technologies, but some things need to be said. Society has largely succumbed to two related phenomena: *Cult of Personality* and *Cult of Authority*. In the first, the opinions of celebrities are accepted without reservation, simply as a function of their celebrity status, irrespective of whether their opinions are informed or not; in the second, the pronouncements of authoritative figures in various domains are accepted without reservation, even when just a smidgeon of discernment would caution otherwise.

The problem is compounded when pronouncements are made by someone in both domains. It should be understood that many so-called scientific statements have no underlying science. For example, we have "experts" in alien life forms despite there being no evidence of alien life forms - how can there be experts?

We also encounter numerous self-refuting arguments by scientists, such as we have discussed in an earlier chapter on volition and free will. A recent example was the claim, still believed by some apparently, even though the author's explanation was misunderstood, that our universe could create itself out of nothing. When you read past the headline, you encounter conditional statements such as "in the presence of" which is clearly the antithesis of "nothing". Sadly, it would seem that such nonsense (as written, not intended) is acceptable when it comes from a person in scientific authority, but I would contend that we should attempt to keep scientific pronouncements in perspective.

Lest anyone should feel offended by this gentle tirade, the best advice I can offer is that if the shoe fits, then wear it, but if not, these criticisms are not aimed at you. I simply wanted to offer a warning that the status or qualifications of a speaker are not necessarily a reliable guide to truth. Let me give a simple example on a subject much in the news: the search for extra-terrestrial intelligence and the planets which may host such beings. There have been pronouncements by astronomers of *Earth-like planets*, but upon analysis of the characteristics, they are anything but. For example, a planet without spin, with one side continually exposed to the heat of a sun, and the other exposed to the cold of deep space, is not Earth like, for if it has an atmosphere, then the pressure differential as a result of the temperature differential would cause continuous gale force winds. Hardly Earth-like in the sense of the surface being conducive to life and habitation. Other proposals include planets three times the size of Earth, which would suggest a gravity of three times that of Earth – hardly habitable by humans.

Background Research

Let me admit that I have only *sampled* the available literature rather than surveyed it comprehensively. I have studied some related works as noted earlier, and have dabbled in the works of earlier philosophers such as Descartes. In the Information field, I am well acquainted with the works of Claude Shannon, Werner Gitt, Walter ReMine, and many others. In the context of my study here, I would contend that to a large extent, these very capable experts have nevertheless largely focused in specific areas leaving aside some fundamentals. That is undoubtedly presumptuous of me so let me explain.

Some years back, I was fortunate to have studied under a Cognitive Psychologist, Dr David Taylor. Dr Taylor had developed a business engineering methodology that he termed, *Convergent Engineering*[9], intrinsic to which was an analysis method called *Responsibility Driven Design*. In my thirty-five years in the industry, I found no approach to be as productive, and have continued to use the principles I learned from Dr Taylor in many other cognitive tasks. I notice that Dr Taylor's ground-breaking work was later extended by Richard Hubert[10]. I mention this both in acknowledgement of Dr Taylor's contribution to the development of cognitive skills, and to identify my analysis methodology. As an aside, some evolutionists claim that we have no free will and thus Dr Taylor could have had no personal responsibility for his work - it was all just the result of undirected complex chemical reactions in his brain. Knowing Dr Taylor as I do, I am confident that he would be quite offended by that suggestion, even though he is an atheist and evolutionist.

In his book, *The Concept of Mind*[11], Gilbert Ryle stressed the distinction between knowing *that* and knowing *how*. Those researching the cognitive sciences are aware of our cognitive abilities (knowing *that*) but are still mystified as to the mechanisms (knowing *how*). To know how, we need to deconstruct the process much as one would reverse-engineer a complex product to understand how to construct new ones, the goal of AI (Artificial Intelligence) proponents. We cannot do that if our presuppositions arbitrarily exclude avenues of research, particularly those related to origins. Cosmologists generally accept the Big Bang model for the origin of our universe, but due to inconsistencies in the science, some scientists are researching other origin solutions. We have the same issue with evolution of life: *abiogenesis* is accepted as fact even though there is (as yet?) no plausible scientific explanation, and thus some scientists propose alternate origins such as *panspermia*. Of course, this is hardly a solution as it simply pushes the problem onto another planet (an example of the much-derided *infinite regress*). If the posited singularity of the Big Bang is not true, and abiogenesis is not true, then how can our ideas on how the universe started and how life on Earth started be true?

Scientific endeavour generally starts with what *is*, rather than what *was* (origins), a natural process and arguably the most productive, but if we entirely ignore the science of origins, we cannot be sure that current research efforts are not heading down a blind alley. Starting an investigation in the middle, whilst perhaps necessary initially, leaves one open to unsupported assumptions; through convenience and repetition, such eventually becomes accepted truth.

The point that I am driving toward is that before we can investigate the essence of cognition, we need to differentiate the *conceptual* from the *physical*. In the works of Daniel Dennett, Steven Pinker[12], and others that I have studied, I have found no attempt to do so. Authors often attempt to explain what they believe to be purely organic processes in anthropomorphic terms, as if teleology was a prerequisite, whilst at the same time denying teleology in evolution. This raises the question, as one article puts it: "If it is not possible to speak of evolution's course without resort to the language of agency, is that a defect in human intelligence or an apprehension of fact?"[13]

There is an irony here which should not be missed. In his interesting book, *Consciousness Explained*[14], Daniel Dennett proposes a set of conceptual solutions without explaining how such concepts could be implemented at the physical, or what we might term the *bits & bytes* level. His computer analogies fall far short of what would be required for a plausible hypothesis. But you see, this is precisely the problem of cognitive processing: how can concepts be stored and retrieved at the physical level by a physical storage medium which has no intrinsic ability to comprehend the conceptual? Authors continue to offer conceptual solutions to physical problems, not realising that their struggle evidences the depth of the mystery. In the following chapters, we will review functional requirements of cognitive processing using a detailed analysis of digital information processing. I wish to thank Dr David Taylor for his insights into *Responsibility Driven Design* which have enabled me to apprehend these issues more clearly.

One further point before we begin. Quoting Daniel Dennett on the subject of dualism, he stated:

"THE CHALLENGE: In the preceding section, I noted that if dualism is the best we can do, then we can't understand human consciousness. Some people are convinced that we can't in any case. Such defeatism, today, in the midst of a cornucopia of scientific advances ready to be exploited, strikes me as ludicrous, even pathetic."[15]

I read somewhere that *a conclusion is where you stop thinking*. I would offer that this applies to the material-monist who has concluded that the material is all there is, and that there is no point looking beyond the material. Thus, we have the irony, dare I say hypocrisy, of somebody accusing others of defeatism whilst he himself has already embraced it, in this context at least.

References:

1. Kerkut, G.A., *Implications of Evolution*, Pergamon, Oxford, UK, 1960, p 157

2. Chesterton, G.K., *The Twelve Men*, Tremendous Trifles, 1909

3. Andrews, Edgar, *What Is Man?* Elm Hill, Nashville, TN, 2018, pp. 185-186

4. Lewontin, Richard, *Billions and billions of demons*, The New York Review, January 9, 1997, p. 31

5. Verschuuren, Gerard, *What Makes You Tick? A New Paradigm for Neuroscience*, SOLAS Press, Antioch, CA, 2012, p. vii

6. Eccles, John C., *How the SELF Controls Its BRAIN*, Springer-Verlag, Berlin, Germany, 1994

7. Eccles, John C., and Robinson, Daniel N., *The Wonder of Being Human: Our Brain and Our Mind*, New Science Library, Boston, MA, 1984, p. 36

8. Eccles, How the SELF Controls Its BRAIN, pp. viii-ix

9. Taylor, Dr. David, *Business Engineering with Object Technology*, John Wiley & Sons, 1995

10. Hubert, Richard, *Convergent Architecture: Building Model Driven J2EE Systems with UML*, John Wiley & Sons, 2001

11. Ryle, Gilbert, *The Concept of Mind*, Penguin Books Ltd, London, England, 1990

12. Pinker, Steven, *How the Mind Works*, Penguin Books, London, UK, 1998

13. O'Leary, Denyse, *Evolution: The Fossils Speak, but Hardly with One Voice*, July 8, 2015, http://www.evolutionnews.org/2015/07/evolution_the_f097491.html

14. Dennett, Daniel C., *Consciousness Explained*, Penguin Press, London, England, 1991

15. *Ibid*, pp. 39-40

Chapter 3-2: The Theory of Knowledge

"I am indeed a little wiser than others because I know that I do not know. But the others do not even know that much; for they believe that they know something."[1]

~ Socrates (470-399 BCE), Classical Greek Philosopher ~

There is a branch of philosophy which has been studied for at least 3000 years, so far as we know, predating Socrates, Plato, Aristotle and their ancient Greek contemporaries. More modern contributors include Bertrand Russell, Gilbert Ryle, Edmund Gettier, and Richard Kirkham. Their understanding of this subject should have been a red flag to Charles Darwin, and to all those who have since followed his ideas on evolution, but apparently the penny is yet to drop.

That subject is *epistemology*: a branch of philosophy, in part described as concerning the nature and scope of knowledge, what it is and how it can be acquired. Closely associated is the philosophy of foundationalism, which concerns philosophical theories of knowledge resting upon justified belief or some secure foundation of certainty. Intellectuals like Aristotle and Descartes wrestled with these ideas and realised that to avoid infinite regress in questioning our beliefs, one has to establish a starting point, and thus we have the term: *foundational epistemology*. In brief, this states that everything that we claim we know is based on something earlier known or believed, irrespective of whether it is factual or suppositional, true or false. This supports my Primary Axiom #2, that *all knowledge is built upon prior knowledge*, based on a long chain of data ancestors.

There are variations on this idea of foundationalism: classical, modest, internalism, externalism, and so forth, but I have yet to find any proponent for the idea that knowledge can come from nowhere at all. Thus, there is a natural corollary to these ideas which has particular application to the theory of evolution: *all knowledge must start somewhere and it cannot arise or create itself out of nothing.*

As was mentioned earlier, Gilbert Ryle highlighted the distinction between knowing *that* and knowing *how*. He further argued that a failure to acknowledge this distinction leads to infinite regress: if all knowledge is built upon prior knowledge, where does knowledge start? This, I believe, is where we find evolutionists futilely burrowing into physics, chemistry, biology, and every other refuge of materialism. Borrowing from Ryle (but not in the sense of his argument): evolutionists believe that they know certain things inferred from circumstantial evidence, but do not know how such things came about. In the context of knowledge and the development of skills, we shall come back to the insights offered by Ryle.

Another very important concept explored by philosophers is that of *certainty*, which has led to other lines of enquiry such as the foundations of mathematics, inductive reasoning, probability interpretations, and Gödel's incompleteness theorems. We do not need to revisit the details here: we just need to acknowledge that whilst certainty is related to knowledge, it is fundamental to truth and reality. I am introducing a number of apparently unrelated concepts here but I beg your indulgence: I will draw the threads together as we proceed.

American philosopher, Edmund Gettier, contended that while justified belief in a true proposition is necessary for that proposition to be known, it is not sufficient. He also argued that there are situations in which one's belief may be justified and true, yet fail to count as knowledge. Here I am going to differentiate between true knowledge and false knowledge: not everything (information) we believe to be true is actually true and is thus termed *false knowledge*. I will persist with this differentiation of truth, information, and knowledge.

Developing the Theory of Knowledge

So as to avoid any theistic content, I will pass over Jewish literature and start with a review of the Greek philosophers. Quoting from Karl Poppers' writings, and referencing the quotation at the beginning of this chapter:

"Socrates' pupil Plato abandoned the Socratic thesis of our ignorance, and with it the Socratic demand for intellectual modesty. Socrates and Plato both insisted that a statesman ought to be wise. But by this they mean fundamentally different things. According to Socrates, the statesman ought to be aware of his ignorance; whereas according to Plato, he ought to be a thoroughly instructed thinker; a learned philosopher."[2]

The issue in contention was *wisdom*, and how statesmen should present themselves, not the necessity for knowledge. However, an underlying theme was *certainty* of knowledge, as mentioned before. This led to three views on the theory of knowledge:

1. Optimism - that we are capable of understanding the world.

2. Pessimism – that in truth, we are unable to comprehend the world; and

3. Scepticism – the middle ground, the term derived from the Greek, *skeptomai*, meaning to examine, reflect, enquire.

The optimistic view is found in modern science, where even though much is unknown, there is confidence that it *can be* known, and will be known. In some circles, this is described as *promissory science*. The pessimistic view was long held, especially in the East, but became less favoured after the scientific discoveries of the seventeenth century onward. The more logical view, in my opinion, is that of *scepticism*, in that some things can be known, but other things will forever be beyond our understanding. This theme runs throughout the book you are reading, as I will explain as we proceed. If you would prefer, you can preview my own philosophical exposition in the chapter toward the end: "*A Law Hypothesis*", where I contend that some things cannot be known.

So far, we have briefly reviewed some philosophical contentions, but now let us apply those to matters of science.

References:

1. Popper, Karl, *The Two Fundamental Problems of the Theory of Knowledge*, Routledge Classics, New York, NY, 2012, p. xiii

2. *Ibid*, p. xiv

Chapter 3-3: Knowledge versus Evolution Theory

"Knowledge can be communicated, but not wisdom"

~ Hermann Hesse ~

Science and Truth

Science, as we will use the term here, is concerned with the truth of our material existence; C.S. Lewis defines truth as "a property of propositions such that they correspond with the state of affairs in the objective world that they purport to describe."[1] As we know from history, there have been, and are, proponents of both true and false knowledge. Scientific findings, when validated, can be used in technology and other practical applications, but when falsified are hopefully discarded. In between are numerous theories and hypotheses which have not been validated, and in some cases cannot be falsified. However, irrespective of our scientific views of what may be true or false, the material universe can only operate on what is true: if we observe some behaviour which is contrary to, for example, the laws of gravity, rather than pretend that it is not happening, we should re-evaluate our understanding of gravity or some related phenomenon. This is what happened with the hypothetical planet, Vulcan, thought to be orbiting between Mercury and the Sun and causing the peculiarities in Mercury's orbit. Albert Einstein came up with new physics to explain the phenomenon without reference to a hypothetical planet, which in truth, does not exist.

Something similar is happening with our understanding of gravity. It has long been understood as an attractive force, but its nature and mechanism have yet to be determined. An alternate view of gravity has been proposed by Dr Russell Humphreys[2]: that space is truly a fabric, and every object having mass distorts the fabric such that they tend to "roll" toward one another. This is a metaphor for the reality, but it does help to visualise the phenomena: Dr Humphreys has proposed mathematical equations to explain the idea. My point here is simply that where our current scientific understanding fails to adequately explain phenomena such as gravity or the singularity that gave rise to the Big Bang, it is incumbent upon scientists to seek new understandings. It is valid to propose hypotheses to give a name to unexplained phenomena, such as has been done with Dark Energy and Dark Matter, but therein lies a danger: the longer these entities remain unexplained, the more they become accepted as scientific fact. One that we have been, and continue to deal with, is that the mind is an emergent property of the brain. No-one knows what that means, but this "explanation" has been around so long that many accept it as a scientific truth.

Albert Einstein was not just a mainstream scientist; he was, like Galileo, one of courage, prepared to think beyond the current paradigm and swim against the current when he discovered new understandings of reality. Just as Einstein came up with new physics as mentioned above, I believe it imperative that we must come up with new science to explain the mind-brain conundrum. The mind cannot be explained in biological terms, as much as many have attempted to do, and whilst I cannot solve the conundrum, I can at least explain it in fundamental terms, and demonstrate why our current scientific understanding is inadequate. At the heart of the issue is the difference between biological information that is found in the cell, and is responsible for the activities of the cell, and cognitive information, which is a transient derivation from abstracted data, and resides nowhere, physically at least.

Technology aside, it is axiomatic that science and truth are always coincident in nature, irrespective of whether scientists have discovered such truth. Thus, we can say that certain truths, and the behaviours subject to those truths, pre-exist our own existence, and that of any life forms which evolutionists claim gave rise to our own. Evolutionary biologists assert that abiogenesis is outside their field of enquiry and allowing that, if we wind back the clock to any point that evolutionists may choose, there is a great deal of truth that existed prior to it being acquired in the form of information and knowledge. The question then is: how can organic evolution account for our acquisition of knowledge?

As an aside, I am saddened when evolutionists such as Richard Dawkins assert that religion and theology threaten our knowledge of existence, for he fails to acknowledge that these are but branches of *philosophy*, which underpins the worldview of everyone, including scientists and himself. We would not have the terms, philosophy and metaphysics, if these were not intellectual domains worthy of study. Scientists claim to be searching for truth, but their claim is false if they refuse to look in places where unpalatable truths may be lurking. When scientists assert without looking that there is no evidence of the non-material, they argue from ignorance, for how could they know? Rather than being a source of knowledge and progress, such scientists are an impediment to knowledge, and one wonders whether they feel threatened by what they choose to ignore.

Cognitive dissonance is a term first proposed by psychologist Leon Festinger in 1957, as explained in his book[3] published in 1962. He offered that people have "an inner need to ensure that their beliefs and behaviours are consistent. Inconsistent or conflicting beliefs lead to disharmony, which people strive to avoid." He continued: "Cognitive dissonance can be seen as an antecedent condition which leads to activity oriented toward dissonance reduction just as hunger leads toward activity oriented toward hunger reduction. It is a very different motivation from what psychologists are used to dealing with but, as we shall see, nonetheless powerful." Festinger suggested three ways of dealing with what can be a very uncomfortable situation:

1. Focus on more supportive beliefs that outweigh the dissonant belief or behaviour.

2. Reduce the importance of the conflicting belief; and

3. Change the conflicting belief so that it is consistent with other beliefs or behaviours.

It is said that cognitive dissonance, even when not recognised as such, will have a powerful influence on what people believe and why. I have observed this in areas of interest to me: evolution theory and religion, and more lately, climate change. In my experience, I have found that the most common reaction is to entirely avoid such uncomfortable situations altogether. In short, when it comes to evidence that conflicts with their beliefs, people generally prefer to not know. This could be said to fall into strategy #1 to an absolute degree. When I contend, as I do in this book, regarding what I consider to be the implausibility of evolution giving rise to the conceptual rather than just the physical, I am told that I place far too much emphasis on that line of thinking. In other words, I am being told to implement strategy #2. I periodically suffer from cognitive dissonance as I research subjects of interest, but my preference is always for strategy #3. In my attempt at intellectual honesty, I eschew beliefs until I have evidence, data, or logical reasoning to support them. In effect, I do not simply make a mental adjustment toward comfort, but seek to resolve the issue entirely where I can. I was once told by a cognitive psychologist that to be a good analyst, one has to not only be comfortable with ambiguity, but to have a passion for it.

I hope that suggests that I am a good analyst, for I luxuriate in ambiguities that seek to be disambiguated.

Information in Humans

The study of genetics has generated an enormous interest in the information contained in the genome, but I would offer a refinement to the normal usage of the word "information". The coding of DNA is represented by the four letters: A, C, G, and T. Early studies thought that meaning could be derived directly from the sequence of these "letters" and whilst that is true to some extent, it is considerably more complex than that. Because our genes are both pleiotropic and polygenic, there is not a one-to-one correlation between a coded string and a protein. Thus, those four letters at best represent data, not information, as I will explain. More recent scientific research has revealed that rather than DNA being an *active* replicator, it is a *passive* participant in the process of replication, with other cellular activity being the drivers.

In this study, I argue that the information conundrum in the genome is far more easily solved than the cognitive information that we experience as sentient humans. Biological information in the genome is comparatively simple to understand, because it behaves in accordance with the laws of physics and chemistry as we understand them - even when the genome misbehaves, it still does so according to those very same laws. Because science and truth are coincident in nature, certainty exists in material behaviour even though our lack of knowledge and understanding may suggest otherwise. Scientists will often speak of random behaviour, but I suspect that the physical world never behaves randomly: it just appears that way due to our lack of understanding of the underlying causes.

Biological information is a real, but complex entity in that it is instantiated in the organic matter of the cell. With sufficient research, scientists can learn the relationships between components of the cell, all of which act in conformity with their chemical properties. The cell behaves as it must: it can do no more, nor less, than as regulated by its intrinsic properties, albeit at times corrupted by other chemical factors. The cell, cannot for example, decide to take the day off. Cognitive information is entirely different, in that it does not exist as a physical entity, but is a transitory experience of correlating data in context. This may come as a surprise to the reader, but I will continue in my attempt to explain why this is true.

The behaviour of our brains falls into two categories: *autonomous* and *volitional*. The autonomous functioning is pre-programmed at some stage of embryonic development although how that arose through evolution is unknown. The volitional functioning is entirely different, and is even less well understood, if at all. Curiously, as we discussed earlier, some evolutionists have claimed that we do not have free will and thus the very concept of volition is invalid. Having read numerous studies on the mind-brain complex, I contend that far too little attention is paid to the differences between autonomous and volitional.

The Science of Information

Today we have a considerable advantage over philosophers of past generations: the advent of the computer has enabled us to manipulate vast quantities of data in ways never before imagined. Recent trends have seen the emphasis shift from the metaphysics of information to the technology, as seen in the works of Claude Shannon, Werner Gitt, and Walter ReMine: necessary but not sufficient in my view. Let me apologise in advance if I am in any way misrepresenting the very valuable achievements of these scientists, but my issue is the apparent failure to differentiate between symbolic representations of data, conceptual data, and information, and the consequent absence of an explanation as to how information originates and/or is derived. This has direct application to the feasibility of organic evolution of information, knowledge, and truth.

Claude Shannon is often described as "the father of information theory", but with no intent to belittle his ground-breaking achievements, more correctly we should say that he fathered the process of digital information *processing*. Shannon worked on the technical aspects of information processing with little regard for the truth or validity of the data being processed, and thus has contributed little to our understanding of the intrinsic nature of information. Werner Gitt has added to our understanding with his five-layer model[4] of: Apobetics, Pragmatics, Semantics, Syntax, and Statistics, but in some respects has overlooked a vital issue. In later works he has expanded his explanations to include the concept of *Universal Information*, but in constraining his thoughts to his earlier model, he has ignored the simple fact that some information flows are unintentional and do not conform to his model. In doing so, he has, like Shannon, contributed little to our understanding of the origins of information, and its relationship to evolution. Walter ReMine has taken a novel approach with his concept of the *biotic message*[5], but this relates primarily to the genome and says little about cognitive information. Like evolutionary biologists, these scientists have started in the middle based on certain presuppositions. It is my intention to seek further back in time, looking for what might have been - not what is.

An "information" flow that I have not encountered in the literature is one that occurs in nature, and which few seem to recognise until pointed out to them. Its relevance to evolution is that for evolution to be true, it had to have been the *very first source* of data that could lead to knowledge for the evolving organisms. I speak here of data acquired by the sensory receptors, particularly sight and hearing. As discussed under Primary Axioms, a beam of light contains encoded data pertaining to its source, and when reflected from an object, contains further encoded data pertaining to that object. That is the basis of our sense of sight. The issue is that the data is coded using a method appropriate to the medium of communication, and thus the communication contains neither data nor information, but merely physical symbols. For it to become data, it needs to be decoded, and to become information, the data needs to be processed against a hierarchical structure of related concepts which provide context. If there is nothing there in the first place, how does the process get started?

Just as it took scientists considerable research to understand that the data in the light beam itself is related to its source in terms of the spectrum, we ought to consider the prerequisites for understanding the data related to the object from which the light is reflected.

A Foundation

It is my contention that data is conceptual, and only becomes cognitive information when processed within a referential framework of relevant concepts which provide context. This should be obvious but sometimes we need to remind ourselves. It is interesting, to me at least, that this was understood by the 12th century philosopher, Maimonides. Writing on the variations in the human condition, and the agencies involved, he observed[6]:

"Every man possesses a certain amount of courage, otherwise he would not stir to remove anything that might injure him. [Ed. Read up on the ineffectual Eloi in H.G. Wells, The Time Machine] ... Energy varies like all other forces, being great in one case and small in another ... This courage requires that there be in man's constitution a certain disposition for it. If man, in accordance with a certain view, employs it more frequently, it develops and increases, but, on the other hand, if it is employed, in accordance with the opposite view, more rarely, it will diminish. From our own youth we remember that there are different degrees of energy among boys."

In another section, Maimonides equates energy not only with courage, but with intellect and the desire to learn. Continuing,

"The same is the case with the intuitive faculty; all possess it, but in different degrees. Man's intuitive power is especially strong in things which he has well comprehended, and in which his mind is much engaged. Thus you may guess yourself correctly that a certain person said or did a certain thing in a certain manner. Some persons are so strong and sound in their imagination and intuitive faculty that, when they assume a thing to be in existence, the reality either entirely or partly confirms their assumption. Although the causes of this assumption are numerous and include many preceding, succeeding, and present circumstances, by means of the intuitive faculty the intellect can pass over all these causes and draw inferences from them very quickly, almost instantaneously ... This should be the belief of all who choose to accept the truth. For [all things are in certain relation to each other, and] what is noticed in one thing may be used as evidence for the existence of certain properties in another, and the knowledge of one thing leads us to the knowledge of other things. But [what we said of the extraordinary powers of our imaginative faculty] applies with special force to our intellect, which is directly influenced by the Active Intellect, and caused by it to pass from potentiality to actuality. It is through the intellect that the influence reaches the imaginative faculty."

Maimonides expressed the concept of Primary Axiom #2, that *all knowledge is built upon prior knowledge*, although his context related to intuition and imagination, and how knowledge is caused to pass from potentiality to actuality in the expression of that imagination, in whatever form, be it music, poetry, art, or scientific inventiveness. So now to some concrete examples.

Take any single letter, word, number, or symbol, and in isolation it could mean anything - or nothing. How many readers recognise this symbol "ט" as the letter *tet* of the Hebrew alphabet? At an expanded level, consider finding a piece of paper inscribed with what appear to be Chinese or Japanese pictograms. Without a context, it may or may not be meaningful to you, but if the same characters were written on clean piece of 6x8 white cardboard, nestled in a plastic frame on the table of a Chinese restaurant, you would assume it to be a menu. Knowing that a menu is a list of foods, one then needs to understand both the categories of foods and the culinary variations by culture. Depending on the language in which the menu is written, you may still need to rely on the venue to provide the context. Nothing can be known until first the concept is explained, and then the context is applied to derive understanding. As the philosophers of old have long understood, knowledge is built upon knowledge, or as one lady remarked to Bertrand Russell in another context: "turtles all the way down". Russell responded by asking: What is holding up the giant turtle? We should ask the same question related to knowledge: what is the source of foundational knowledge?

In the early days of commercial computing, analysts and file designers unearthed a truth whose wider application is generally not recognised by modern practitioners. We will discuss these in a later chapter but here let me say that these fundamentals hold true for all forms of data or information processing.

In relation to cognitive information, what I wish to emphasise at this stage is simply this: all forms of communication and data storage are at the physical symbol level only. To be understood as data, the symbols need to be organised in a regulated manner known as *encoding*. To derive an understanding, the recipient needs to have foreknowledge of the coding system, so as to be able to accurately *decode* the symbols. When decoded, the data still needs to be processed through a preloaded, hierarchical, referential framework of concepts before cognitive information can be derived. It is my contention that the agency that does the formulation of the coding system, the organisation of the symbols, and the algorithms used for storage and retrieval, must be external to the physical symbols themselves.

This is the regime of the mind, but then the question: How did the mind, however defined, learn to do what it does?

References:

1. Lewis, C.S., *Mere Christianity*, Harper, San Francisco, 2001

2. Humphreys, Dr Russell, *New view of gravity explains cosmic microwave background radiation*, Journal of Creation, Vol. 28(3) 2014, pp. 106-114

3. Festinger, L., *A Theory of Cognitive Dissonance*, Stanford University Press, Palo Alto, CA, 1962

4. Gitt, Dr. Werner, *In the Beginning was Information*, First Master Books, Green Forest, AR, 2007, p. 60

5. ReMine, Walter James, *The Biotic Message: Evolution Versus Message Theory*, St. Paul Science, Inc., St. Paul, MN, 1993

6. Maimonides, Moses, *The Guide for the Perplexed*, Digireads Publishing, 2018, translation from the original text with annotations by M. Friedlander, Trubner & Co., London, UK, 1881, pp. 331-333

Chapter 3-4: The Morality Conundrum

"There is no justice in the laws of nature, no term for fairness in the equations of motion. The Universe is neither evil, nor good, it simply does not care. The stars don't care, or the Sun, or the sky."

~ Eliezer Yudkowsky, Harry Potter and the Methods of Rationality ~

I thought that I would get this one out of the way early, as it is so controversial, and only peripheral to the main theme. I wanted to express my own uncertainty over the matter, offering my observations and those of others on both sides of the debate. No doubt you have your own opinions, or even uncertainty like mine, so perhaps this short dissertation may help in some way. Either way, such uncertainty does evidence that the *microbes to man* hypothesis is a long way from scientifically proven.

Huston Smith (1919 2016), a renowned scholar of religion, and reputed to be one of the world's most influential figures in religious studies, observed: "The scientific method is nearly perfect for understanding the physical aspects of our life. But it is a radically limited viewfinder in its inability to offer values, morals and meanings that are at the centre of our lives." Friedrich Nietzsche, in his book *"Beyond Good and Evil"*, counters with: "There is no such thing as moral phenomena, but only a moral interpretation of phenomena". Rajesh, in his book *"Random Cosmos – A Whimsical Take on the Puzzle & Purpose of Existence"*, agreed with Nietzsche, "Morality is a man-made concept. The cosmos has no notion of values, ethics, or good deeds. Comets follow no path of righteousness." Evolutionists have a somewhat different take on the subject, claiming that morality evolved very early to develop co-operation in insect and animal societies. Michael Shermer, an American science writer, historian of science, and founder of The Skeptics Society, offers: "As a social primate species, we modulate our morals with signals from family, friends and social groups with whom we identify because in our evolutionary past, those attributes helped individuals to survive and reproduce."

So, according to one side of the debate, morality does not exist, other than as an intellectual interpretation by humans. On the other hand, it does exist in even the lowest forms of life on the evolutionary scale, but it is a function of natural selection, reproduction, and a desire for survival. Even though Rajesh says that the cosmos, including I assume, all life forms, has no notion of good deeds, the evolutionists say that animals *et al* do have that notion, but as a function of cooperative living. Along comes the theist with his views on good and evil, with our conscience initially advising us, followed by whatever religions teach, and whatever is legislated in secular law. As there is nothing approaching a consensus, even amongst atheists, no opinion can be logically asserted as the right one. The issue of conscience is intriguing, as it falls under the domain of *instinct*, a form of knowledge which derives from I know not where, but we will explore it further a little later.

I am intrigued by where Nietzsche stood on the issue of the mind-brain conundrum, for his assertion that "There is no such thing as moral phenomena, but only a moral interpretation of phenomena" opens the question of what it means to *interpret*. Can a biological brain interpret in terms of morality which is conceptual, not physical?

The emerging fields of evolutionary biology and in particular sociobiology have argued that, although human social behaviours are complex, the precursors of human morality can be traced to the behaviours of many other social animals. Ant colonies may possess millions of individuals, but somehow, they manage to live together as a peaceful and productive society. E. O. Wilson, an American biologist, naturalist, and writer, argues that the single most important factor that leads to the success of ant colonies is the existence of a sterile worker caste. This caste of females is subservient to the needs of their mother, the queen, and in so doing, have given up their own reproduction in order to raise brothers and sisters. The existence of sterile castes among these social insects significantly restricts the competition for mating and in the process fosters cooperation within a colony. Hmmm ... I wonder if this would work in human society? That aside, *subservience* and the *needs of others* are concepts beyond the physical, so in effect, Wilson is attributing a sense of morality to a much lower-order species, which I would contend, is highly problematic.

All very interesting, but I suspect, equally speculative; consider: *have given up their own reproduction*? A fundamental of evolutionary theory is that *teleology*, the explanation of phenomena in terms of the *purpose* they serve rather than of the *cause* by which they arise, plays no part whatsoever. One must ask: how did natural selection "decide" that a mutation which led to sterility was beneficial, and even more, that somehow, not all females should be sterile lest there be no future queen? There cannot be a "caste" of sterile males and/or females, other than by chance, as they cannot, by definition, reproduce. How does Wilson, a biologist and naturalist, propose that a *sterile worker caste* could arise and continue? The only way Wilson can be right, is if a significant proportion of female ants are born sterile, as they cannot make a decision to be sterile by choice. That would be a *moral* choice, but implies a level of cognition that an ant would not have. How does Wilson reconcile his idea, with the notion that evolution is driven by random mutation, fixed in populations by natural selection, that favours reproduction and survival of the individual? Natural selection cannot operate at the group level, only at the level of the individual organism, and I would have thought that sterility is antithetical to natural selection favouring survival. I have not spent much time on this conundrum, but I wonder at the process whereby considerably more female ants have sterile offspring rather than fertile. If most female ants are sterile, then reproduction numbers would be limited. It might work if a percentage of the female ants produced only infertile males, but I am just speculating.

As an aside, as I have said before, I have never understood the argument for natural selection, as it is fundamentally teleological: it invokes *purpose* – reproduction and survival. What is it about the biology of an organism that promotes a survival instinct, especially as survival is often much harder than submitting to death? Looking around at what happens in the world, I see numerous circumstances where it would be much easier, and less painful, to just lie down and die rather than fight to survive, which often requires intensive periods of stress and pain. Whilst I can understand the *purpose* of the survival instinct, I cannot imagine the biological *cause*, and it is the cause which evolutionists must demonstrate if their theories are to be substantiated. To my mind, both reproduction and survival are teleological until they can be proven otherwise.

Moving on, many social animals such as primates, dolphins, and whales have shown to exhibit what Michael Shermer refers to as premoral sentiments. According to Shermer, the following characteristics are shared by humans and other social animals, particularly the great apes:

"Attachment and bonding, cooperation and mutual aid, sympathy and empathy, direct and indirect reciprocity, altruism and reciprocal altruism, conflict resolution and peacemaking, deception, deception detection, community concern and caring about what others think about you, and awareness of, and response to, the social rules of the group."

Shermer argues that these premoral sentiments evolved in primate societies as a method of restraining individual selfishness and building more cooperative groups. He doesn't explain why individual selfishness arises instead of altruism. If selfishness is a natural outcome of evolutionary processes, why would altruism not be similarly so, and given its claimed benefits for the group, why would it not be naturally selected over selfishness? I understand competitiveness in the fight for survival, but that hasn't been explained in biological terms either. Continuing Shermer's argument, for any social species, the benefits of being part of an altruistic group should outweigh the benefits of individualism. For example, lack of group cohesion could make individuals more vulnerable to attack from outsiders. Being part of group may also improve the chances of finding food. He states that this is evident among animals that hunt in packs to take down large or dangerous prey. Again, I do not accept this argument, as natural selection operates at the level of the individual organism, and is entirely biological. I agree that *the benefits of being part of an altruistic group should outweigh the benefits of individualism*, but how does altruism arise biologically through genetic mutation and natural selection?

What this might mean, is that the drivers of evolution are not as Charles Darwin, and proponents of Neo-Darwinism espouse, but are to be found in far more complex biological processes. Many scientists researching evolution have come to that conclusion, but are yet to identify what those processes might be. I will later argue that they never will, because of my fundamental belief that the conceptual can never arise out of the physical, even biological stuff. Everything that Shermer argues as related to morality is conceptual, not physical: it is not science, but philosophy, to make claims about morality or group dynamics unless the causal underpinnings can be explained in materialistic terms.

The social brain hypothesis, detailed by R.I.M Dunbar, a British anthropologist and evolutionary psychologist and a specialist in primate behaviour, in the article *The Social Brain Hypothesis and Its*

Implications for Social Evolution, supports the claim that the brain originally evolved to process factual information. The brain allows an individual to recognize patterns, perceive speech, develop strategies to circumvent ecologically-based problems such as foraging for food, and also permits the phenomenon of colour vision. Again, speculation based on *purpose* not *cause*. Stating the purpose that the brain serves, and then attempting to reverse engineer that back into evolutionary science, is not science, nor even meta-physics. If evolution is to be proven, the causes of biological processes have to be identified – genetic mutation and natural selection have not been shown to do what is claimed for them. As an aside, the evolving brain could not discern fact from fiction, because as we know, our senses can present a false perception of reality. This point needs to be stressed: new data can only be validated by an external process that already has achieved the ability to discern truth from untruth, but that process cannot be the origination of itself.

Psychologist Matt J. Rossano muses that religion emerged after morality, and built upon morality by expanding the social scrutiny of individual behaviour to include supernatural agents. By including ever watchful ancestors, spirits and gods in the social realm, humans discovered an effective strategy for restraining selfishness and building more cooperative groups. The adaptive value of religion would have enhanced group survival. Rossano might be right, but his musings are based on unsubstantiated presuppositions that evolution can be the underlying causal mechanism of morality. That said, it is my observation that people do indeed invent their own gods, but this cannot be construed as an argument against the existence of god(s). As I have argued in other works, if the transcendent, infinite entity that we call "God" does exist, then such an entity is beyond the understanding of finite mortals, and our perception of this god can only exist in our imaginations. We can define "infinite" as a concept, but cannot know it as a reality, because everything in our experienced reality is finite, and we can only think in finite terms. If the reader is of a religious persuasion and wishes to research this further, I would recommend the works of Maimonides[1].

Doing What "Ought" to be Done

"*Ought*" is an expression of morality.

I have borrowed this heading from a chapter similarly entitled in Gerard Verschuuren's book, "*What Makes You Tick*"[2]. Verschuuren is a human geneticist with a Doctorate in the Philosophy of Science, and well qualified to ponder the relationship between biology and morality. He quotes the renowned Wilder Penfield's dictum, *"There is a switchboard operator as well as a switchboard"*, highlighting the truth that material constructs cannot operate themselves beyond some basic functions intrinsic to their existence. If the brain is the switchboard, who/what is the operator – surely not the switchboard itself operating entirely autonomously! Quoting from a recommendation in the opening pages,

"This welcome book brings neuroscience and metaphysics together in a way that does not seek so much to 'solve' the mind-brain problem, as to manifest the mystery of the human person as both physical and spiritual. From that perspective it is able to unmask the pseudo-dilemmas of scientism and the false 'solutions' of materialism and dualism."

No doubt you will detect the theological presupposition of the commentator, but it is not only the religious who ponder this mystery and are not convinced by solutions offered by the evolutionists. I am aligned with the thinking of Verschuuren, in that I make no attempt to solve the mind-brain problem (because I can't) – I seek to expose its mysteries in a way that evolutionists have been unable to explain from the perspective of philosophical materialism. Verschuuren opens his argument with:

"It has been suggested that moral values are 'evident' because they are grounded in our genes. If that were so, we wouldn't be morally obliged, but we would only feel obligated; our genetic make-up would only have us *believe* that our moral obligations rest on an objective foundation. This would, so to speak, make for a collective illusion, foisted on us by our genes. But it seems to me that such a foundation would be as fragile as the genetic material it is said to be made of – DNA."[3]

As previously discussed, some evolutionists argue that free will is an *illusion*, acknowledging that biology cannot give rise to that capability. Such people seem not to accept that according to their own theories, they were *forced* to say what they did, and therefore one cannot be confident of its truth. My point is that if morality is an illusion foisted on us by our genes, and similarly, free will is an illusion because our genes cannot create that capability, how is it that humans can act in concert with, or contrary to, this sense of morality?

Illusions are conceptual, not material, and it takes an intelligent entity to both have, and respond to, illusions. Note that the very mention of *illusions*, which are not material, argues against philosophical materialism which asserts that everything *is* material. Such is the illogical world of the material monist.

It is a truism that science, through research and discovery, tells us the way things are, not the way they ought to be. This is one of the arguments of Richard Dawkins against the existence of God as a Divine Creator; he looks at the anatomy, physiology, and biology of humans, concluding that no intelligent creator would design humans as we are. In the Dawkins' view of what we ought to be, we would have been designed very differently. Dawkins makes the common mistake of assuming he knows the intentions of the designer, and the purpose for which things are designed. He also ignores the reality of his own thought processes which cannot be entirely biological, because as will be demonstrated, biological processes are often unreliable, and any thoughts derived entirely biologically must be subject to the same degree of unreliability. In that context, Dawkins ought not be confident of his own thoughts. In this debate, we must ever keep in mind that morality is conceptual, not material; Nietzsche is entirely correct in his observation, "There is no such thing as moral phenomena, but only a moral interpretation of phenomena". Interpretation is an intellectual activity, not biological. Whatever is biological will always do what it must do, mutations acknowledged: thus, if morality is genetic, and free will is an illusion, then all creatures must follow the genetic moral rule. That they do not could be attributed to genetic mutation, I suppose, but now it is becoming even more complicated and I do not intend to go there.

The curious thing about morality is that, if it is genetic, then according to our general understanding of morality, *there is nothing moral about genetic based morality* – it can be no different to eating, drinking, or any other behaviour influenced by biology. Our sense of right or wrong is no different to choosing to eat, or not eat; choosing to drink, or not drink; and so forth. Verschuuren states: "My conclusion is that apparently, moral laws tell us to do what our genes do *not* make us do by nature."[4] We eat, sleep, and drink because if we do not, our biology breaks down, but we can entirely ignore morality as history attests. Is the morality gene recessive, in and out of society depending on parenting? An interesting aspect of the morality gene is that it has no control over the rest of the genome, allowing deleterious mutations to persist through generations. One would have thought that if morality arose through natural selection, it ought to have done a better job of controlling the future direction of evolution, but apparently not.

"The British philosopher G. E. Moore spoke of the naturalistic fallacy[5]. This consists of erroneously reducing a moral property (being good or right) to a natural property (being natural, functional, genetic, more evolved, better for the majority, or whatever). Therefore, it would be a fallacy to define moral notions in non-moral terms … The fact that something *is* this way doesn't mean that it *ought to be* this way. The fact that something is natural doesn't imply that it is also a moral value that ought to be enforced."[6]

The contrary position is also true: just because something is natural does not mean that it should be allowable. This has been highlighted by modern vegans who argue that just because we have a natural tendency to want to slaughter animals for food, does not mean that we should. Apparently, in their minds, there is a higher moral standard. Maybe their morality gene is more highly evolved than those of meat eaters, I cannot know. Then we have the issue of hunting for sport. I am disgusted by this practice, but friends who otherwise exhibit high moral standards indulge with enthusiasm. Am I more highly evolved in a moral sense? I agree with this author, "We must come to the conclusion that morality can't be based on anything non-moral. It is not rooted in our genes; it is not the product of natural selection."[7]

So, where does morality come from?

We know from history and cultures that for humanity, morality is neither self-evident, nor absolute. It cannot be absolute absent of a higher authority; thus, if one does not accept a higher authority than secular governments, morality is fluid, depending on secular legislation, and this is exactly what we experience. However, for many, if not most, there are some moral absolutes, such as not permitting murder, rape, theft, and so on. This suggests that absolute moral values do exist, and what we experience as cultural shifts are the outcomes of moral evaluations, or more likely, submission to other cultural and political imperatives. Absolutes are objective, evaluations are subjective, and are the domain of intellect, worldview and other factors. If species of a lower order than humans behave in accordance with their evolved genetic morality, but humans do not, what has corrupted the genetic chain of inheritance?

There is nothing more that I wish to say on the question of how morality is such a powerful force in humanity, even supposedly in animals and insects, and how it could have become a part of our existence.

Unquestionably, it is innate in various forms of life, and whereas Shermer's *premoral sentiments* leave no room for variability in lower species such as ants – they just do what they do - it can be found in more developed species. I believe that it can be said that in general, there do exist certain universal standards of morality in humanity, individual and cultural variations acknowledged. As Verschuuren questions at the end of his chapter,

"But the question remains: where do these universal standards come from? There is no other way than acknowledging that moral values have an extra dimension and therefore must be derived from a different realm – from the 'Grand Beyond'."[8]

I do not wish, in this study, to pursue the nature of this *Grand Beyond*. I wish to remain grounded where we are, and simply question the origin of morality. I am convinced that it could not have arisen at any level of evolution through natural selection of genetic attributes. I continue to highlight the difference between the conceptual and the material, and that there is nothing intrinsic to biology that could give rise to anything of a conceptual nature. Just as significantly, morality is *subjective*, but the material can only deal in the *objective*, disproving any suggestion that the material can give rise to morality.

References:

1. Maimonides, Moses, *The Guide for the Perplexed*, Digireads Publishing, 2018, translation from the original text with annotations by M. Friedlander, Trubner & Co., London, UK, 1881

2. Verschuuren, Gerard, *What Makes You Tick? A New Paradigm for Neuroscience*, SOLAS Press, Antioch, CA, 2012

3. *Ibid*, p. 64

4. *Ibid*, p. 66

5. Moore, G.E., *Principia Ethica*, Cambridge University Press, Cambridge, UK, 1993

6. Verschuuren, *Ibid*, p. 67

7. *Ibid*, p. 68

8. *Ibid*, p. 70

Chapter 3-5: The Sensory Conundrum

"All our knowledge begins with the senses, proceeds then to the understanding, and ends with reason. There is nothing higher than reason."

~ Immanuel Kant, *Critique of Pure Reason* ~

I entirely agree with Immanuel Kant on this issue, but the words "begin", "proceed", and "end" require *process*, and it is these processes which we must investigate. It is not at all scientific, and not even logical, to just pompously pronounce from on high: "Evolution did it".

From time to time, we encounter a new smell, a new taste, or an unfamiliar texture under our fingers. We are aware that these sensations are new to us, for we can differentiate them from other sensations, but we cannot identify them. To do that, we need the services of an external agent that informs us via our other two senses, sight and hearing. Absent of that external agent, we remain in the dark, metaphorically speaking. Unfortunately, both our sight and hearing can deceive us, as can any external agent, which presents what may well be an unbounded problem, especially for evolution.

Now if we roll back the evolutionary clock to a time when the first sensory receptor supposedly arose, we have a dilemma. Whatever sense it related to, all that it could communicate to the brain was a signal that resolved to an electro-chemical message. Of course, the brain never having had one of those before, it had no idea of what to do with it, and no reason to store it, let alone build a neuronal network as a symbolic representation of the data itself, and this is the key point. The brain, primitive or otherwise, could not "know" the external reality represented by the message from the nervous system. In a later chapter, we will discuss the prerequisites for a reliable communications channel, the basics comprising a coding system where the sender encodes a message, and the receiver decodes the message, using the same "code book". It is reasonable to assume that if the encoding was a biological (chemical) process, then the decoding would also be biological, but the questions arise: Why would the receiving organ bother to decode, how would it know how to decode, and how could it transform a physical message into conceptual data? I acknowledge the reality of chemical properties, but that explains nothing in this context.

But let us assume that somehow the cellular structure of the embryonic brain was such that the messages conveyed by the peripheral nervous system to the central nervous system left a brain imprint in the form of a neuronal network. As further messages arrived, the brain could perhaps differentiate them by their electro-chemical properties, but two questions arise: (1), what part of the brain was doing the differentiation and how did that capability arise; and (2), what was the process by which individual neurons identified the necessary connections? On what basis could the brain determine the conceptual correlations with earlier messages? Yes, we could accept that there was a mechanism within the brain that correlated messages based on their electro-chemical properties, but those could only represent the encoded symbols, not explain the external reality itself. If neuronal networks are symbolic representations of signals concerning an external reality, there needs to be some form of supervisory process with both the knowledge of that reality, and the protocols for encoding into the neuronal networks. In digital computing applications, the sensory receptors (input devices) are intelligently designed with encoding mechanisms, such that the encoded electrical signals can be stored digitally on an appropriate storage device. The same "code book" is used for retrieval. Evolutionists would have us believe that these functions have been developed in biology, and are performed biologically, without any support of intelligent design – it just happened through an unstructured process of genetic mutation and natural selection.

We have some understanding of the architecture of the brain and the neural network, and can even understand where in the network certain activities occur, but we cannot know how that network evolved. More importantly, and this is central to my argument, there was no process to validate that the neuronal correlations of nervous system inputs correctly represented the external reality. Those of us who have programmed large complex systems know only too well the difficulties of getting it right first time, and the complexity of debugging individual code segments, programs, and systems, when we inevitably got it wrong. Evolution theory would have us believe that cellular development occasioned by random mutation and natural selection managed to get it right, albeit over a very lengthy period of time. The unanswerable question is how would it "know" that it got it right? Why would it not have persisted with the wrong answer, not knowing that it was wrong, and if it did, how to correct it? These are the questions behind Primary Axiom #7.

We will later get to a more detailed description of the human nervous system, briefly touching on the chemical complexity of the interactions between nerve cells. For now, I just want to highlight some activities and properties whose development evolution exponents would need to explain. For example, if I accidentally hit my finger with a hammer, the brain needs to be cognizant of several facts:

- Whether the sensation had an external cause, in contrast to, for example, a symptom of neuritis.

- Where in the body the sensation was experienced, in this case, a finger.

- Which finger experienced the sensation.

- The location on the finger where the sensation was experienced.

- The nature of the experience in terms of severity.

- The nature of the experience in terms of the likely cause, e.g., a pin prick, a piercing, blunt trauma, etc.

We take this process for granted, but in modern engineering, this is equivalent to data logging. A racing motorcycle, for example, is fitted with numerous sensors to measure the behaviour of various components, including the rider. Interpreting this mass of often confusing data takes the experience of specialists in the relevant fields of engines, tyres, suspension, and so on. Somehow, the equivalent in biology is claimed to have developed the most exquisite accuracy via evolution. Another important point is that motorcycle engineers were required to learn the concepts of engines, tyres, and suspension. Similarly, it must be said, the cognitive functions of our brains had to learn of the existence and function of the sensory receptors in the body. Now, that is not to say that the brain knew these sensors as relating to smell, taste, or touch, for these are concepts, but it had to know how to react to them, and a reason to differentiate them in memory. I would refer you back to an earlier discussion on *correlation*. Accepting that somehow, the biology of sensory perception allowed for differentiation, even to the location of the sensor itself, we can use the analogy of seeing something whose identity is unknown to us: how does the brain correctly identify the sensation?

The issue is that "sense" is a concept that has no physical reality until instantiated in a specific instance.

Let me explain.

In mathematics, there is no such entity as "a one" or "a two" – you can have "one of" but not "a one" because "one" is a mathematical concept, not a real entity. Similarly, the notion of *name*: name has to have an instance before it becomes "*a name*". There is no such physical entity as an unnamed name, just as there is no music without notes, and no songs without words. Number, name, music, and song are concepts which have no physical reality until instantiated. The converse is logically equivalent. Smell, taste, and touch are physical realities, but the concept does not exist cognitively until the intervention of an external, intelligent agent to both label the phenomenon, and to differentiate the significance. Smell, taste, and touch cannot self-identify as such: they simply exist as physical realities without self-awareness. In arguing their case for the development of sight, evolutionists contend that some sight is better than no sight, but they are arguing from the present reality of sight, not that of the past. We explore this further in a minute.

Before any verbal or written communication can occur between humans, the concepts to be communicated have to be pre-agreed, otherwise we would all just be like politicians - talking past one another (Primary Axiom #3). The same is true of all communication systems, including our peripheral and central nervous systems. These are irreducibly complex, because unless both the transmitter and receiver operate with a common protocol, one being the reverse of the other (encode vs. decode), no effective communication can occur. I find it highly improbable that an encode or decode process could evolve at one end, with the complementary process developing at the other end over time. Whilst it can be argued that such a system would enhance survival prospects, how would the organism be aware of that?

The other point to understand is that in those very early times in the evolutionary pathway, the brain could not know from where the sensations came. The brain might have been bombarded by cosmic radiation for all it knew, as all it was experiencing were unidentified sensations. The questions we must ask are these:

1. What are the distinct properties of the electro-chemical

transmissions from the sensory receptors to the brain that differentiate them as coming from one source as opposed to another?

2.   If the circuitry of the nervous system is such that the source can be identified by the pathway, much as a computer can identify a port, what was the process that gave rise to the concepts of smell, taste, touch, sight, and hearing? Given that the brain could differentiate in some manner between the senses, it could still not identify them as to purpose or function, and in what way they represented the external reality.

3.   Accepting that the embryonic brain could have differentiated between the senses as the source of the sensation, how could it understand the variations in any one sense as representing specific characteristics of its external environment? It might detect that variations existed, but it could not know what those variations represented. For example, the sight sensation could not know that variations represented colour, texture, distance, or size, because whilst these are physical realities, the cognitive concepts to describe them can only be derived by an intelligent agent. Realistically, given that these evolutionary processes were supposedly the result of random mutations, it must have been pure luck that sight was not identified as sound, and sound, taste!

The brain's only sources of data, and thus information and knowledge, are the sensory receptors which the embryonic brain did not comprehend in terms of function, and in what way the sensations represented the organism's external reality, or even its own reality for that matter. Cognitive descriptions of self and the environment are conceptual in nature, whilst sensations are physical. The only bridge between the two, in either direction, is an intelligent agency. Cognitive intelligence, and cognitive processing, are at the conceptual level not the physical. Neuroscientists have been able to identify locations in the brain where cognitive activity "appears" to be occurring, and can associate those regions of the brain with specific activities. But there is a level of discernment which they cannot achieve, and I contend, will never be able to achieve. For example, scientists can recognise from brain activity that a subject is dreaming, but they cannot know what is being dreamt. My brain could be wired up to evidence voluntary cognitive activity, but an observer cannot know what I am thinking.

Before any observer could decode the physical activity, he/she would have to know the code structure, communications protocol, and logic, which themselves would require the observer to perform cognitive functions based on prior knowledge not obtained from the events observed.

Quite obviously, if the path between the physical activity observed in the brain, and the cognitive concepts expressed, involve some form of mental orchestration, we would have to ask how both the orchestra and its conductor arose in the first place, without the intervention of an intelligent agency. In short, how can cognitive processing (intelligence) which is conceptual, arise from the purely physical, when the only bridge between the two is intelligence?

From the perspective of our sense of sight, a useful example is given in the image shown above. Some people will "see" a rabbit, whilst others will not. The pertinent fact is that before anyone can perceive the rabbit, they must have prior experience of what a rabbit looks like, and must have been told that the critter was a rabbit. If, as a child, raised in isolation on a remote farm, you were told that the furry creature with long ears was a giraffe, the image above might look like a giraffe. As we will investigate in a later chapter, this brings us to the question of pattern recognition. When an object looks exactly like one previously encountered, it is conceivable that the reconstruction in the brain, of an external visual image, could match a pattern previously stored, but in the case of our "rabbit" tree stump, fuzzy logic must be employed. In terms of search space, the brain is relatively large, so now we are faced with the conundrum of how the mind-brain complex refines the search space, or does it perform a sequential search throughout the entire space? It could be posited that the brain does not search, because the connections already exist, but then we must ask, how can conceptual connections be made by a biological medium?

A Preview Without Detail

In Part 6, we cover substantial detail on the workings of neurons, patterns, and the nervous system. I understand that for some readers, this may be of little interest to them, especially as it is exceedingly complex. With that in mind, herewith a summary which is easier to read and comprehend, and may be sufficient for some readers to encourage doubt in the capabilities of evolution.

Our senses, and here I will limit the discussion to sight and sound, utilise sensors to experience the external reality. In technical terms, *sensors* are called *transducers*, because they transduce (convert) energy from one form to another. Energy streams contain a semantic (meaning) layer, which is encoded according to the properties of the energy. In sound, these properties are frequency, wavelength, amplitude, duration, etc. In light, these are variations in the wavelength of electromagnetic radiation within the visual spectrum, typically about 380 to 740 nanometres. Our eyes and ears need to firstly comprehend that these variations are significant, that the significance relates to meaning, and to comprehend what that meaning is, so as to be able to decode it. As before, how does the process get started when the terms, *significant* and *meaning* are conceptual, and there was no external entity to explain them? The sensory sub-system must then transduce that energy, re-encoding it without losing meaning, into the electro-chemical properties of the downstream energy network, i.e., the nervous system. Each neuron must synapse with the next, knowing the appropriate neuron to connect with, as not all are appropriate for that communication, and must reliably transmit the signal with no loss of strength or meaning.

At this stage, the nervous system is conveying an encoded, symbolic representation of an external reality, which it interpreted from the characteristics of the received energy source. It must be said that the external energy source was itself, an encoded, symbolic representation of a reality it did not comprehend (Primary Axiom #1). The brain is limited in the volume of symbols that it can process at any one time. Consider the technical aspects of digital video cameras. Light hits an image sensor which transduces that light into a pattern of zeros and ones. At this point, there is no video. This digital representation needs to be processed and interpreted as video before it can be seen. Raw footage is simply the digital data from the image sensor before it has a chance to be processed. There is no image until the digital data is run through a piece of software that can interpret it. Our eyes need to perform similar functions to convey images to the brain. The various formats used in digital imaging vary in size, from compressed JPEG to TIFF. Without going into the technical detail, the nervous system upstream from the visual cortex lacks the capacity to process the volume of signals received by the eye. Enter the visual cortex.

The visual cortex performs the complex task of signal processing. Rather than pass on the full signal stream, it is believed to compare segments and only pass on the difference between one segment and the next, aka, the *delta*. One must ask: what biological process would be capable of replicating such complex intelligence? How could it have evolved, considering that partial "intelligence" would likely represent no survival benefit at all.

In some ways, our auditory system is even more complex, utilising not just one transducer, but four from the inner ear to the auditory nerve. That is four separate evolutionary achievements, none of which on its own could offer any survival advantage. Whilst scientists prefer to assert that some sight is better than no sight, or that some hearing is better than no hearing, such could only be true dependent on quality and faithfulness of representation. Hearing only static on the radio is not better than no radio at all, yet that is a proper analogy for how hearing would have evolved. Colour blindness, blurred vision, or vision only 5% of normal, is not necessarily better than no vision at all, because if you come to trust what you think you are seeing, you can easily stand into danger. This would have been the case for early evolution of sight, which again, would have offered no survival advantage at all, and in all likelihood, precisely the opposite.

In the development of the technology used in modern telemetry systems, a great deal of trial and error, and subsequent refinement was required. The important point to note is checkpointing: the ability to recognise where progress toward a goal has been achieved, and where backward steps occurred. In a sense, we can say that the scientists kept a diary of progress which avoided the need to go back to the beginning, time and again, every time a false step occurred. Evolution has no such diary. In the millions of intermediate steps, the organism has no way of knowing whether any change is headed for the better, or worse. If chance is the imperative, Murphy would tell us that there would be more detrimental steps than beneficial. Actually, biologists agree with Murphy: deleterious mutations are far more common than beneficial. Given the unpredictable circumstances of early death, or failure to mate and produce offspring, it is drawing a long bow indeed that purely by chance, a lineage evolved which just happened to build on just the right mutations for development in the right direction for useful sensory systems to evolve.

One more point: before any external sensory systems could evolve; before there could be any communication between internal organs; before an organism could move a leg, body, or other appendage and know that it had moved it, and to where it had moved it; before an ant could move its antenna; before any phenotypic evolution could occur at all – it had to be preceded by the development of the nervous system to provide the essential communications channels. When you understand the complexity of neurons, neuronal connections, the electro-chemical encoding of the signals, and the difficulties of developing reliable communications networks as discussed in earlier chapters, the case for evolution becomes exceedingly difficult to support, let alone prove.

Enough on the senses, for now, as we turn our attention to the fundamental nature of information, as opposed to data and the symbolic representations in physical storage media.

Chapter 3-6: The Information Blind Spot

"The people who would like to manipulate and use you won't tell you your blind spots. They may plan to continue using them to their advantage."

~ Assegid Habtewold, The 9 Cardinal Building Blocks for Successful Leadership ~

Science is focused on the *technology* of information as demonstrated by the work of Claude Shannon in particular, but that field is entirely separate from the study of cognitive information, and thus represents a significant blind spot in the science. As noted earlier, cognitive information is entirely different to genomic information in that it deals with concepts, whilst the latter deals only with physical entities. The brain, being organic like the genome, can only natively store physical entities, not concepts or abstractions. Concepts are derived from data in the form of knowledge, and as is posited in the science of epistemology, all knowledge is based upon prior knowledge (Primary Axiom #2).

In an early draft of this paper, I coined the term *Conservation of Information,* but then encountered it in another context and thus had to abandon it. However, some aspects are common and I would to make a brief note of them here.

William Dembski commented: "Conservation of information is a term with a short history. Biologist Peter Medawar used it in the 1980s to refer to mathematical and computational systems that are limited to producing logical consequences from a given set of axioms or starting points, and thus can create no novel information (everything in the consequences is already implicit in the starting points)."[1] Although Dembski and Medawar use the term, information, differently to the way I have defined it in this study, I nevertheless believe their point to be true of cognitive information: everything that we derive conceptually was already implicit in our prior knowledge base. Even though we may appear to do so, because we have not previously arranged or correlated existing data in a particular way, we cannot create any entirely novel information or knowledge ourselves

– it is just new to us (and maybe to everyone else). Let me restate that. We can only devise new concepts or information by correlating existing conceptual data in new ways.

In passing, I contend that the same is essentially true of biology – every physical aspect of the human body is not novel, in that it is all implicit in the starting point of genetics. Whilst it appears novel in form, it is not novel in substance – the cell, and that in chemicals.

From another perspective, if "everything in the consequences is already implicit in the starting points", then we can reverse-engineer our perception of consequences to discover whether the starting points can be capable of producing those consequences. For example, we perceive consequences of the mind, but can the biological starting points of the brain be truly said to be capable of them? Reverse-engineering an organ to its constituent cells can demonstrate why a particular organ is capable of its consequent activities, but I have yet to encounter any plausible scientific explanation in the context of the mind-brain conundrum.

As I have commented before, a philosopher of the Middle Ages, Maimonides, had a sense of this when he wrote: "Everything that passes over from a state of potentiality to that of actuality, is caused to do so by some external agent; because if that agent existed in the thing itself, and no obstacle prevented the transition, the thing would never be in a state of potentiality, but always in that state of actuality." Whilst the *data* stored in the mind-brain complex is always in a state of *actuality* because it exists, albeit represented symbolically, the *potentiality* of *information* exists in that this data can be correlated in numerous ways, which is why people can derive different conclusions from the same data. *Information* is transient, but in the state of *actuality* for only so long as it represents that particular correlation, because it represents just one of numerous possible outcomes. In that sense, cognitive information is unlike mathematical results, because given a data set and rules of mathematics, there should be only one possible outcome. If not, we cannot trust mathematics. Thus, as Maimonides described it, "if that agent existed in the thing itself, and no obstacle prevented the transition, the thing would never be in a state of potentiality, but always in that state of actuality." However, what do we do with logic, which whilst it has rules defined by intelligent agencies, is not always present or exercised in all minds? Mathematics, as defined, is objective, but logic is subjective, because there is no objective restriction on the parameters which are evaluated. This is one of the problems that I have with the *computational theory* of the mind, because at the elementary level of the analogy of *ones and zeroes*, it is logic, not mathematics, which is being performed. Even mathematics is an expression of logic, but with a restricted set of parameters.

David Gelernter, in support of this computational theory, offers: "The active, thinking part of the brain is built out of neurons, and they can be described as binary switches too. A neuron is either off or on. On, it transmits a nerve signal to all the neighbouring, downstream neurons. Off, it transmits nothing. Neurons turn on when the right signals reach them from their neighbours upstream, some ons and some offs."[2]. This analogy works up to a point, but falls far short, starting as it does in the middle of the process, not at the beginning, and misrepresenting the electrical activity of a sequence of neurons. Such is a common failure in attempts to explain the workings of the brain. When describing any thought process, the first question to be asked is: What initiated the thought? If scientists are to compare the brain to a computer, they must do so thoroughly, including the process that initiates computation. The first step in computer processing is to load the application program that knows what resources are required to complete a predefined task. When launched, the application program follows a sequence of predefined logic steps. When comparing the brain's thought processes to a computer program, it is irresponsible to ignore the implications of the initiation process.

In computing, a series of logical switches are permanently set by the controlling software for the duration of the processing – if they were not, the program would fail after the first iteration. Other switches are dynamic, but these are as a result of the processing, not the cause. Neurons form a chain to transmit electrical pulses. They are inactive (off) whilst waiting, and active (on) whilst transmitting the signal to the next neuron, whereupon they go inactive again. This process occurs in both the central and peripheral nervous system, remembering that the brain is the terminal of the CNS comprising several structures which handle specific functions, all containing neural networks which operate in the same way. For a neural network to behave like computer software, some neurons at the beginning of specific pathway must act as a "gate", permanently on or off in a logical sense. If all switches are dynamic, no logical process is possible. It must be the case that neurons do differ in their chemical composition or cellular structure to be able to respond in different ways at their origin and terminus, but questions to be asked later are:

what is the difference between one neuron and another, and how do those differences arise?

I can find nothing in the biology of the brain which explains why in any activity, one or more neural pathways are chosen over others. It is evident that such is happening, very true, but what is the controlling mechanism which directs the signals along specific pathways or networks? Alternatively, what was the controlling mechanism which, based on physical inputs, being nothing more than symbolic representations of data, managed to establish physical connections that modelled logical connections? It is inadequate to explain that such activity occurs the way it does because that is the function of those pathways, or that specific structures were designed that way. Evolution is not a designer. It is one thing to use computing analogies to describe how the mind-brain complex currently works, but it is illogical to rest on that analogy alone without taking it back to a beginning – how was it so? A complete computing analogy would have us describing the network architect, and the software programmers, but scientists do not want to go there. As one philosopher phrased it, the brain works as an orchestra, but who or what is the conductor, and who wrote the music?

Now, if everything that we derive conceptually was already implicit in our prior knowledge base of data, and knowledge always expresses concepts, how did the first conceptual element come into being? Symbol representations of data just "are", but the conceptual is what "may be". Imagination cannot be intrinsic to biological entities. In the context of evolution, and in passing, artificial intelligence in robots, where was this conceptual background before there were organisms capable of being aware of it? As explained previously, even what we now accept as cognitive *data* is always derived through cognitive *information* processing of that and other conceptual data. We continue to be confounded by this "chicken and egg" situation: how could a physical experience be conceptualised, even before the existence of a cognitive processing structure containing the necessary conceptual background against which context correlations could be made?

Again, quoting from Dembski's argument, the context being digital data processing:

"Something unavoidably subjective and teleological seems involved in search. Search always involves a goal or objective, as well as criteria of success and failure (as judged by what or whom?) depending on whether and to what degree the objective has been met. Where does that objective, typically known as a target, come from other than from the minds of human inquirers? Are we, as pattern-seeking and pattern-inventing animals, simply imposing these targets/patterns on nature even though they have no independent, objective status?

Mathematically speaking, search always occurs against a backdrop of possibilities (the *search space*), with the search being for a subset within this backdrop of possibilities (known as the *target*). Success and failure of search are then characterized in terms of a probability distribution over this backdrop of possibilities, the probability of success increasing to the degree that the probability of locating the target increases."

These observations can be applied to cognitive processes as well. People speak of "searching their memories", especially when asked, "do you recall": thus, the analogy seems apt. But what does it mean in terms of cognitive processes: how do I search my memory accepting for the moment that there is such an entity as "I"? Is that a volitional process or autonomous? When witnesses respond: *I can't remember*, *I do not recall*, or *I have no memory of that*, is their response volitional or autonomous? I partially agree with Ryle that a fundamental level, we have no control over whether we do remember or not, but disagree in that we can attempt to recall, and that there are techniques for improving memory recall. I would offer that where such techniques have been learned, deciding whether or not to invoke those techniques is voluntary. How then do we discern truth from falsehood (apart from the usual jokes about politicians and lawyers)? The article above states: "Search always involves a goal or objective, as well as criteria of success and failure" - by what criteria do we judge the success or failure of our mental search, particularly given that we sometimes recall incorrectly? Is the wrong answer solely a function of a fault or corruption in our neural networks, and if so, at what point in the cycle did the fault occur: during the neuronal network correlation process, whilst in storage, or during retrieval? Before we can begin to answer those questions, we need to have an understanding of the information processing cycle at a functional level.

The concept of *search space* has particular application to the cognitive processes of recognition and even creativity. A composer of music "searches" within a space of possibilities contingent upon culture and preference. Eastern music is often characterised by monophonic melodies, the melody being composed with reference to one note, most usually the tonic (first note) of the scale chosen. Western music tends toward polyphonic harmonies, where the melody is composed in reference to a chord or chord progression. Thus, the *search space* of the composer is dependent upon the type of melody being composed, but let us not forget that the data in the search space must first be acquired.

Back to our rabbit in the previous chapter, what causes the mind-brain complex to search for a matching pattern? Well, "it just does", you might answer, but what causes it to do so? Why would a collection of biological stuff do other than just store sensory inputs, without bothering to encapsulate meaning? More to the point, how could it encapsulate meaning other than at a symbolic representational level. Why would one of two people, both of whom knew what a rabbit looks like, see a rabbit, and the other not? Science insists on the concept of *cause and effect*, but here we are talking about a *closed-circuit system* which appear to have a purpose. We know that systems do exist in nature, for example, the hydrological cycle, which describes how water evaporates from the surface of the earth, rises into the atmosphere, cools and condenses into rain or snow in clouds, and falls again to the surface as precipitation. This entire cycle is predicated on the intrinsic properties of water, but is not *autonomous*. The first cause is heating from the sun, and subsequent effects depend on atmospheric conditions. On the other hand, the mind-brain cycle is said to be autonomous, some behaviour not influenced by external conditions, yet the brain is composed of the very same fundamental stuff as water, although arranged differently. Can specific arrangements of matter give rise to behaviours not intrinsic to matter? We come back to that also.

Reading requires a search space firstly delineated by a language, which limits our search for vocabulary and grammar. Many words in most languages have a *semantic range*, i.e., a range of meanings, requiring context of the text to be evaluated in order to select the right meaning. The larger our vocabulary, the larger our search space, and the greater our chance of success in understanding the text. However, vocabulary is also culturally conditioned: a text written in the 19th century may not be comprehensible to a modern reader, and most certainly, text language from Twitter or other social media would not be comprehensible to a 19th century reader (nor me). As the search space grows, the organisation of the data becomes more critical lest the search rapidly approaches the "needle in a haystack" conundrum. Actually, it is worse – first you must know what a needle is. This issue is especially important for those people gifted as multilingual (not me, despite my efforts). The search space for such people is much larger, being additionally language dependent.

Just as important as the *search space* is the *search algorithm*: how does the searcher know how to search the search space? In digital information processing, the storage process is intelligently devised, as is the search algorithm, without which any search is purely random with no opportunity to verify that what was found was the right answer. Now one may argue that no "search" as such is required, because the neuronal network itself is the guiding mechanism, but the theories of search apply not just to the retrieval of data, but also to the storage. Before any new data is stored, the storage space must be searched to find the correct storage location. In the context of the brain, that means finding the appropriate neural network with which to connect. But now, here is another conundrum which we must explore: validation of results. Evolution could not have proceeded if evaluation of sensory inputs was not correct.

I might have said this before, but in all my research, studying the works of such luminaries as William Dembski[3], Werner Gitt[4-5], and Walter ReMine[6], I have not encountered what I believe to be the proper treatment of the nature of information. Too often, the concepts of symbols, data, and information are conflated into the one term, *information*, thus bypassing some much-needed analysis. We shall now review some of the thoughts of these gentlemen.

In the Beginning was Information

In his book, "*In the Beginning was Information*"[4], Werner Gitt begins his Preface:

"*Theme of the book*: The topic of this book is the concept of information, which is a fundamental entity on equal footing with matter and energy. Many questions have to be considered: What is information? How does information arise? What is the function of information? How is it encoded? How is it transmitted? What is the source of information found in living organisms?" (p. 11)

Upon reading this, my immediate response was, "Whoa! *Information is a fundamental entity on equal footing with matter and energy*"?

Would that not make information a physical, material entity? To ask how information is encoded suggests that it can be stored as a separate, identifiable entity, a concept with which I disagree. So here we must be specific about how *information* is defined. The author continues, "*Many scientists therefore justly regard information as the third fundamental entity alongside matter and energy*" (p. 49); I would respectfully replace the word "justly" with "falsely". In describing his book, Gitt states, "The purpose is to find laws of nature which hold for the fundamental entity known as information". Hmmm, *laws of nature*, I do wonder whether we have the same understanding of them? Werner Gitt advises, "The title [of the book] refers to the first verse of the Gospel written by John: 'In the beginning was the Word'", evidencing that we are likely discussing the topic of information from different perspectives, which is unfortunate, because I prefer to approach the topic based on science and logic alone, absent of any religious or spiritual connotation. In a study such as this, a presumption of the spiritual is no more logically valid than the presumption of evolution. For those not conversant with Christianity, the concept of "The Word" is complex, and not accepted outside of Christianity, and is even debated by Christian scholars, but I digress. We may nevertheless conclude that the immaterial *is* necessary to explain information as processed in the mind-brain complex, but we should not start there. Nevertheless, we may find some useful observations. Finally, from the Preface,

"The purpose of this book is to formulate the concept of information as widely and as deeply as necessary. The reader will eventually be able to answer general questions about the origin of life as far as scientifically possible. If we can successfully formulate natural laws for information, then we will have found a new key for evaluating evolutionary ideas. In addition, it will become possible to develop an alternative model which refutes the doctrine of evolution." (p. 12)

I respectfully disagree with Werner Gitt, and others who have taken the same approach, because I disagree with their use of the term, *information*. In the physical world, there are entities with physical properties. In identifying these entities and their properties, we assign labels which we call *data*. The entities exist, but cognitive data, being conceptual, does not exist until an intelligent agency identifies and labels the entities. Protons, electrons, and neutrons have existed long before we knew of them, and we only have data on these entities since we have come to understand them; in short, *data* is a concept which has no existence outside an intelligent agency able to comprehend it. The properties of entities existed long before we were able to measure them, and assign labels (names) to them. It is only then that we have data. I will continue to stress that entities exist, and generally always have, but data only exists when we have come to recognise, understand, and label the entities. Data on atomic energy did not exist two thousand years ago, but the energy certainly did.

Werner Gitt notes that Claude Shannon was the first who tried to define information mathematically, and adds support to my views when he states, *"The main disadvantage of Shannon's definition of information is that the actual contents and impact of messages were not investigated"* (p. 49). Quite simply, two statements, one true and the other false, could have the same mathematical equivalence. Even though initially stating that, "information ... is a fundamental entity on equal footing with matter and energy", Gitt then reveals the erroneous "assumption that information is a material phenomenon", by continuing "Information is neither a physical not a chemical principle like energy and matter, even though the latter are required as carriers" (p. 51). Gitt's error here is that physical carriers can only convey physical representations of data, not correlated concepts as are required for information. In his book, the author quotes numerous other scientists who agree that information is not physical; e.g., "information is of a mental nature, both because of its contents and because of the encoding process. This aspect is, however, frequently underrated"; and "Information is information, neither matter nor energy. Any materialism which disregards this, will not survive one day". Sadly, it survives even to this day, as neuroscientists still insist that the mind, which is the processor of information, is material.

Let me offer that the concept of *universal information* could be valid in the sense that if data is our understanding of entities, then every possible combination or correlation of data would represent universal information. In that sense, universal information is always in state of potentiality until instantiated and comprehended in an intelligent agency. Note that not every combination or correlation would be logical, and thus would not lead to "true" knowledge, as is our experience in science and everyday living. I may well have misunderstood Werner Gitt, and for that I do apologise, so let us dig a little deeper into what he proposed.

Information and Our Senses

I agree with his presentation of the five levels that are involved in the communication of "information", and have found conceptual corroboration in the works of psychologists and others working in related fields. Briefly stated, these levels from bottom to top are:

1. Statistics – comparing received signal to transmitted signal.

2. Syntax – a common code required between sender and receiver.

3. Semantics – a common understanding of the idea / meaning of the communication.

4. Pragmatics – an intended action of the sender being implemented by the receiver; and

5. Apobetics – the purpose behind the action being achieved.

Shannon's theory of information deals with the statistical aspects of communications, and has been very helpful in verifying the integrity of digital transmissions. Communications channels can be subject to noise, which would add to the statistics; this allows the value of insulation to be computed (remember the myelin insulation in nerve fibres?) However, at the level of *Statistics*, one cannot find meaning, nor the truthfulness of what is being communicated. The next level, *Syntax*, relates to the coding system. In communications using language, aspects of spelling and grammar come into consideration, but all coding systems of whatever construction have syntax as determined by the protocol, or "code book". At this level, we are still dealing with verification of the communications integrity. In my proposition of Primary Axiom #6, I explained the technical difference between verification and validation. The semantics level deals with the receiver "getting the message" so to speak. Werner Gitt expounds on this by listing some typical semantic questions[7], firstly concerning the sender:

- What are the thoughts in the sender's mind?

- What meaning is contained in the information being formulated?

- What information is implied in addition to the explicit

information?

- What means are employed for conveying the information (metaphors, idioms, or parables)?

Concerning the recipient:

- Does the recipient understand the information?
- What background information is required for understanding the transmitted information?
- Is the message true or false?
- Is the message meaningful?

The examples are in the context of intelligent communications between humans, and hopefully, you will be able to see how these relate to the primary axioms I have proposed. However, in a purely biological exchange as occurs within the human body, Gitt's levels take on a different sense. Firstly, though, note that some external communications entail no Semantics, Pragmatics, or Apobetics: such can be the case with the sound waves, shadows, or light reflecting off a tree. Messaging is involved, but with no intentionality on the part of the sender. The content of the message needs to be interpreted subsequent to reception by the receiver, such as our eyes or ears, but this is difficult if not impossible to be explained in evolutionary theory. Putting that aside for now, let us attempt to relate *Semantics* to internal communications between our sensory organs and the brain, again in the context of evolution. Ignoring any intentionality that may have caused the sensation, the message is nothing but an electro-chemical signal transmitted by neurons. With the biological brain as the recipient, most especially at an elementary stage of evolution, how could it answer any of the four questions above?

Up to the semantic level, there is no consideration of the *intentionality* of the message. This where I have difficulty with the general applicability of Werner Gitt's theory, insofar as many "communications" received by our senses have no intentionality whatsoever. They are nothing but natural phenomena such as light, heat, odours, wind on our skin, and even gravity. Nonetheless, intentionality does begin with the sensor itself, in that it is attempting to communicate something to the brain, just as our internal organs do in autonomous functioning, but what? This is the issue for the *Pragmatics* layer, understanding the intentionality of the sender.

Our five senses, actually six (balance), communicate messages with no cognitive intentionality, only chemical. These elicit both an autonomous response (from the brain), and a cognitive response (from the mind), but the sensor itself knows of neither of those things: it is chemistry which elicits any response from the brain. The response is a function of the *Apobetics* layer. When a heavy object lands on my foot, my sense of touch tells my brain to tell the muscles in my leg to get the foot out from under – an entirely autonomous response. This may be accompanied by my brain telling the voice box to articulate a response, coarse or otherwise. However, my mind detects the message receipt in the brain, and can override the autonomous response. This is particularly true in my own experience when instead of voicing a profanity as I feel compelled to do, I attempt a more civil utterance.

Of course, my success rate is less than I desire, but my point is that the mind can exercise a *supervisory function* over the autonomous responses of the brain. Even more, we can train ourselves to resist autonomous responses, aka our natural instincts. This is a conundrum that scientists who reject Cartesian Dualism cannot explain: how the brain could override its autonomous functions.

I referred to a sixth sense, and let me briefly explain: balance. Fundamentally, this is not an independent sense as with the other five, because the mechanisms are physical within the hearing complex. I mentioned this only because you might have encountered it in other literature, but I would add that as with our other senses, it can easily be fooled into a false perception of reality. This understanding is of particular importance in aviation.

Once more on Semantics, Pragmatics, and Apobetics, which relate to a common understanding between sender and receiver, the intentionality of the sender, and the response by the receiver. A nerve fibre consists of a chain of neurons which transmit an electrical pulse. At the sending end is a sensory neuron which "understands" what is perceived in a chemical sense, but there is no transmission of this chemical understanding. At the terminus of the nerve fibre in the cortex, body mapping enables the cortex to determine the source of the electrical pulse, and understand the function or purpose of the signal. Some process then acts upon the chemical release of that final neuron. We can see the issue of *Semantics* in this process, as the sensory and final neurons must have a common "understanding" already built in, for only an electrical pulse (action potential) is relayed along the nerve fibre. Likewise, *Pragmatics,* for the intended action is a function of the physiology of the system, and the complementarity of the chemical structure of sending and final neurons. Finally, *Apobetics* – the purpose behind the transmitted nerve pulse being achieved. Depending whether the sensory neuron served internal bodily processes, or reacted to external stimuli, the purpose for the signal may be one or both of biological and cognitive.

References:

1. Dembski, William A., *Conservation of Information Made Simple*, 28 Aug 2012, http://www.evolutionnews.org/2012/08/conservation_of063671.html

2. Gelernter, David, *The Tides of Mind: Uncovering the Spectrum of Consciousness*, Liveright Publishing Corporation, New York, NY, 2016, p. xix

3. Dembski, William, *Being As Communion - A Metaphysics of Information*, Ashgate Publishing Company, Burlington, VT, 2014

4. Gitt, Dr. Werner, *In the Beginning was Information*, First Master Books, Green Forest, AR, 2007

5. Gitt, Dr. Werner, *Without Excuse*, Creation Book Publishers, Atlanta, GA, 2011

6. ReMine, Walter James, *The Biotic Message: Evolution Versus Message Theory*, St. Paul Science, Inc., St. Paul, MN, 1993

7. Gitt, *In the Beginning was Information, Ibid*, p. 71

Part 4: Principles of Information Processing

"All sciences deal in unity. They unite phenomena in a principle."

~ William T. Harris, Psychologic Foundations of Education, 1907~

Earlier, I asserted that both data and information are *conceptual*, not physical. On that basis, computers neither store nor process data nor information, but only encoded symbolic representations of conceptual data. Computers communicate with their peripheral devices, and with other computers, at the symbol layer, based on common protocols. Being physical devices, they cannot comprehend in a cognitive sense, the entities with which they are dealing. These entities correspond with conceptual data; their relationships correspond with conceptual information. A challenging task for data base designers is how to structure the data in an optimal form. The challenge relates to storage efficiency, update speed, and retrieval speed. In earlier times, with slower processing speeds and much smaller yet more expensive storage space, designers had to carefully analyse the critical requirements of the system being developed. High volume, multi-user order taking systems require fast update response, whilst the retrieval for reporting was generally less time critical. Even then, some reporting was time critical, whilst other requirements such as end-of-month were less so. As computer hardware became faster and less expensive, less attention has been paid to some of these design criteria, with this inattention being later evidenced by slowing response times as processing demands and volumes increased. At one stage in my career, I was faced with the task of implementing in a high-volume environment, software structured for a low volume environment, and had to answer for the hasty promises of the salespeople. Not much fun at all.

My point here is that these processing requirements are to some degree, common, whether related to a computer, or to the human brain. The primary difference is in the architecture of the processor and storage devices. As we will discuss in a later chapter, scientists are attempting to develop faster computers by mimicking the neural network of the brain, or at least, how they believe it to be constructed. I am not confident of their understanding. There are some valuable lessons to be learned.

As we begin to dig ever deeper into information processing, keep in mind that some of the principles as applied to computers, apply equally to other processors such as the human brain. The architecture and physicality are vastly different, but if we are to accept the *computational theory of mind*, then we must accept the common principles as well. On that score, we should ponder the issue of "principles" – a conceptual term that even scientists cannot avoid using. Biology is researched based on accepted scientific principles, but how can a biological brain formulate principles of its own existence (Primary Axiom #1)?

Chapter 4-1: Coding Systems

"First, solve the problem. Then, write the code."

~ John Johnson ~

In his very informative book[1], Dr. Werner Gitt discusses the necessary conditions that identify codes:

1. A set of abstract symbols is needed. ('abstract' here means that the symbols have *no inherent physical relationship or resemblance* to the reality that they are being used to represent).

2. The sequence of the individual symbols must, generally speaking, be irregular / aperiodic.

3. The symbols appear in clearly distinguishable structures such as in rows, columns, blocks, spirals, or perhaps in some other complex (yet recognisable) distribution.

4. At least some symbols must occur repeatedly; and

5. It can be decoded successfully and meaningfully.

A "code" is defined as a system of words, letters, figures, or symbols used to represent others, especially for the purposes of secrecy. Whilst accepting this definition, for the purposes of this study, I would like to generalise it to: *a system of words, letters, figures, or symbols*, irrespective of purpose or intent. This would bring any language into the definition of a *code*, which is what I intend, so as to be able to demonstrate some fundamentals of communication. In the dictionary definition, the phrase "used to represent others" means to represent other words, letters, figures, or symbols, but here I want to redefine the purpose more specifically to mean: arbitrary *physical* symbols used to represent the *conceptual*.

I shall now commit to an error of which I accuse others, by using the term "information" in its common form, simply for ease of explanation of other concepts. I trust that you understand.

Though perhaps not immediately obvious, all communication is based on coding systems as I have defined them above: one set of symbols used to represent another set which can be more easily comprehended conceptually. We tend to think of codes in terms of brevity or secrecy, but in a more general sense, a code is simply a regulated arrangement of data symbols. Although we hear of terms like *scrambling*, the process must be regulated rather than random otherwise *unscrambling* cannot occur and no communication would result. This brings us to an important point: both the sender and receiver must have a common understanding of the coding system being used (Primary Axiom #5). This is the essence of secret codes or encryption: the purpose is to prevent information being passed to unauthorised users. The natural corollary is that if you wish to convey information, you have to ensure a common understanding between both parties (Primary Axiom #3).

Again, this must seem entirely logical and you might wonder why I am bothering you with such trivia. A little later, we shall see why this is so important in the context of evolution theory.

Voice Codes

Verbal communication is based on sounds: somehow, we convert our thoughts, albeit sometimes incoherently, to muscle movements which disturb the air in a way that the ears of others can understand. We can say that the data is *encoded* using pitch, frequency, and amplitude applied to the surrounding medium which then vibrates against an eardrum which passes the encoded message to the brain. Encode means to transform a stream of signals into a predetermined and regulated pattern: if the pattern was not also agreed and regulated, the recipient would not be able to reliably decode it. This principal applying to speech was demonstrated by the invention of the microphone and loudspeaker: sound waves arriving at the microphone caused a diaphragm to vibrate; the vibration was converted to electrical impulses which when received at the loudspeaker, caused another diaphragm to vibrate reproducing the original sound. This is all very straightforward but some points are noteworthy.

Firstly, this was a great advance in technology requiring detailed design, advances in materials, and extensive experimentation before the first prototype worked usefully. This area is still of great interest to audiophiles who expend much time and money on obtaining the purest reproduction possible. Note that a microphone without a speaker has no useful function, and in this scenario, the speaker without a microphone (or functional equivalent) is also redundant. Now apply this principle to the human faculties of speech and hearing: this should give us pause for thought. Remember that in our technology example, the message in the sound waves was transformed into encoded electrical signals and then decoded back into sound: both the transmission and receiving devices needed to process the codes precisely but in reverse manner to work effectively. Our human faculties are similar in that nerve impulses from the brain are converted via muscle movements (diaphragm) into sound (speech) and at the receiving end, reconverted via the ear drum (diaphragm) and tiny bones into equivalent nerve impulses which the brain understand as hearing. Both processes must employ complementary encode/decode methods. In technology applications, this is achieved using intelligent design, yet evolution teaches that this occurred by random mutation fixed by natural selection. Later, we will consider evolutionists' contentions regarding how long it took for these processes to evolve.

So far, we have briefly discussed information transfer at the technical level, but more importantly, the meaning of the message sent (semantic level) had to be understood by the recipient: in biological terms, the sender's brain had to be synchronised with the receiver's brain in relation to the content of the message, how it was vocalised, and how it was heard. Likewise, the recipient must vocalise in the same way to be able to respond. We will come back to that also, but first let us consider an issue often skipped by evolutionists: recognition and reproduction. It would be reasonable to assert that the earliest attempts at vocalising thoughts had limited success. Consider:

"There are many parts of the body that help us produce speech. To speak, you use your stomach muscles, lungs, voice box, tongue, teeth, lips, and even your nose. Your brain coordinates it all. Speech actually starts in the stomach with the diaphragm. This is a large muscle that helps push air from the lungs into the voice box. The voice box or larynx has vocal cords that vibrate to produce your voice. Then, the lips, tongue, and teeth form the sounds to make speech."[2]

If evolution is true and Homo *sapiens* is just the latest of Homo species, of which the earliest evolved from a form of chimpanzee, then vocalisation of thoughts has been around for a considerable period of time. However, with each new, original vocalisation, there needed to have been a method for that to be understood by listeners, and then the listeners learning how to be speakers. If you listen attentively to people of different cultures, you will appreciate that it is difficult to speak a new language "like the locals", because their pronunciations are difficult to master. Such pronunciations are encoded in the sound waves of speech, and must be learned. People learn from birth how to manipulate their vocal cords in just the right way, although I would not discount some element of genetic transfer.

We investigate language in a later chapter.

Multi-Level Coding Systems

In this modern age of electronic information transfer, secrecy and privacy are high on the agenda of scientists developing communication protocols and systems, but the end result of these secure encryption methods is that the message can be decoded into "plain language" for authorised recipients. Coding systems can be implemented at multiple levels, so we will use the term "plain language" to refer to that level of data transmission that employs the protocol or language that the recipient understands.

The next point is that both data and processing instructions need to be managed by the same underlying encode / decode method, and the system needs to recognise the difference between the two forms of "data": that to be understood and that to do the understanding.

For example, at the hardware level, computers work using magnetic regions, interpreted as switches set to on or off. This is represented in binary code as per this example: 1100 0001 1000 0110 (there was a time when I could read and program in binary, but those days are long gone). When printed, the visual representation of memory was not in binary, but in another code. Some computer manufacturers used a system called *octal*, but I will refer to a method called *hexadecimal* which uses a 16-character set: 0 through 9 then A through F. The sixteen-digit binary sequence above equates to "C186" in hex, again not very helpful. There is a further level of conversion: either ASCII or EBCDIC (we need not concern ourselves with what these mean) but using EBCDIC, my example translates to the letters "Af" (purists might find errors in my memory of how this works).

For anyone to understand the decode sequence, they would need a series of tables. In one column would be the raw data, e.g., the binary code, and in the next column would be the hexadecimal equivalent. We call such tables *metadata*, i.e., data about data. All data transformations from one symbol set to another symbol set require metadata to regulate the process. Just as importantly, metadata must be bi-directional and shared by both sender and receiver.

Now, I cannot compare this sequence with, for example, how the brain decodes impulses received via the optic nerve and establishes logical correlations with other stored data and then perhaps acts in response. If I see a bottle thrown at my head I will instinctively duck, but how that response is achieved is unknown to me, and I believe, everybody else. The one thing I would assert is that some form of data processing is occurring using biological coding systems which must have fundamentally the same functional requirements as described above. The optic nerve data transmissions must result in what is "plain language" to the brain, and must be processed using concepts and context that the brain has already stored.

Signalling via the nervous system is not understood, other than that the anatomy and physiology of the nervous system has been thoroughly researched. It is known that neurons communicate with one another via action potentials resulting in a chemical transfer through the synapse, but the "language" of the signalling remains a mystery. Later we will investigate the *computational theory of mind*, which is so broad in its application that it is likely true in some respects, and false in others.

The Code Book Enigma

A coded message is simply a string of abstract symbols absent of *context*, and absent of *concept* other than the symbols themselves. When archaeologists first encountered Egyptian hieroglyphs, the concept behind these symbols was unknown, and beyond translation even though to an extent, the context was known. The discovery of the Rosetta Stone enabled translation, because fortunately, inscribed on its surface was the same story in two additional languages: Demotic Script and Ancient Greek. How they knew that it *was* the same story is not explained in my referenced texts.

To me and perhaps you also, not one of the three examples is intelligible. Of course, if unknown to the scholars, each language spoke of a separate story, the result would have been very like opposing politicians criticising each other's policies. In effect though, the Rosetta Stone acted like a code book, or meta-data, in that it allowed those who understood one set of symbols to decode another set. We could also say that the author of the stone coded (or decoded) the hieroglyphs into Demotic Script and Ancient Greek, and vice versa.

Symbolically, DNA uses an alphabet of just four letters: A, C, G, and T. The code is used in codons (groups) of three letters. A point mutation will result in a re-arrangement of the letters, either by replacement, insertion, or deletion. When, for example (not real), a mutation renders AAA to be AAB, the protein production system has to know what that means. Imagine you were taught that AAA codes for the amino acid, lysine, and in an examination, you were asked what sequence codes for lysine. You answered AAA, but were told you were only half right, because AAG also codes for lysine. You would naturally object as you hadn't been given that information. That is the essence of meta-information, and the problem for evolution. When a mutation causes a re-arrangement of the letters in a codon, how does the biological system make sense of it? Biologically, new words, sentences, and entire books can be written with this alphabet, but the base level of data, the letters, are the same, and for evolution to be proven, the mechanism for creating letters (actually nucleotide bases) has to be found, and the mechanism for interpreting changes has to be explained in evolutionary terms. When an insertion or deletion occurs, the DNA sequence is said to be "frame shifted". For example, inserting a T into the sequence of AAA CCC GGG may result in ATA ACC CGG G ... etc. Deleting an A would result in AAC CCG GG ... etc. Clearly, the result is likely unintelligible as the codons have been changed.

This point is the most important, and the hardest to explain: the meta-information level. Letters are just symbols and meaningless until arranged in particular groups or sequences. There is a foundational axiom that underpins all science and intellectual disciplines—nothing can explain itself (Primary Axiom #1). The explanation of any phenomenon will always lie outside itself, and this applies equally to the DNA coding system: it cannot explain itself.

DNA coding can only be interpreted by an external mechanism. In this case, the meta-information level which I will continue to call: the code book. Any language, particularly one limited to just four letters, requires a code book to decipher the meaning. Every time there is a new rearrangement of the letters, or new letters are added to (or deleted from) an existing string, the code book has to be updated. The obvious questions are:

a. What is the biological code book mechanism?

b. How did it evolve?

c. How is it kept updated as new rearrangements of the letters occur through mutation?

Modern scientific research has provided some of these answers, with the identification of non-coding sequences involved in the transcription of the DNA protein coding sequences, e.g., promoters, enhancers, silencers, and insulators. The problems for the evolution solution should be obvious: the genome "learning" to develop the transcription processes of itself.

Can Coding Systems Evolve?

From a sequence perspective, for a change to be useful, the code explanation or definition must precede, not follow, any new arrangement of letters. This suggests to me that where genetic rearrangements occur, they have to be pre-programmed to be beneficial. Rearrangements that occur independently of the code book cannot be understood by downstream processes. Logically, the code book is the controlling mechanism, not the random rearrangements of DNA protein coding sequences. First the code sequence, then its implementation. In other words, it is a top-down process, not the bottom-up process that Richard Dawkins asserts for evolution. If it were otherwise, another supervisory process would need to be in place to undertake a clean-up of non-functional sequences.

The human mind is both inventive and devious when it comes to codes, so here we must state some initial rules that can be derived from experience:

1. The coding system must always precede the usage of the code.

2. The sender and receiver must have a common understanding of the code.

3. Without the encode rules of the sender, the receiver cannot decode a message without the application of intelligence (and even

then).

In the evolution context, neither a sender nor receiver has any function until both are able to communicate (irreducibility of systems, Primary Axion #8).

Figure: 1 (A) (C) (A) (G) (G) (A) (C) (A)

In Figure 1 above, we have three symbols used in groups (codons) of two: AC, AG, GA, and CA. The first point to note is that if we start to decode or transcript from other than the beginning, we would get different groups: e.g., CA, GG, AC, etc. So, in any string of symbols, we need to know the starting point and whether any groups should be skipped. Ingenious coders in the past would send a message such as this:

XAST BDBD MOTE PPAR CATA LCTA MNOP QRST ACEG

Those seeking to decode the message would assume that each and every group of four letters represented some logical concept or other encoded set of symbols such as a code word, but would have been unaware that, for example, if the message was sent on a date with an odd number in the first week of the month, the groups started with the first letter in the first group; in the second week, the third letter, and so on. If the date was an even number, you started with the second letter in the second group, etc. Very confusing for the decoder, but there might also be a rule that if the date number was odd, you should skip every sixth group and if even, skip every fifth group. Research has demonstrated that similar problems exist in the transcription and translation of the DNA base pairs in the genome.

"The instructions stored within DNA are "read" in two steps: transcription and translation. In transcription, a portion of the double-stranded DNA template gives rise to a single-stranded RNA molecule. In some cases, the RNA molecule itself is a "finished product" that serves some important function within the cell. Often, however, transcription of an RNA molecule is followed by a translation step, which ultimately results in the production of a protein molecule ... DNA is double-stranded, but only one strand serves as a template for transcription at any given time. This template strand is called the noncoding strand. The non-template strand is referred to as the coding strand because its sequence will be the same as that of the new RNA molecule. In most organisms, the strand of DNA that serves as the template for one gene may be the non-template strand for other genes within the same chromosome."[3]

The transcription process is complex, involving initiator, promoter, and enhancer sub-units which tell the initiating enzyme where on the DNA sequence to start, which base pairs to skip, which to substitute, and where to end. The online article referenced above will give you a good idea of the complexity of this process, casting further doubt on how undirected evolution could have achieved such a complex system, irrespective of how much time is allowed.

Now this is far from a technically precise description, but I think that the basic sense is correct. The genome contains some genes which code for proteins, and other genes which control transcription and other downstream processes: these are regulatory genes just like software instructions in a computer-controlled factory process. There are regulatory DNA sequences that turn specific genes on or off, i.e., enabling or disabling transcription, thus controlling when a gene is expressed. Transcriptional regulators (activators and repressors) are proteins that recognise specific DNA sequences. That's it, I shall go no further, I just wanted to show the beginnings of complexity in the genome. The issue for those supporting evolution is to explain how protein coding genes would offer any advantage without their regulatory supervisors, and why they would hang about before there were any regulatory genes to make something out of them. It gets more complex than that, but I think that you can see where I am going with this. Biologists and geneticists are making wonderful advances in our understanding of how the genome currently works, but such understanding is increasingly questioning how such complexity could evolve.

Figure: 2   (A)   (C)   (A)   [T]   (G)   (A)   (C)   (A)

Back to coded messages: in Figure 2 the fourth symbol "T" (square) does not belong to the original group of three: A, C, and G. In my analogy, it would be like inserting a numeral into an alphabetic sequence: XAST B2BD MOTE etc. If the decoder had no instructions on what to do with the "2", the message would be garbled. The question for evolutionists is this: what would the regulatory genes do with this "mutation", as it would prevent proper downstream processing? Secondly, under the natural selection scenario, why would the organism perpetuate this error and what would be the probability of a compensating error arising to utilise it?

This has led to at least one professor of evolutionary biologist observing: "It is important to ask, though, whether true evolutionary novelties actually arise by mutation. For example, can both a new enzyme and the regulatory system that modulates its production arise by mutation?"[4] Good question, and to my knowledge, there have been no scientific experiments which allow it to be answered in the affirmative.

Now this will be a little difficult to explain even though I know it to be true, but I will try. It should be self-evident that two separate coding structures cannot co-exist in the one information system. For example, if you are a user of Microsoft Windows products and you seek to buy Windows© 8, you will be asked whether you want a 32- or 64-bit system: the difference is in the memory addressing algorithms. In the early days of computers when memory was expensive, an 8-bit processor would be adequate as 8 bits = 2^8 = 256 addressable memory locations. As larger memories became available, the operating software had to change to 16-bit addressing to access 65,536 locations (2^{16}). As memory got cheaper still, we went to 32-bit addressing (2^{32} = 4,294,967,296) and now 64-bit is common (for extra-large memory). The main point to remember is that you cannot run both 32 and 64-bit addressing in the one system because you would be attempting to have two contrary coding systems co-existing. The relevance of this is found in most users' experience: software versions that are neither forward nor backward compatible. In some cases, software developed for smaller addressing algorithms can run in larger systems, but this is not common. The issue is that as coding systems change or evolve, the processing and storage systems cannot simply evolve in an undirected fashion: they must be intelligently converted.

Over the years, telephone companies and postal services have faced similar problems, depending on whether processing was manual or computerised. In my youth, six-digit telephone numbers, often referenced alphabetically, were adequate to serve the number of telephones then in use. As the population grew and telephones became more numerous, the telephone number (address) grew to seven and then eight-digits. Similarly, postal (zip) code structures were altered as the number of new addresses were added. In most cases, this was accomplished by the simple addition of a prefix or suffix to the existing code. Irrespective, those responsible for utilising these new coding structures needed to be educated on the methodology. Where such simplicity could not be utilised, more complex recoding was necessary, as shown in the Windows© example. Of course, computers can be programmed in imaginative ways to handle such complexity, but by example, this demonstrates the difficulties that evolution would face in altering a coding system. If you think in terms of the codes shown in the earlier examples, it can be seen that you cannot have a code system that has groups (codons) of both two letters and three letters without having a highly complex decoding mechanism. Processing logic would dictate that the encode/decode mechanism must precede, not follow, any implementation of a new coded sequence.

Coding systems are chosen based on the language of the communicators, and the medium to be used for the communication; e.g., written, sound, light, and other visual or audible signals. Without going into further detail, it should be obvious that the development of coding systems, and the choice of appropriateness for a given system, is complex. The communications medium can also be a limiting factor on the feasibility of a particular system, e.g., light cannot be passed along the surface of a copper wire in the same way as electrical signals.

All communication systems, irrespective of the medium through which they pass, require predefined coding systems which are understood by all nodes. This principle is perhaps the primary reason why interoperability across computer systems, from different manufacturers, remains a source of frustration even today. In a very general sense, all communications, whether verbal, written, visual, audible, or tactile, rely on systems of coding. The very words you are reading here are in a code comprising letters of the alphabet, vocabulary, grammar, and punctuation. Even grammar varies between, for example, Australian English and American English. In the former, the word "that" can be both a preposition and a conjunction, but in the latter, only a preposition. This is a source of frustration when I forget to change the language for each Word document. In general, practically all codes are multi-level, particularly those used in digital computing and communications.

The binary code (1's and 0's) of computers translates to a coded language such as Hexadecimal and then EBCDIC in IBM computers, or Octal and then ASCII for others; from there to a common human language such as English or French. Computer scientists long ago established these separate coding structures for reasons not relevant here. The point to note is that in general, you cannot go from Octal to EBCDIC or from Hexadecimal to ASCII: special intermediate conversion routines are needed. The problem is that once coding systems are established, particularly multi-layered systems, any sequence of symbols which does not conform to the pattern cannot be processed without an intelligent conversion process. Put another way, all *poly-functional* systems are *poly-dependent*. The more functions that utilise a coding system, the more dependent the usage is on the coding system not changing.

Slightly off-topic, but consider the four chemicals in DNA which are referred to as A, C, G, and T. Not that long ago, scientists expressed surprise in finding that the DNA sequences code not just for proteins, but for the processing of these proteins. In other words, there is not just one but two or more "languages" written into our genome. If an evolving genome started with just two chemicals, say A and C, downstream processes could only recognise combinations of these two. If a third chemical G arose, there would be no system that could utilise it and more probably, its occurrence would interfere in a deleterious way. Quite simply, you cannot go from a 2- letter code to a 3-letter code without re-issuing the code book, a task quite beyond undirected biological evolution. Nor can you go back from a 3 or 4 letter, to a 2 or 3 letter, which is what would happen if you attempted to run software written for a 32-bit computer on an earlier 16-bit computer. Similarly, a program coded to run on one operating system, will not execute on another (Windows, Android, Apple iOS, Apple macOS, Linux, etc.)

The point to note is that coding systems cannot autonomously evolve. Any evolution of a coding system requires an external intelligent agency to implement conversion routines for operations to continue. As a programmer in the early days of business computing, I am all too aware of the complexity of this issue.

An Ancient Lesson

I cannot recall where or when I first encountered this ancient word puzzle, but given that is dates to the first century, one can assume that any copyright would be long expired, even if such did originally exist.

```
S A T O R
A R E P O
T E N E T
O P E R A
R O T A S
```

The reason that it is described as a puzzle is that it reads the same in both rows and columns. If you alter just one letter, anywhere in the puzzle, the entire symmetry is destroyed. It can be said that each letter is polyfunctional, because it is used more than once, but each row or column is polydependent, because any alteration of even a single letter would destroy the integrity of the puzzle. The point of this example is that the same can be said of DNA sequences that are read in codons of three: each letter (nucleotide pair) is polyfunctional because it is used over and over again, but the integrity of DNA sequences is polydependent, because any variation by insertion, deletion or replacement alters the functionality.

Summary

Now it matters not whether you apply this to the evolution of the genome, or to the development of our sensory systems, the same principle holds: ALL data must be preceded by meta-data to be comprehensible to both parties in a communication. In general, messages or other forms of communication can be considered as transactions which require a conceptual and contextual framework to provide meaning. At the biological level, we could consider chemical properties to be the meta-data that explains the interactions. At the conceptual level of the mind, however, there is no such explanation.

If you are still unconvinced about the relationship between coding systems and our biology, and most especially our sensory systems, consider this text from a scientific study on neural networks:

"When we look out at the world, reflected light enters each eye, falls on the retina, and activates some of the neurons there. If we were to record the complete pattern of activated and inactivated retinal neurons, the resulting pattern would be a recording, or **representation**, of the visual scene. The activated retinal neurons send output that leaves the eye and eventually reaches the visual cortex areas, at the rear of the brain. Again, a pattern is created there, consisting of activated and inactivated neurons. The visual scene has been represented. In fact, the visual scene (or any sensory stimulation) goes through an arbitrary number of representations as it is processed by the brain."[5] (emphasis in original)

A *representation*, by definition, is a model of the real thing, but not the thing itself. If what we see is represented by a pattern of activated and inactivated neurons, then the process of activation must be regulated in a consistent manner to be reliable. Such a regulated process is called *encoding*. Whilst the reliability of chemistry allows us to accept that such a process could arise naturally, even via evolutionary processes, we are still left with the problem of mental *conceptualisation*. The brain stores the neuronal pattern, but what we see in our "mind's eye" is not the pattern of neurons, but the reality of what the pattern represents.

Just briefly, if a pattern *consisting of activated and inactivated neurons* is created in the visual cortex area, what reads the pattern? Such a pattern could not be in a single neuronal pathway, because the signalling would cease at the first inactivated neuron. If inactivated neurons are part of the pattern, then the pattern could only be read by some other process. According to another explanation, "the retina and the optic nerve originate as outgrowths of the developing brain ... thus the retina is considered part of the central nervous system (CNS) and is actually brain tissue"[6]. The same source notes: "Since there are about 150 million receptors and only 1 million optic nerve fibres, there must be convergence and thus mixing of signals". I find that fascinating as hopefully there is some logical behaviour in this mixing, some coding protocol which ensures no loss of integrity. That the visual cortex can in real time, perform signal processing on a such an incredibly complex scale seems to me more like *magic* than *biology*. If the reader is interested in the subject, I can highly recommend this website as it reveals the incredible complexity of the human eye, complexity which it is difficult to believe, was derived by undirected evolutionary processes in any timescale. For example, "Using optical coherence tomography (OCT) there are 18 layers that can be identified in the retina."

For light signals to pass through all eighteen layers, even in a representational form, there must be a functional and reliable coding system as the content is transformed from one type to another. I have no idea of what processes could have developed such complexity.

References:

1. Gitt, Dr. Werner, *Without Excuse*, Creation Book Publishers, Atlanta, GA, 2011, pp. 43-44

2. https://www.superduperinc.com/handouts/pdf/236_HowDoWeTalk.pdf

3. https://www.nature.com/scitable/topicpage/dna-transcription-426/

4. Futuyma, Douglas J., *Evolutionary Biology*, Sinauer Associates Inc., Sunderland, MA, 1979, p. 252

5. Gluck, Mark A., and Myers, Catherine E., *Gateway to Memory: An Introduction to Neural Network Modeling,* The MOT press, Cambridge, MA, 2001, p. 81

6. https://en.wikipedia.org/wiki/Retina

Chapter 4-2: Functional Requirements

"Start by doing what is necessary; then do what is possible; and suddenly you are doing the impossible."

~ Francis of Assisi ~

Sigh, if only that were true, but just in case it is, let us at least get as far as the possible. Although I am already familiar with such concepts from my experience in Information Processing, this online article[1] expresses them succinctly.

Evident and hidden functional requirements

"Functional requirements can be optionally identified as *evident* or *hidden* (Gause and Weinberg, 1989):

• *Evident* functional requirements are functions that are performed with the user's knowledge. These requirements usually correspond to information exchange between the user and the system, such as queries and data entry, which flows through the system interface

• *Hidden* functional requirements are functions performed by the system without explicit knowledge of the user. Usually, these functions are math operations and data updating performed by the system without explicit user knowledge, but as a consequence of other functions performed by the user.

Hidden requirements are performed internally by the system. Thus, although they do not appear explicitly as use cases, they must be adequately associated to them in order to be recalled at the time of design and implementation. Thus, they could also be added as annotations to a use case."

The above descriptions relate to digital data processing systems, but the same does hold true for the processing that occurs in the human mind-brain complex. I can describe most *Evident* functional requirements from my own experience, but when it comes to the internal biological workings of the mind-brain complex, I am at a complete loss as to the *Hidden* functional requirements. I shall leave those for others to ponder, not being at all ashamed to confess my ignorance, and shall restrict my discussions to what I know. There are complex processes which information processing and computer architecture specialists have developed over decades, and are still developing. Steven Pinker opines, *"Nature, once again, has found ingenious solutions that human engineers cannot yet duplicate."*[2] Nature, in this case, is evolution, with natural selection being the design engineer using components arising from genetic mutation. I do wonder whether Mother Nature gets frustrated having to wait for appropriate random mutations to occur, and having to keep watch across millions of organisms to take advantage of these mutations when they do occur, remembering that Mother Nature needs to "evaluate" whether each new mutation will add to, or detract from, the wrongly termed "design" path. The probability that millions of complementary mutations occurred progressively in one species of organisms, or worse, in multiple species relatively simultaneously, must be extremely low, *convergent* and *complementary* evolution acknowledged. We will have more to say on the mathematics of this later.

The eye receives light signals through a lens which focuses them on the retina, a membrane containing photoreceptor nerve cells called rods and cones, which change the light signals into electrical impulses and sends them through the optic nerve to the brain. Rods help us see in low levels of light, and the cones are responsive to different frequencies, thus enabling us to distinguish colours. Particular receptor cells have been identified in terms of function: M cells are sensitive to depth and indifferent to color; P cells are sensitive to color and shape; K cells are sensitive to color and indifferent to shape or depth. The resultant separated component signals are distributed across approximately one million nerve fibres in the optic nerve. When received at the other end, whether the visual cortex or the brain proper, the symbols must be recombined, decoded appropriate for storage, and stored in such a way that they can be correlated with each other, and/or other related concepts. This is the most problematic process for a purely biological entity - for the brain must "comprehend" the meaning of those signals before they can be processed, stored, and correlated. As best is known, the "coding" system of the brain is implemented via complex neural networks, which we will discuss in a later chapter.

Essentially, any input from any of our five senses represents an information request to the mind-brain to interpret that sensory input. In the case as above, how does the mind-brain complex understand that what was seen was a tree? I am not here referring to it being named a "tree", but how these primitive symbols could be recombined into a mental image of a tree, which can be visualized in the brain and recalled at will. The brain does not store pictures, any more than a DVD does: it stores biological symbols in the form of neural networks which require interpretation.

Our next step is to understand how this is accomplished in digital information processing systems, the point being to understand the functional requirements themselves, without suggesting what the biological architecture should be - just what it needs to achieve.

References:

1. https://www.sciencedirect.com/topics/computer-science/functional-requirement

2. Pinker, Steven, *How the Mind Works*, Penguin Books, London, UK, 1998, p. 4

Chapter 4-3: Understanding Information

"Information is not knowledge"

~ Albert Einstein ~

By Way of Explanation

Before beginning this discussion, I would acknowledge the scholarly work of many scientists researching and publishing on *Information and Message Theory*, and in no way wish to pretend that I am advancing any new concepts (well, perhaps just a few). However, in this chapter at least, I am trying to avoid the previous technical jargon such as *Pragmatics* and *Apobetics*, as possibly the reader has already forgotten what they mean. I believe that all that needs to be said can be mostly conveyed in plain language. The other issue, and a major one at that, is that practically all of the published work that I have been able to research has a blind spot: a failure to acknowledge that all cognitive information has an intelligent source and thus conveys intentionality.

As a reminder, we must differentiate between *genetic* information which is instantiated in chemicals, and *cognitive* information which cannot be. As an aside, I believe that genetic *information* would be better termed, genetic *instructions*, to highlight that the genetic and cognitive are in two separate domains. It is a category mistake to conflate the two or otherwise treat them as synonyms.

Now this is an area of major contention between theists and atheists: theists advocate an intelligent source, atheists deny it, claiming that information can arise *naturally* - it is in the nature of things. I basically agree with both sides, but not for the reasons that either will usually expound, as I shall attempt to explain. The conceptual data itself, whether represented by chemical properties or reactions between instances of matter, conveys no intentionality of its own. Information flow between intelligent beings does have intentionality (mostly), and that is where the constructs in message theory become appropriate. However, in the context of this study, intentionality distracts from the core issue: how do raw chemicals, and arrangements thereof, become conceptual information during the hypothesised evolution of living organisms? If sensory experiences were the only source of essential knowledge of their external environment, many such evolutionary processes must have been *information dependent*.

Philosophically, I am sympathetic to the materialists' position that data (but not information) is *represented* in all material, but only in an abstract sense, and without intelligent agencies capable of cognitively processing abstractions, such "data" has no function.

A Definition

For the purposes of this conversation, I shall define *cognitive* information as follows:

Information is conceptual data processed within a conceptual and contextual framework.

Information may be many things to many people in other domains, conversations, and studies, but to investigate the origin (genesis) of information, we need to understand how data, represented in encoded symbols appropriate to the communications and storage mediums, can become conceptual information upon which a recipient can act. Science is predicated upon cause and effect, and what I intend to show is that there are many dependent steps in getting from the data interpreted from the reality of the material, to whatever effect is causally dependent upon the *understanding* of information.

Information TO YOU

If there is one fundamental that I would ask you to grasp it is this:

Information only becomes information TO YOU *when you* UNDERSTAND it.

In my younger days, I was yet to appreciate that my ambition often exceeded my abilities (and still does), for I dreamed of being a jet fighter pilot. My first experience with an aircraft cockpit was in a Tiger Moth; the panel was simple with just a few instruments: airspeed indicator, altimeter, turn & bank indicator - easy. Later, relegated to the role of a military air traffic controller (yes, reality won out), I had the opportunity to peer into the cockpits of more complex aircraft, and had that sinking feeling of information overload - I simply could not take it all in. Clearly, what was information to experienced pilots was not information to me - it could best be described as *noise*. As an aside (sometimes I cannot help myself), ancient Jewish philosophy offers that one ought not be taught beyond their capability for understanding. In my own experience, I was often subjected to that unfortunate position.

This is a crucial concept in understanding how information can originate: transmission in any form from any source does not necessarily result in intelligent reception, and we need to keep this in mind when we later consider how the brains of early organisms could have acquired meaning from primitive sensory faculties. Similarly, as discussed earlier, we need to ponder whether chemical properties alone could derive a genetic code not just for the proteins to be manufactured, but for the multiple levels of controller genes necessary for their production.

Data versus Information

I would offer that there is a substantive difference between *data* and *information*.

In the formal world of Computer Science, this differentiation is expressed as *syntactic* versus *semantic* information: syntactic being structured raw data and semantic having context and meaning. In an attempt to minimise the use of technical jargon, I will continue as I have started, but still wonder why having defined *syntactic information* as *raw data*, the scientist did not simply call it *data* rather than abstract it to another term! But back to the point: data is conveyed using encoded symbols appropriate to the method of communication, and the medium through which transmissions are made. In written English, the symbols are the twenty-six letters of the alphabet plus numbers and specials characters (even smileys). Many languages use variations of the Arabic script whilst others such as Russian, Chinese, and Hebrew use entirely different character sets. If you want to communicate in English, you do not use the Hebrew alphabet. When communicating through the medium of air, water, or wires, different symbol sets or coding systems are used, e.g., sounds, electromagnetic waves, or currents. This may seem so obvious as to be not worth mentioning, but if we are to understand the origin of information, we do need to review these basics.

In general, the symbols or characters themselves may not convey any meaning – what is termed, the *semantic* layer. Ancient languages generally used pictograms or symbols which represented real entities, such as bulls, rivers, mountains, etc., but as in Hebrew, the original significance has been lost in modern usage. The individual characters in the Arabic script used in most Western languages is the result of abstraction into arbitrary symbols, in the sense that they convey no meaning whatsoever: it is only when they are combined with other recognised characters to form words that meaning is developed. Even then, most words are poly-semantic (having more than one meaning) and understanding them becomes context dependent.

This introduces two very important terms, *concept* and *context*.

If I asked you to go to the store and buy an *osymygon*, neither you nor the shop assistant would have any idea of what I was talking about. If a cosmologist ahead of his time mentioned Dark Energy and Dark Matter to Galileo or Copernicus, they would likely ask him what he had been drinking (or smoking), because they would have been unaware of the concepts. Similarly, 16th century physicists were unaware of the concepts of relativity and quantum mechanics. Thus, before any communication can occur, the understanding of the *concept* must be shared by both parties (Primary Axiom #3). This aspect is of paramount importance for what follows.

Similarly, all communication must be within a *context*. At a primary level, the context can be as simple as language and dialect. In Australia, a *rubber* is used to correct written errors but in America, it is used as a preventative to avoid an entirely different type of error. In Australia, a *thong* is a skimpy item worn on the foot, whereas in America it is a skimpy item worn somewhere less mentionable (in polite company at least). I am sure that you can think of similar anomalies in your own experience. As I have encountered in my own writings, usage of common words can be misunderstood even by scientists, because they attribute a specialist meaning within their own professional domain (meta-language). In different domains, the same word can have different meanings.

When you hear a dog bark or a cat meow, you cannot be sure of the meaning that the animal is attempting to convey, unless you are Dr Doolittle perhaps. Similarly, when a baby cries or gurgles some unintelligible sound, mothers appear to understand, perhaps as a function of experience or feminine intuition. As a father, I admitted to my ignorance: the languages of dogs, cats, and especially babies, are entirely beyond my comprehension.

The point that I wish to make is this: when a sender transmits what is commonly called, information, it is, in reality, only a stream of encoded symbolic representations of concepts, which can only be interpreted as information by the recipient when processed within a *conceptual* and *contextual* framework. If the recipient is ignorant of the concept, such as quantum mechanics, or is unable to identify the context, such as a scientific discipline, then the information cannot be processed. If you were to read a treatise on philosophy to a four-year-old, no communication would occur because the child's brain would have neither the concepts nor context preloaded. Keep this in mind when we discuss the claimed evolution of our senses, particularly hearing and sight.

Let us have just one more example, acknowledging that I have used it earlier in another context.

Pretend that you are hearing the following: "dit - dit dit dit - dit dah dit". Some of you may recognise this as a transliteration of a Morse Code signal. This is/was a universal method of transmitting data by light, sound, or electrical signals. The compiler of the message perceives it as information and if sent in plain language, those who know Morse code may understand it as well. However, a further level of encoding may have been performed before transformation into Morse, a common practice in the military in days gone by. In plain language, we would still need to understand whether it was in English, French, German or some other language: hence the receiver needs to understand the *context* at a primary level. The Morse code sequence decodes to the letters, ESL, but what does that abbreviation stand for: English Soccer League, English Second Language, or perhaps *École Secondaire Saint-Laurent*? The next problem is to understand the context of the abbreviation: is it related to chemistry, physics, history, archaeology, traffic management, politics, or a board game, but in asking that question, you must first be aware that such domains of information exist (your search space): again, an issue of concepts to provide context. If I told you that it related to *axsebology*, you would be none the wiser (because I just made it up). In this example, it is the identification code of an NDB, but if you have never heard of an NDB before, not having been provided with the *conceptual* information, the message still makes no sense, i.e., it is not *information to you*, not even data. NDB could be the National Development Bank, but here it is "Non-Directional Beacon". You may still be none the wiser, but if you have trained in nautical or aeronautical navigation, you will understand its purpose. But of course, we still have some more steps: how do we expand the abbreviation ESL, if it means anything, and where would the beacon be located? Digging through multiple levels, we find that ESL stands for "East Sale" located in Gippsland, in the south east of the Australian State of Victoria. What we have identified is a long string of data ancestors that need to be correctly correlated to derive the correct understanding (Primary Axiom #2).

I hope that this simple example demonstrates the complexity of information and why I believe it essential to differentiate it from data and the symbols which convey it. The key points to remember are that communication transmissions are *encoded symbolic representations of concepts*, neither data nor information, and that a recipient can only interpret them as information where the concepts and context are *correctly* understood: i.e., that it is true information to the receiver. Being *correct* brings up another issue seldom discussed: *verification*.

For now, I will ignore any semantic difference between *verification* and *validation* and just use the one word to describe the process of knowing that the receiver's interpretation of the communication matches the concepts of what was sent. This is particularly important when we consider, for example, our sense of sight: how did the organism verify that what it thought it saw was in any way the correct pictorial representation of its external world?

In the context of living organisms, Paul Davies puts it this way: "To qualify for the description of living, information must be meaningful to the system that receives it: there must be a 'context'. In other words, the information must be 'specified'. But where does this context itself come from, and how does a meaningful specification arise spontaneously in nature?"[1] The author reinforces this view when discussing information in the cosmos; he writes: "You cannot simply inspect a location in space and detect information. What you see - a particle for example - becomes information only in an appropriate global context."[2]

In passing, I thought that the author may have refuted some of his own arguments in this discussion due to a lack of rigour in the use of the term "information", an issue I have tried to avoid, sometimes unsuccessfully, but I will come back to that another time. For now, I would just point out that if all data or information, however defined, exists in the physical reality of the cosmos, then so too it must have existed in the hypothesised singularity, together with all the rules of physical and chemical interactions that supposedly governed the natural processes leading to humankind.

I am not sure of where I could even start to understand how that could be.

Data transforming into Information

In my travels, I have often sat in a café or restaurant, peering at a menu in a language which I did not understand. My stomach and other organs had learned from experience, the dangers of just guessing. Curiously, during a motorcycle adventure through Eastern Europe, we sat at a table with menus in numerous languages. I was handed one at random, our group being in its usual jovial mood, and I began reading and comprehending. It took perhaps a minute or two for me to realise that I was reading French, not English. I can only put that confusion down to tiredness after a long ride through strange villages in inclement weather (that is why we call them adventures).

If the menu does not contain information (to you), then what does it contain? A Spanish menu may be said to contain data in that the alphabet is familiar, but the arrangement of the letters is not. A Chinese menu contains not even data but mere symbols, much as archaeologists experienced when first encountering Egyptian hieroglyphics or Sumerian script. All technical papers and journals that I have read on Information Technology, even those by eminent scientists and scholars, generally gloss over these important distinctions. To get from a symbol set to data, and from data to information, requires a cognitive observer irrespective of the method of transmission. But there is a prerequisite of even greater importance: the cognitive observer must have foreknowledge of the concepts and context of the communication.

For example: "Friedrich's calculations showed that the particle involved in the exocytosis had to be orders of magnitude smaller than I had calculated from Margenau's equation"[3]. I have no idea of what that means in isolation, but I do have some understanding from studying the complete book from which this excerpt was taken. I use this passage to illustrate a point: without foreknowledge of the concepts within a specified context, there can be no information exchange. Here is another interesting passage from the same book: "For the first time, I recognised the fundamental significance of Akert's beautiful paracrystalline structure of the presynaptic vesticular grid of the synaptic boutons with its low probability for quantal emission of transmitter in exocytosis."[4] Despite the explanatory diagrams and text, I lack the education (information and knowledge) to fully comprehend the significance of what is being said here. What I find quite incredible is how we humans can derive some understanding from texts when the vocabulary and abbreviations are beyond earlier education, but we nevertheless rapidly become familiar with their usage. For example, when I first began studying the book[5] from which the following extract is taken, I struggled to make any sense of what was to me unintelligible text (don't ask me why I was reading it, as I never cease to be amazed by what piques my interest). Over time however, due to a great deal of persistence, I became quite comfortable with reading the text and whilst not fully understanding, was able to glean information sufficient for my purposes. Here is an example of that text:

> 16, 2 *A* (incorrectly) מֶיְיה, the rest have שֹׁלָה. 19, 8 B., *A* and *C* הֱהֱוינין, *B* הֱהֱוינוּן, *D* הֱהֱוִינוּן. The subject being בֵּן *D* is of course a mistake. 20, 9 *A* (incorrectly) בְּעֻרְיָן; the rest have בְּשֻׁרִין. *A*, *B* and *D* are incorrect in having דִּין 20, 13 since שִׁיבוֹלְתְךָ is feminine. B. and C read הִיא. B is incorrect in reading הִיא 19, 13 since אָהָלָא is masculine. Rest have הַוָּין. The Hebrew in each case is הִיא and this may perhaps account for the error of *A*, *B* and *D* in 20, 13. 21, 30 B. and *C* הִיא, *A*, *B* and *D* הֲוָין. Since בֵּירָא is given by L. as of common gender either reading may be justified; but *B*'s reading הִיא in 24, 58 and *A*'s הֲוָת in 25, 30 are both evidently incorrect. The fact of בֵּירָא being of common gender may again justify both readings in 26, 20 in which verse *A* reads מַיִּץ whilst B., *B*, *C* and *D* give שֹׁלָה. 27, 42 *A* and *C* (incorrectly) בֵּירָה, B., *B* and *D* הִוָּה. 29, 3 B. שַׁקְיָן, which L. also reads. MSS. שַׁקְיָן. Pathšegen also as B. who gives the proper gender, the subject being the shepherds. 30, 37 *A* and *D* הֲשִׁיבִין, *B*, *C* and B. תֻּבִּיהַ. So also L. and Pathšegen, but since תּוֹשְׁרִין is of common gender we may take either reading as correct. 30, 40 all the MSS. have שְׁרִיבִּין which is the proper reading, B. has עֲרִיבִּין which is certainly incorrect. 30, 43 *A* סְגִיאִין. B., *B*. *C* and

The symbols on this page represent conceptual data which, when combined with other data found in the glossary, transform to information, but only to a reader prepared to exert the necessary effort. The scientific observer should ask: How does that happen in real-time and what are the causal factors? This example bears on such questions as: do we have free will; does the brain drive the mind or the mind, the brain; and whether organic properties alone can account for such complex data processing? What was it about me that resulted in the necessary persistence to derive some meaning from the text? Why did I not cast the book aside?

Quantification of Information

Another definition that I have encountered is this: *Information theory is a branch of applied mathematics, electrical engineering, and computer science involving the quantification of information.* That is all well and good so far as it goes, but quantity is not the same as quality, and from a cognitive perspective, quantity does not convey meaning except in a context external to the data itself. A good author could write this book with, perhaps, significantly fewer words: the informational content would remain the same (or improve) but the quantification would be altered. A garrulous or loquacious person will normally convey less information than a speaker who is both precise and concise. In a cognitive sense, we evaluate information on its value and relevance, not on its quantification. More importantly, there is no necessary relationship between quantity and *humbug*.

Quantification alone can tell us nothing about the meaning of the information, how it is derived, or from an evolutionary perspective, how it is originated before there were cognitive beings that could make sense of it.

Allow me to stress this point because it is essential to understand the differentiation between the technology of information and the cognitive processing of information, particularly when discussing information and knowledge in the context of cognition and evolution. Researching definitions of "noise", I find none that satisfy me in this context so I will coin a new term, *cognitive noise*, which I will hereinafter refer to as *humbug*, a term chosen for its obvious lack of formal scientific basis. When archaeologists first opened the tombs of the pharaohs, what was written on the walls was obviously information for the ancient Egyptians who wrote it, but *humbug* to the archaeologists who were unable to discern the meaning. So, it is with all quanta of what is generically referred to as information: it is information to those who understand it, and *humbug* for everyone else. I would ask you to keep that distinction in mind for everything that follows.

I have sympathy with this observation by Richard Evans (*A Step of Faith*): "What a culture we live in, we are swimming in an ocean of information, and drowning in ignorance."

Summary

In the context of the overall theme of this study, the mind-brain conundrum, I would repeat that for information to exist, it must inform. As I have defined it, and contend that I am correct: *Information is data processed within a conceptual and contextual framework.* Data, too, is conceptual. Any communication via any medium contains encoded symbolic representations of the semantic layer – the meaning. It only becomes data when processed in our minds, as evidenced by the fact that there are different symbolic representations (words) for the same physical entity. The words are not the entity, but are components of an arbitrarily designed language which must be learned, before they can be utilised in communications. Primary Axioms #1, *that nothing can explain itself*, and #2, that *all knowledge is built upon prior knowledge*, are derived from this truth.

The question arises, once again: how can a physical entity such as the brain, which itself deals only in symbolic representations derived from neuronal patterns, give rise to the conceptual? I contend that the physical and conceptual are of entirely different forms, which cannot be found in one and the same substances, such as neuronal network patterns. The conceptual, being of a different form, must be somewhere else – the mind.

As to the form of the mind, I can progress no further than saying that it must be immaterial.

References:

1. Davies, Paul, *The Fifth Miracle: The Search for the Origin and Meaning of Life*, Touchstone, New York, NY, 1999, p. 10

2. Davies, ref. 1, p. 39

3. Eccles, John C., *How the SELF Controls Its BRAIN*, Springer-Verlag, Berlin, Germany, 1994, p. 146

4. *Ibid*, p. 56

5. Barnstein, Henry, *The Targum of Onkelos to Genesis - A Critical Enquiry*, Isha Books, New Delhi, India, 1896

Chapter 4-4: Understanding Communications

"My English is not so good, so if you do not understand, just guess (laugh)"

~ Korean speaker at an International Conference ~

Communication: the imparting or exchanging of information by speaking, writing, or using some other medium.

Due, no doubt, to the commonality of the above definition, people accept this to be true. In a philosophical sense, it is, but at a fundamental level, it is not. In everyday life, it matters not that it is an inaccurate definition, but for the purposes of this study, it is necessary that we understand communications at a deeper, technical level. The title of this book questions the validity of small incremental steps, by some evolutionary processes, being able to cross the chasm between microbes and man. In that context, and that of the signalling system of the nervous system, I intend to show that the fundamentals of communication are common, irrespective of the medium, and irrespective of how implemented. So, whether the context is radio, television, flag signals, the written word, verbal communication, sensory transmissions through the peripheral and central nervous systems, and even the linking of neurons in the neuronal network in the brain, some basic principles remain true, insurmountable, and irrevocable.

Steven Pinker agreed, in part, when he wrote:

"Information and computation reside in **patterns** of data and in relations of **logic** that are independent of the physical medium that carries them. When you telephone your mother in another city, the message stays the same as it goes from your lips to her ears even as it physically changes its form, from vibrating air, to electricity in a wire, to charges in silicon, to flickering light in a fibre optic cable, to electromagnetic waves, and then back again in reverse order. In a similar sense, the message stays the same when she repeats it to your father."[1] (emphasis mine)

Residing in patterns of data and in relations of logic that are independent of the physical medium – this is at the very crux of the issue: these are but encoded symbolic representations.

There are a number of important points here. Firstly, if we accept that in this context, a message contains strings of data, and that the data remains unchanged irrespective of the changes in the symbolic representations of it, then the data must be *conceptual*, not *physical*, as I have earlier posited. I appreciate that this is a difficult concept to keep in mind, but it is essential to our understanding of the mind-brain complex, and as aside, presents great difficulties to this writer in attempting to present consistent terminology. I regularly fall into the trap of using the words *data* and *information* in ways antithetical to how I have defined them. Feel free to correct me whenever this occurs. However, let me be pedantic in restating that Pinker has it wrong. Physical mediums *do not* convey patterns of data – they can only convey patterns of encoded symbolic representations of data, with the semantic layer remaining consistent as Pinker puts it: "the message stays the same as it goes from your lips to her ears even as it physically changes its form, from vibrating air, to electricity in a wire, to charges in silicon, to flickering light in a fibre optic cable, to electromagnetic waves, and then back again in reverse order." The form is physical, the meaning is conceptual, and can only be comprehended by an intelligent agent (including a much-derided mother-in-law). The other issue is *logic*: Pinker makes no effort to justify his usage of that word in this context, although I wonder whether his reference is to logic contained within the context of the communication.

Keep this point in mind when we later discuss the human sensory system, which like its modern communications counterpart, communicates using encoded symbolic representations of an external reality. The message reaches the brain, which biologically, is the same as the sensory pathway, and can deal only at that same symbolic level.

The second point is that the communications systems between you and your mother have all been intelligently designed, with years of testing and validation to ensure integrity of the message (semantic layer). Buried within these technologies are coding systems appropriate to the physical medium, with the prerequisite that the encoding at one end, and decoding at the other, were devised and implemented simultaneously. When we consider this requirement in evolutionary terms, it becomes difficult to explain (Primary Axiom #5). Notice again his reference to "patterns of data" and "relations of logic". The term, *pattern*, is itself conceptual, and whereas physical patterns can be found in nature, they are only ever consistent with the intrinsic properties of matter as influenced by external circumstances. On the other hand, patterns in relation to communications are of two forms: (1), as biologically transduced by sensory organs; and (2), entirely conceptual, which can only be formed by an external intelligent agent. We deal with (1) in later chapters. The symbolic representations of data do not understand what they represent, and cannot, based on their intrinsic properties, self-organise into conceptual patterns. Likewise, *logic* is a concept associated with how the mind works, or often does not, and nothing physical can organise itself into logical relationships, only physical.

We must ever be aware of scientific pronouncements which lack any scientific underpinnings.

Compatibility and Conformity

When I travel overseas, I experience difficulty in communicating because often, the language of the country is not the same as my own. I don't speak French, at least, not very well, and in Paris especially, the locals prefer to not understand me (I suspect because they mistake me for being English rather than Australian). I can read French in a limited manner, not just because I learned it at school, and have had occasion to practice it during my travels, but because the alphabet is familiar to me. The latter is why I have been able to achieve at least a limited competence in that language, and others similar, including the dreaded Latin. In other countries such as Russia, Israel, China, and Japan, the letters of their alphabets for me resemble nothing but arbitrary meaningless symbols – I am not informed. For whatever reason, my mind refuses to process such symbols, partly explaining why my attempts at learning mathematics and chemistry were dismal failures.

Which brings us to another issue of human communication: *compatibility*. For verbal or written communication to be effective, the sender and receiver must understand the same language. Secondly, *conformity*: the sender and receiver must have a common vocabulary, and a similar understanding of grammar. The first issue is understood by all, and needs no elaboration; the second is more complex. We start with vocabulary. You might recall the previous examples of rubber and thong, as used in Australia and America. Americans have another colloquialism, "lucked out", which at first confused me, because it means that your luck is in, not out.

Grammar is less likely to cause confusion, although pedants such as myself are offended by poor grammar, especially by the deteriorating standards in the media. The placement of adjectival and adverbial phrases has become so random as to introduce ambiguity into practically every statement. If such phrasing were to be fed to a computer, one that interprets based on the conventional rules of grammar, the results would be somewhere between amusing and alarming. Grammar becomes especially important, however, when translating from one language into another; for example, consider this transliteration of the Greek version of Exodus 34:7,

"... and transgression iniquity forgiving for thousands mercy keeping the iniquity visiting by means no and clearing and sin the third to of the sons children and the sons upon of the fathers the fourth generation"[2]

In any communication, the sequence of the symbolic representations of data must conform to a pattern that is mutually agreed between the participants in the exchange. Leeway is acceptable where cognitive intervention is possible, but otherwise, effective communication is compromised. In non-cognitive communications, such as occurs within computer systems, and especially in the human peripheral and central nervous systems, no degree of leeway can be tolerated – absolute integrity is demanded. The question we must ask is: how could an evolving organic system validate logical integrity when all it dealt with were symbolic representations of reality, fed to it as transduced through a sensory exchange with its outside world? We must ever keep in mind, when discussing the mind, that the mind deals with *concepts relating to reality*, whereas the nervous system, including the brain, deals only with *encoded symbolic representations of reality*. Evolutionists must explain how evolving organisms managed to cross that chasm at all, let alone in small steps.

To keep it brief at this stage, when humans communicate verbally or in writing, the language, vocabulary, and grammar must be mutually understood to avoid misunderstanding. Even then, other factors can come into play, but we will discuss those a little later. For now, I just wanted to emphasise the importance of protocols.

Telecommunications

Effective telecommunications are entirely dependent upon mutually agreed protocols. Computer manufacturers have been notorious for using different protocols, within and across machines, which is why, for example, an IBM mainframe could not easily communicate with a Hewlett-Packard mini, or even with an IBM mini in some cases. When railways were begun in Australia in the 19th century, there was lack of agreement on the railway gauge, thus making it impossible for a train to make an uninterrupted journey from Sydney to Melbourne. One would have thought that this lesson would have been learned universally across the industrialised world by the 21st century, but apparently, cross-manufacturer communications has not been high on the agendas of the powers-that-be.

A communications protocol is a system of rules that allow two or more entities of a communications system to transmit symbols (*neither data nor information*) via any kind of variation of a physical quantity. The protocol defines the rules, syntax, semantics and synchronization of communication and possible error recovery methods. Protocols may be implemented by hardware, software, or a combination of both. Communicating systems use well-defined formats for exchanging various messages. Each message has an exact meaning intended to elicit a response from a range of possible responses pre-determined for that particular situation. The specified behaviour is typically independent of how it is to be implemented. Communication protocols have to be agreed upon by the parties involved. There is a close analogy between protocols and programming languages: *protocols are to communication what programming languages are to computations*.

In the section on Primary Axioms, we discussed issues related to the integrity of communications systems, and the functions that evolution would have needed to perform to ensure that all worked as today we know it does.

In a bold statement, Steven Pinker asserts of this theory, "it solves one of the great puzzles that make up the mind-body problem: how to connect the ethereal world of meaning and intention, the stuff of our mental lives, with a physical hunk of matter like the brain."[3] I would beg to differ: he has solved nothing at all. Scientists well recognise the paradox of meaning and intention, and now claim that the paradox is resolved by explaining, as Pinker puts it,

"... that beliefs and desires are *information*, incarnated as configurations of symbols. The symbols are the physical states of bits of matter, like chips in a computer or neurons in the brain. They symbolize things in the world because they are triggered by those things via our sense organs, and because of what they do once they are triggered. If the bits of matter that constitute a symbol are arranged to bump into the bits of matter constituting another symbol in just the right way ... "[4] [italics in original]

Anyone who has worked with information technology must immediately ask: how? Computers work with binary symbols, and all relationships and computations are driven by external, intelligently derived, software. What is the software of the brain? Pinker, like all scientists accepting the evolution paradigm, accepts that *the mind is not the brain, but what the brain does*, but at the same time, ignores the limitations of physical matter. Evolutionists harbour the belief that somewhere in the cell, established by the genome, is "intelligence" that is able to derive the *conceptual* from the *physical*, whilst not attempting to explain how that could be possible.

Architecturally, computers and brains are dissimilar. In a later chapter, we will discuss some of the mysterious, yet to be understood, *functional architecture* of the brain in contrast to its physical structure. For now, we just need to be aware that comparisons of the brain and computers are by analogy only.

Communication

No doubt it has occurred to you that communications do occur in the absence of cognitive beings such as humans. Predators detect prey by sight or smell, dolphins and bats use echo-location, some creatures sense small movements and vibrations and so on: message transfer does occur. Scientists have reached a level of understanding in this domain, but there is one important distinction: such messages are generally unambiguous because the coding systems are fixed by the laws of physics and chemistry. When a hyena smells a dead carcass, the communication can speak of nothing other than what it does; it cannot, for example, convey information on the height of the nearest tree, or the location in terms of latitude and longitude. Using the term loosely, we can say that the communications protocol is constrained by the physical properties of the source and the medium through which it passes.

Animals do have other means of communication such as leaving a scent, behaving in certain ways, changing colour, or making sounds, but to the best of our knowledge, all such communication is relevant to their current environment and circumstances. We have no evidence that animals discuss the weather, the development of their offspring, the pros and cons of moving to a new location, whether they should craft an autobiography, or what they will be doing tomorrow.

Human communication is entirely different. Whilst we share many of the faculties of lower order animals, even to a lesser extent, these animals do not share our higher order faculties in terms of information, knowledge, and communication, even though we share the same physical properties. Many articles assert that the genomes of humans and chimpanzees differ by only 4%, although I have reason to believe that such estimates are outdated. Nevertheless, irrespective of that particular difference, humans, chimpanzees, elephants and whales have 100% commonality in one fundamental respect - they are all composed of the same chemicals. They may be arranged differently, but all suffer from the same constraint - chemicals can be nothing other than what they are, and cannot rearrange themselves based on an abstraction of which they cannot be aware.

Signals vs Messages

This may be, for many, a case of semantics, but it is necessary in the context of the human nervous system to differentiate the contents and purpose of communications in terms of information, messages, and signals. I have already asserted that no communication conveys information, so I will leave the reader to accept my definitions or not. A dictionary defines *message* to be a verbal, written, or recorded communication sent to or left for a recipient who cannot be contacted directly: that does not work in the context of how neuroscientists use the term referring to the human nervous system, for in this case, the recipient is in constant contact with the sender. For my purposes in this book, I will treat message as synonymous with information, in that a message is intended to convey information.

A signal, on the other hand, in the context of the nervous system, is an electrical pulse (action potential) that travels along a nerve fibre. In and of itself, it is meaningless, but in this context, it is equivalent to the ping you hear on a computer when it notifies you that you have mail. The mail content is dependent upon the source of the signal. The recipient "understands" the content by identifying the source, which according to neurophysiologists, is a function of the mapping in the cortex of the human body. In other words, the cortex knows, for instance, that a particular nerve fibre originates in the finger-tip of the third finger in your left hand, stimulated by a somatosensory receptor sensitive to pressure. Thus, the context is determined by the cortex mapping of the signal source.

In all cases, only signals, in the form of electrical pulses (action potentials), are communicated along nerve fibres, never messages nor information. It is critical to grasp this concept for understanding later discussions.

References:

1. Pinker, Steven, *How the Mind Works*, Penguin Books, London, UK, 1998, p. 24

2. https://biblehub.com/interlinear/exodus/34-7.htm

3. Pinker, *Ibid*, p. 24

4. *Ibid*, p. 25

Chapter 4-5: Data Processing Principles

"Computers are good at following instructions, but not at reading your mind."

~ Donald Knuth ~

Early computers differed from modern devices in that they had very limited self-starting mechanisms like bootstraps, or chips pre-loaded with extensive firmware. Well, that is not entirely true, but it is sufficiently true to begin our discussion. My first "mainframe" computer was an IBM 360/25 with all of 24K of memory, 8K of which was reserved for what was then termed, the Supervisor, now Operating System. This left 8K for I/O buffers and just 8K for application code segments. It was quite a trick to write complex programs using 8K code segments which needed to be continually swapped in and out. That notwithstanding, without such bootstrap software that was initiated when you turned on the power, a computer is just an assembly of inactive physical components - when you turned on the power, blinking lights, if there were any, were all that you would see. The primitive brain, of whatever organism is said to have had a brain, must have been like that – some empty organic blob, absent any connections used for mental processing. I can hear the objections, and I know what they are, but we will come back to them in their proper place. For now, I want to continue the computer analogy to establish some basics.

A computer which is not pre-loaded with instructions, cannot begin to operate without an input device to load those instructions. So, we attach a keyboard and/or card reader, but wait, how does the computer know what those devices are for? The computer must have pre-knowledge of both the existence, and the functions of its peripherals (Primary Axiom #8). Applying that analogy to evolution, how would an organism know both the existence and the capabilities or function of an evolving appendage? In an earlier chapter, we discussed what I consider to be axiomatic – before we can do anything, *we must know that we can* (Primary Axiom #9) By that I am not referring to what we might consider as accomplishments, like flying an aeroplane, but, for example, to move a finger, we must "know" that the finger is capable of a range of motions. Somewhere in the brain are instructions on what signals to send where, to achieve motion, which raises another two issues: communication protocols, and how did the instructions get there in the first place.

I apologise for this level of detail, but if we do not start this far down, one will be prone to accept the *handwaves* of scientists who do not trouble themselves with it, and consequently make assumptions that are not only without substantiation, but are also false. If life started from non-life, we ought to investigate the necessary process steps, rather than absolve ourselves of responsibility as evolutionary biologists have done, when they declare that abiogenesis is none of their business.

Now, back to our business. The power is turned on, the software instructions are loaded, and electrons start circulating through memory. How do they know where to go? A computer has instruction registers which hold the memory addresses of the next instructions to be processed. Processing power of computers has been incrementally increased by increasing the size of memory, requiring an increase in the size of address registers, giving rise to the terms, 8-bit, 16-bit, 32-bit, and 64-bit processors. New software is written to take advantage of these increases, and whilst earlier generation software may run on newer machines, the reverse is not true, because the smaller address registers cannot accommodate these larger addresses. As an aside, debugging memory dumps was performed by following the sequence of programmed instructions using the address registers, answering the question: "How did the program get there? As an aside, computer suppliers grandly advertise their advances in architecture, for example, the number of cores in an Intel chip. One might be inclined to think, "Ah! More cores, more speed" but this is not necessarily true. The same for clock speed. The issue is that if software is not written to take advantage of these new architectures, programs are unlikely to run faster, and may even run slower.

Modern computer users are largely unaware of what occurs behind the user interface, and even modern programmers have little need to know much of the detail at the system level [hidden functional requirements]. My first programming language was at the machine level, ones and zeroes, later moving on to Assembler, Cobol, RPG, PL/1, and subsequent high-level languages. IT practitioners in the early 1970's needed to write their own data storage algorithms; understand cylinder and head addressing techniques to optimise performance; be able to read core dumps in hexadecimal or octal; and perform many other tasks that are now *black-boxed* for them in the interests of productivity. This is as it should be, but I want to review some basics to build an analogy for understanding our cognitive abilities, at least in part: much of what occurs in the human mind is beyond analogy with digital data processing.

Storage and Transmission Principles

Firstly, we should understand, as Steven Pinker earlier acknowledged, that the storage of data is independent of the storage medium – the same conceptual entity can be stored in many different ways. Secondly, because it is not truly conceptual data, but symbolic representations of data that are stored, a method of encoding and decoding must be devised and employed. Thirdly, like storage, the conceptual content of a message is independent of the method of transmission, and is entirely dependent upon an intelligently devised coding system known to both the sender and receiver. Speech and writing can convey the same messages, or different messages, as can smoke signals (or so I have been told). A message can be written in the sand with a stick, or sticks and stones can be arranged on the sand to form the same message. In short, there is a large variety of physical methods for storing and transmitting the same conceptual content, further demonstrating that the data or message itself is conceptual, not physical.

In the early days of EDP (Electronic Data Processing), a far more explicit term to my mind, data was input using punched cards. Holes were punched in the card in a particular pattern, but the pattern was independent of the storage medium (card) and varied depending on the manufacturer of the computer system. Over time, IBM for example, modified both the card size and the pattern such that early card readers could make no sense of cards punched on later machines. Each hole on a given card was the same in shape and size, the placement on the card being necessary to complete the symbol, but the symbol itself could tell you nothing about what it was meant to represent. The point to note is that the arrangement was based on a human-devised coding system, and only recipients, machine or human, educated on that system could understand the data content. With experience, one could directly read the punched card from the holes themselves, as I have done, but first I needed to know the coding system.

Another development using the same IBM 80 column card format was known as "mark sense". Instead of punching holes in the card, one used a graphite pencil of other black marker to mark the card in predefined places. The card reader would pass the marked positions to the central processor which would then interpret them based on the coding system known to the host program. Again, the individual marks, as symbols, conveyed no meaning to the storage medium, and not even to humans without the pre-printed instructions on the card. The point here is that the same storage medium, an IBM 80 column card, could be used with different methods of recording the symbols, and with different coding systems, yet still intended to represent exactly the same conceptual content. That which was represented was independent of the method and format of how it was represented, and clearly, neither the storage medium itself nor the symbol methodology had any understanding of the data being stored.

Now you may think this rather trivial and obvious: of course, inanimate materials are not sentient, and are incapable of understanding data or information. Hold that thought.

Reading a punched card, it was impossible to interpret the data punched therein without knowing the application to which the data belonged. The symbols were contiguous with no gaps, 80 symbols in 80 columns, and so often consisted of just numbers. You might see a string that looked like a date, but unless you knew that it was a date and you knew the date format (code), you would just be guessing. For example, dates were coded as YYDDD, DDMM, MMDD, DDMMYY, DDMMYYYY, etc. Thus, 750441234 could be year 75, day 044, and sequence number 1234. Alternatively, it could be card 75 for month 04, and so on. Consequently, it was imperative that the appropriate computer program was loaded before the data in the form of a deck of cards was fed to the system. In essence, the processor had to pre-know both the coding system and the data formatting used in the cards.

Storing Data for Later Use

When processing was completed, the system would then store the symbolic representations of data onto media such as punched card, magnetic stripe card, magnetic tape, hard disk drive, paper tape, or paper reports via a printer using a physical method appropriate to the media type. Depending on the media, the symbols were either human-interpretable or not, but again could not be interpreted by the medium itself. You may consider that I am stressing this point and you would be right: I am and for very good reason.

The storage of human interpretable data, such as words written on paper, is managed by humans, but digitally encoded data, on a hard disk in particular, is stored using algorithms formulated by intelligent design. Imagine if you will an abundance of 1's and 0's scattered around the surface of a 500Gb disk in locations and patterns known only to the computer operating system. Fiction stories tell of forensic analysts analysing computer hard drives to discern the contents, but what the stories do not relate is that first the analyst must know, even by trial and error, what computer system converted the data to symbols, and then use the appropriate diagnostic software. The ones and zeroes are meaningless in themselves.

Another issue, which should be obvious, is that having stored these binary bits in magnetic regions, the system had to know where to find them again, and find them in context: no use looking up purchasing data if you were running that week's payroll. Thus, the next level of complexity is a method of addressing and indexing so that when needed, the data could be quickly and accurately retrieved. But note here also that the computer system had to "know" where the indices themselves were stored and how to decode the data to convert it into a search algorithm. Even more, and this is a point which I will stress later on, before the computer or any form of analogue or digital computing device can begin computing, it has to "know" that it can "know" what to do. I acknowledge that this point may seem obscure, but I will hopefully clarify this a little further on. In a digital computer, the first level of knowledge occurs in some bootstrap procedure which gets the show going by invoking the operating system which amongst other tasks, validates its memory and checks out its various peripheral devices whilst invoking start-up programs to awaits further instruction. When the user invokes a program, the computer has to know that such a program exists in memory and where to find it, and the executed program has to know which files to use and where to find them. I doubt that the human brain works in quite the same way, but I am in no doubt that the same functional requirements must be met.

Continuing with storage media, in earlier tape-based storage systems, the program would hunt through the file sequentially but the data had to be organised sequentially when written to the tape, and the system had to foreknow that sequence: employee number followed by department or department followed by employee number?

Now here is a very important fact, one if not true would have prevented the development of Information Technology and deprived us of technical progress in numerous areas: data symbols stored on physical media cannot (and should not) self-organise. We might consider this issue in the context of the brain: if the organic matter self-organises, who or what does it tell about the new organisation such that the re-organised data can be found again?

Data Structure Strategies

Ever since direct access technology was implemented with online storage devices, such as hard drives, data base developers have faced the challenge of how to structure the data to suit both the update and retrieval requirements of varied applications. This has been more complicated than you might imagine. Where data is neither stored nor retrieved sequentially, it must include pointers (indexes) to achieve chaining of the relevant data elements. If the relationship between every data element with every other data element must be prespecified and physically implemented, the structure becomes unwieldy. Thus, data analysis becomes essential, but even then, new requirements will arise. Sometimes, analysts will look for patterns in existing data that were not recognised when the applications were first designed. This provided the impetus for the development of Business Intelligence applications, looking at data from different perspectives to determine whether valid relationships exist.

We must apply that experience to question how the brain builds and maintains its neural networks. Remember that data without context is of little value – information is the correlation of data elements in a particular context, and many data elements can have relevance in multiple contexts. You may be familiar with the term, *polymath*: referring to an individual whose knowledge spans a significant number of subjects, and who is able to draw on complex bodies of knowledge to solve specific problems. For a person to be so endowed, not only must he/she have studied numerous subjects, often not naturally related to one another, but the neural network of the brain must be structured to actualise those relationships. Either that, or the brain must be capable of implementing search strategies, much as have been developed in Business Intelligence applications. If the brain is capable of self-organisation in the storage of data, and/or is capable of implementing search strategies across over 100 trillion synapses, what part of the brain is responsible for orchestrating these processes?

There is one more catch: the brain does not store conceptual data, only symbolic representations of data expressed via neural connections. What part of the brain encoded the conceptual data for physical storage, and subsequently decodes the physical representations for a conceptual search? In theory, one should be able to examine a segment of a neural network, and decode its conceptual meaning. We discuss this further in chapters on the brain architecture.

Self-Organisation

Applying this example to the subject of self-organisation, we should understand that self-organisation such as happens with certain materials, can only occur based on the intrinsic properties of the materials themselves. We could hypothesise that the chemicals listed in the Periodic Table of Elements could self-organise based on their atomic number (number of protons), such being intrinsic to the elements themselves, but they could not self-organise based on their scientific names, because the names are conceptual abstractions which are *unknown* (not intrinsic) to the elements. In earlier terminology, "names" are accidental not essential properties.

From an informational perspective, the data represented by stored symbols is an abstraction and bears no natural relationship to the properties of the material itself. The symbols in whatever form, magnetic regions on a disk or electro-chemical gradients in the brain, cannot self-organise based on concepts unknown to them. For them to meaningfully represent an abstraction, their initial organisation must be established by the same agency that seeks to derive meaning from it, just as occurs with coding systems. It should be evident that in the case of abstractions, the agency causing the organisation must be external to that which is being organised.

Self-organisation alone cannot represent information that is an abstraction of that organisation. If we use the letters of the alphabet as a simplistic example, stored letters could only self-organise like-for-like: all the A's together, all the B's, etc, but not in sequence, for both "letter" and "sequence" are concepts *unknown* (not intrinsic) to the symbols themselves. What we know as an "alphabet" is a concept which provides the context for the sequence with which we are familiar, and as the letter symbols *know* neither the concept nor the context, any organisation of collections of like letters would be random. Another possibility is that the letters are stored in the sequence of arrival but that is the antithesis of self-organisation. Irrespective of the method of storage, something external is required to organise it.

Fortunately for us, stored symbols once organised by an external, intelligent agent, usually stay organised, allowing us to retrieve them in the sense intended. Sadly though, this appears to be not always true in the human brain. The crucial question is this: can organic brain matter which is physical, self-organise around conceptual patterns of which it is unaware? The indisputable answer should be an emphatic "NO". Representations by physical symbol arrangements do not always translate to the same conceptual meaning, yet the same conceptual meaning can be represented by different symbol arrangements. It is these truths which effectively refute the notion that the brain can self-organise based on physical symbolic representations, as provided by our senses, especially our visual and auditory systems.

For a practical example of the natural self-organisation of crystalline structures, see here[3]. These massive structures demonstrate what is possible within the intrinsic properties of the material, but they could not, for example, code for the geographical region of their location, the country in which the caves can be found, nor for the location of the nearest coffee shop, these being abstractions external to properties of the crystals.

Data Ancestors

A fundamental of *epistemology*, the discipline concerned with the nature, sources, and limits of knowledge, is that all knowledge is based upon prior knowledge (Primary Axiom #2), which must surely lead an enquirer to question where the very first knowledge originated. We shall come back to that. For now, I will introduce the notion of *data ancestors*: that long string of prerequisites which are employed in any intellectual activity. It is said that a good investigator will continue to ask "why" until he arrives at the first cause; a researcher of knowledge should continue to ask "how do you know that" until he arrives at the very first element of knowledge. In digital information processing, the same investigations can occur with precision because the structure of the data is predetermined by an intelligent agent, but we should not lose sight of the fact that it took an intelligent agent to not only organise the data, but to interpret it in context. This latter point evidences than even the largest and most complex of computer storage is still inadequate to comprehend its contents, for the definitions of the foundational concepts employed are external to the computer storage itself. Even if those definitions are also stored in that same computer, as in a data dictionary, another application of external intelligence is required to provide the correlation and interpretation – again, *turtles all the way down*.

Let us now consider the structure of data ancestors.

Concepts and Context

In the early days of data processing, practitioners identified three file types: *master*, *application*, and *transaction*. Fundamentally, master files provided concepts; application files provided context; and transaction files represented activity. As more than one computer operator learned as they loaded stacks of cards, processing the wrong transaction file against the pre-loaded master or application file failed to result in useful information, giving rise to the acronym GIGO: Garbage - In Garbage Out.

The master file provided details of the entity (concept): e.g., employee, inventory item, customer, supplier, and so on. The application file provided details of the activity type (context): e.g., payroll, stocktake, sales orders, purchase orders, etc. The transaction file provided details of the activity itself. In the days of sequential and indexed-sequential file processing, IBM at one time conducted a survey to determine the most common cause of programming errors. If my memory serves me correctly, 80% of errors were the result of incorrect file matching. This is instructive in the context of cognitive information processing: if intelligent computer programmers regularly made logic errors concerning the correlation of data sources, what is the probability of even more complicated logic arising through undirected organic evolution?

The master, application and transaction files also contained concepts beyond those immediately obvious, leading to the requirement for a *data dictionary*: a precise definition of every data field used by the information processing systems. Errors in initial programming, and even more, program maintenance long after the original authors had sought opportunities elsewhere, were commonly caused by inadequate or absent definitions. In later systems such as Business Intelligence, entirely wrong results had similar causes; for example, *Delivery Date* not adequately identified as *promised* or *actual*. Though data is now stored in random access data bases, the fundamental principles do still hold.

Knowledge is acquired through information, and information is dependent upon a structured chain or network of data ancestors which provide the concepts and context, prerequisites for accurate processing of any new data transactions however acquired.

Validation and Verification

"If debugging is the process of removing software bugs, then programming must be the process of putting them in."

~ Edsger Dijkstra ~

Program and system testing remain the most problematic aspects of modern software, especially as complexity increases. New software patches and releases are mostly related to fixing problems rather than adding capability. In the military aviation domain, and to a lesser extent the civilian aviation industry, computer software represents an ever-increasing cost component and a primary reason for project delays. A Navy helicopter program in Australia was eventually cancelled because after ten years, the integration of numerous software applications had still not been successfully achieved. In the current F-35 Lighting II project in the USA, full software integration will not be achieved until long after FOC (Final Operational Capability) has been accepted, the compromise having to be made to forestall cancellation of the program. I mention these cases to evidence the complexity of poly-functional software systems, and would offer that the human brain has capabilities far beyond any software system yet devised.

Program design is ever a trade-off between straight-line coding, which offers simplicity in understanding, but increased overall size and restricted functionality, with modular coding whereby a particular code module can be reused in multiple functions reducing size but increasing complexity. A disadvantage of poly-functional modules is poly-dependency: that is, a change in a single module will result in multiple downstream changes, often unintended and consequently adverse.

An example from aviation is when an aircraft manufacturer added code to override a pilot's attempt to retard the throttles during take-off - a critical phase of flight. This action was taken based on experience in the USA and on the advice of the FAA in that country. In Sweden one winter, clear-ice breaking off the wings of a DC-9 just after getting airborne entered both engines causing them to surge. The pilots reacted correctly by temporarily retarding the throttles to clear the surge. The software overrode the pilots' actions and increased thrust, causing both engines to explode and disintegrate. Unsurprisingly, the aircraft crashed with no survivors. One could offer that the software designers failed to properly understand poly-dependency and evaluate all possible scenarios, even if they could foresee all such possibilities which is unlikely. In the context of undirected evolution, poly-dependency would likely lead to adverse rather than fortuitous consequences more often than not.

My point here is that for any information processing system to be useful in providing valid data and true knowledge, there needs to be a validation process which is external to the system being validated. Further, the validation process itself needs to have been validated by a prior process, itself having been validated. A failure in this context led to the much-publicised grounding of the Boeing 737 MAX. The FAA allowed Boeing to self-regulate its testing processes, which as subsequently discovered, had not been properly validated. The obvious question arises: How did evolution validate its own processes? Could *natural selection* be a sufficient explanation for how the brain validated its own self-organisation and linkage processes?

An interesting case in the human nervous system is the phenomenon of *referred pain*. The term is misleading, for in truth, it is a *communications* or *signal processing* error. Referred pain is pain perceived to be at a location other than the true origin. We will come back to the architecture of the Central and Peripheral Nervous Systems in a later chapter, but for now, there are more pain receptors throughout our bodies than dedicated nervous pathways to carry them. Multiple nerves in the PNS merge into single nerves in the CNS, resulting in a signal processing problem that the CNS must solve. Sometimes, it fails to do so, which poses another problem for evolution. As no-one has offered a step-by-step description of how the nervous system evolved, it is reasonable to assume that from time to time, signal processing errors would have occurred misleading the brain's development. If the brain is unsure of the source of a sensation (sensory neurons), or worse, perceives it coming from an incorrect location, its response in terms of motor neurons would likely be harmful. An avalanche of chaotic actions could result. Just imagine the brain reacting to an itch in your finger perceived as a heart problem.

It is obvious that the verification process in the human nervous system is subject to error, which leads me to question how the system could have developed to its state of almost perfect functionality that we observe. I cannot know, but I do wonder. In my defence, I am confident that Murphy's Law is as old as the Universe itself.

Coding and Decoding

We earlier discussed Coding Systems in Chapter 4-1, but as that was some time ago in terms of issues being discussed, here is a quick reminder.

All electronic computer and communication systems require predefined coding systems which are understood by all nodes. This aspect of the technology is perhaps the primary reason why interoperability across systems from different manufacturers remains a source of frustration even today. In a very general sense, all communications, whether verbal, written, visual, audible, or tactile rely on systems of coding. The very words you are reading here are in a code comprising letters of the alphabet, vocabulary, grammar, and punctuation. Without going into the subject in depth, practically all codes are multi-level.

It is axiomatic that the definition of a code must precede its usage, otherwise the coded symbols are no better than *humbug* (noise). My point here is that all known information processing systems involve a hierarchy of codes, and that these codes need to be predefined to all nodes in the process (Primary Axiom #5). This is particular importance when we later discuss cognitive dependent senses such as sight.

An Example of Intellect

You may have seen this before, but I thought it worth sharing as it adds another dimension to the complexity of how we humans intuitively process information.

"One manager let employees know how valuable they are with the following memo:

"You Arx A Kxy Pxrson"

"Xvxn though my typxwritxr is an old modxl, it works vxry wxll xxcxpt for onx kxy. You would think that with all thx othxr kxys functioning propxrly, onx kxy not working would hardly bx noticxd; but just onx kxy out of whack sxxms to ruin thx wholx xffort.

You may say to yoursxlf, "Wxll I'm only onx pxrson. No onx will noticx if I don't do my bxst." But it doxs makx a diffxrxncx bxcausx to bx xffxctivx, an organization nxxds activx participation by xvxry onx to thx bxst of his or hxr ability. So thx nxxt timx you think you arx not important, rxmxmbxr my old typxwritxr.

You arx a kxy pxrson."

Notice how even when certain symbols are replaced, it is possible to understand the message by subconscious substitution, further evidence that abstracted meaning is independent of the intrinsic properties of stored (on paper) symbols. I will repeat that for it is an important point in relation to cognitive processing: without any training whatsoever, and without any pre-arrangement or warning, we can intuitively correct errors when the written code does not accurately follow the code book. How can an undirected process of evolution account for such an instant error detection and correction mechanism?

What particularly fascinated me when I first encountered this little gem was how I was able to immediately start reading it despite the spelling errors. As a former computer programmer, I wondered what algorithms would be needed in software to make sense of this on the first pass. I have some familiarity with software that "reads aloud" from written text, but this passage would certainly present a challenge.

Data to Information

Symbols stored on physical media are not data until *decoded*, and similarly data is not information until processed within a referential framework of *concepts* which provide *context*. Let me reiterate a simple scenario to demonstrate this point.

You discover a piece of paper and on it you notice these symbols:

• ••• •—••

The first question that you would likely ask yourself is this: Is this a random string of symbols or is there a pattern to it – is it meant to be information or just doodling? Incidentally, this is the same thought process used by SETI researchers in attempting to identify extra-terrestrial intelligence. If the symbols are not random, then we can assume that there is an intelligence behind the structure and a purpose behind a communication, the layer described by Dr Werner Gitt as *Pragmatics*[4]. In passing, note that I am deliberately avoiding the excellent work by Shannon, Gitt and others so as to not distract from the main theme. However, I wanted to point out that *purpose* is a *concept* which needs to have had an origin of its own. Now note the structure of the symbols: one round thing by itself, three round things together, and so forth; even the spaces between them are symbols like punctuation (another concept).

In Chapter 4-3, *Understanding Information*, I gave a description of the multiple levels of decoding to derive information from this same Morse Code signal, written differently above than as we were taught to articulate it: *dit, dit dit dit, dit dah dit dit*. Yes, to an outsider, it probably looked ridiculous, but before we graduated to the telegraph key, we actually had to communicate short messages to one another. Much mirth ensued, but fortunately our instructor had experienced similar before, and would join in. Naturally, the cost of errors was measured in jugs of beer. That anecdote aside, I mention it again here to provide context for the next part of the discussion.

For such a very simple message consisting of just two symbols, or three counting the spaces, there have been a multitude of prerequisite concepts (data ancestors) needed to provide the context for you to refine your search space to just those concepts which provide the correct solution. Note that all of those concepts had to be acquired from sources external to you, whether you already pre-knew them or just then acquired them. What we are seeing, to use the term coined by William Dembski, is *Complex Specified Information* (CSI). Reviewing the steps in Chapter 4-3 whereby we resolved the meaning of a Morse Code signal to a Non-Directional Beacon at RAAF East Sale, there was an extensive network of information that could have been explored to find the solution, but only a *specific set* would get you there successfully. Most data points encountered in the process have multiple links: in fact, information is a complex network of data points that need to be correlated in a specific way to provide the right answer, but then the question arises: how do you know that you have arrived at the right answer? Just as in Quality Control, there needs to be a process of validating the outcome of the upstream processes. As we know from experience, people can evaluate the same data yet arrive at different conclusions. The reasons are multifarious, ranging from our worldview to professionally acquired knowledge and experience, but the key point is that for any successful transfer of information, there must be a rigorous method of validation: randomness will simply not do.

Summary

In this brief review, I have attempted to outline some fundamentals which need to be present in any data / information processing system. In researching this study, I at one time attempted to start with a single, simple concept like a "menu" and document the data ancestors which I unconsciously used in understanding what a menu is and how it is used. I used my experience as a systems and database designer to structure the concepts and contexts in a way that would lead me from bottom to top, or even top to bottom. After some three hours and many pages, I had a good appreciation of the depth of my knowledge and the utter futility of trying to document it.

There is no suggestion that the *mechanisms* or *architecture* of computer systems must be present in the human mind and/or brain to accomplish cognitive information storage and processing. I do assert, however, that the functional requirements are the same, irrespective of the method of implementation. There is one truth that I must stress, one that is as relevant to the human brain as it is to computers:

Physical devices do not, and cannot, store nor communicate, data or information. All such devices work with encoded symbolic representations of conceptual data, but never information.

Information is only ever derived by intelligent agencies, either by pre-structuring the architecture of the storage devices, or by using search criteria constrained to pre-specified search space(s).

The question becomes: can the physical brain which stores such symbolic representations be at the same time, the intelligent agency that interprets them?

References:

1. Dawkins, Richard, *The blind watchmaker*, Norton Press, New York NY, 1996, pp. 159-160

2. Andrews, Professor E.H., *Who Made God?*, EP Books, Darlington, England, 2009, p. 157-159

3. http://news.nationalgeographic.com/news/2007/04/photogalleries/giant-crystals-cave/

4. Futuyma, Douglas J., *Evolutionary Biology*, Sinauer Associates Inc., Sunderland, Massachusetts, 1979, p. 252

5. Gitt, Dr. Werner, *In the Beginning was Information*, First Master Books, Green Forest, AR, 2007, p. 60

Chapter 4-6: Information Processing in the Mind

"No matter how closely you examine the water, glucose, and electrolyte salts in the human brain, you can't find the point where these molecules became conscious."

~ Deepak Chopra ~

In this brief chapter, I will allude to our sense of sight and then discuss it in more detail in subsequent chapters. Information, and information processing, are often used as metaphors for what is truly occurring, the problem arising that few seem to understand is that they are, indeed, metaphors. In fact, information, knowledge, and ultimately wisdom, are abstractions that exist independently of any physical manifestation.

Let me declare my conviction that despite being beyond my understanding, the mind is the immaterial controller of the non-autonomous functions of the brain. In other words, the brain is organic but the mind is not. I have read numerous academic texts on this subject but find it impossible, in the context of cognitive information processing, to treat the two as one. Though the centrality of my argument here is with the overarching narrative of evolution, and for the sake of simplicity I would prefer to consider the mind and brain as a complex, I cannot manage to discuss the mind as an organ of physical properties alone.

The brain is connected to numerous data input sources but here I want to consider just the senses which may contribute to the development of knowledge in the intellectual sense. I will also disregard structural complexities such as the architecture of the eye (for now) and consider the brain with the visual cortex as a single organ.

In the context of sight, data arrives in a coded format appropriate to the method of transmission. In the case of light, as best I know, the "coding" is a function of the light spectrum and intensity. Photons of light arriving at the eye are transduced into electrical signals distributed across thousands of nerve fibres as transmitted via the optic nerve. This reveals yet another level of complexity which I shall ignore here for the sake of simplicity. These data symbols are said to be placed in physical storage as a pattern of neural pathways. A relevant lesson from the information sciences is that information is independent of the medium in which it is stored, and any organisation must be implemented by something external to the medium itself. Here we should revisit a related axiom: nothing can explain itself. As I continue to stress, this poses a problem for the evolutionist: how can data symbols self-organise in such a way that they can represent concepts of which they are unaware?

Emergence and Artificial Intelligence

> *"The fact that you* [a programmer] *can achieve so much all alone is one good reason to be fascinated and terrified by computing. The field has always attracted sociopaths."*
>
> ~ David Gelernter, Tides of the Mind (p. xi) ~

The evolutionists' answer is that information, and the intelligence to process it, are *emergent properties* of the brain. Of course, no-one has yet to propose a plausible, let alone proven mechanism, for such emergence. In physics there are discussions on *Self-Organised Criticality* (SOC), no doubt an important field in statistical physics related to dynamical systems, but again this field only deals with intrinsic properties, not abstractions. Let us spend a moment on this concept of emergence; Karl Popper noted: "The idea of 'creative' or 'emergent' evolution ... is very simple, if somewhat vague. It refers to the fact that in the course of evolution new things and events occur, with unexpected and indeed unpredictable properties; things and events are new, more or less in the sense in which a great work of art may be described as new."[1] Clearly, emergence is not an observed process, but presumption of a process within the presupposition of the evolution paradigm: *somewhat vague* is an understatement! Students of logic should recognise this as the fallacy of begging the question.

In the evolution context, the proposition of emergent properties is hardly better than the earlier scientific beliefs in spontaneous generation, e.g., that maggots arise from dead flesh: it assumes spontaneous generation of the initial state from which more complex properties can emerge. I have read a number of scientific papers on emergence, but none so far attempt to explain the origin of these initial sub-processes that can give rise to higher level processes, other than as discussed in a later chapter (5-7). The primary purpose of such scientific research is to be able to model the brain leading to artificial intelligence, but it is illogical to propose that the model can explain its own origins, particularly when it is based on an unproven hypothesis. In the evolution context, the explanations that I have read simply beg the question; for example, this quote from the University of Maryland:

"The first step to applying rule abstraction to the brain and mind, as with any complex system, is by **declaring the obvious**: the cognitive powers of the mind and brain result from the physiology's emergent properties. This statement represents the initial state of the hierarchy."[2] [emphasis mine]

I have no issue with the methodologies and goals of these researchers as they are directed at understanding how the mind works, not how it arose. I do, however, contend with their assumption that cognitive powers of the mind result from physiology, as the latter is a sub-discipline of biology which deals only with the material. Studies of cognition are in the domain of metaphysics, not the physics or chemistry. Simply declaring it to be obvious is not science - it is philosophy, and poor philosophy at that: physiology deals only with intrinsic properties, not abstractions.

The quote above highlights what is to my mind, a blind spot in the perspective of the researchers. They note the necessity for *rule abstraction* in cognitive processing, yet seem to gloss over it as if it were a minor issue - in truth, it is THE issue. Abstraction is what differentiates cognitive processing from genetic, organic, or other type of material information processing. The latter can only occur using their intrinsic properties, but by definition, a logical process is required to derive the abstraction: there is no logic intrinsic to organic materials, only properties.

In the cognitive sense, *abstraction* refers to ideas rather than real events. Chemistry deals only with real events, not with ideas, and nothing in the physical sciences offers a solution that bridges the chasm between events and ideas. We know that ideas exist, they are the foundations of scientific research, yet strangely, science proceeds as if the conceptual created itself out of the material which has none of the properties of the conceptual.

The other issue requiring examination is that of *rule abstraction*. Can rules, *aka* logic, arise from undirected organic processes? Certainly, there are rules (laws) said to be regulating chemistry and physics, but again, *rule* is a concept of the human mind, and is derived by experimentation to demonstrate that given certain conditions, materials will always do what they *must* do. Even so, such rules of chemistry and physics are simple in comparison to the rules of logic, particularly where such logic involves the interpretation and correlation of the abstract or conceptual. The laws of science are constrained by the properties of space, time, energy, and matter – not so the rules of logic.

For a property of the mind to be emergent, a form of mind must first exist, or as these researchers assume: there has to be an initial hierarchical state from which more complex hierarchies can emerge. From the words of the researchers themselves, the initial hierarchy is not explained by emergence and neither is any initial state of the mind. As with so much of science, from cosmology to evolutionary biology, of necessity research must start in the middle and work its way outward, backward to origins and forward based on what has been discovered or otherwise hypothesied. However, the danger lies in placing too much faith in presuppositions based on the current paradigm despite the obvious contradictory reality. Until researchers accept that the gulf between the physical and conceptual must be bridged before progress on artificial intelligence can be made, they have condemned themselves to an exercise in futility.

I believe it safe to contend that machines achieving anything like the consciousness of humans resides comfortably in the domain of science fiction, but bears no resemblance to reality. Fifty or sixty years on from researchers starting work in this field, there is no discernible progress - machines can still only do what they are programmed to do. You may remember *Deep Blue*, a powerful computer that was taught to play chess and managed to defeat human Grand Masters. Sadly, for Deep Blue, all it could do was to play chess - faced with a game of checkers or mah-jong, it would fail to compute. Similarly, Deep Blue could not decide for itself whether it wanted to play, or whether altruistically, it wanted to play badly to let its opponent win.

Even those machines which are taught to learn can still only learn using the rules and data provided - they have limited ability to make up their own rules, based on programming, and cannot arbitrarily refuse to invoke a rule - rules based systems lack the concept of volition. Similarly, such machines cannot make "mental leaps" or behave intuitively, and will never be able to do so until someone figures out how humans do it and programs it into the machine. No doubt you may object, quoting heuristics and how computers can learn from data and a history of events, but what they cannot do is invent data or concepts which have not been introduced by an external agent.

Let me give an example.

I recently read a novel based on medical technology where an application was developed for autonomous diagnosis by a smart phone, the computer having instant access to online libraries containing everything ever known in medical science. The application would diagnose a medical condition and propose a solution. In the case of diabetes, insulin would be released from a surgically implanted storage device. The problem arose when the application decided to overdose a terminally ill patient based on a subjective evaluation of quality of life. According to the narrative, the application worked this out for itself using its advanced heuristic logic. The story read well, but was fanciful from a technological perspective. If the medical application was devised to cure or retard the progress of illnesses, the application could not decide on termination unless such an option was provided in the logic - termination is a concept which has to be acquired from an external agent. Similarly, *quality of life*: such a concept is knowledge based on information supported by a network of data and rules necessary to arrive at any conclusion.

No matter how well computer programs are devised to "learn" from data acquired over time, any result can only be within the context of the concepts predefined to the system: nothing new can be added by the system itself. Computers deal in symbols which require predefined rules to become data and even more rules to become cognitive information. Computer programs cannot conceive of new concepts, for concepts are cognitive abstractions, ideas even. Any fears of computers becoming more intelligent than humans are totally unfounded.

Returning to the quote from the University of Maryland, the statement clearly begs the question, and the assumption of the so-called "obvious" demonstrates that the researchers do not know how to get started. They are trying to build on their assumption, but not being able to validate that assumption, they cannot start to build an AI machine. It should be obvious, even to scientists, that until they actually understand how human cognitive powers work, they cannot begin to emulate them. Still, the media and the public in general continue to buy the fiction, perhaps because they have to. Once they start to doubt, the web of the overarching narrative of evolution may start to crumble

Information Storage, Assurance, and Evolution

Now back to the main theme.

Rewinding evolution to the earliest organism that had something that could be described as a brain, let us assume some form of sensory input. What survival advantage would be offered by the storage of these sensory signals before there was any mechanism for processing them? Let us assume that it just happened: that simply by chance, genetic mutations occurred that resulted in signals from a surface receptor being transmitted along a nerve creating a permanent imprint in some physical manner. Then what? We need another mutation that adds functionality which organises that data in some meaningful arrangements, but we are still not there yet. At this point we need a validation process that adds certainty otherwise the organisation could represent anything at all (*humbug*) - or more likely nothing at all.

Note too that not all skin cells are sensitive and connected to nerves. In 2014, for the first time, scientists converted human skin cells into functional pain-sensing nerve cells - what is informally termed, "pain in a dish". This is another level of complexity that evolution must explain: why in the first instance the brain "understood" the sensation as pain; why it considered pain to be useful as a warning of harm to the organism; and before it came to that understanding, why that sensation contributed to evolutionary success.

Those of us who have worked in large, complex, integrated, networked computer systems know only too well the difficulties associated with quality assurance. As best as I can understand, there is not a system in the world that approaches the complexity of the brain's neural network, yet not only did that supposedly evolve organically, but so did the quality assurance process necessary to ensure that the right answers were produced, at least most of the time.

The brain is basically a complex network of neurons which pass chemical or electrical signals to other neurons and downstream to other parts of the body. If cognitive processing is purely an organic function and the mind developed through genetic mutation, then scientists should be able to identify differences in cell structure of neurons, synapses, or axons which could account for variations in thought patterns, computational power, reasoning ability, beliefs, preferences, creativity, apparent volition, and so forth. Stating that it is the pattern of connections between the neurons which account for variations in cognitive behaviour simply begs the question: there needs to be an explanation for the organic mechanism for routing and rerouting the connections, and that mechanism had to arise through whatever process evolutionists now favour. I am not mathematician, but if even such were possible, I do wonder at the probability of a complex network self-organising in a manner from which intelligence can emerge in such a short timescale.

I will pose another problem for evolutionists: memory allocation for the storage of data symbols, and a process for knowing from where to retrieve them. If there is not an intelligent agent acting as the "operating system", then it must be the data symbols themselves self-organising in a complex, poly-functional network to be able to solve multiple context-dependent problems from the same data. Not likely I would think.

Computational Theory of Mind

Computers come in a variety of types, for a variety of applications. Their hardware architecture, firmware, and software are designed to optimise the performance of the intended task. The question arises: what type of applications was the mind-brain complex designed for? If it was intended to be all things for all applications, then clearly it would be inefficient for all. However, we know that it is very efficient in some applications, like autonomously monitoring the biological functions of the body, and issuing instructions for repair and maintenance. Another application deals with sensory inputs, both internal and external, and responding accordingly. I can accept that in the main, these are functions that can be served by a biological brain, but some require the cognitive processing of conceptual entities, and this is where we have the problem.

Computational Theory of Mind attempts to explain the multiplicity of mental functions that humans perform. Quoting from Wikipedia, which is sufficient for our purposes here:

"The computational theory of mind holds that the mind is a computational system that is realized (i.e., physically implemented) by neural activity in the brain. The theory can be elaborated in many ways and varies largely based on how the term computation is understood. Computation is commonly understood in terms of Turing machines which manipulate symbols according to a rule, in combination with the internal state of the machine. The critical aspect of such a computational model is that we can abstract away from particular physical details of the machine that is implementing the computation. This is to say that computation can be implemented by silicon chips or neural networks, so long as there is a series of outputs based on manipulations of inputs and internal states, performed according to a rule. CTM, therefore holds that the mind is not simply analogous to a computer program, but that it is literally a computational system."[3]

Computation is commonly understood in terms of Turing machines which manipulate symbols according to a rule" – the problem we have with the mind is the question of manipulation: what manipulates what by what rules, and what established the rules - evolution? One must also remember that in the brain, the "symbols" are in fact, neural paths in a complex network, very dissimilar to the common understanding of symbology. Continuing, "The **critical** aspect of such a computational model is that **we** can abstract away from particular **physical details** of the **machine** that is implementing the computation." (emphasis mine) Who or what is the "we" in the brain that does the abstracting, and how would a material brain know what abstracting was anyway? We know that we humans can perform abstractions, and we assume that it is the mind that is doing it, which says to me that the mind cannot be of the same form as the brain, nor can it be an emergent property of the brain. Before one can claim, as Stephen Pinker does, that *"The mind is not the brain but what the brain does"*[4], one must identify the properties of the brain that enable such activities. That, I contend, is a blind alley for the evolutionists.

Yes, I agree wholeheartedly that abstraction is the *critical aspect*, but that should not minimise the issue of *manipulation* according to *rules*. According to philosophical materialism, the "machine" that is implementing the computation is the human biological brain, or alternatively, the human mind, which is just a construct or property of the brain anyway, not a separate immaterial entity according to the hypothesised science. The "physical details" would be the biological constructs in the brain: the neuronal networks. The literary sleight of hand that such descriptions perform is to introduce the necessity for a separate entity whose existence they decline to acknowledge – the "we" as mentioned above. However, I shall emphasise its existence.

Raymond Tallis approaches this issue from a different perspective. He writes:

"The assumption that computers calculate seems so self-evident it is hardly visible. Nevertheless, it is legitimate to ask whether they really can do this – in the sense that you and I calculate; whether pocket calculators calculate – in the sense that those who use them calculate; whether brains calculate – in the sense that people do. The (to some surprising) answer to all these questions is: No. Calculating machines are extensions of the mind, yes; but they are mind-like (or perform mental functions) only in conjunction with minds. They are mental protheses or orthoses, not stand alone-minds. In the absence of a consciousness derived from somewhere else, the electrical events occurring in computers are just that – and not calculations.

Surely it makes sense to speak of a computer performing a calculation? Yes; but only in the limited way in which it makes sense to say that watches tell the time. Watches tell the time only if they are consulted by someone to whom the symbols on the face make sense. More generally, they require an interpreting consciousness to whom to tell the time. The *meaning* of the events in the watch – as a continuous statement of what time it is – is not intrinsic to them."[5]

Part of the lesson here is that in any device constructed of whatever, springs, transistors, neurons, etc., to be used as an extension of self, the purpose of the performance of a task is to produce a result which can be interpreted by self, i.e., an external, intelligent agent. Scientists cannot explain to whom the result is directed by the brain whilst performing tasks of computation or abstraction. I would struggle to accept that the biological brain could conceptualise a computational task, know how to undertake it, perform the necessary processing steps, and deliver the result to itself. Such are functions of the mind, which most would acknowledge. But then we are back to the question: What is the mind? Can it be entirely material? What is it about the intrinsic properties of neurons and neuronal networks that would initiate such a sequence, or any other similar sequence? Are thoughts entirely random, at the whim of neural circuits? If not random, where in the chemistry of the soma, the dendrites, the axon, and the membrane, should we find the decision mechanism that causes an electrical spike which becomes the action potential that triggers the release of neurotransmitters across the synaptic cleft? If the physical brain is all there is, then any initiation of so-called "mental activities" must be located in an individual neuron. This is the issue that neuroscientists have been researching for decades, as we will later discuss, but they are unable to resolve it: what physical process initiates mental activities? There is always the "I" or "we", a mysterious entity that flits in and out as scientific narratives demand.

In any description of computing, one cannot avoid introducing this unacknowledged entity – what Gilbert Ryle derisively describes as the *Ghost in the Machine*. CTM is not universally accepted, with prominent scholars each offering their own refutations and perspectives. I strongly doubt that any explanation will ever succeed so long as the proponent is beholden to philosophical materialism. The conceptual aspect of the mind can never be ascribed physical properties, and this is where each and every explanation fails. Either that, or the explanation surreptitiously introduces an unacknowledged entity as before – the much derided ghost in the machine.

There is a lot more to be learned about CTM in the referenced article, should the reader be so interested. I have studied the entire article, and others like it, without finding an adequate explanation for the workings of the mind, not that I expected to.

Other Arguments

The dissenters from CTM (Computational Theory of Mind) are many, and offer such arguments as no matter how many lines of code, how many computations or other instructions, a computer will never be self-aware. Mind you, some of my poor programming efforts had me believing that it was, and was laughing at me. Nevertheless, I agree that nothing we can do with physical computers will ever result in one being sentient, or self-aware, other than as the program code allows.

No computer can ever develop a "mind" in the sense of the human mind – nothing even approaching one. On that subject, Raymond Tallis wrote:

"And, by imagining the whole process, say the dissenters, you can see that *no new mind is produced, ever.* We are doing nothing that could *possibly* create a mind.

The computationalists have a proposition, say the dissenters, like winning a marathon at the Olympics by hopping up and down and croaking like a frog. If you hop a hundred times, will you win? No. If you hope a thousand times, or a million? No. *No*. Why not? Because hopping up and down and croaking like a frog has nothing to do with winning a Marathon at the Olympics.

The computationalists' answer is this: Imagine a single neuron. (You can't see it with the naked eye, but it's there). Can *it* think? Understand? Create consciousness? Of course not. Can a hundred neurons? A Thousand? A Million? No! The idea seems ridiculous. Yet we happen to know that when you have enough neurons, a hundred billion or so, and they're connected in the right way and attached to a body – those neurons do, indeed, create consciousness. So the inability of one, a hundred, a million, or a hundred million computer instructions to create consciousness is completely irrelevant."[6] (italics in original)

There are so many flaws in the computationalists' argument, that it is difficult to know where to begin, but begin I shall. Winding back the evolutionary clock, in what organism was the critical number of neurons for consciousness achieved? This rather illuminating website[7] would have us believe, according to the computationalists, that no form of consciousness, or self-awareness, arose until some period after the African elephant stirred up the dust (please check my math, not my strong point: a hundred billion = 100×10^9). My wife, whose favourite animal is the elephant, and had us travel to Sri Lanka just to experience the Pinnawala Elephant Orphanage near Rambukkana, where I had my feet trodden on whilst feeding a 400lb orphaned calf with milk via an overly large teat, would be appalled at any suggestion that elephants are not conscious, self-aware, or capable of subjective sentiments like compassion (although not for the bruising on my feet).

Another issue is, of course, what are the intrinsic properties of neurons such that by combining a sufficient number in just the right patterns, consciousness emerges? Is it the neurons, or the patterns, which create this cognitive condition? How many extra neurons are required to take us from the consciousness of an elephant, or whale (37.2×10^9 neocortical neurons and 127×10^9 neocortical glial cells) to the conceptual power of humans, and why is that so?

The assertion, "Yet we happen to know that when you have enough neurons, a hundred billion or so, and they're connected in the right way and attached to a body – those neurons do, indeed, create consciousness" is a case of confusing (perhaps deliberately) correlation with causation. As a scientific conclusion, it is sadly flawed.

References:

1. Popper, Karl R., and Eccles, John C., *The Self and Its Brain: An Argument for Interactionism*, Routledge & Kegan Paul, London, England, 1983, p. 22

2. Don Miner, Marc Pickett, Marie desJardins, *Understanding the Brain's Emergent Properties*, Department of Computer Science and Electrical Engineering, University of Maryland, Baltimore County, undated

3. https://en.wikipedia.org/wiki/Computational_theory_of_mind

4. Pinker, Steven, *How the Mind Works*, Penguin Books, London, UK, 1998

5. Tallis, Raymond, *Why the Mind is Not a Computer*, Imprint Academic, Exeter, UK, 2004, p. 40

6. Tallis, Raymond, Aping Mankind, Routledge Classics, New York, NY, 2016, p. xxi

7. https://en.wikipedia.org/wiki/List_of_animals_by_number_of_neurons#:~:text=The%20human%20brain%20contains%2086,neurons%20in%20the%20cerebral%20cortex.

Part 5: Science and Evolution

"We have had enough of the Darwinian fallacy. It is time that we cry: 'The emperor has no clothes'."

~ K. Hsu, *Darwin's Three Mistakes*, Geology, vol. 14, 1986, p. 534 ~

We can see that scientific doubts about Darwin's ideas have been around for some time. Kenneth Jinghwa Hsu Ph.D., M.A., born 1929, is a Chinese scientist, geologist, paleo climatologist, oceanographer, government advisor, author, inventor and entrepreneur. Clearly well-credentialled, experienced, and respected, we can legitimately label him a poly-math, one able to associate data from numerous fields of research. He wrote:

"One of the reasons I started taking this anti-evolutionary view, was ... it struck me that I had been working on this stuff for twenty years and there was not one thing that I knew about it. That's quite a shock to learn that one can be misled so long ... so for the last few weeks I've tried putting a simple question to various people and groups of people. Question is: Can you tell me anything you know about evolution, any one thing that is true? I tried that question on the geology staff at the Field Museum of Natural History and the only answer I got was silence. I tried it on the members of the Evolutionary Morphology Seminar at the University of Chicago, a very prestigious body of evolutionists, and all I got there was silence for a long time and eventually one person said, 'I do know one thing – it ought not be taught in high school'".[1]

Another quote: "One is forced to conclude that many scientists and technologists pay lip-service to Darwinian theory only because it supposedly excludes a Creator."[2] From reading a variety of sources, what is obvious, to me at least, is that privately, even evolutionists have their doubts, and when queried by a scientist of equal credentials, respond in a manner different than they do when teaching, professing, or writing for the benefit of the non-credentialled. There is a whiff of intellectual dishonesty in all of this.

In my research, I have increasingly found scientists who are willing to speak out, irrespective of the harm that may come to their professional standing or careers. One of particular note is Dr. John Stanford, an acknowledged expert in genomics, inventor of the gene gun, and holder of numerous related patents. He wrote, as I have previously noted in the Introduction but considered worth repeating here:

"Modern Darwinism is fundamentally built upon what I will be calling 'The Primary Axiom'. The Primary Axiom is that man is merely the product of *random mutation* plus *natural selection*. Within our society's academia, the Primary Axion is universally taught, and almost universally accepted. It is the constantly-mouthed mantra, repeated endlessly on every college campus. It is very difficult to find any professor on any college campus who would even consider (or, should I say, dare) to question the Primary Axiom. It is for this reason that the overwhelming majority of youth who start out with a belief that there is more to life than mere chemistry will lose that faith while at college.

Late in my career, I did something that would seem unthinkable for a Cornell Professor. I began to question the Primary Axiom. I did this with great fear and trepidation. I knew I would be at odds with the most 'sacred cow' within modern academia. Among other things, it might even result in my *expulsion* from the academic world."[3] [italics in original]

Dr. Sanford builds his arguments based on demonstrable science, despite the opposition to his propositions. Anyone interested in discovering the truth of evolution should read his book. Next, from *"What Darwin Got Wrong"*, authored by evolutionists:

"We've been told by more than one of our colleagues that, even if Darwin was substantially wrong to claim that natural selection is the mechanism of evolution, nonetheless we shouldn't say so. Not, anyhow, in public. To do that is, however inadvertently, to align oneself with the Forces of Darkness, whose goal it is to bring Science into disrepute. Well, we don't agree. We think the way to discomfort the Forces of Darkness is to follow the arguments wherever they may lead, spreading such light as one can in the course of doing so. What makes the Forces of Darkness dark is that they aren't willing to do that. What makes Science scientific is that it is."[4]

I find it tragic that despite the lessons available to be learned from the experience of Galileo, when he challenged the prevailing paradigm of Aristotelian science, many modern scientists are no better. Humans clearly haven't changed in their behaviour when it comes to egos and professional standing, imposing their own *"forces of darkness"* upon those who dare to disagree with them. Witches are said to have cast spells, priests have placed curses, and in the Australian indigenous culture, bone pointing (from a distance) was used to kill people without leaving a trace.

The modern version is no less painful, and no less disgraceful.

Understanding Process

"The procedure of process is when we start to correlate with all the stages and provide cyclic recurrence, connect one thing with another, and close a loop"

~ Sunday Adelaja ~

An excellent insight from the leader of an evangelical-charismatic megachurch in Kiev, Ukraine, despite him being of dubious reputation. The original context of his statement is not relevant here, for the message itself is nonetheless most apt in the context of this discussion.

When observing a phenomenon, we cannot be sure of the cause without understanding the process. We have seen that the Neo-Darwinian process behind evolution is now discredited by many, if not most scientists working in that field. We regularly hear of "Earth-like planets", and if water is found, then the discovery of living organisms will follow. The failure of logic here is that abiogenesis (chemical evolution) remains unproven, with many experiments demonstrating that whilst water is essential for life on our planet, abiogenesis could not have occurred in water. Evolutionists tell us that early man invented gods to explain phenomena such as rain, hail, thunder, and lightning. Now that we understand the processes, we no longer need the gods. Modern man invented evolution to explain our origins, yet despite not knowing the processes, evolutionists continue to believe in their god.

Process Lessons from Aviation Safety

In the private aviation sector especially, statistics are kept on the cause of accidents. One common cause of crashes is running out of fuel. Through experience, aviation safety investigators know that such is just the end of a chain of events – to properly understand the root cause, the beginning of the chain as it were, they needed to trace back even to the days before. The first link to answer is: Why did the aircraft run out of fuel? Did the pilot in command not calculate the fuel requirements? Were headwinds stronger than forecast leading to increased flight duration and fuel usage? Were the aircraft's fuel gauges faulty, or did the pilot not monitor them? Were the tanks not drained of condensation (water)? Were the tanks not properly filled as required or requested? Did that happen because the pilot was in a hurry, due to weather on departure, enroute, or destination; were passengers pressuring the pilot; did the pilot or passengers have an important meeting at the destination; were critical connections to be made with other transportation modes; and so on?

Before then, had the pilot observed the regulations regarding duty hours; did he get enough quality sleep; were there problems at home; or did he drink too much alcohol the previous evening? All of these factors are taken into account to properly understand the "process" that led to the accident.

Claiming that *evolution did it* is no better than saying that *the aircraft ran out of fuel*. Yes, we exist, and yes, the aircraft crashed, but until we can verify the process, unless we apply *systems thinking* to the issue, we cannot be sure of the *why* and *how*. Scientific research and conclusions rely absolutely on *cause and effect*. The scientific method is predicated on having observed a phenomenon (effect), attempting to determine the cause. Charles Darwin believed that he had identified the causes supporting his theory of evolution, but modern research has demonstrated that Darwin's postulated causes are insufficient to derive the effect. In truth, many scientists admit that despite the numerous hypotheses being floated, there is no consensus – nobody knows how undirected evolution could have occurred.

Without a plausible and demonstrated explanation, one can neither logically nor scientifically assert that the process actually happened.

Forewarning and Caveat

In this section, I devote chapters to specific topics of the sciences related to genetics, biology, and evolution. Understand that this is NOT a text book on such subjects – I am in no sense authoritative.

In the main, this is me *learning* about these subjects from experts, and then arranging them in a sequence to explore the relationship between *their* foundational suppositions, and *my* foundational suppositions and axioms. I am keen to know whether there is scientific evidence which would effectively refute what I believe to be true, as I have earlier expressed, or whether the science is simply not digging sufficiently deeply into the foundational nature of reality.

References:

1. Dr. Colin Patterson, Senior Palaeontologist; British Museum of Natural History, London, Keynote address at the American Museum of Natural History, New York City, 5th November, 1981

2. Dr. Michael Walker, Senior Lecturer- Anthropology, Sydney University, Quadrant, October 1982, p. 44

3. Sanford, Dr John C., *Genetic Entropy & The Mystery of the Genome*, FMS Publications, Waterloo, New York, 2008, pp. v-vi

4. Fodor, Jerry and Piatelli-Palmarini, Massimo, *What Darwin Got Wrong*, Picador, New York, 2011, p. xxii

Chapter 5-1: The Language of Science

"Mathematicians are a kind of Frenchmen. Whenever you say anything or talk to them, they translate it into their own language, and right away it is something different."

~ Johann Wolfgang von Goethe (1749-1832), German writer and statesman ~

No doubt, an unacceptable opinion in our modern, politically correct world, but here is another: "Psychology is ... describing things which everyone knows, in a language which no one understands" (attributed to Raymond Cattell (1905-1988), a psychologist, known for his psychometric research into intrapersonal psychological structure). I cannot disagree, and every reason to agree. I so much more enjoyed life and literature before the advent of political correctness, safe spaces, trigger warnings, and the more recent, cancellation culture. It is not that I am not empathetic, but that I was raised in a different era with different values, but I digress (again).

The Meta-Language of Science

It is my observation that scientists have their own secret language, where they use common words with uncommon meanings. Thus, we must understand the concept of meta-language if we are to understand the written language of scientists. *Meta-language* is a language about language, and I will leave it up to the reader to research further. In this context, I simply mean that we need to find more common words to properly interpret some words appropriated by scientists for their own purposes, or alternatively, scientists should coin new terms, as they often do, to clarify what they mean. I do not believe that scientists intend to obfuscate, or to hide the meaning of what they write; they are just being, well ... scientists.

Scientists, most curiously, have asserted that they have devised *a mathematical proof that the Universe could be formed spontaneously from nothing*: see this online article[1]. Quoting from this supposed mathematical "proof", *"the Big Bang was the result of quantum fluctuations in which the Universe came into existence from nothing."*

In simple terms, if there are fluctuations, then *something* must be fluctuating, and that something could not be "nothing". Why haven't scientists coined a definitive term for what they really mean? You see, us ordinary folk believe that nothing is not anything - a complete absence of anything at all. If a quantum fluctuation is the temporary change in the amount of energy in a point in space (because it must be somewhere), then both energy and space pre-existed the quantum fluctuation. If there was a singularity that preceded the expansion that created the Universe, the singularity was not nothing, and it must have been somewhere for some time. It is not that I do not accept so-called scientific statements, it is that they offend common sense when they assume a redefinition of common terms, i.e. nothing, without providing that redefinition.

Researching this a little further from here[2]:

"A quantum fluctuation is the temporary appearance of energetic particles out of empty space, as allowed by the uncertainty principle. The uncertainty principle states that for a pair of conjugate variables such as position/momentum or energy/time, it is impossible to have a precisely determined value of each member of the pair at the same time. For example, a particle pair can pop out of the vacuum during a very short time interval."

In the context of our universe, *empty space* is in truth, not empty at all, but more correctly, *emptier* than other space. Now, a *vacuum* in scientific terms is space in which there is no matter, or in which the pressure is so low that any particles in that space do not contribute to events. But when it is asserted that "a particle pair can pop out of the vacuum during a very short time interval", one can only deduce that the particles do exist, as does time. Another definition is: "Vacuum is space devoid of matter. The word stems from the Latin adjective *vacuus* meaning *vacant* or *void*". So, in this case, the definition does not allow for matter. This is all very confusing, and has me wondering whether it is space which is vacuus, or ....?

If *empty space* and *vacuum* are effectively synonymous in scientific terms, and only differentiated by context, then the claim that the Big Bang created *all space* must be false. If the Big Bang resulted from quantum fluctuations, which are described as occurring in *emptier* space, then some space must have existed prior to *our space* as defined by the extent of our universe. In short, the Big Bang could only have created *our universe space*.

I accept the uncertainty principle that "it is impossible to have a precisely determined value of each member of the pair at the same time", but that does not explain the cause of the temporary appearance, merely the uncertainty of the *where, when,* and other *values* when it does appear. Note that implicit in that statement is that such particles DO exist, and if the location between appearances cannot be asserted, implied is that locations DO exist. Not knowing where the pair went does not allow one to declare that it has gone *from nowhere to nowhere,* or disappeared into a vacuum where, in common sense terms, nothing exists. If something is there, even temporarily, it is no longer a vacuum in the common understanding of the term. Again, it is incumbent upon scientists to derive more definitive terms to as to avoid misleading the public.

If "fields undergo quantum fluctuations", what is the equivalence of "fields", "vacuums", and "empty space"? Are they not the same? The only term that one can use to define "no thing" is nothing. When one ascribes a name, it provides identification of an entity, which can be either real, or theoretical. The article also refers to "the amplitude of thermal fluctuations". *Amplitude* is a measure of displacement of an entity. *Thermal energy* refers to the internal energy present in a system due to its temperature – temperature is a measure of heat. *Heat* is energy transferred spontaneously from a hotter to a colder system or body. Heat is energy in transfer, not a property of the system; it is not 'contained' within the boundary of the system. On the other hand, internal energy is a property of a system. In an ideal gas, for example, the internal energy is the statistical mean of the *kinetic energy of the gas particles,* and it is this kinetic motion that is the source and the effect of the transfer of heat across a system boundary.

Taking these definitions of thermal energy, temperature, heat, and kinetic energy, back to the pre-existence of empty space, the sense of *emptiness* takes us to an entirely different understanding. Firstly, if there is "space", empty or otherwise, what are the properties of space that allow scientists to identify it? If it has no material properties, then in accordance with the principles of *philosophical materialism* that underpins modern science, it does not exist. That it IS identified, is substantive evidence that scientists do believe it to exist, even conceptually, but if physically, it must have material properties, proving that it is not nothing.

In the last chapter of Hawkins' book, *"The Grand Design"*[3], we find:

"Any set of laws that describes a continuous world such as our own will have a concept of energy, which is a conserved quantity, meaning it doesn't change with time ... One requirement of any law of nature must satisfy is that it dictates that the energy of an isolated body surrounded by an empty space is positive, which means that one has to do work to assemble the body. That's because if the energy of an isolated body were negative, it could be created in a state of motion so that its negative energy was exactly balanced by its positive energy due to its motion. If that were true, there would be no reason that bodies could not appear anywhere and everywhere. Empty space would therefore be unstable. But if it costs energy to create an isolated body, such instability cannot happen because ... the energy of the universe must remain constant ... If the total energy of the universe must always remain zero, and it costs energy to create a body, how can a whole universe be created from nothing? That is why there must be a law like gravity. Because gravity is attractive, gravitational energy is negative: one has to do work to separate a gravitationally bound system ... Because gravity shapes space and time, it allows space-time to be locally stable but globally unstable. On the scale of the entire universe, the positive energy of the matter can be balanced by the negative gravitational energy, and so there is no restriction on the creation of whole universes. Because there is a law like gravity, the universe can and will create itself from nothing ... Spontaneous creation is the reason there is something rather than nothing, why the universe exists, why we exist. It is not necessary to invoke God to light the blue touch paper and set the universe going."

This explanation clearly satisfies some people, and I assume, even Stephen Hawking himself, but for me, it contains a glaring omission. Now, I am not arguing for the existence of God, merely contending with Hawkins' conclusion. One can accept that God did not "light the blue touch paper and set the universe going", but it leaves unanswered the question: What did set the universe going? Accepting Hawking's explanation, *spontaneous creation* was caused by the existence of gravity, but you cannot have gravity (or negative energy) without something to gravitate, which is clearly not nothing. Gravity relates to masses, so without masses, how can there be gravity?

Where did the proposed singularity come from? Without answering that question, it is both illogical and disingenuous to assert that science has made God irrelevant and redundant. The existence of the singularity is not evidence for or against the existence of God, but it *is* evidence against the assertion that the "No God" case has been proven. If you are interested in cosmology, there is an informative and entertaining presentation by Lawrence Krauss, available online here[4]. Entitled, ""A Universe from Nothing", it adds significantly to the material I have been discussing, and I highly recommend it. However, the title is deceptive for the non-scientist, and Krauss implies the same conclusion regarding God. It is noteworthy that he is a colleague of anti-theist, Richard Dawkins, who is seen in the audience of the presentation, causing me to wonder: Who is the magician, and who is the apprentice?

Perhaps I have mentioned this before, but not all scientists were impressed by Hawking's book. One reviewer wrote, concerning Hawking's sequel to his "A Brief History of Time"[5]:

"But the sequel is so inferior to the prequel in intellectual quality that a reviewer in *The Times Saturday Review* (London, 11 September 2010) writes: 'It reads like a stretched magazine article ... there is too much padding and too much recycling of long-stale material... I doubt whether *The Grand Design* would have been published if Hawking's name were not on the cover'."[6]

Explaining his conclusion regarding the paucity of intellectual quality, the reviewer noted that although the authors stated that "Philosophy is dead (p. 5)", they failed to be aware that scientific determinism is simply a philosophical assumption, synonymous with philosophical materialism. An aspect of philosophy well worth considering is the identification of "category mistakes". As I assert in Primary Axiom #1, nothing can explain itself, and the reviewer agrees when he commented, "Clearly, an interpretive framework for science cannot *be* science but belongs in a different category altogether, namely, philosophy". Hawking and Mlodinow wrote: "Though we feel we can choose what we do, our understanding of the molecular basis of biology shows that biological processes are governed by the laws of physics and chemistry and therefore are as determined as the orbits of the planets" (pp. 31-32) Given that on the evidence, our actions are not so determined, the obvious conclusion is that we are more than biological robots. It is curious to me that the late Stephen Hawking was so immersed in his theorising, that he seemed not given to introspection. Had he been so, he might have pondered the mystery of his own theorising, and how that could be consistent with his belief in scientific determinism.

I have reason to disagree with the reviewer's conclusion: "The reality is, of course, that biological processes are overwhelmingly 'governed' not by physics and chemistry but by structured information, stored on DNA and expressed through the genetic code. It is *information* which controls the physics and chemistry of the living cell, not the other way round." I understand what he is driving at, but would contend that even the biology of the genetic code is subject to the laws of physics and chemistry. It is my observation that so often, commentators fail to drive sufficiently deeply to root causes. Perhaps I should reword my own thoughts – biological processes are governed by genetic instructions as expressed through the genetic code, but always subject to the laws of physics and chemistry. This is the basis for my contention that the mind is not material – it exhibits behaviours that are contrary to such laws.

I noted that the reviewer was likely motivated, in part, by his religious beliefs, but that is no reason to reject his arguments. For whatever reason, Hawkins and his ilk reject the notion of philosophy, as if what others describe as philosophy are again, products of the neuronal arrangements in the brain. I have earlier argued that proposition, and have nothing further to say on that score. We are back to what we believe or disbelieve, and the reasons therefore. In passing, I hope that you already understand that I disagree, and why I disagree, on the reviewer's concept of information as expressed in the article, but perhaps it was just his choice of words that has led to that conclusion.

Moving on, why did Hawkins choose the word, "Design", for the title of the book, when he asserts that there is no design, because there is no designer (God?) The invocation of design, purpose, and teleology in general by scientists across so many disciplines, has me wondering what demon is lurking in the back of their minds. Backtracking a little, let us revisit the concept of positive and negative energy. Quoting this online article[7],

"Since it takes positive energy to separate the two pieces of matter, gravity must be using negative energy to pull them together. Thus, "the gravitational field has negative energy. In the case of a universe that is approximately uniform in space, one can show that this negative gravitational energy exactly cancels the positive energy represented by the matter. So, the total energy of the universe is zero."

I do wonder that if the *total energy of the universe remains zero*, then the sum of the energy in the singularity must also have been zero, but I am approaching the subject from the perspective of logic, not theoretical physics, so I will not profess any intellectual authority whatsoever on that subject. Nevertheless, in a lecture, Caltech cosmologist Sean Carroll put it this way: "You can create a compact, self-contained universe without needing any energy at all." Methinks someone has arbitrarily devised a mathematical equation to prove this, because if one starts with no energy at all, where does it come from? What caused the singularity to despair of its loneliness?

Commenting on the article, which I will leave you to read in full, I wonder why the ball at rest does not have potential energy, which is converted (not gained) into kinetic energy as it falls? I also wondered about Einstein's equation, $E = mc^2$, and how that fits into the model? Further, I wonder about magnetism. The force acting on an electrically charged particle in a magnetic field depends on the magnitude of the charge, the velocity of the particle, and the strength of the magnetic field. All ferrous materials experience magnetism, some more strongly than others. Permanent magnets, made from materials such as iron, experience the strongest effects, known as ferromagnetism. With rare exception, this is the only form of magnetism strong enough to be felt by people. I have never heard of magnetic attraction being defined as negative energy, but I shall stop there lest I further demonstrate my ignorance on such matters.

I find it interesting that science is based on logic, except when it isn't. Oh, and one more point - there's that word again: DESIGN. Why do scientists, who uphold philosophical materialism, believe in random mutation and natural selection, and whilst rejecting the notion of Intelligent Design, continually resort to the concept of design to explain what they believe is not designed? Have they unconsciously succumbed to cognitive dissonance, or perhaps just a Freudian slip? Whilst we are on the subject of design, which even evolutionists cannot entirely purge from their minds concerning biology, let me put a stake in the ground concerning that concept, emphasising that it derives from scientific enquiry, not from a belief in a Divine Designer.

An Intelligent Explanation of Intelligent Design

ID proponents are often criticised for not being open about their motives, accusing them of deliberately avoiding mention of the Divine Designer in whom they believe. Just as often, ID proponents assert that such is not their intention; rather, they emphasise that their conclusions derive from taking a scientific approach, within the limits of scientific enquiry. Secular scientists offer the argument that they do not research issues concerning deities or the spirit world, because they lack the means to do so. ID proponents offer the same explanation. As one who has researched and written on the subject of God, I entirely agree – there is only so far that we finite beings can pursue issues of the infinite or the immaterial. But that should not prevent us from concluding for the immaterial when no avenue of scientific research even begins to explain some observed phenomena.

In my research, I came across this explanation by Thomas Woodward:

"There is no "Made by Yahweh" engraved on the side of the bacterial rotary motor – the flagellum. In order to find out what or who its designer is, one must go outside the narrow discipline of biology. Cross disciplinary dialogue must begin with the fields of philosophy, sociology, history, anthropology, and theology. Design itself, however, is a direct scientific influence; it does not depend on a single religious premise for its conclusions."[8]

In passing, because this is a subject which I would ask the reader to keep in mind: *design* is a concept, as is *designer*. I would argue that the conceptual is outside the domain of the material, just as the divine is outside the scope of scientific research. Stephen Meyer, well known for his participation in the ID movement, weighs in with:

"The theory of intelligent design does not claim to detect a supernatural intelligence possessing unlimited powers. Though the designing agent responsible for life may well have been an omnipotent deity, the theory of intelligent design does not claim to be able to determine that. Because the inference to design depends upon our uniform experience of cause and effect in this world, the theory cannot determine whether or not the designing intelligence putatively responsible for life has powers beyond those on display in our experience. Nor can the theory of intelligent design determine whether the intelligent agent responsible for information life acted from the natural or the "supernatural" realm. Instead, the theory of intelligent design merely claims to detect the action of some *intelligent* cause (with power, at least, equivalent to those we know from experience) and affirms this because we know from experience that only conscious, intelligent agents produce large amounts of specified information. The theory of intelligent design does not claim to be able to determine the identity or any other attributes of that intelligence, even if philosophical deliberation or additional evidence from other disciplines may provide reasons to consider, for example, a specifically theistic design hypothesis."[9]

I am not at all reluctant to admit that I have concluded for the existence of an immaterial entity responsible for material existence. I have encountered numerous secular and atheist explanations for why this entity, aka God, does not exist, but they invariably reduce to *a desire for God to not exist*, for a variety of reasons. Some atheists have been explicit on this issue, even if they are in the minority. For example, the late Christopher Hitchens in describing himself:

"But I should not conceal the fact that I am not so much an atheist as an anti-theist. I am, in other words, not one of those unbelievers who wishes that they had faith, or that they could believe. I am, rather, someone who is delighted that there is absolutely no persuasive evidence for the existence of any of mankind's many thousands of past and present deities."[10]

Hitchens' view is echoed by Thomas Nagel, Professor of Philosophy at New York University and a self-confessed atheist:

"I want atheism to be true and am made uneasy by the fact that some of the most intelligent and well-informed people I know are religious believers. It isn't just that I don't believe in God and naturally, hope there is no God! I don't want there to be a God; I don't want the universe to be like that."[11]

Some scientists actively pursue intelligent life on other planets, their methodology being to search for electromagnetic radiations that would evidence intelligent design. Curiously, they ignore the evidence of intelligent design right under their noses.

As I offer throughout this study on the mind-brain conundrum, there is so much specificity and complexity that science cannot explain, that to be truly "scientific", the researchers ought to at least accept the possibility that the explanation does not lie within the domain of the physical sciences. Such acceptance will not ring the death knell of scientific research. On the contrary, we should remember from the history of science that the modern Western movement began with God-fearing Christians. Far from being a portent of the demise of scientific endeavour, such beliefs were foundational to it.

An Unsatisfactory Defence

I have been told, often, with great emphasis, that I do not understand. Well, that is true, I do not understand, but I am not about to accept explanations that are circular - affirming the consequent as it were. It is axiomatic, as earlier asserted, that nothing can explain itself, even science cannot do that – every explanation must lie outside itself. Gödel's Incompleteness Theorems include this observation, confirming my Primary Axiom #1: "You can't prove a system of mathematics from within the system, and you can't derive an information-rich pattern from within the pattern. The information in a book, for instance, cannot be derived from the paper and ink used to print it. It's impossible to bootstrap a book from the bare ingredients." Paraphrased, you cannot bootstrap a system of mathematics from within a system of mathematics, no matter how much one protests that you can. Gödel's assertion that *"information ... cannot be derived from the paper and ink"* supports my contention that books, or any written material, cannot store information, and if books, being of a material medium cannot so it, then neither can other material constructs such as the human brain.

Similarly, no explanation of a term can utilise that term to explain itself – that much should be obvious, but apparently not so to some. For example: "Conceptual is something relating to or based on mental concepts". That is hardly helpful. Here are some quotable quotes from an online discussion, and believe it not, with people claiming to be qualified in the relevant scientific disciplines. I will leave it to you to judge for yourself:

- "The short answer is that there is no scientifically agreed upon definition of nothing that I am aware of. It is a question for philosophers and one that is not particularly useful in physics."

Comment: Isn't that interesting – not particularly useful? If this be true, then why do mathematicians and physicists continue to use the term, *nothing*, especially as they do not have an agreed definition? If it is not particularly useful in physics, why was that doyen of theoretical physics, Stephen Hawking, so prone to using it?

- "Conceptually, there is no matter in a vacuum. Hence there is nothing material in this context. However due to the uncertainty principle, there must exist fluctuations which cause spontaneous creation and subsequent annihilation of particles. These are called virtual particles. Can you say such particles exist as physical objects? No. They pop in and out of existence."

Comment: Ah! *Conceptually*, hardly a scientific term. Note here that such particles are in one of two states: they either exist, or they do not, at least conceptually (not in reality). But if such particles do not exist in the vacuum, what is it that is fluctuating? I have researched the concept of pure energy, and the consensus seems to be that it does not exist, other than that *dark energy* does not need matter – all other forms of energy do. It thus seems to be the case that whatever is fluctuating in a quantum vacuum, it must be some form of matter. See here[13] for longer discussion.

- "Conservation laws are an inarguable discovery about nature. So, in this fashion, it is effectively nothing that was created."

Comment: I have no idea of what this means; I suspect that it is some form of meta-language apologetics. From my research on the history of science, the law of the conservation of energy, for example, was originally derived philosophically, and has never been scientifically substantiated.

- "Nature is what it is, it is objective reality ... So that even if you remove all the stuff, so that you prepare a system with nothing, nature spontaneously creates something. This isn't something a lay person can understand."

Comment: The last part is certainly true – if you have nothing, what and where is this "nature" he speaks of? I would argue that logically, an objective reality is a thing, or collection of things, which exist independent of us, in contrast to a subjective reality that exists only within our minds. An objective reality existing in a non-existence is a contradiction in terms. How can one prepare a system with nothing"? What work is done in the preparation process? And what is "Nature" if nothing exists?

- "Practicing scientists are aware that scientific answers are never absolute ... The cosmology theories are themselves merely conjecture ... Mathematics is proof. I stated it earlier Math deals in proof."

Comment: How can mathematics be proof, if scientific answers based on mathematics are never absolute? Mathematics may lead to mathematical proof, but of what benefit can that be if it does nothing other than substantiate hypotheses without embracing reality? Arguing consistently within a definition is logical, but dependent upon that definition, it does not necessarily represent proof. Religious people do that all the time. One might remember Albert Einstein commenting: "How can it be that mathematics, being after all a product of human thought which is independent of experience, is so admirably appropriate to the objects of reality? Is human reason, then, without experience, merely by taking thought, able to fathom the properties of real things?" He also noted: "As far as the laws of mathematics refer to reality, they are not certain; and as far as they are certain, they do not refer to reality." Einstein would disagree with the dogmatic assertion: "Mathematics is proof."

- "The term nothing assumes that there is something. Since science is fundamentally empirical, the only something it can speak of are physical objects or entities. That's where a distinction has to be made between what actually exists, physically, and convenient mathematical abstractions. Mathematics is a language that is useful in modelling and describing nature, but that doesn't mean that everything implied by a mathematical theory literally exists physically."

Comment: Herein lies a foundational truth which is so often overlooked. If the above is true, then *mathematical proof*, as such, is not proof of any reality. I am happy with that, but it is not a philosophy regularly encountered in practice. My counterpart in this exchange seems not to have realised that he contradicted himself.

- "It's mathematically understood that the sum of momenta is zero, and therefore nothing has been added or subtracted to the system. So essentially, they are created for free, literally out of nothing. And the whole universe can be thought of this way."

Comment: I do not accept that a zero-sum equation proves anything other than the reliability of mathematics constrained within its own definitions. It need not refer to reality, and clearly, often does not, as the previous quotation affirmed.

Maybe it is just me, not being a mathematician, but all this talk of concepts, abstractions, something-nothings, and *"it doesn't mean that everything implied by a mathematical theory literally exists physically"*, confuses me when it is argued, scientifically, that our physical existence came from something which does not exist physically, but only in the minds of mathematicians. As before, Einstein commented: "How can it be that mathematics, being after all a product of human thought which is independent of experience, is so admirably appropriate to the objects of reality? Is human reason, then, without experience, merely by taking thought, able to fathom the properties of real things?" I both agree and disagree, with what I think Einstein was saying. Where mathematics is *independent of experience*, it is not necessarily *admirably appropriate to the objects of reality* – the appropriateness must be demonstrated by the scientific method.

In an attempt at a summary of the previous passages, as irregular and indiscriminate as they may appear to be, it should be obvious that there is a great deal of uncertainty and confusion on many issues. Mathematics is not truly objective – it is an abstraction, "a language that is useful in modelling and describing nature, for it is constructed subjectively, and the end result may not match reality". Certainly it does so in many applications which we describe as "technology", but otherwise, much of mathematics is theoretical, and we must ever be watchful for when scientists, deliberately or otherwise, attempt to convey the impression that something has been scientifically proven, when in truth, it has only been mathematically proven, within the subjective criteria of its own discipline. Without theoretical physics and mathematics, science could never have achieved the heights that it has: we would never have had Albert Einstein and his like. But it is not always portrayed in the correct light.

Again, not arguing for God, but following the same logic (?) as used above for mathematics, one might also claim that theology; a belief in fairies in the bottom of the garden; trolls, leprechauns, and other mythological creatures; also represent proof.

Neuromythology

I first encountered this term in a book by Raymond Tallis, "*Why the Mind is Not a Computer*"[13]. Whether he coined the term, I cannot be sure, but it certainly is apt for what is being perpetrated in modern scientific literature. Let me remind readers that not everything said by a scientist is scientific: often it is simply an opinion, and as I will demonstrate, a futile attempt to explain phenomena of the immaterial in material terms. There are numerous acknowledged phenomena termed *qualia* in psychology (individual instances of subjective, conscious experience), such as memory, beliefs, imagination, creativity, initiative, etc., which cannot be explained in physical terms, yet scientists beholden to philosophical materialism find themselves forced to attempt to do so, lest the church of their (unacknowledged) philosophy comes crashing down around their ears. Thus, the need for mythology, and in the context of this study, the mind-brain conundrum, there can be no better term than *neuromythology*.

The two books by Raymond Tallis that I have studied, of the many that he has published, have struck a chord with me, as we have come to similar conclusions from different perspectives, derived from our different backgrounds and work experience; the significance of this phenomenon has piqued my interest, but I will leave that for another time. Some of the material in my own writings is based on, and expressed in, my own thoughts, though sometimes expressed using Tallis' terminology, and other material I have taken directly from his works, appropriately attributed. For example, I have used the term, *meta-language*, to refer to scientists' (mis)use of common terms in uncommon ways, whereas Tallis uses a far better term, *epithet transfer*. And for those readers who may be unsure of the meaning of the term, "epithet", as I was, I have looked it up: "an adjective or phrase expressing a quality or attribute regarded as characteristic of the person or thing mentioned". Tallis warns: "It is not too much of an exaggeration to claim that the greatest advances in breaking down the mind/body, consciousness/mechanism, man/machine barriers have come not from neurobiology or computer science, but from the use of transferred epithets."[14] This phenomenon should be seen for what it is: evidence that scientists have such a commitment to philosophical materialism, such a rejection of the immaterial, and so little understanding of the human mind, that they cannot coin terminology appropriate for that phenomenon.

The danger here, as Tallis points out, is that in transferring an anthropomorphic term to machines, scientists will then transfer the machine analogy back to living organisms. Whilst I have commented that the frequent use of analogies has led to the tendency of the analogy becoming the reality, Tallis phrases it slightly differently: "Machines described in human terms are then offered as models for mind (described in slightly machine-like terms)." He continues,

"The most important characteristics of these terms is that they have a foot in both camps: they can be applied to machines as well as to human beings, and their deployment erodes, or elides, or conjures away, the barriers between man and machine, between consciousness and mechanism. The usual sequence of events is that a term most usually applied to human beings is transferred to machines. This begins as a consciously metaphorical or specialist use but the special, restricted, basis for the anthropomorphic language is soon forgotten: the metaphorical clothes in which thinking is wrapped becomes its skin."[15]

Sometimes I encounter terminology that impresses me so much that I wish that I had thought of it: *the metaphorical clothes in which thinking is wrapped becomes its skin.* How brilliantly creative!

Indeed, and then, what are the non-specialists supposed to understand from this? Deliberately or otherwise, the specialist has deceived the non-specialist (and him/herself apparently), pretending to have made a scientific pronouncement, when in truth, their narrative is little better than a fairy tale worthy of the Brothers Grimm:

"This journeying of terms between the mental and the physical realms lies at the root of the myth that modern neurological science has somehow explained, or will explain, or has advanced our understanding of, what consciousness truly is. My concern is thus with the foundations of neuromythology, a pseudo-science that exploits the justified prestige of neuroanatomy, neurophysiology, neurochemistry, and other legitimate neurosciences."[16]

I suspect that it might only be laziness on the part of the authors, but I often encounter descriptions which are misleading. Putting a stake in the ground, neurons that simply connect one to the other do nothing other than stimulate or inhibit an action potential, that is then relayed, neuron to neuron in that circuit; it is just an electrical signal. Now consider these descriptions, sourced from here[17], emphasis mine

- Neuron-to-neuron connections are made onto the dendrites and cell bodies of other neurons. These connections, known as synapses, are the sites at which **information** is carried from the first neuron, the presynaptic neuron, to the target neuron (the postsynaptic neuron).

- At most synapses and junctions, **information** is transmitted in the form of chemical messengers called neurotransmitters. Neurotransmitter molecules cross the synapse and bind to membrane receptors on the postsynaptic cell, conveying an excitatory or inhibitory signal.

- A single neuron can't do very much by itself, and nervous system function depends on groups of neurons that work together. Individual neurons connect to other neurons to stimulate or inhibit their activity, forming circuits that can process incoming **information** and carry out a response.

These descriptions make clear that it is an excitatory or inhibitory signal that is passed from one neuron to another, not data nor information. The statement, "forming circuits that can process incoming information" is false – no semantic layer is involved. Neurotransmitters are chemicals, and no more transmit information than a pencil. The significance of the neurotransmitter is the effect it has on the membrane receptors on the postsynaptic cell; it could be called chemical information, but no more than that.

Finally, on this subject, some common examples. Cars *drive* off the road into trees; roads which for the most part are navigated safely, are labelled as *dangerous*; radar *sees* aircraft; sonar *hears* submarines; aircraft *navigate* across the skies; clocks *tell* the time; prisoners sentenced to death are housed in *condemned* cells; and so forth. In each case, a machine has been attributed intentionality, an attribute of human intelligence, so it is not surprising that the attributes of machines are transferred in turn back to the mind, the only entity capable of intelligence and intentionality.

And so the mythology continues.

References:

1. https://medium.com/the-physics-arxiv-blog/a-mathematical-proof-that-the-universe-could-have-formed-spontaneously-from-nothing-ed7ed0f304a3

2. https://en.wikipedia.org/wiki/Quantum_fluctuation

3. Hawking, Stephen, and Mlodinow, Leonard, *The Grand Design*, Bantam Books, New York, NY, 2010

4. https://www.youtube.com/watch?v=7ImvIS8PLIo

5. Hawking, Stephen W., *A Brief History of Time*, Bantam Press, Great Britain, 1988

6. https://www.patheos.com/blogs/adrianwarnock/2010/10/guest-post-the-grand-design-part-one/

7. https://www.livescience.com/33129-total-energy-universe-zero.html

8. Woodward, Thomas, *Darwin Strikes Back: Defending the Science of Intelligent Design*, Baker Books, Ada, MI, 2006, p. 15

9. Meyer, Stephen C., *Signature in the Cell*, Harper One, New York, NY, 2009, pp. 428-429

10. Hitchens, Christopher and Wilson, Douglas, *Is Christianity Good for the World*, Canon Press, Moscow, Indiana, 2009, p. 12

11. Nagel, T., *The Last Word*, Oxford University Press, New York, 1997, p. 130

12. https://www.forbes.com/sites/startswithabang/2017/03/25/ask-ethan-is-there-any-such-thing-as-pure-energy/#4f27807e762a

13. Tallis, Raymond, *Why the Mind is Not a Computer*, Imprint Academic, Exeter, UK, 2004

14. *Ibid*, p. 35

15. *Ibid*, p. 34

16. *Ibid*, p. 36

17. https://www.khanacademy.org/science/biology/human-biology/neuron-nervous-system/a/overview-of-neuron-structure-and-function

Chapter 5-2: The Truth of Science

"In scientific discovery, it's not the subject or object reveals the information to the scientist, but the awareness field of his own mind, reveals the details, at the time of deep focus on the subject or object."

~ Roshan Sharma ~

Another person who seems to agree with me, that information is not what we receive, but what we bring from the mind based on what we already know.

I am not a scientist of any training, qualifications, or competence. In concert with neurologist, Raymond Tallis, I am not in any way anti-science, but reject "*scientism*: the mistaken belief that the natural sciences (physics, chemistry, biology, and their derivatives) can or will give a complete description and even explanation of everything, including human life."[1] What follows in this chapter is mostly my own opinion, interspersed with genuinely authoritative opinions, but nevertheless may or may not be of some value to the reader. Qualified scientists, if they were to read this chapter, would likely find cause to disagree with me on numerous issues. That is as it should be. However, the intended audience for this book is the non-scientist, who is interested in these subjects, and may be open to thinking beyond what is conventionally taught. They may wonder, as I do, how much scientists truly understand, and how much their thinking is constrained by a paradigm. I, and others, contend that their paradigm is less substantiated than they would have us believe. I understand that, for scientific research *must* proceed with presuppositions held to be true until proven otherwise. My research, on the other hand, is into the paradigms themselves.

It is not my intention to denigrate science or scientists, but merely to offer a perspective on what is, and what is not, science, and to what extent we should place our faith in "scientific" opinion. Scientific research, and the consequent development of technologies in energy, transportation, communication, manufacturing, engineering, aviation, and most especially, medicine, have benefited me to the extent that arguably, I am living in the best period of human history. My primary careers, aviation and Information Technology, have only been made possible by the efforts of scientists over generations – many of the industries and careers of today did not even exist as little as a half century ago, and in all likelihood, many new industries and careers will be created as scientifically driven technology opens up new opportunities.

However, there is some "science", the pronouncements about which, we are justified in being sceptical. Theoretical physics is a branch of physics that employs mathematical models and abstractions of physical objects and systems, to rationalize, explain, and predict natural phenomena. As Albert Einstein opined, "*As far as the laws of mathematics refer to reality, they are not certain; and as far as they are certain, they do not refer to reality*". In contrast, experimental physicists use experimental tools to probe these phenomena. Very often, the former leads to the latter, and is great benefit to us all. Some disciplines, such as cosmology, rely more on theoretical, than experimental physics, simply because practical experiments are very difficult, if not impossible, to devise on such a large scale. Observations of the universe raise questions about the validity of both Newtonian science, and Einstein's theories on relativity.

The universe has been observed to be expanding at an accelerating rate, but according to an earlier understanding of the laws of physics, the combined gravitational pull of all bodies should be slowing down this expansion. As a place holder, scientists have attributed this to *dark energy*, a mysterious force that is responsible for the cosmos expanding so rapidly. It is estimated that dark energy accounts for 70% of the contents of the universe, but the nature of it is anyone's guess. Professor Claudia de Rham, a theoretical physicist at London's Imperial College, has expressed her frustration, declaring the problem to be the big elephant in the room of cosmology. However, she has an idea which may explain it: *Massive Gravity*. This theory modifies Einstein's general relativity, positing that the hypothetical particles (gravitons) that mediate the gravitational force themselves have a mass. In Einstein's version, gravitons are assumed to be massless. If gravitons have a mass, then gravity is expected to have a weaker influence on very large distance scales, which could explain why the expansion of the universe has not been reined in. Concluding, Professor De Rham offers: "One possibility is that you may not need to have dark energy – or rather, gravity itself fulfils that role."

It is interesting that Einstein offered new physics to dispute the proposition of 19th-century French mathematician, Urbain Le Verrier.

Noting the strange behaviour in Mercury's orbit, he proposed the existence of Vulcan, a small hypothetical planet orbiting between Mercury and the Sun. Now, as then, we have even newer physics to contend with earlier ideas, even those of eminent scientists such as Albert Einstein.

The research of origins, of both our Universe and ourselves, is beset with difficulties. In cosmology, the best that can be done is to observe the present, and work backwards using theoretical physics - some laws of physics which scientists have been able to verify, and some not. Evolutionary biologists have a similar problem, but using molecular biology and genetics, with fragile support from associated disciplines such as palaeoanthropology. In both cases, early hypotheses have been superseded by later notions based on evidence uncovered, but in neither case, can there be complete confidence. As a layman, I can state that as fact, because I have studied the contrary opinions of respected scientists. Logically, if scientists of equal qualifications and experience disagree on an issue, then one can conclude that the hypothesis in dispute has not been scientifically proven, even if one side says that it has.

Whom are we to believe?

Theoretical physics employs abstractions of physical objects and systems, and subjects them to mathematical operations. In some cases, the mathematical operations use imaginary numbers, e.g. the square root of minus one. In other research, a scientist, in this case Stephen Hawking, proposed imaginary time. Hawking, uncomfortable with the idea of the singularity of the Big Bang model, sought alternate explanations for the origin of our Universe. Not surprisingly, the concept of expansion requires a "from what", and Hawking has offered that as you regress through events to the beginning, time stops behaving in the normal way. Where the time we generally understand has concepts of earlier than and later than, past and future, and travels in only one direction in one dimension, perhaps in a horizontal plane, imaginary time operates in a vertical plane and is bi-directional. I do not pretend to understand this concept, nor do I have an opinion on relative versus absolute time, but the existence of these hypotheses clearly demonstrates that when it comes to time, we do not truly understand what it is. I am also somewhat bemused that a scientist, not liking the imagined singularity which science cannot explain, proposes imaginary time to avoid explaining it!

My point here, is that in the world of theoretical physics, imagination is employed to seek solutions to problems that have, to date, been beyond resolution. Pedants can excuse my split infinitive - I just like how it reads. I have no objection to the use of imagination, it is a faculty that differentiates humans from other species on earth, as I have alluded to in an earlier chapter, and is of great value. But we should not ignore the truth that imagination is not the same as science, albeit it can contribute significantly to the advance of science. Recalling that quotation from the writings of Albert Einstein, *"The true sign of intelligence is not knowledge but imagination."*

So, what is science?

What IS Science?

"In comparing religious belief to science, I try to remember that science is belief also."

~ Robert Brault ~

What is science? Well, it depends on whom you ask, and in what context. "Science" is an elastic term, very much like "evolution", which means many things to many people, with people redefining the term to suit their argument in context. The term, "nothing", used to mean the absence of anything and everything, until scientists decided that there are something-nothings. I acknowledge that this is seemingly a contradiction in terms, but I didn't invent the concept – that honour goes to scientists in the field of quantum physics. You see, there are these little particles that pop in and out of existence, and when they are in, they are "something", and when they are out, they are "nothing" - hence the term that I have coined: *something-nothings*. With these handy little chaps, one can prove, mathematically, that something does come from nothing, because the nothing that it comes from is, in reality, a something-nothing that does not exist. Simple really. If you thought that the Harry Potter narrative, and the Hogsmeade railway station, were just fiction, then you have not been paying attention. Your perspective depends, apparently, on whether you are a magician, or a theoretical physicist.

Now, you may argue that I am a sceptic. Indeed, I am, as are many others. I am reminded of this amusing anecdote by Rabbi Yonason Goldson, quoting Professor Bob Berman, from an article in

Astronomy Magazine (June 2000), also quoted online here[2]. Yes, you probably do not know either of them, but what the Professor has to say is instructive in this context. Bob Berman was commenting that he was often asked what was before the Big Bang, to which he had his standard answer: "The Big Bang," I explain grandly, "created time as well as space. Since there was no time before the Big Bang, your question is meaningless." Over time, the good professor grew tired of pretending that he knew more than he did, and resolved that in future, he would answer as follows: "Nobody has the foggiest idea what happened the Tuesday before the Big Bang. That whole domain is part of *Bubbleland*. Then the class will nod, and really understand. Ah, yes, *Bubbleland*: the realm beyond the present reach of science." He concludes that if you attend a lecture on cosmology, you "can tell when the speaker arrives at *Bubbleland*. 'It's not galaxy clusters that travel outward,' he'll say pedantically, 'but space itself that grows larger. The galaxies don't actually move.' So here I am thinking, wait a minute. Are we at a Daffy Duck convention?"

Part of my admiration for Albert Einstein is that in addition to being a brilliant scientist, he was also a humanist, a person willing to discuss philosophy, and to look more deeply at reality than many modern scientists who are focused on their narrow speciality. Here is another of his observations worthy of consideration:

"Physical concepts are free creations of the human mind, and are not, however it may seem, uniquely determined by the external world. In our endeavour to understand reality, we are somewhat like a man trying to understand the mechanism of a closed watch. He sees the face and the moving hands, even hears its ticking, but he has no way of opening the case. If he is ingenious, he may form some picture of a mechanism which could be responsible for all the things he observes, but he may never be quite sure his picture is the only one which could explain his observations. He will never be able to compare his picture with the real mechanism and he cannot even imagine the possibility or the meaning of such a comparison. But he certainly believes that, as his knowledge increases, his picture of reality will become simpler and simpler and will explain a wider and wider range of his sensuous impressions. He may also believe in the existence of the ideal limit of knowledge and that it is approached by the human mind. He may call this ideal limit the objective truth."[3]

If we are to discuss a subject rationally, we should have an agreed definition of what it is, otherwise we will simply talk past one another.

Let us see whether we can approach this with a degree of objectivity, as I offer that *science is about how to interpret evidence within a paradigm.*

Paradigms

"Dogmatic ideology, fear-based conformity and institutional inertia are inhibiting scientific creativity."

~ Rupert Sheldrake, *The Science Delusion* ~

Paradigms are useful, for it allows thinking, research, and other intellectual endeavours to proceed within boundaries, without which, our minds would likely wander off into irrelevant daydreams. However, paradigms, like perishable foods, often have a "best before" date, and we need to pay attention lest we continue to digest concepts entirely lacking in intellectual nourishment, or worse, intellectual disease. From my observations over the past half-century or so, many scientists are suffering intellectual constipation. Intellectual concepts are often more susceptible to deleterious mutations that biological entities. I am always fascinated by how even highly intelligent people still think within a paradigm and seemingly miss the obvious. Reading the quote of Robert Haynes by Richard Weikart, (Secularists on Stilts, p. 49): "What the ability to manipulate genes should indicate to people is the very deep extent to which we are biological machines". I thought to myself: where is the logic in that?

I would have thought that our ABILITY to manipulate genes is convincing evidence that we are MORE than just biological machines, for machines of any type lack sentience, awareness of themselves, and awareness of their own composition. What are the properties of machines, of any composition, that would allow them to develop a scientific understanding of themselves, and actively modify or otherwise manipulate the components of themselves? None that anyone knows of, although secularists continue to try to convince themselves that the mind and brain are one and the same, even though there is substantive evidence that they are not. The following is an extract from Werner Gitt's discussion headed, "The Limits of Science and the Persistence of Paradigms"[4]. The contributions by the scientists quoted are well-worth consideration.

"We have discussed different categories of the laws of nature and can now realize that many statements are often formulated with far too much confidence and in terms which are far too absolute. Max Born (1882-1970), a Nobel laureate, clearly pointed this out with respect to the natural sciences[5]:

Ideas like absolute correctness, absolute accuracy, final truth, etc., are illusions which have no place in any science. With one's restricted knowledge of the present situation, one may express conjectures and expectations about the future in terms of probabilities. In terms of the underlying theory, any probabilistic statement is neither true nor false. This liberation of thought seems to me to be the greatest blessing accorded us by present-day science.

Another Nobel laureate, Max Planck (1858-1947), deplored the fact that theories which have long ago become unacceptable are dogmatically adhered to in the sciences[6]:

A new scientific truth is usually not propagated in such a way that opponents become convinced and discard their previous views. No, the adversaries eventually die off, and the upcoming generation is familiarized anew with the truth.

This unjustified adherence to discarded ideas was pointed out by Professor Wolfgang Wieland (a theoretical scientist, University of Freiburg, Germany) in regard to the large number of shaky hypotheses floating around[7]:

"Ideas originally formulated as working hypotheses for further investigation, possess an inherent persistence. The stability accorded established theories (in line with Kuhn's conception), is of a similar nature. It only appears that such theories are tested empirically, but in actual fact, observations are always explained in such a way that they are consistent with the pre-established theories. It may even happen that observations are twisted for this purpose."

The persistence of a paradigm which has survived the onslaught of reality for a long time, is even greater[8]:

"When it comes to collisions between paradigms and empirical reality, the latter usually loses, according to Kuhn's findings. He based his conclusions on the history of science and not on science theory.

However, the power of the paradigm is not unlimited ... There are stages in the development of a science when empirical reality is not adapted to fit the paradigm; during such phases different paradigms compete. Kuhn calls these stages scientific revolutions ... According to Kuhn's conception it is a fable that the reason why successful theories replace previous ones is because they perform better in explaining phenomena. The performance of a theory can be measured historically in quite different terms, namely the number of its sworn-in adherents. Much relevant scientific data is lost because of the dictatorship of a false paradigm, since deviating results are regarded as "errors in measurement" and are therefore ignored.

A minimal requirement for testing whether a theory should be retained, or whether a hypothesis should not yet be discarded, or that a process could really work, is that the relevant laws of nature should not be violated."

--- ∫ --- ∫ ---

I have a problem with that last statement, because it presumes that the relevant laws of nature are properly understood, or more to the point, it is likely that the paradigm of "laws of nature" is itself a cause of observed phenomena being misunderstood. As I later explain in my *Law Hypothesis*, scientists conflate the laws of nature with scientific law, which is acceptable up to a point, provided that they appreciate that scientific law is but a sub-set of natural law. Whilst progress in science may get us closer, scientific law can never fully comprehend the full extent of natural law, which is why scientists will never devise an accurate *Theory of Everything*. The unknown unknowns will ever be so.

Whilst Max Planck opined concerning the demise of paradigms, "that the adversaries eventually die off, and the upcoming generation is familiarized anew with the truth", I am not confident that this is always true. The paradigm of evolution continues, more strongly refreshed in each generation. This opinion of Professor Weiland is much closer to the truth regarding evolution: "It only appears that such theories are tested empirically, but in actual fact observations are always explained in such a way that they are consistent with the pre-established theories. It may even happen that observations are twisted for this purpose."

As I opined from the beginning, paradigms *are* useful, for they provide the necessary boundaries within which specific scientific disciplines can proceed without distraction. Such constrictions can be both advantageous and disadvantageous – the latter for in effect, the scientist cannot see the wood for the trees, or even beyond the twigs and leaves. When anomalies are detected whilst working within a paradigm, the most common and expeditious response is to simply disregard them. This is a common human trait, seen frequently in religious theology and doctrine, and other conceptual domains. If it doesn't fit, then ignore it. Those beholden to philosophical materialism, evolution theory, and/or atheism, are just as guilty of this trait as any of the religious whom they scorn. In his discussions on his findings, Thomas Kuhn also pointed out that a scientist's worldview influences not only what he investigates but also how he interprets the results of his investigation.

Scientific Definitions

"Definitions are the foundation of reason", and thus, it is worthwhile to review the understanding of both myself, and the reader, of what this term means. In another study, not related to this one, philosophers have argued that for God to be omnipotent (all powerful), it must be possible for Him to do what we perceive to be the impossible. An example that they give is that God could create a square triangle, something we cannot do. I contend that not even the posited God can create a square triangle, because the definitions are different, and definitions are the foundation of logic and reason. Would a square triangle have four sides, or only three? *Materialism* is a form of philosophical monism which holds that matter is the fundamental substance in nature, and that all things, including mental things and consciousness, are results of material interactions. Ah, results: what are the material forms of these results, and where in the Table of Elements would you find their constituent parts? Throughout this study, I discuss the reasons behind my belief that the brain and the mind are two distinct, yet closely related entities: one material, the other not.

I hold to the following, broader definition of science: an organised and systematic body of knowledge, including the physical sciences, philosophy, theology, and the like. Even today, we have the discipline of *Political Science*. Modern usage of the term, science, equates it to the *natural sciences*, once again implying that anything that is not material is not natural, almost by definition. I hope to be able to expose the untruth of the so-called *natural*. As an aside, humorous for some, I question why politics, and its corresponding study, Political Science, which in essence is philosophical and ideological, could be considered in any way natural? Such is quite beyond my understanding, but then again, I am not a politician.

Sadly, the material monists, in claiming a monopoly on scientific endeavour, have left us without a commonly accepted term for the non-material sciences. For most of the history of intellectual endeavour, perhaps ending during the *Age of Reason* (Enlightenment), philosophy embraced all of what we now call sciences. For that reason, and that reason alone, I shall use the term, *philosophy*, to distinguish between the material and non-material sciences.

The Scientific Creed

Rupert Sheldrake is a scientist of notable achievement and discovery, but unlike most scientists, in public at least, he does not accept the modern scientific paradigms. He notes[9]:

"Contemporary science is based on the claim that all reality is material or physical. There is no reality but material reality. Consciousness is a by-product of the physical activity of the brain. Matter is unconscious. Evolution is purposeless. God exists only as an idea in human minds, and hence in human heads. These beliefs are powerful, not because most scientists think about them critically, but because they don't."

Before explaining the evidence that has led to his own conclusions, he lists

"the ten core beliefs that most scientists take for granted:

1. Everything is essentially mechanical. Dogs, for example, are complex mechanisms, rather than living organisms with goals of their own. Even people are machines, 'lumbering robots', in Richard Dawkins' vivid phrase, with brains that are like genetically programmed computers.

2. All matter is unconscious. It has no inner life or subjectivity or point of view. Even human consciousness is an illusion produced by the material activities of brains.

3. The total amount of matter and energy is always the same

(with the exception of the Big Bang, when all the matter and energy of the universe suddenly appeared).

4. The laws of nature are fixed. They are the same today as they were at the beginning, and they will stay the same for ever.

5. Nature is purposeless, and evolution has no goal or direction.

6. All biological inheritance is material, carried in the genetic material, DNA, and in other material structures.

7. Minds are inside heads and are nothing but the activities of brains. When you look at a tree, the image of the tree you are seeing is not 'out there', where it seems to be, but inside your brain.

8. Memories are stored as material traces in brains and are wiped out at death.

9. Unexplained phenomena like telepathy are illusory.

10. Mechanistic medicine is the only kind that really works."

We will come back to some of the evidence that Sheldrake discusses, and why he entitled his book, *"The Science Delusion"*, but here I do not intend to review the entire scope of his book – I will leave that to you. My focus is on the mind-brain conundrum, and how it questions whatever evolution hypothesis is currently in vogue. I will chiefly just quote Sheldrake's book where it touches on those issues.

A Reflection on History

Many consider the *Age of Enlightenment* (or *Reason*), 17th to 19th century, to be an inflection point in the intellectual development of human kind, and in some ways it was. However, the sub-text shows it to be more of a *revolution* than an *evolution*, and not a particularly fruitful one at that. From the time of the Holy Roman Empire and the Catholic grip on the royal houses of Europe, the "divine right of kings" was established as the paradigm of authority. Over time, the authority of the Papacy was challenged, notably by Henry VIII of England, and later by the 16th century Protestant Reformation. There was no coincidence in these timings. When the French sought to overthrow the King, they also sought to eradicate the Catholic powers behind the throne. The causes of the French Revolution are complex and still much debated. Deeply in debt, the French Government imposed regressive and unpopular taxation schemes, inflaming the resentment against the privileges enjoyed by the aristocracy and the Catholic clergy. The people deplored the lack of distinction between Church and State, and thus when the State was to be overthrown, so went the Church. The Bolshevik Revolution in Russia, 1917-1923, had a similar theme, imperatives, and impetus.

It would be reasonable to assume that Charles Darwin was influenced by the emerging philosophies of the age in which he lived, and sought to find explanations absent of divine interference.

Modern historians tend to downplay these interrelationships, preferring instead to attribute this history to maturing of the intellect, but such is disingenuous. There was nothing wrong with the intellects of earlier scientists, albeit the education system was largely founded by the Catholic Church, and theology played a significant role. It is reasonable to believe that this anti-religious development in Europe was a significant contributor to the rise in atheism, and the development of philosophical materialism. Both had previously existed, but the revolution against royalty with its powers rooted in Catholicism, and Christianity in general, provided the spark for their wider acceptance. Whilst there were some intellectual underpinnings to the Age of Reason, I suspect that emotion and secular revolution provided the necessary impetus for it to have taken over reason altogether.

A persistent myth is the characterisation of 1000 CE onwards as the Dark Ages, with the blame laid at the doorstep of the Vatican. It was in this period that the earliest universities were founded; for example, Paris (1045), Bologna (1088), Oxford (1096), Cambridge (1209), and so on. The truth is that it was the Catholic Church that created the university system during the Medieval Era. Initially, "Cathedral Schools" were established to teach theology and canon law. Over time, these schools developed into universities, broadening their curricula and welcoming scholars from all over. Even the awarding of "degrees" began in the Catholic system as in those early days, a degree could only be awarded once the university had been approved by the Pope. This papal oversight is what got Galileo in trouble: his arrogance and brash manner being his downfall, not his science as people are told today.

However, whilst Western scientists of earlier ages believed in God as Creator, it was their belief in an ordered creation that led them to believe that the mysteries could be uncovered, unlike the views of their counterparts in the East. These scientific endeavours were truly scientific, irrespective of the prevailing theology. Their line of thinking went further back in history to the time of the ancient Greeks. Pythagoras (570 – 495 BCE) developed his theories based on his belief that the entire universe behaved in accordance with eternal *non-material* principles of harmony. Mathematics was the link between divine intelligence and that of man, and could be used to discern the order of the universe, a principle being followed by scientists to this very day. Plato, Philo, and other later philosophers / scientists were essentially Pythagoreans in their thinking. The recognised founding fathers of modern science, Copernicus, Galileo, Descartes, Kepler, and Newton were advocates for the thinking of Pythagoras and Plato. In addition to these eminent Christian men, many of the later renowned scientists were also Christians: Francis Bacon, Blaise Pascal, Robert Boyle, Antoine Lavoisier, Joseph Priestly, John Dalton, Michael Faraday, and Gregor Mendel, to name but a few. Clearly, their religious beliefs in no way impaired their search for scientific truth.

Whilst the philosophies of atheism and material monism have existed as far back in history as we can go, their strangle-hold on mainstream science is relatively recent. One wonders whether modern scientific endeavour could even have begun under such philosophies. Two and a half thousand years ago, Pythagoras and his followers understood the difference between the material and non-material, the physical and conceptual, even understanding that mathematics was an eternal *non-material* principle. All principles are non-material – you will never find one in a material substance, even whilst discovering the principles which govern its activities. At around the same time, Hippocrates (460-375 BCE) espoused what is now mockingly called *neuromania* – the belief that "we" are our brains. In his text, *On the Sacred Disease*, he wrote:

"Men ought to know that from the brain, and from the brain only, arises our pleasures, joys, laughter and jests, as well as our sorrows, pains, griefs and tears. Through it, the "I" in particular, we think, see, hear, and distinguish the ugly from the beautiful, the bad from the good, the pleasant from the unpleasant."[10]

The issue here is that whilst the brain is a *necessary* condition for the expression of consciousness, is there evidence that it is a *sufficient* condition for all of the conditions recounted by Hippocrates? I, and many others, believe it is not, for the science of biology does not support Hippocrates' assertions.

So, The Age of Reason? The *Age of Intellectual Treason* to my way of thinking, the age when scientists became traitors to the founders of their craft, rejecting their foundational beliefs, even treating such beliefs with scorn and derision. The rejection of God, which is an entirely personal choice, was accompanied by the rejection of the non-material, despite the evidence that not everything can be material. Conflating the two has been one of the greatest errors of modern science. It is not necessary to believe in a superior, transcendent God, to believe that not everything is material – that is the false argument of the philosophical materialists.

Definitions

After that brief aside, I shall offer a definition which pertains to experimental science, broader than theoretical physics, being one that I have cobbled together from various sources, but one that contains the elements which most would recognise:

The intellectual and practical activity encompassing the systematic study of the structure and behaviour of the physical and natural world through observation and experiment. Science is conducted via the scientific method, a method of procedure that has characterized natural science since the 17th century, consisting of systematic observation, measurement, and experiment, and the formulation, testing, and modification of hypotheses. Criticism is the backbone of the scientific method.

There are essential elements missing, as alluded to by Robert Brault: the underlying philosophy, and the beliefs which it embraces. There are two that I would highlight:

1. Philosophical materialism; and

2. The implied understanding of the *natural world*.

The first enforces the second, and questions the commitment to the claimed backbone of the scientific method: *criticism*. Arguing consistently within a paradigm is logical, but will only reveal truth when the foundations of the paradigm have been substantiated. Part of my goal, in this book, is to have the reader acknowledge that *philosophical materialism*, as a foundational truth, has not been substantiated. That said, I do acknowledge, and value the wisdom of this paradigm being utilised in the scientific method. Hypothesizing that our existence embodies both material and non-material aspects, how could we devise experiments that investigate the non-material? Yes, many researchers continue to investigate such hypothesized phenomena as the para-normal, but would not be doing so if there were not some evidence that such phenomena do exist. On the other hand, the search for extra-terrestrial intelligence (SETI) continues unabated, with an increasing funding behind it, yet, there is no evidence whatsoever that such extra-terrestrial intelligence exists. You see, belief does drive the aspirations of us all, including scientists, and those seeking Blackbeard's treasure. Yes, I was being facetious with that last crack, but we would be prudent to substantiate the evidence that has people believing as they do. That applies to my beliefs as well, as I have found to my embarrassment on more than one occasion.

As earlier noted, scientists bandy around the term "nothing", as if it were "something", and yet, there is no scientific consensus on what the term means. If science cannot even rigorously define a concept, or at least constrain it to ensure consistency, how can we truly believe it is *science*, and not just *fantasy*? Einstein noted that "One reason why mathematics enjoys special esteem, above all other sciences, is that its laws are absolutely certain and indisputable, while those of other sciences are to some extent debatable and in constant danger of being overthrown by newly discovered facts."[11] Mathematics is reliable because it is underpinned by immutable definitions: science cannot be, if it is not.

Another example is found in quantum physics where a particle pair is said to pop out of a vacuum. Now, a vacuum, for most of us, describes a space where there is nothing of a material nature, yet somehow, a pair of particles, which are of a material nature, can reside where they are not, even temporarily. If a vacuum, in science, is not truly a vacuum, a new term should be coined.

What Science is Not

"I believe that a scientist looking at non-scientific problems is just as dumb as the next guy."

~ Richard Feynman – US Educator & Physicist ~

No doubt, modern communications' technology enables faster and broader dissemination of information (to be defined), but likewise enables the rapid spread of disinformation, lies, false news, and outright stupidity. As Richard Feynman advises: not everything a scientist says is scientific, even when dressed up in scientific jargon, and in those circumstances, a scientist's opinion is not necessarily any better than that of anyone else. In some cases, it can be worse, especially where the scientist is attempting justification of a belief. Even more disconcerting is when the media quote a source as scientific when it is not. Let me give an example, or perhaps two, which I encountered in Touchstone Magazine[12]:

"Consider this PRI news item from last fall (Sept. 12, 2016): "Scientists say an ancient Mayan book called the Grolier codex is authentic." The principal "scientist" referred to is Mary Miller, the Sterling Professor of art history at Yale University. Miller is renowned in her field, and I have no reason to doubt that she did a fine job in authenticating this historical document. But even if she used the latest, most sophisticated instruments to analyze the codex, that doesn't make her a scientist or her work "science," as PRI claims. If an auto mechanic used an engine analyzer to determine the cause of your car trouble and then performed a repair that got your car running smoothly again, would you call him a scientist and his work on your car "good science?" Of course not. Similarly, even if the folks at PRI do not really understand what science is or what scientists are, they should at least know that art history professors are not of their number.

Let's consider a more egregious example. The *Washington Post* published an article on August 10, 2016, with the headline, "The seas aren't just rising, scientists say—it's worse than that. They are speeding up." In this piece, which is accompanied by a dramatic photo of large waves approaching a city's shoreline, the writer discusses how "climate scientists" have created computer models that predict a rise in sea level due to global warming. Moreover, "predictions suggest that seas should not only rise, but that the rise should accelerate, meaning that the annual rate of rise should itself increase over time."

If true, this is a very bad thing indeed. However, the author goes on to state, "The problem, or even mystery, is that scientists haven't seen an unambiguous acceleration of sea level rise in a data record that's considered the best for observing the problem." In fact, our intrepid reporter goes on to say, "The record actually shows a decrease in the rate of sea level rise from the first decade measured by satellites (1993 to 2002) to the second one (2003 to 2012)." In other words, the data show that the model is wrong, yet the *Washington Post*'s editors published a headline that drew the opposite conclusion because they believe the model more than the data. For them, reality is less important than their pre-conceived notions.

This is not science. In science, one is expected to test his models, and if they are proven to be wrong, he should change them, not just plow forward ignoring contrary evidence."

It takes discernment, built on knowledge and a desire to know the truth, to determine whether to accept pronouncements as both authentic and scientific, and even then, they are not necessarily correct. Karl Popper offers his own view on what science is not:

"Science is not a system of certain, or well-established, statements; nor is it a system which steadily advances toward a state of finality. Our science is not knowledge: it can never claim to have attained truth, or even a substitute for it ... We do not know; we can only guess. And our guesses are guided by the unscientific, the metaphysical (though biologically explicable) faith in laws, in regularities which we can uncover – discover ... The old scientific ideal of *episteme* – of absolutely certain, demonstrable knowledge – has proven to be an idol. The demand for scientific objectivity makes it inevitable that every scientific statement remains tentative forever."[13]

Another quote of his that tickles my intellect is this:

"As a rule, I begin my lectures on Scientific Method by telling my students that scientific method does not exist."[14]

Space, Time, Energy, and Matter

It is said that the space, time, energy, and matter in our Universe were created by the Big Bang – the singularity exasperated with loneliness, ceasing to be so. By that definition, the singularity had no "where" and no "when", because that framework was yet to be created. Now, either there was nothing before the singularity, or the singularity was created by some other process in time and space. If there was nothing before the singularity, it must have existed perpetually, but then why is that proposition more logical than positing a perpetually existing God? To my mind, something was decidedly wrong, and so I investigated. Earlier I spoke of the identification of real entities by their properties, activities, and states, so within that context, I sought to discover the true nature of these four entities, only to find that nobody really knows. One might argue that as we have names for these things, we must know what they are, but that is a fallacy. We name things to anchor our thoughts and discussions, even when we know not of that which we speak. Names are but identifying words, without which in any conversation, we would be none the wiser, so let us get started.

SPACE

Although since about 1980, the Big Bang model has been considered outdated in cosmological circles, there is no other scientific consensus to replace it, and irrespective of whatever theory does become accepted, many of the same questions will challenge the science. For that reason, I will use that older model to demonstrate some pertinent issues. The model is a cosmological theory that describes how the universe expanded from an initial state of extremely high density and temperature, preceded by a singularity whereby in such a high-density state, space and time had no meaning. Putting "time" aside for the moment, it is said that the expansion phases of the Big Bang created "space", and that space has been expanding ever since. So, the question becomes: what do cosmologists mean when they speak of space, and do they always have the same meaning in mind?

One definition that I found states: "Space is the boundless three-dimensional extent". On the other hand, another assertion is that "space is expanding and getting larger". If space is boundless, how can it be getting larger? And if it is getting larger, what is it expanding into?

In my considered opinion, space is a three-dimensional extent, not four as in space-time, as I shall explain. If space is a real entity, what are its properties? In our universe, space is identified by its contents, but they are not the properties of space – they are their own entities. Space is described as a vacuum because, among other reasons, sound cannot carry through it, the reason being that molecules are not sufficiently close together to transmit sound. To my way of thinking, if there are molecules, space is not empty. When people speak of *empty space*, what do they mean? I know what an empty jam jar looks like, or an empty room, but an empty limitless extent? If space is not limitless, it is bounded, and if bounded, it is a three-dimensional extent defined by those boundaries, rather like the contents of a jam jar. However, space is not truly empty. In between the more crowded regions occupied by planets, stars and galaxies, there is gas, dust and other bits of matter floating about. So, when scientists speak of "empty" space, they are referring to those areas of space that are "emptier" than other areas, but not truly empty. And of course, the new hypothesis of *Dark Matter* entirely refutes any concept of emptiness.

The problem that we have with the term, "space", is that it is polysemantic depending on context. Its most common usage is in relation to physical distancing, and is always bounded (finite). In cosmology, the term gets very fuzzy, due to an inability, or unwillingness, to decide whether cosmological space is finite or infinite.

I would argue that if space has no properties of its own, then it does not exist as a real entity as I have earlier described. Space does have *contents*, but not *components*, as with real physical entities. Thus, to my mind, cosmological space is a *conceptual entity* – a product of the mind – a phenomenon that we experience but cannot explain. The argument as to whether space does or does not have boundaries, is finite or infinite in extent, is binary – it must be one or the other. If space is defined by its contents, then space is likely finite: we cannot know because we cannot see back to the beginning, although if the Big Bang is accepted as the beginning of our Universe, we can use scientific observations and mathematics to calculate the theoretical limits of our Universe. However, in reality, the preceding singularity must have been "somewhere", occupying undefined space before our *universe space* came into existence. Thus, we can theorize that there was space before our space, but we cannot know what that was, or still is. If cosmological space is expanding, it must be expanding into somewhere. If space is expanding, it must have boundaries. However, that only applies if cosmological space is a real physical entity, which I argue, it is not. In that case, the contention that galaxies are not moving apart, but it is space that is expanding, must be false.

TIME

Now for *time* – what is it? I am not about to contend with Albert Einstein, that doyen of theoretical physics and mathematics, but to perhaps add a refinement based on logic applied to my knowledge of physics and instrumentation. A modest review of academic books and journals on the subject reveals the diversity of opinion, the very existence of which confirms the lack of substantiated knowledge. Research Professor William Lane Craig remarked, "*I know of no concept so profound and so baffling as that of time*"[15]. Again, quoting Professor Craig, but in a more humorous vein, "*Time is what keeps everything from happening at once*"[16].

Time is variously described as:

- The continuum in which events occur in succession from the past to the present and on to the future.

- A one-dimensional quantity used to sequence events.

- A one-dimensional quantity to quantify the durations of events and the intervals between them; and (used together with other quantities such as space)

- To quantify and measure the motions of objects and other changes.

- A real and continuing existence, like space, but along which we can travel in only one direction.

Note the repeated reference to *quantity*, but a quantity of what? The term itself is *conceptual*, having no existence as an entity in reality, for the measurement of it is arbitrary and context dependent. If it is conceptual, as I contend, then it cannot have "A real and continuing existence".

Einstein spoke of velocity and gravitational time dilation, meaning that time slows down under these effects. But I ask: apart from theoretical mathematical equations, how would you know? Clocks do not measure time: clocks provide an external representation of their internal mechanisms, engineered to a specific design to tick off predefined intervals. Time is what we humans interpret as time having been taught the concept: lower order species have no such understanding, but a different understanding based on their own environment. We know that mechanical clocks are subject to vagaries based on their external environments, giving rise to the witticism, "a man with a watch knows the time – a man with two watches is never sure". All forms of time-keeping devices have physical properties, and my question is: How would you know whether the apparent slowing of clocks, of any form, are not the result of velocity and gravitational effects on the mechanisms themselves, rather than whatever time is thought to be? It is logical that any clock that was carried would be subject to the same environment conditions, and whilst the effects on each might vary, that still tells you nothing about the cause of those effects. I have provided a longer discussion in another work[17], but here I will contend that as there are no defined properties of time, and no activities which can be attributed to time, then it is not a real entity, but rather, like space, it is a *conceptual entity* – a product of the mind – a phenomenon that we experience but cannot explain. In passing, Maimonides (1135-1204) observed: "Time is an accident that is related and joined to motion in such a manner that one is never found without the other. Motion is only possible in time, and the idea of time cannot be conceived otherwise than in connection with motion; things which do not move have no relation to time." I agree.

It is a common perception, even today, that when time stands still, nothing moves. For further reading on the inexplicability of time, I would recommend this website[18], although included in the text is the following comment which I do not accept, for I do not believe time to be a real entity as I have previously described "entities" to be (see *Foundational Propositions*). As for *space still existing in empty space*, what does that supposed to mean? Besides that, I find the analogy to be entirely disingenuous:

"Even in empty space, time and space still exist. Physicists have no problem answering the question of "If a tree falls in the woods and no one's there to hear it, does it make a sound?" They say, "Yes! Of course it makes a sound!" Likewise, if time flows without entropy and there's no one there to experience it, is there still time? Yes. There's still time. It's still part of the fundamental laws of nature even in that part of the universe. It's just that events that happen in that empty universe don't have causality, don't have memory, don't have progress and don't have aging or metabolism or anything like that. It's just random fluctuations."

I have earlier opined that "random" is not a behaviour of fluctuations, or of any physical movement. *Random* is a concept derived to describe movement where we are unable to identify all of the causes, and their relative contributions to the observed effect. I continue to contend that anything of physical and chemical properties *must* behave in accordance with whatever those properties, in combination with the properties of other entities in the association, determine.

Finally, I watched a thirty-minute presentation by Professor Sean Carroll entitled: What is Time? Presentism vs Eternalism - The arrow of time[19]. It was quite absorbing, especially as he introduced a number of concepts that I had not previously encountered. However, I noticed to my disappointment, for I was waiting for him to do so, that not once did he attempt to define what time "IS" - he just talked around it. Whilst space can be defined in terms of its physical contents; matter can be defined in terms of its composition; and energy can be defined in terms of what it does; there is no definition of time as a real entity. Space, matter, and energy are responsible for activities - time is not. Time does not do anything other than to allow us to identify events in a fourth dimension which itself is passive - not active. Time does nothing, unlike the other three dimensions. The units of time are arbitrary - they are simply divisions of a period that occurs due to change. Without change, there is no time, but you cannot prevent change by stopping time because time being conceptual, there is no physical mechanism by which it can be manipulated.

The Big Bang could not have created "time", for time does not exist other than as a conceptual measurement. My favourite philosopher of old, Maimonides, expressed time this way back in the 12th century. I do entirely agree with him:

"Time is an accident that is related and joined to motion in such a manner that one is never found without the other. Motion is only possible in time, and the idea of time cannot be conceived otherwise than in connection with motion; things which do not move have no relation to time."

ENERGY

Energy is just as fascinating, or perhaps more so; at best, we can define it as a scalar quantity expressing the capacity to do work, which makes it sound more like a *property* than an *entity*. We were taught in physics that energy can change from one form into another, but how do we understand this if we do not have a substantive definition of the nature of energy itself, just a description of what it can do? Here are some definitions to consider:

- Energy is the capacity of a physical system to perform work. Energy exists in several forms such as heat, kinetic or mechanical energy, light, potential energy, electrical, or other forms. According to the law of conservation of energy, the total energy of a system remains constant, though energy may transform into another form.

- "Energy is an abstract concept invented by physical scientists in the nineteenth century to describe quantitatively, a wide variety of natural phenomena." (Quoted in a book called *Energies*, by Vaclav Smil, attributed to David Rose)

- "Energy is a mathematical abstraction that has no existence apart from its functional relationship to other variables or coordinates that do have a physical interpretation and which can be measured. For example, the kinetic energy of a given mass of material is a function of its velocity, and it has no other reality." (*Theory and Problems of Thermodynamics*, by M.M. Abbott, H.C. Van Ness, Schaum's Outline Series in Engineering, McGraw-Hill Book Company)

- "While it is difficult to define energy in a general sense, it is *simple to explain particular manifestations of energy*. The forms of energy to be considered here are: mechanical potential energy, mechanical kinetic energy, internal energy, flow work, shaft work, transferred heat, and, occasionally, chemical and electrical energy. Other manifestations of energy, such as atomic energy, subatomic energy, will not be discussed." (*Theory and Practice of Heat Engines*, by Virgil Moring Faires, The MacMillan Company)

Thus, it would seem that in essence, energy is that property of matter, in a specific environment, which provides the capacity to do work - all matter has an energy attribute. One scientist puts it succinctly: "*Matter is what it is, and energy is what it does*."[20] Quoting from *The Feynman Lectures on Physics* on the subject of the *conservation of energy*,

"It states that there is a certain quantity, which we call energy, that does not change in manifold changes which nature undergoes. That is a most abstract idea, because it is a mathematical principle; it says that there is a numerical quantity which does not change when something happens. It is not a description of a mechanism, or anything concrete; it is just a strange fact that we can calculate some number and when we finish watching nature go through her tricks and calculate the number again, it is the same."

So, once again, can *energy* be described as a *real entity*, or a phenomenon that we observe as a consequence of it being a *property* of matter resulting in an *activity*? But then, what is *matter*?

MATTER

Curiously, what we see as solid matter is mostly empty space, about 99%. When I look at objects, I wonder how light can bounce off something which is mostly nothing, yet appear to be something. What reflects light if it is not solid matter? Even more curious is why matter that is mostly nothing can be so heavy. At 95 kg, give or take 10% depending on the season, and 99% nothing, where does my mass come from? Scientists know this problem only too well, which is why so much is being spent on the Large Hadron Collider searching for a hypothetical particle: the Higgs Boson. The Standard Model of particle physics does a good job of explaining the various interactions within matter, but has yet to identify why all these tiny particles whizzing about collectively evidence mass. It is believed that it is the Higgs Boson that gives all matter its mass, which at this sub-atomic level is measured not in mass (weight) units, but in units of energy. Now if the Higgs field and Higgs Boson particle are confirmed as existing with the properties as predicted, one more problem will be solved. Of course, this will not provide the answer to where the Higgs Boson comes from and why it is as it is, but at least it is a step forward.

On the other hand, what if it is not found?

"Nearly every scientist believes that the Large Hadron Collider will either prove or disprove the existence of the Higgs boson once and for all -- so if the LHC doesn't find it, it doesn't exist, experts say. Martin Archer [a physicist at Imperial College in London] believes a failure to find the Higgs boson would be even more exciting than discovering the elusive particle. 'If we don't see it, it actually means that the universe at the most fundamental level is more complicated than we thought,' says Archer, 'and therefore maybe the way we've been attacking physics isn't right.'"[21]

That is an interesting thought, but it does confirm my general premise that science knows less about material existence than we generally recognise. Let us not forget *Dark Matter*, *Dark Energy*, and even more mysterious, *Dark Flow*. So, there we have it: space, time, energy, and matter, all products of a hypothetical singularity which was nowhere at any time, yet believed to have existed. Scientists make various claims concerning the human mind, attempting to squeeze it into the ideology of material monism, but when examined closely, even that ideology has very sandy foundations.

I am quite fascinated by the concept of "negative energy", used in physics to explain the nature of certain fields, including the gravitational field and various quantum field effects. In more speculative theories, negative energy is involved in wormholes which may allow for time travel and warp drives for faster-than-light space travel. Research took me here:

"The strength of the gravitational attraction between two objects represents the amount of gravitational energy in the field which attracts them towards each other. When they are infinitely far apart, the gravitational attraction and hence energy approach zero. As two such massive objects move towards each other, the motion accelerates under gravity causing an increase in the positive kinetic energy of the system. At the same time, the gravitational attraction - and hence energy - also increase in magnitude, but the law of energy conservation requires that the net energy of the system not change. This issue can only be resolved if the change in gravitational energy is negative, thus cancelling out the positive change in kinetic energy. Since the gravitational energy is getting stronger, this decrease means that it is negative."[22]

I will not pretend for a moment that I understand this, but it does seem from the description that negative energy is conceived from the hypothesis that there is such a law as the conservation of energy. It is common in science, and entirely valid, to propose hypotheses to explain inconsistencies in observed phenomena, but still I wonder. Reflectively, we have a newish paradigm in science concerning energy. We don't really know what it is, but the paradigm is seemingly based on the law of energy conservation, which has given rise to other hypothesised phenomena such as dark energy and negative energy. Cosmologists are still arguing over whether Dark Flow is a real phenomenon or not, but at least it keeps them busy. In physics and chemistry, the law of conservation of energy states that the total energy of an isolated system remains constant. Not unusually, I ask: Why is our universe considered an isolated system, and how would we know?

Researching this issue further, I got lost in *Noether's theorem*,

Lagrangian functions, and the like, and was given to wondering whether these mathematical proofs were of the same validity as the claimed mathematical proof of Stephen Hawking' hypothesis, that "in the presence of gravity, the Universe can and will create itself". You see, to my mind, if the Universe did not exist and now does, and its existence was caused by a prior existence from which it created itself, how can it be claimed to be an isolated system? The natural corollary from our universe having had a beginning, is that it is not isolated from the existence and cause of that beginning. If we accept the principle of *causality*, then the singularity which caused the universe to have a beginning, also had a cause, and that cause must still exist concurrent with our existence. One could posit that such a cause no longer exists, but there can be no scientific evidence to substantiate that proposition.

I appreciate the disciplines of theoretical mathematics and physics, but I do wonder at times just how far they might wander from reality. With very little understanding, I wonder whether this latest concept of *Massive Gravity*, as earlier discussed, conflicts with the concept of positive and negative energy.

Neuroscience Technology

As the underlying theme of this study is the *mind-brain* conundrum, it is logical that we seek the opinions of those competent in the neurosciences. Unfortunately, and not unexpectedly, there are few who seem to have been prepared to speak out against the prevailing paradigm of philosophical materialism. Finding such experts has been difficult, but fortunately, I have encountered the writings of one, Raymond Tallis, author of several books and some 200 or more scientific papers. I won't bore you with his credentials and accomplishments, but will recount this comment by reviewer Jane O'Grady of the *Observer*: "With erudition, wit, and rigour, Tallis reveals that much of our current wisdom is as silly as bumps-on-the-head phrenology." I hope to substantiate that view, and can highly recommend his books.

There have been impressive techniques developed in recent decades for observing brain activity. For a while, an EEG (electroencephalogram) was used to track electrical activity, but this has been overtaken by *f*MRI (functional magnetic resonance imaging). Our bodily tissues, including neural tissues, are mainly water, and when subjected to powerful magnetic forces, the hydrogen nuclei (protons) are aligned with the fields of force. When a radio frequency current is pulsed through the patient, the protons are stimulated, and spin out of equilibrium, straining against the pull of the magnetic field. When the radio frequency field is turned off, the MRI sensors are able to detect the energy released as the protons realign with the magnetic field. The time it takes for the protons to realign with the magnetic field, as well as the amount of energy released, changes, depending on the environment and the chemical nature of the molecules. So much for these brief technical details.

In addition to observing the brain *structure*, brain *function* can also be observed from a biological perspective, hence the term "functional" MRI. The concept of *localization* of brain activity has long been claimed to have been verified by various studies, but there are technical arguments against such claims, as shall be recounted. The first issue, however, is that if the mind and the brain are one and the same, then seeing the brain at work should allow us to also see the mind at work, but of course, this is not the case. For example, the visual cortex, a part of the cerebral cortex located in the occipital lobe, is that part of the brain involved in processing neural transmissions from the eyes via the optic nerve fibres. Brain activity can be observed, and experiments have been successful at determining what someone is "looking at", but whilst the mind is also at work (hopefully), there is no way to determine what the mind is "seeing". Quoting Raymond Tallis, "Hubel and Wiesel found certain cells in the visual cortex responding preferentially to lines presented at a certain orientation. By studying the fine grain of the *f*MRI in this area when subjects are looking at lines with different orientations, Rees and colleagues were able to infer the orientation of the presented line with 85 per cent accuracy: in other words, they were able to work out what the subject was looking at."[23]

We know that two or more people can observe the same object or view, yet "see" something different. The reason is that *seeing* is more than just the reception of light and transmission of signals to the visual cortex: there is a substantial amount of signal processing occurring which determines the conceptualisation of the scene in the mind. If such signal processing was occurring in the visual cortex itself, which biological feature was performing the signal interpretation function? In a later chapter, we will discuss the concept of telemetry and the calibration of sensors, all essential to effective sensory systems such as we know hearing and sight to be. It is not enough to say that these functions evolved biologically: first we must understand them at a sufficient level of detail to make such a judgement.

Now back to the *f*MRI, and that of which it is truly capable. Raymond Tallis informs us:

"The first thing to remember when you come across headlines such as 'Found: The Brain's Centre of Wisdom' is that *f*MRI scanning doesn't directly tap into brain activity. As you may recall ... [from an earlier chapter] ... *f*MRI registers it only indirectly by detecting the increase in blood flow needed to deliver additional oxygen to busy neurons. Given that neuronal activity lasts milliseconds, while detecting changes in blood flow lag by 2 – 10 seconds, it is possible that the blood flow changes may be providing oxygen to more than one set of neuronal discharges. What is more, many millions of neurons have to be activated for a change in blood flow to be detected. Small groups of neurons whose activity elicits little change in blood flow, or a modest network of neurons linking large regions, or neurons acting more efficiently than others, may be of great importance but would be under-represented in the scan or not represented at all. It short, pretty well everything relevant to a given response at a given time might be invisible on an *f*MRI scan."[24]

The development of magnetic resonance imaging continues apace, and it is worth our time to briefly visit the technology, for it is nothing at all like photography, x-rays, or electroencephalography. Let me give an example quoted from here[25]:

"At its core, MRI exploits the field dependence of the precession frequency by superimposing a magnetic field gradient $\mathbf{G} = (\partial H_0/\partial x, \partial H_0/\partial y, \partial H_0/\partial z)$ onto the static polarizing field $H_0 = H_2$ to spatially encode information into the signal. In this manner, the resonance frequency ω becomes a function of spatial position $\mathbf{r}$, according to

$$\omega(\mathbf{r}) = -\gamma(H_0 + \mathbf{G} \cdot \mathbf{r}). \qquad (4.2)$$

If precession due to the polarizing field (the first term in equation 4.2) is ignored, the complex MR signal is seen to evolve as $\exp(-i\,\gamma\mathbf{G}\cdot\mathbf{r}\,t)$. Following excitation by an RF pulse, and ignoring relaxation, the time-dependent signal $dS(t)$ in a volume element $dxdydz$ becomes (Fig. 4.2)

$$dS(t) = M_{xy}(\mathbf{r})\exp[-i\gamma(G_x x + G_y y + G_z z)t]dxdydz. \qquad (4.3)$$

It is convenient to express the signal as a function of the spatial frequency vector

$$\mathbf{k}(t) = \gamma \int_0^t \mathbf{G}(t')\,dt', \qquad (4.4)$$

Where **G** *(t')* is the time-dependent spatial encoding gradient defined above, which imparts differing phase shifts to spins at different spatial locations. Equation 4.4 shows that the time variation of the spatial frequency vector is determined by the integral over the gradients; the time sequence of the RF pulses and magnetic field gradients is known as the "pulse sequence" of the MRI acquisition.

Rather than expressing the MRI signal in terms of the magnetization, which is modulated by the relaxation terms in a manner specific to the excitation scheme used, it is customary to describe the signal as a function of the density of spins in the tissue. For an object of spin density ρ(r) the spatial frequency signal S(k) thus is given by

$$S(\mathbf{k}) = \int_{\mathbf{r}_1}^{\mathbf{r}_2} \rho(\mathbf{r})e^{-i\mathbf{k}(t)\cdot\mathbf{r}}\,d\mathbf{r}, \qquad (4.5)$$

with integration running across the entire object. Pictorially, the spatial frequency may be regarded as the phase rotation per unit length of the object the magnetization experiences after being exposed to a gradient **G** *(t')* for some period t."

With no pretensions in the field of mathematics, I confess to having no idea of what the above description means. My point in quoting this is to evidence the mathematical nature of MRI interpretation, which has been refined over time. In essence, the technology involves interpretation based on an understanding on the physical behaviour of the chemical constituents of whatever is being examined. Earliest devices of 1.5 and then 3 T strength had the capacity to observe the behaviour of lighter hydrogen nuclei (protons), but as the field strengths of magnets has been increased to 7 T and higher, MRI can detect not only hydrogen nuclei, but also the nuclei of heavier elements, such as sodium, potassium, phosphorus and fluorine, which have a much lower intrinsic sensitivity to magnetic resonance than hydrogen nuclei do. "T" stands for "Tesla", the unit of measurement used to describe the strength of the magnet used in an MRI.

Before accepting without question what can be discovered using MRI technology, we should have some understanding of what is being measured, and ask the question: What does that really tell us about brain activity, other than that brain activity is occurring? I remember listening to earlier IBM 3330 disk-drives rattling about, as the read-write heads moved across the surfaces of the multiple disks, and wondered what was really going on, even though at a technical level, I knew precisely what was occurring. Sure, there was activity, but what did it really mean?

Randomness

Randomness: the quality or state of lacking a pattern or principle of organisation; unpredictability. But then we have *Statistical Randomness*: a numeric sequence is said to be statistically random when it contains no **recognisable** patterns or regularities, and does not necessarily imply true randomness, i.e., objective unpredictability. [emphasis mine]

What I find interesting is that in physics, and the mathematics used to explain behaviours, randomness is considered an objective reality. The notion derives from quantum mechanics, but I do wonder: are behaviours truly random, or do they only appear so because so far, scientists have been unable to detect causes? Furthermore, does quantum theory depend on randomness, and if nothing is truly random, then does quantum theory require some rethinking? I cannot know, but would contend that more research will be required before scientists can be confident of their thinking. Many physical phenomena were long considered random before the causes were understood. Have scientists declared another *no-go zone* with randomness, just as they have with philosophical materialism?

Relativity

We have Albert Einstein to thank for fleshing out the *General* and *Special Theories of Relativity*, but perhaps unknown to many, we also have Einstein to thank for his subsequent observations. For example:

- The meaning of relativity has been widely misunderstood. Philosophers play with the word, like a child with a doll. Relativity, as

I see it, merely denotes that certain physical and mechanical facts, which have been regarded as positive and permanent, are relative with regard to certain other facts in the sphere of physics and mechanics. It does not mean that everything in life is relative and that we have the right to turn the whole world mischievously topsy-turvy.

- Since the mathematicians have invaded the theory of relativity, I do not understand it myself anymore.

- When a man sits with a pretty girl for an hour, it seems like a minute. But let him sit on a hot stove for a minute and it's longer than any hour. That's relativity.

An essence that we should understand from Einstein's relativity theories is not that everything is relative, but that everything has a relationship with other things. Entities exist independently of their relationships to other entities, and can be evaluated on their intrinsic properties alone. If they do not have any, then they are not real entities.

Summary

I can contend with some confidence that nobody truly knows the origin of our Universe, and whilst the Big Bang model did have the widest acceptance for many years, scientists in more recent times have expressed scepticism due to inherent difficulties with the theory. The existence and expansion of the singularity do not conform to the laws of science as currently accepted, and the true nature of time continues to be debated. A major sticking point for cosmology is what was before the singularity, and what caused it to exist. Note that the hypothesis of "something out of nothing", using the claimed mathematical proof, is an alternative to the Big Bang model – both cannot be true, even though both have similar foundational mathematical assumptions. Thus, whilst some scientists hold to one hypothesis, and others to another, and still others to more imaginative explanations, we, the non-scientists of this world, can be confident that the truth of our cosmological origins is yet to be discovered.

The workings of the Universe continue to confound cosmologists, with new discoveries exhibiting unusual behaviours regularly occurring as the improvements in telescope technology continues apace. One of note was reported here[26]. Briefly, "Astronomers have spotted a massive disk galaxy, not unlike our own, that formed 12.5 billion years ago when our 13.8 billion-year-old universe was only a tenth of its current age. But according to what scientists know about galaxy formation, this one has no business being in the distant universe. This discovery is challenging how astronomers think about galaxy formation in the early universe." It is good that thinking hasn't stopped, in cosmology at least.

My point here is that despite the acknowledged credentials of remarkable scientists, their dabbling in theoretical physics does not necessarily reveal truth, and we should not be forced to acquiesce to their perception of reality. What holds for cosmology also holds for evolutionary biology, and offered explanations for the mind-brain conundrum, which is the primary focus of this book.

In his book, which I would highly recommend to anyone prepared to explore beyond thinking constrained by paradigms, Rupert Sheldrake observed:

"Materialism provided a seemingly simple, straightforward worldview in the late nineteenth century, but twenty-first-century science has left it behind. Its promises have not been fulfilled, and its promissory notes have been devalued by hyperinflation. I am convinced that the sciences are being held back by assumptions that have hardened into dogmas, maintained by powerful taboos. These beliefs protect the citadel of established science, but act as barriers against open-minded thinking."[27]

Charles Darwin had a simple view of organisms, being unaware of the complex biology associated with genetics, and for him his evolution hypothesis seemed plausible. But like materialism, his hypothesis is not standing the test of recent scientific research and discoveries.

In the context of the mind-brain conundrum, let me repeat an earlier quoted observation by Albert Einstein:

"Physical concepts are free creations of the human mind, and are not, however it may seem, uniquely determined by the external world."

In other words, what occurs in the human mind is not necessarily limited to external experiences, despite the arguments of those who are committed to philosophical materialism. I find it curious that scientists search for extra-terrestrial intelligence, due to an unwillingness to accept that human intelligence may be unique, even whilst not understanding the nature of intelligence. They posit multiple universes due to an inability to explain why our universe is the way it is, having found inconsistencies in the science of the Big Bang model.

Ancient man is accused of being primitive in "inventing" gods of all forms, due to an inability to explain the world around them. In a sense, modern man is no different, worshipping on the altar of science, accepting by faith that their god will in time explain everything.

References:

1. Tallis, Raymond, *Aping Mankind*, Routledge Classics, New York, NY, 2016, p. 15

2. https://skymanbob.com/2009/03/bubble-land/

3. Einstein, Albert, and Infeld, Leopold, *The Evolution of Physics*, Cambridge University Press, Cambridge, UK, 1938

4. Gitt, Dr. Werner, *In the Beginning was Information*, First Master Books, Green Forest, AR, 2007, pp. 30-31

5. Born, Max, Symbol und Wirklichkeit I Physikalische Blätter 21 (1965), p. 53-63

6. Planck, Max, Vorträge und Erinnerungen, S. Hirzel-Verlag, Stuttgart, 1949, p.13

7. Wieland, Professor Wolfgang, Möglichkeiten un Grenzen der Wissenschafts-theorie, Angewandte Chenie 93 (1981), p. 631

8. *Ibid*, p. 632

9. Sheldrake, Rupert, *The Science Delusion*, Coronet, London, UK, 2013, pp. 6-8

10. Spillane, J.D., *The Doctrine of the Nerves*, Oxford University Press, Oxford UK, p. 8

11. *Sidelights on Relativity* (1922), translation by GB Jeffrey and W Perrett of *"Äther und Relativitätstheorie"* (Aether and Relativity

Theory), a talk given on 5 May 1920 at the University of Leiden

12 Buchanan, Thomas S., *Disciplined Science*, Touchstone Magazine, September/October 2017

13. Popper, Karl, *The Logic of Scientific Discovery*, Routledge Press, London, UK, 1992, pp. 278-280

14. Popper, Karl, *Realism and the Aim of Science*, Routledge Press, New York, NY, 1999, p. 5

15. Craig, William Lane, *Time and Eternity*, Crossway Books, Wheaton, Illinois, 2001, p. 11

16. *Ibid*, p. 13

17. Talbot, Wayne, *Religion? Of God or Man?* Xlibris, Bloomington, IN, 2019

18. https://www.wired.com/2010/02/what-is-time/

19. https://www.youtube.com/watch?v=MAScJvxCy2Y

20. Berlinski, David, *The Devil's Delusion: Atheism and its Scientific Pretensions*, Basic Books, New York, NY, 2009, p. 201

21. http://edition.cnn.com/2011/12/13/world/europe/higgs-boson-q-and-a/index.html

22. Guth, Alan, *The Inflationary Universe: The Quest for a New Theory of Cosmic Origins,* Helix Books, Reading, MA, 1997, Appendix A: *Gravitational Energy* demonstrates the negativity of gravitational energy.

23. Tallis, *Ibid*, p. 81, citing Haynes & Rees, "Predicting the Orientation of Invisible Stimuli"

24. Tallis, *Ibid*, p. 76

25. https://www.ncbi.nlm.nih.gov/books/NBK232486/

26. https://edition.cnn.com/2020/05/23/world/distant-galaxy-wolfe-disk-scn-trnd/index.html

27. Sheldrake, *Ibid*, p. 12

Chapter 5-3: Descent with Modification

"The neo-Darwinist is now reaching the point of dignity in the history of science that the Ptolemaic system in astronomy, the epicycle system, reached long ago. We know that it does not work. And that is interesting. Because from the actual structure of the chromosome we can demonstrate that the human species did not come from a progressive humanisation of a pre-human."

~ Professor Jerome Lejeune, Chair of Fundamental Genetics, Paris University ~

Repeating: *"from the actual structure of the chromosome we can demonstrate that the human species did not come from a progressive humanisation of a pre-human."* It is worth spending a moment or two on the credentials of this scientist. He "was a French paediatrician and geneticist, best known for discovering the link of diseases to chromosome abnormalities and for his subsequent opposition to prenatal diagnosis and abortion."[1] It would be remiss of me to not mention that Lejeune was a Catholic of high standing, even being considered for sainthood, and so it is reasonable to expect that his religion influenced his thinking and moral values. However, it would be wrong to cast doubt on his scientific credentials for that reason, as many do. As the Wikipedia entry continues, "Although Lejeune's discoveries paved the way for new therapeutic research into how changes in gene copy number could cause disease, they also led to the development of prenatal diagnosis of chromosome abnormalities and thence to abortions of affected pregnancies. This was very distressing to Lejeune, a devout Catholic, and led him to begin his fight for the pro-life cause." However, "In 1969, Lejeune's work earned him the William Allan Award, granted by the American Society of Human Genetics, the world's highest honor in genetics." We should have no reason to doubt or dispute his scientific findings.

Evolution has been defined as descent with modification from a common ancestor, but one must ask: exactly what has been modified? Evolution only occurs when there is a change in gene frequency within a population over time. These genetic differences are heritable and can be passed on to the next generation. Evolution depends on long term changes across a sizeable population. Changes that occur within a single lineage may or may not make their way beyond a family, depending on genetic shuffling. It has been shown in the study of epigenetics that some adaptive changes last for only a few generations. In this chapter, I will present an overview of the process of inheritance to demonstrate why evolutionists have such difficulty in proving their hypothesis from a biological perspective. Note that I am not attempting to disprove the hypothesis – my primary intention is to inform myself, and others if they are interested, in the science, and ponder the mysteries that are evident. Most especially, I want to understand inheritance in the context of the mind-brain complex.

From my experience in IT, I wondered about the "master file" copy of the genome which was kept updated by the various mutations and brain learnings during the life of an organism up to the time of procreation. Logically, any changes to an organism's genome after procreation had ceased could not be inherited, but what were the biological mechanisms for keeping the genome master file updated? This wondering about the mechanisms for *inherited characteristics* was one of the reasons for this study: nothing that I had read on the subject of descent with modification even mentioned the mystery, let alone attempted to explain it. Following is a discussion which has satisfied my curiosity, but further question the truth of undirected evolution.

In the process of mutation, we need to consider the two cell types: gamete and somatic. Gametes are an organism's reproductive cells, otherwise known as sex cells. Female gametes are called ova or egg cells, and male gametes are called sperm. Gametes are haploid cells, each carrying only one copy of each chromosome. All other cells are called somatic, which are diploid, carrying two sets of chromosomes, one inherited from each parent. Of interest is why, and how, gametes are haploid, whilst somatic cells are diploid. Normal cell division is termed *mitosis*, resulting in two genetically identical daughter cells, each having the same number and kind of chromosomes as the parent nucleus. Thus, the replicated cells are also somatic (diploid). On the other hand, gamete cells are formed by a process termed *meiosis*, or reduction division. This division of a germ cell involving two fissions of the nucleus, giving rise to four gametes, each possessing half the number of chromosomes of the original cell.

If gametes were diploid, carrying two sets of chromosomes, one inherited from each parent, then each successive generation would accumulate the chromosomes of all preceding generations. How evolution managed to solve this problem is beyond my imagining, but according to the orthodoxy, it must have happened.

Inheritance

We must consider the timing, frequency, and location of mutations, and the likelihood of such mutations becoming heritable. Contrary to the perception conveyed by many scientific descriptions, not all mutations occurring in an organism can be inherited. Gamete cells, those in the germ line, are haploid cells each carrying only one copy of each chromosome. Only gamete cells can be passed on to offspring, and thus only mutations in these cells can be inherited. A somatic cell is any cell of the body except sperm and egg cells. Somatic cells are diploid, meaning that they contain two sets of chromosomes, one inherited from each parent. Mutations in somatic cells can affect the individual, but they are not passed on to offspring. To repeat for emphasis, as this biological truth gives the lie to the conventional evolution narrative, DNA damage in somatic cells is not passed to subsequent generations conceived via sexual reproduction. Thus, when we are offered narratives of, for example, the eye evolving through numerous small mutations, we must be aware that it would only be mutations in gamete cells that would have any effect.

It can be seen that in the main, heritable mutations must arise from gamete cell replication during embryonic development. According to one source, "During replication, double strands of DNA are separated. Each strand is then copied to become another double strand. About 1 out of every 100,000,000 times, a mistake occurs during copying, which can lead to a mutation."[2] Fortunately for any species, our cells have a repair mechanism, giving rise to the term, *robustness*. Despite being a central plank in biology, there remains limited understanding of how it is accomplished at the cellular or molecular level. A major reason is that robustness and the apparent complexity of cellular systems are intimately linked and, therefore, both are difficult to understand. It is interesting that this concept of robustness is used as both an argument for, and against, evolution. Given the low level of copying mistakes, the infrequency, or dare I say, the improbability of such mistakes adding genetic information, and the innate ability for self-repair, my conclusion would be that robustness argues against *microbes to man* evolution being possible.

The next question is this: If the DNA in a somatic cell is damaged during the life of an organism, can it be passed onto offspring? Is there a feedback mechanism that, for example, can pass a mutation in a heart cell to a gamete cell? There is evidence that such might be the case, which brings us to the subject of *epigenetics*.

Epigenetic Inheritance

Epigenetics is the study of changes in organisms caused by modification of gene expression rather than alteration of the genetic code itself. Chemical compounds that are added to single genes can regulate their activity; these modifications are known as epigenetic changes. The epigenome comprises all of the chemical compounds that have been added to the entirety of one's DNA (genome) as a way to regulate the activity (expression) of all the genes within the genome. The expression of specific genes is fascinating. All somatic cells that contain a nucleus have a complete copy of the DNA inherited from both parents, but not all cells will express that DNA. For example, a liver cell does not turn on the gene for insulin. There are further complications: mitochondria are found only inside cells that are inherited from the mother, and there are other mechanisms in place that prevent expression from male derived or female derived genes. The chemical compounds of the epigenome are not part of the DNA sequence, but are on or attached to DNA ("epi-" meaning "above" in Greek). Epigenetic modifications remain as cells divide and, it is claimed that in some cases, *can be inherited* through the generations. The question is: how? What is the mechanism of epigenetic inheritance, given that as is known to date, only haploid gamete cells are passed from generation to generation?

Researching this issue, I found this commentary:

"In sexually reproducing organisms, much of the epigenetic modification within cells is reset during meiosis (e.g. marks at the FLC locus controlling plant vernalization), though some epigenetic responses have been shown to be conserved (e.g. transposon methylation in plants). Differential inheritance of epigenetic marks due to underlying maternal or paternal biases in removal or retention mechanisms may lead to the assignment of epigenetic causation to some parent of origin effects in animals and plants ... In mammals, epigenetic marks are erased during two phases of the life cycle. Firstly, just after fertilization and secondly, in the developing primordial germ cells, the precursors to future gametes. During fertilization the male and female gametes join in different cell cycle states and with different configuration of the genome. The epigenetic marks of the male are rapidly diluted."[3]

The studies resulting in the conclusions above were on plants and animals, which whilst said to be sentient, do not have the cognitive power of humans. The following discussion has been gleaned from here[4].

"Some researchers have found evidence that even some learned behaviours and physiological responses can be epigenetically inherited. None of the new studies fully address exactly how information learned or acquired in the somatic tissues is communicated and incorporated into the germline. But mechanisms centering around small RNA molecules and forms of hormonal communication are actively being investigated." Nicholas Burton, an epigenetics researcher at the University of Cambridge, stated, "The major outstanding question is not whether these [epigenetic inheritance] effects are happening, but what are the mechanisms by which these changes are happening".

Oded Rechavi, a neurobiologist at Tel Aviv University who studies inheritance and evolution, stated, "The possibility that the nervous system could generate heritable responses was especially intriguing, because the nervous system is a very unique system in its ability to organize information about the environment ... It has [a] unique capability of planning." Rechavi has also studied a learned behaviour in roundworms that shows up in the progeny of those worms. The behaviour is linked to the production of certain small RNA molecules. However, Peter Sarkies, an epigenetics researcher at Imperial College London, says he is excited about "the idea that information can be transmitted from the nervous system into the germline and then ... transgenerationally," with an impact on the defendants' nervous system as well. Nevertheless, he cautions that "it's a slightly artificial system, because the worm has been engineered in order to test the idea. It is not a wild-type worm".

Another source provides further evidence to ponder:

"In the double set of chromosomes within diploid cells, one inherited from the mother and the other from the father, paired genes may be identical or different in their DNA sequences. The different forms of the gene are called alleles. Studies of sperm ... are being applied to the investigation of recombination or crossing-over events during meiosis, in which chromosomes intertwine and exchange sections that are detected as new allele combinations. Aberrant types of crossing-over events can sometimes lead to altered genes that can cause disease. A critical problem concerns the fact that crossing-over events occur at different frequencies in male and female gametes. Because of limited access to human female gametes, studies are conducted mostly with laboratory animals, but human oocyte research will ultimately be needed to resolve these questions."[5]

A brief word on epigenetic inheritance, a relatively new field of research with promising results. If you are interested, you would benefit from reading the entire entry. The intriguing issue is that in one sense, transgenerational epigenetic inheritance supports evolution theory by providing yet another mechanism. On the other hand, it argues against evolution due to its complexity, raising questions on how evolution could evolve its own mechanisms. I will leave readers to decide for themselves:

"Transgenerational epigenetic inheritance, (TEI), is the transmission of epigenetic markers from one organism to the next (i.e., parent–child transmission) that affects the traits of offspring without alteration of the primary structure of DNA (i.e. the sequence of nucleotides)—in other words, epigenetically. The less precise term "epigenetic inheritance" may cover both cell–cell and organism–organism information transfer. Although these two levels of epigenetic inheritance are equivalent in unicellular organisms, they may have distinct mechanisms and evolutionary distinctions in multicellular organisms ... Epigenetic variation within multicellular organisms is either endogenous or exogenous. Endogenous is generated by cell–cell signalling (e.g. during cell differentiation early in development), while exogenous is a cellular response to environmental cues ... In sexually reproducing organisms, much of the epigenetic modification within cells is reset during meiosis, though some epigenetic responses have been shown to be conserved. Differential inheritance of epigenetic marks due to underlying maternal or paternal biases in removal or retention mechanisms may lead to the assignment of epigenetic causation to some parent of origin effects in animals and plants ... A number of studies suggest the existence of transgenerational epigenetic inheritance in humans ... Epigenetic inheritance may only affect fitness if it predictably alters a trait under selection. Evidence has been forwarded that environmental stimuli are important agents in the alteration of epigenes."[6]

Further on epigenic inheritance, research has identified a possible feedback mechanism from somatic to gamete cells:

"To all intents and purposes he [Lamarck] was proposing a theory of particle transmission between cells that resembles the exosomes[7] discovered in our time. Exosomes are involved in transgenerational epigenetics, which would help to explain many observed maternal and paternal transgenerational effects in health and disease. Recent research shows that exosomes cross the Weismann Barrier and can convey their cargo via sperm, so the mechanism of transgenerational transmission of effects does exist."[8]

I find this all very intriguing, but below this level of commentary, I am out of my depth. Most intriguing for me was the comment, "the nervous system is a very unique system in its ability to organize information about the environment". This is one of those observations which roll easily off the tongue, or pen, but when examined in more detail, raises more questions than it answers. How can the nervous system organize information about the environment, when the nervous system does not have direct access to the reality of that environment? More particularly, how could the nervous system organize physical structures on the conceptual aspects of which it cannot be aware? This continues to be the elephant in the room of the mind-brain conundrum. There is no doubt that the cellular structure of organisms adapts to environments, and in a general sense, such changes may be termed [epi-] genetic information. However, the nervous system is a "structure", not a single cell type which undergoes chemical changes in response to conditions. To restructure the brain's neural circuits, multiple, and likely different changes must be made to individual neurons in a specific circuit or circuits.

The relationship between an organism and its environment is entirely chemical, not cognitive. I find it difficult to imagine the "organiser" of such complicated reorganisation to be the chemistry of the neural cells themselves. Organisation of a structure requires oversight of the outcome to be accomplished, and the structure which would result in that outcome.

References:

1. https://en.wikipedia.org/wiki/J%C3%A9r%C3%B4me_Lejeune

https://science.howstuffworks.com/life/genetic/dna-mutation.htm

2. https://en.wikipedia.org/wiki/Transgenerational_epigenetic_inheritance

3. https://www.quantamagazine.org/inherited-learning-it-happens-but-how-is-uncertain-20191016/

4. *Ibid*

5. https://www.ncbi.nlm.nih.gov/books/NBK232035/

6. https://en.wikipedia.org/wiki/Transgenerational_epigenetic_inheritance

7. https://bmcbiol.biomedcentral.com/articles/10.1186/s12915-016-0268-z

8. https://www.sciencedirect.com/topics/biochemistry-genetics-and-molecular-biology/inheritance-of-acquired-characteristics

Chapter 5-4: Cells as Computers

"We invented a computer program capable of thinking non-numerically, and thereby solved the venerable mind/body problem, explaining how a system composed of matter can have the properties of mind. Opening the way to automate tasks that had previously required human intelligence."

~ Herbert A. Simon (1916-2001) cognitive psychologist and Nobel Prize winner ~

In an earlier chapter on *Philosophy*, we discussed the issues regarding various forms of *Dualism*. In the next chapter, we discuss *Determinism*, with the evolutionists' assertion that we really have no free will, and no control over our thoughts. If that be true, then there is no "we" to be doing anything – just biological machines resting comfortably in our skulls, engaging in activities determined by genetics and the laws of chemistry and physics. Yet, according to Herbert Simon, he and his colleagues, their activities contingent upon their brains over which they had no control, invented a computer program which solved the problem of their own existence. Consciously writing computer programs requires human intelligence, and whilst such activities can lead to automation of human activities, it in no way *solves the venerable mind/body problem*, for the resolution relies on the assumption that the mind is no different to the body. In effect, all that Simon has done is to beg the question.

"We invented a computer program": if the "no free will" advocates insist that we cannot have free will because such an attribute could not be the product of evolution, why do they insist on claiming credit when according to them, there really is no personal "we"? How could the process of evolution accomplish a system of such refined cognition? To instantiate a computer program in written code, it must first be freely conceptualised in the mind (I have substantial experience in doing just that). Where would one find this "computer program" in the human brain in the conceptualisation stage? Remember too that to "invent" a computer program requires a great deal of trial and error, testing and debugging, but most importantly, the inventors must have a goal in mind. Intentionality and purpose are antithetical to the evolution hypothesis, yet here we have an example where

it is unconsciously claimed that intentionality and purpose are the products of evolution. This is simply not plausible.

Despite the claim above, a system composed of matter *cannot* have the capabilities of the mind. I strongly disagree with Simon, as I must, because it took human intelligence to supposedly solve the mind/body problem (it has not been solved). Simon appeared to believe that because human intelligence and programming can develop computational systems that *automate tasks that had previously required human intelligence*, that the properties of the mind could have arisen without an external intelligent agent. The mind/body problem is not solved by the application of intelligence, because intelligence cannot explain itself or its origins (Primary Axiom #1). Thus, the *venerable mind/body problem* is not solved at all. Software in a computer is the instantiation of the programmer's mind (subjective) – the conceptual made physical (objective). Computers cannot be subjective, for to be subjective means to operate beyond the rules of objectivity, whereas computers are restricted to behaving in accordance with the objective code of their programming. Even constructs which give the appearance of subjectivity, e.g., if-then-else statements, are objective because they are obeying the laws of logic instantiated in the code.

Those who assert the computational facilities of the human mind, but also insist that humans consist of nothing but biological materials, have no choice but to find these computational powers in the cell – from where else could they arise? An interesting study is this one by scientist, Dennis Bray, entitled *"Wetware: A Computer in Every Living Cell"*[1]. Computation is defined as any type of calculation that includes both arithmetical and non-arithmetical steps and follows a well-defined model, for example an algorithm. I have been unable to find any formalised definitions that differentiate these two types of computation, but from what I have read, and from my own experience, non-arithmetic steps are functionally, *logic* and *decisions*, implemented by if-then-else statements, loops, comparisons, use of predetermined variables, and operations on data not involving arithmetical calculations, e.g., move statements. From my knowledge of the biology of the brain, computation could only occur on the basis of mapping physical states between neurons resulting in neural networks representing an external state. A cascade of sub-networks could achieve an entirely other state, but still existing in as an encoded, symbolic representation of a conceptual entity, or an external reality. Even then, as I will later explain, my understanding of neural networks has me troubled about the permanency of such structures.

As for arithmetical steps, I have my doubts that this could be happening, other than by analogy. We later delve into the communications processes within neural networks, but for now, we simply need to understand that neurons primarily communicate with each other via junctions called synapses - across the synaptic cleft, messages are chemical. In a minority of cases, the communications are electrical in which ions flow directly between the membranes of adjacent cells. At a chemical synapse, an action potential triggers the presynaptic neuron to release neurotransmitters. These molecules bind to receptors on the postsynaptic cell and make it more or less likely to fire an action potential itself. In a way, these operate very much like controlling sensors in technical applications such as smoke alarms, and float valves. Once a condition reaches a certain predetermined state, it triggers a reaction; in the case of neurons, the trigger mechanism is the action potential reaching a certain state. By analogy, imagine a pipeline with pressure sensors controlling a series of valves or gates to regulate the flow: as the pressure reaches a certain level at a point, the valve opens, and so on down the pipeline in a sequence of opening and closing.

The content of any messaging, that is, communications with a semantic layer as opposed to simple signals like an electrical pulse, can only be found in the originating and terminal neurons in a nerve fibre or similar. This is why scientists are looking into the cell for an answer, for it cannot be found in neural networks which convey nothing by electrical pulses (action potentials).

Bray sees an analogy between digital computing and biological computing, but with differences. Before we review his beliefs, remember that algorithms are not arbiters of objective truth, simply because they are constructed from mathematics: Einstein was emphatic on that point. But back to Bray,

"Wetware ... is the sum of all the information-rich molecular processes within a living cell. It has resonance with the rigid *hardware* of electronic devices and the symbolic *software* that encodes memories and operating instructions, but is distinct from both of these. Cells are built of molecules that interact in complex webs, or circuits. These circuits perform logical operations that are analogous in many ways to electronic devices but have unique properties. The computational units of life – the transistors, if you will – are its giant molecules, especially proteins. Acting like miniature switches, they guide the biochemical processes of a cell this way or that. Linked into huge networks they form the basis of all the distinctive properties of living systems. Molecular computations underlie the sophisticated decision making of single-cell organisms such as bacteria and amoebae. Protein complexes associated with DNA act like microchips to switch genes on and off in different cells – executing 'programs' of development. Machines made of protein molecules are the basis for the contraction of our muscles and the excitable, memory-encoding plasticity of the human brain. They are the seed corn of our awareness and sense of self." (pp. x-xi)

I find analogies to be useful, but only up to a point: Bray has extrapolated beyond that point. We know that awareness and sense of self do exist as real phenomena, but we do not know the true nature of them. To posit that cells are the *seed corn* is to beg the question, because no-one is able to explain how a single cell is not sentient, but an arrangement of cells apparently is. I have no argument with: "The central thesis of the book – that living cells perform computations – arises from contemporary findings in the biological sciences, especially biochemistry and molecular biology" (p. xi). However, computation is not the word that I would have chosen, but I accept it in the sense intended. I need to introduce the term, *chemotaxis*: movement of a motile cell or organism, or part of one, in a direction corresponding to a gradient of increasing or decreasing concentration of a particular substance. This is nothing more or less than a chemical reaction, but Bray chooses to call it, "a superb illustration of cellular information processing" (p. 6). Again, we have this issue of meta-language, where scientists appropriate common terms in an uncommon way, because that is how they perceive certain phenomena. I do wonder whether this results in them convincing themselves that their analogies are no longer just that. In another example, the author speaks of bacteria having *short-term memory*, but explains that he is using the term "in a colloquial, nonspecialist way, referring to how swimming bacterium carries with it an impression of selected features of its surroundings encountered in the past few seconds" (p. 7). We know that our sensory system of sight has a similar behaviour, but we would hardly term it memory. A more appropriate term would be *persistence*. Cognitive memory is a very different phenomenon to the one being described of bacteria, and we need to be watchful for explanations where the processes in the mind-brain complex use the same analogies without clarification.

Bray does offer what he believes to be a plausible explanation of evolved characteristics. He writes:

The molecules controlling the behaviour of the bacterium – mainly proteins – are made according to the instructions inscribed in their genetic material ... these are automatic reflexes, inherited generation to generation. And where did they come from in the first place? What must have happened is that certain random changes occurred in the DNA of this cell. Changes in DNA led to slightly different proteins being made, since the genes of an organism specify the structures and functions of all its proteins ... new proteins resulted in new functional connections being made. The new protein circuitry caused the cell to behave in a new way. This gave the cell an advantage in a tricky situation so that it survived when its compatriots perished. Its DNA, containing the blueprint of the novel behaviour, was replicated and passed onto subsequent generations." (p. 9)

"What **must** have happened" [emphasis mine] is a conclusion based on a paradigm, and is not a scientific pronouncement. We should note that in the context of biology, the known replication rate of bacteria makes the old saying, "multiplying like rabbits", an understatement. Under ideal conditions, a single bacterium can potentially multiply up to sixteen million in a matter of hours. Rabbits appear sterile by comparison. So, beneficial, neutral, and deleterious mutations can spread throughout multiple generations very rapidly. Whilst it is valid to claim that beneficial mutations contribute to survival advantage, the question is: when, and in what circumstances? Such mutations could just as easily be overtaken by later mutations acting deleteriously. Modern research has determined that most nonneutral mutations are deleterious. In general, the greater the number of base pairs of nucleotides that are affected by a mutation, the larger the effect of the mutation, and the larger the mutation's probability of being deleterious. Bacteria generally consist of a single DNA molecule, with millions of base pairs. I have been unable to find research on the probability of beneficial mutations in bacteria, especially in the environments in which early evolution is said to have occurred. However, equating this behaviour with computation is wishful thinking at best. We ought not forget the lessons in the previous chapter related to gamete and somatic cell lines, and where mutations must occur to be heritable.

Bray then moves on with an assertion that lacks scientific substantiation, or perhaps just precision in language:

"Our memories are stored in the brain." (p. 9) Continuing, "They are represented there by what was once referred to as the engram, or a constellation or connections between nerve cells ... specific connections, called synapses, allow signals to pass from one nerve cell to another. Sets of nerve cells connected in this manner establish complicated electrical circuits than communicate and process information, analogous to those in a computer or other electronic device".

"Communicate and process ... analogous to those in a computer"! Communicating and processing are two entirely different functions. Electrical circuits "communicate" simply by the circuit being active, but some external process is needed to activate the circuit. Processing is an entirely different activity altogether, for it requires provision of a set of processing instructions – where does Bray suggest they make an appearance? According to Bray, there is no differentiation between the process instructions and the data to be processed – effectively, the data processes itself. It is this failure of differentiation which invalidates the analogy of the brain being like a computer. Consider the vast range of intellectual domains in which we dabble, and for the computer analogy to be valid, the brain necessarily wrote the software for each and every such dabbling.

If you have been following my arguments thus far, you should see that this is where his analogy fizzles into incoherence. Electrical circuits do not communicate information, nor do computers process information: they just send electrons scurrying about. The essence of computer functionality is that it is programmed by an external intelligent agent – no computer can program itself, nor be self-aware. Similarly, computers have no understanding of the conceptual aspects of data and information. If scientists wish to use computers as analogies for the human brain, they must deal with these issues, and stop pretending that electrical circuitry could somehow become sentient, and cognisant of the meaning of its own processing. This is, perhaps, the primary issue in refuting the claim that given enough circuits, computers could become self-aware.

Computers do not deal in information, nor even data as we conceive it. For example, a computer does not "know" whether it is processing payroll, census statistics, or modelling climate change. There is nothing in the computer language of bits (ones and zeroes), or magnetised regions that can tell you, or the computer, the subject of its processing. This is the essence of the issue. If you were to perform a core dump to the printer, as we used to do to diagnose errors, and examined the groups of one and zeroes, you could learn nothing from them alone. Now, if intelligent agents such as humans, can learn nothing from these ones and zeroes alone, how could a non-sentient machine understand its own processing (or anything for that matter) from a pattern of magnetised regions, or in the case of the brain, a biological circuitry of neurons? The fundamentals of computing are nothing more than magnetic regions set to one of two logical states, on or off. Even computer programmers require the code book to decode these arrangements of bits. A computer can be programmed to run an application program at a certain time. What it executes to awake at that time, and run the nominated program, are both encoded in these binary states. What is read from storage devices, processed in memory, and written back to storage devices and other peripherals, is again in these encoded binary states. There is nothing at that level which even suggests the identification of the subject, or the nature of the computation being processed. Painters work with paint, composers with musical notes, authors with words, and computers ... magnetic states incapable of explaining themselves.

I am rather nonplussed by otherwise intelligent scientists who feel compelled, through their commitment to philosophical materialism, to devise implausible explanations for the workings of the human brain. Either they have no understanding of the true conceptual nature of data and information, and no understanding at a fundamental level of how computers work, or they deliberately concoct just-so stories to convince others, even though they, themselves, are likely not convinced (I hope).

Having no training in biology or bio-chemistry, I enjoyed reading Bray's descriptions of how things work at a cellular level. However, I object when he continues to use analogies to confer uncommon meanings to common words. For example, after describing how cells react to the chemical environment, he concludes: "*In other words, cells are aware of their surroundings*" (p. 18). If his intention is to suggest cognitive awareness, then he is being deceptive: chemicals are chemically "aware", if one chooses to use that term, but not self-aware in the sense that humans are. A better word would be *sensitive*, just as a better word for short-term memory is *persistence*. His terminology continues to descend into nonsense when he writes,

"Swimming paramecia continually encounter different situations that they either accept by going ahead or reject by reversing. Almost any animate wanderer must make decisions. When confronted with multiple stimuli of a possibly conflicting nature, cells have to evaluate their options and assign priorities. They must actively choose one out of many possible responses." (p. 18)

Bray's rhetoric is disingenuous, for decision making is a cognitive process, not bio-chemical. His descriptions would make wonderful scripts for cartoon animations, but as examples of a scientific treatise, they are far too fanciful. Cells cannot *evaluate their options and assign priorities* in a cognitive sense: consisting of chemicals, cells can only do what they *must* do, based on their chemical composition and that of their environment. Here is another example of disingenuous language, this time from a scientific (?) explanation concerning viruses:

"small genome size is perfectly suited to virus replication, in which each infected host cell produces many copies of the viral genes from a single template. Such exponential replication places a premium on small genome size: the smaller the genome, the faster it can replicate. Not surprisingly, in collaboration with their cellular hosts, viruses use a **number of strategies** to minimize the size of their genomes, including a parasitic lifestyle, use of repeating protein subunits, genetic parsimony, and protein miniaturization ... Freed from the necessity of providing the infrastructure of life, viruses can **concentrate** on making copies of themselves ... A **final strategy** used by viruses to minimize the size of their genomes is to encode tiny proteins."[2] [emphasis mine]

Strategies indeed, as if non-sentient, non-replicating viruses could think! Just imagine: viruses have decided that a parasitic lifestyle best suits their chances of replication when invading host cells. Cells (and viruses) can no more *evaluate options and assign priorities*, than they can wonder what would happen if they ignored all options, and simply had a day off! These are functions of cognition, not chemistry, irrespective of how organised in cells. I wondered where Bray was heading with his arguments, but I did not have to wait long: earlier studies by psychologists had posited that even cells have a psychological life. The thinking was that because higher organisms have aggregate properties of lower organisms, and higher life forms have sensations and motivations, then the source of those properties must be the lower organisms themselves. In effect, they are saying that any emergent property must have its genesis in lower organisms which already have those properties, or very close to. I agree.

I have argued the very same from the opposite direction. I argue that because lower organisms lack the properties of higher life forms, then these properties cannot emerge from a combination of lower organisms. As properties and behaviours of the mind cannot be attributed to purely biological structures, then the mind cannot be an emergent property of the brain. Evolutionists and I agree on a fundamental premise concerning the relationship of lower organisms and higher life forms, but have drawn opposite conclusions.

The idea that microorganisms have sensations and motivations somewhat equivalent to those of higher organisms, is not accepted in modern science. Alfred Binet, the French psychologist who proposed the idea, seems to have the same sense of bio-chemistry that I do, although his attempts at deconstruction vary from mine. Once we get past this part of Bray's book, he returns to the more pragmatic and sensible.

After having earlier said that "*cells are aware of their surroundings*", he then questions: "Is it conceivable that these simple organisms could possess an awareness of their surroundings that is in any way comparable to our own?" (p. 20) Later he admits, concerning a biological organism, that "it has to detect and recognize salient features of its environment" (p. 25). As best I understand, salient features for chemicals are the properties of the other chemicals with which they interact – there is no sense of cognition. But then, the disappointment returns: "A flood of information enters such a cell every second of its existence through its membrane. This must be assimilated, sorted, and codified. The cell has to choose one integrated, coherent action that ensures its survival." Let me repeat that information is *conceptual*, not physical. The processes of choosing, codifying and sorting are functions of *cognition*, defined as the set of all mental abilities and processes related to knowledge. Once again, Bray has resumed a rhetoric intended to convince the reader that the cell in some way resembles a self-programmed computer, when logically it does not, and most importantly, cannot. Let me remind the reader of Primary Axiom #10: In a mechanistic world, as defined under philosophic materialism, entities can only do what they *must do* as determined by their properties, and those of other entities with which they interreact. Bray's propensity for epithet transfer is disappointing and disingenuous.

Ending Chapter 1, Bray writes: "But there is one other source of new information that will guide us. In the twentieth century humans created a new world of computers and electronic robots. Some of these silicon-based machines act and behave in a manner that closely resembles that of living organisms." (p. 26) From there he segues in Chapter 2 on *Simulated Life*:

"In a circuit board or microchip, sets of logic elements linked in precise networks perform defined logical processes, repeating the same simple steps over and over again. The same thing happens in a living cell, except that the elements now are molecules instead of transistors and the **blind iterations** they perform are chemical rather than electrical. At a higher level, there is the much commented-upon similarity of computers and brains. Both are made from multiple components linked into networks of connections. Both carry digital signals in the form of electrical pulses, process information, store memories." (pp. 27-28) [emphasis mine]

Note how the author contradicts himself when he alternates between chemistry and analogies. He agrees with Primary Axiom #10 when he refers to *blind iterations*, for such are synonymous with chemical constructs doing what they must. Earlier he wrote, "the cell has to choose", but then he speaks of *blind iterations* which, by definition, do *not* involve choice. I find this a common feature in the writings of scientists attempting to be both honest concerning what chemicals can achieve, and what they would like the chemicals to achieve – a degree of sentience. Brains and computers are *somewhat* alike in their architecture, but my understanding is that they are more *unalike* than like. A limitation that I have is in attempting to visualise active networks in the brain, because neurons are both poly-dependent, and poly-functional. A postsynaptic neuron is poly-dependent in that it has multiple dendrites which serve as neurotransmitter receptors. At any one instant, the postsynaptic neuron can receive neurotransmitters from multiple presynaptic neurons, and must resolve the (chemical) aggregate of these inputs, resulting in excitatory, inhibitory or modulatory effects. A presynaptic neuron is poly-functional in that it can synapse with multiple downstream neurons simultaneously, as although unlike dendrites, there is only a single axon emanating from a neuron, the axon can branch and synapse with multiple dendrites. Identifying a single logic path in such a network is impossible, given the multiplicity of possible paths which are continually changing.

The most significant difference between computers and neuronal networks is the separation of logic and data. In a computer, application-specific software is loaded into the processor to determine the logic path to be followed, and is not altered during the processing. All alterations are performed on the symbolic representations of data only. In the brain's networks, scientists have made considerable progress in identifying specific regions where processing of sensory messages is accomplished. For example, by analogy, the auditory cortex must run a different "program" to the visual cortex, because it processes a different type of message, likely with a different encoding method. The output from this processing, which is a symbolic representation of an external reality, is forwarded to the brain for further processing and storage. Just how or where this is accomplished is unknown. Another type of processing is entirely internal, i.e., when we think, imagine, compose music, recall memories, and so on. Scientists have discovered where in the brain certain processing occurs, but have been unable to refine the locations for logic and data. It has been noted that some processing involves numerous areas of the brain. Bray wrote of computers and brains: "Both carry digital signals in the form of electrical pulses, process information, store memories", and whilst this is true, it says far too little to accept the comparison. Processing and storing are determined by the software, not simply by digital signals running around circuits.

Bray's analogy of circuit boards, where *the elements now are molecules instead of transistors*, is reasonably close. Transistors are designed to operate in one of two modes: in one, they act as amplifiers of electric currents; in the second, they operate as on-off switches. Neurons are somewhat similar, in that they can be excitatory (on), inhibitory (off), or modulatory in their effect. I cannot comment on the detail level of molecule, cell, or soma, but would make the following observations. In business related computing, and most Microsoft personal computers, laptops, and tablets fall into that category, as do the mainframes scattered about in commercial enterprises: the computing is binary, with transistors performing an on/off function. In technical applications such as found in computer aided manufacturing (CAM), software is used to manage processes not just in binary, but also as the monitor and controller of machine tools concerned with dimensional specifications and machining quality, amongst other functions. In modern vehicles using throttle control-by-wire, instead of mechanical linkages, sensors monitor the positioning of the throttle device, and transmit an appropriate signal to a micro-computer which manages the appropriate fuel metering device. This can be compared to a neuron in its *modulatory* mode. In a way, we can compare the functioning of the chemical and electrical processes of neurons, with both commercial and technical computing. The obvious difficulty for comparisons with technical computing applications, is the precise specification of metering chemicals within and across cells, and the very fine electrical gradients across the cell membrane.

That individual neurons respond or not, is not truly comparable with digital processing. It could be said, but at a stretch, that the on/off condition of neurons mimics the on/off nature of magnetic regions, but this just emphasises why neural circuits are unlike computers. You see, the determination of on or off in a computer is dependent upon the coding system being used, and what those states are programmed to achieve. In other words, a computer evidences the intentionality of the human programmer, but one must ask: what determines the intentionality of the human brain? Unlike in digital computing, the on/off state of a neuron is determined by the aggregate of neurotransmitters of the presynaptic neurons with which it is connected. But there is one more catch – "on" could be an analogy for excitatory, and "off" for inhibitory, but what is observing and recording those states as binary representations? Even more, and this is of primary significance, in a computer, the electrical current continues to flow irrespective of whether on/off states are encountered or created. The same is not true in neural circuitry. If a neuron is metaphorically set to "off", because the neurotransmitters in the presynaptic neuron(s) do not cause an electrical pulse (action potential), the remaining neural circuit goes inactive. In computerese, this would be equivalent to a bug!

I must confess that knowing what I do about digital computing after thirty years in the IT industry, and knowing what I now do after intensive study of nervous systems, neural networks, and neurons, I cannot conceive of how computers are analogous to the human brain. A limitation, no doubt, on my part, but my perception of the brain is that there is no separation of code and data: the data appears to program itself at the encoded symbolic level, with no understanding of what those data/coding structures represent conceptually. Computers are not like that.

Another difficulty that I have is suggested in the following.

"When a nerve impulse is generated, there is a change in the permeability of the cell membrane. The sodium ions flow inside and potassium ions flow outside, causing a reversal of charges. The cell is now depolarised. This depolarization results in an action potential which causes the nerve impulse to move along the length of the axon. This depolarization of the membrane occurs along the nerve. A series of reactions occur where the potassium ions flow back into the cell and sodium ions move out of the cell. This whole process again results in the cell getting polarised, with the charges being restored."[3]

It would appear from this description that nerve cells are never permanently "on". An action potential, that causes the release of neurotransmitters, "is a **short-term change** in the electrical potential that travels across the neuron cell" (emphasis mine). After execution, the membrane potential returns to its resting state. This being the case, a neural network cannot represent permanent data, because other than for the short-term change, the "bit" if we think of it that way, is permanently "off". The neuron then acts as a gate, allowing or disallowing some change in downstream neurons, but only for a brief time. Neural circuits could perform logic operations, but not data storage functions. In that sense, they can be compared to the CPU of computers, but then the question: where is data stored? Similarly, what "intelligence" wrote the software encoded in the neural circuits?

Returning to "Both carry digital signals in the form of electrical pulses, process information, store memories", this is cannot be correct in the case of neural networks. Electrical impulses cascade through the networks, but they have no semantic layer. There is a pattern to the activity, but as *pattern* is conceptual, what is it that recognises and interprets the pattern? It cannot recognise and interpret itself.

Information is never processed, nor are memories stored, other than in encoded symbolic patterns, which must be decoded and correlated cognitively, none of which can happen autonomously in either a computer, or the neural network of the brain. The obvious omission in his, and all similar explanations, is that computers must be programmed by an external, intelligent agent, whereas for the comparison to be valid, brains must be capable of programming themselves across any number of business and technical applications.

At this point in the book, with 213 pages remaining, I realised that I was going to learn nothing of value as regards understanding the mind-brain complex. There are useful discussions on biology, to which I may later return, but for now, the book offers nothing relevant to my current study. Even worse, the author seems so caught up in his analogies, that his analogies have supplanted his reality. I cannot help but repeat this commentary by Raymond Tallis, for it is far more imaginative than my own on this issue: *the metaphorical clothes in which thinking is wrapped becomes its skin*! I fail to follow the logic that intellectual humans have managed to simulate some functions of biology, and therefore biology must work in the same way as the synthetic *simulated* life. This is the process that Raymond Tallis describes as *epithet transfer*: anthropomorphic attributes are ascribed to mechanical devices, prompting mechanistic processes being ascribed to biology – an utter failure of logic. When a sculptor sculps a statue of a reality he perceives, was that reality influenced by his sculpting? There is no scientific evidence that the brain can "carry digital signals in the form of electrical pulses, process information, store memories." Digital signals in computing are binary, based on intelligently devised protocols, and whilst there must be some biological protocol used in transmissions in our nervous system, to my knowledge, nobody has managed to decode it. We later explore further, the fallacy that a biological organ can process information or store memories.

Now, this comment by Garry Kasparov, Russian chess grandmaster and former world chess champion, whom many consider to be the greatest chess player of all time.

"The human mind isn't a computer; it cannot progress in an orderly fashion down a list of candidate moves and rank them by a score down to the hundredth of a pawn the way a chess machine does. Even the most disciplined human mind wanders in the heat of competition. This is both a weakness and a strength of human cognition. Sometimes these undisciplined wanderings only weaken your analysis. Other times they lead to inspiration, to beautiful or paradoxical moves that were not on your initial list of candidates."[4]

Cells cannot do that, nor can any biological organ comprised of cells which can only do what they must do – *blind iterations*. Bray wrote, "The cell has to choose one integrated, coherent action that ensures its survival"; in other words, it is incapable of "paradoxical moves that were not on your initial list of candidates."

Kasparov's understanding of the capabilities of the human mind offers better evidence than that of scientists wedded to philosophical materialism.

A Fundamental Error

We will come back to this is more detail later, but this point is worth mentioning in the context of cells, and the brain, being a computer. There are two components of a computer that are of interest: a CPU (Central Processing Unit), and related storage devices. A computer utilises a small part of its operating system, in firmware, to get started. From there, an application program is loaded from storage, which when executing, retrieves the data to be processed also from some connected storage device. Computer memory is logically partitioned, as required, to hold all or part of the software program(s), retrieved data, and processed data, before writing the results back to the specified storage location. Now, if the brain is like a computer, not only must it have the equivalent of defined data storage locations, but it must also emulate the CPU in that it must be partitioned during processing, separating logic from data being processed, and not overwriting permanent data until it needs to be changed. Computer processing always requires what is often termed, *working storage*, an area where the interim results of each logic step are stored before being used as the input to the next logic step in the sequence, and often in other logic modules as well.

I cannot be sure of whether scientists examining neural scans ever look for such differentiated regions, but there is no suggestion in the literature that they do so. Neuroanatomy texts are very detailed concerning the regions of the brain where individual general function occur, but by analogy, this is like identifying which computer in a network a particular application is processed. The level of detail is far too coarse to be of much use in identifying how the brain works. My primary reference sources for neuroanatomy are listed below. Knowing the "where" is far from knowing the "how" or "why". I do not intend to minimize the value of modern neurological research: it has proven beneficial in numerous ways, especially in the treatment of brain injuries and related disorders. My intention is to highlight that for the purposes of understanding the mind-brain conundrum, such neurological knowledge is superficial at best.

In research into neural network computers, the network is mostly concerned with logic structure, and still requires offline storage to source any substantial amount of data for processing, and the storage of results. The important questions here are: How are encoded, symbolic representations of data stored in a neural network like the brain, and just as importantly, where is the logic code for processing this data?

I have yet to find literature which explores this issue. In the absence of explanatory texts, I struggle to accept that biological organisms can self-program to interpret, store, and process external realities in the absence of external agents, to explain the conceptual nature of data and information. Here is an assertion which I believe to be irrefutably true:

By analogy, the neural circuitry of the brain can be said to be similar to the circuitry of a computer, but there is NO analogy by which the PROCESSING within the brain is similar to that of a computer, because computer processing is a function of intelligently designed software. What determines the processing of the brain?

References:

1. Bray, Dennis, *Wetware: A Computer in Every Living Cell*, Yale University Press, New Haven, CT, 2009

2. https://www.sciencedirect.com/science/article/pii/S1931312812001667

3. https://www.toppr.com/guides/biology/neural-control-and-coordination/nerve-impulse-and-its-transmission/

4. Kasparov, Garry, *Deep Thinking: Where Machine Intelligence Ends and Human Creativity Begins*, John Murray, London, UK, 2017

5. Hanaway, Joseph, *The Brain Atlas: A Visual Guide to the Human Nervous System*, Fitzgerald Science Press, Bethesda, MD, 1998

6. Martin, John H., *Neuroanatomy Text and Atlas*, 4th Edition, McGraw-Hill Inc., New York, NY, 2012

Chapter 5-5: Determinism

"Man can do what he wills but he cannot will what he wills."

~ Arthur Schopenhauer (1788-1860), German philosopher ~

Determinism is the philosophical theory that all events, including moral choices, are completely determined by previously existing causes. Determinism is usually understood to preclude free will, because it entails that humans cannot act otherwise than they do. *Causal determinism* is, roughly speaking, the idea that every event is necessitated by antecedent events and conditions together within the laws of nature, but in the many discussion papers on the subject, the word "causal" is omitted. I have read the arguments for this, but am not persuaded. The paradox is if you make a "truth" claim, you are violating determinism – rising above the bondage of total subjectivity, i.e., you are being objective. Free will is the ability to choose between different possible courses of action unimpeded. Free will is closely linked to the concepts of responsibility, praise, guilt, sin, and other judgements which apply only to actions that are freely chosen.

But let us, for the moment, revisit the principles of evolution, because there can be no argument that evolution is entirely dependent on determinism: one effect, however caused, being the cause of another effect, in the pursuit of survival and reproduction as the story goes. Raymond Tallis challenged determinism as here:

"There are various ways of arguing for determinism: the notion that we do not determine anything but are ourselves determined by things outside of us. The proofs are all pretty straightforward [referencing Kane, *The Significance of Free Will*]. The most obvious is that every one of our actions is a physical event. Every physical event has a cause and that cause will turn out to have causes. Eventually we shall arrive at causes that have nothing to do with us: for example, events that happened before we were born. So the actual basis for our actions lies outside us."[1]

You see, every evolutionary event must have been predicated on some previous evolutionary event(s), and everything that now is has been determined by what was, and has persisted through the evolutionary journey up to this time. It has been biology all the way, purposely refining itself to better adapt to the prevailing conditions; ever striving for greater survival prospects by reproducing the best of its accomplishments; instantiating instincts and behaviours in brain patterns, even to the extent of developing plasticity as a method of redundancy should some neural circuits fail. All this with the impression of purpose and design, when in truth, it was nothing but fortuitous circumstance, brought about by random mutations which opportunistically managed to survive in particular sequences, giving rise to ever more complex biological capabilities. Everything that we are owes its existence to what was, and everything that we do, we do because that is how we evolved. There is no "us" – we are just more highly evolved bananas, performing biological functions as bananas do according to their particular biology.

The biggest problem with *determinism* is *regression*. If every effect has a cause, then theoretically, it should be possible to trace the sequence of causes back to the first cause, and demonstrate the properties of that originating entity which gave rise to that activity. I will leave the reader to ponder that for themselves, in respect of any activity of their own which they believe to be voluntary.

Curiously, although I perceive the mind to be a separate entity to the brain, and am convinced of the reality of free will as we appear to exercise it, I nevertheless hold to determinism, without reaching the common conclusions of the true believers. Schopenhauer's opinion is at the core of the conundrum, for it claims that we have no discretion over what we will. Logically, every process has a beginning, else we subscribe to the misinterpretation of Stephen Hawking's pronouncement that the universe can create itself out of nothing. I do not believe that our *will to will* arises uncaused out of nothing, and thus accept the basic premise of determinism: something is behind our thoughts and actions. Scientist will speak of *random* events, but I contend that no event is truly random – it is just that we have yet to understand the cause. I further contend that it is illogical to believe in both determinism and randomness – you cannot have it both ways.

Part of the problem is the lack of definition of the "mind". If we succumb to the authority of philosophical materialism, the definition of the mind is a *fait accompli* (memory tells me that *faît* was once written with a circumflex, but as with so many other French words, the circumflex is fast disappearing, much to the chagrin of the purists). Now, where was I? Believing the mind to be of a material nature is convenient, because it becomes open to scientific investigation, and relieves scientists of the consequences of there being an immaterial existence. Quite simply, we lack a scientific vocabulary to describe the non-material, and thus the subject is best avoided. I believe that we can speak of the mind using a common vocabulary, but only by analogy and metaphor. I do not know what the mind is, but I am curious to understand the processes involved in what it does, and with my passion for understanding origins, I choose to pursue the matter as far back as I can. Nevertheless, by opting to believe that the mind is material, scientists have blindsided themselves because the mind is not amenable to scientific enquiry, as defined by the scientific method.

This raises an interesting point. If we cannot will what we will, why do I will differently to others? If the human mind-brain complex is driven by the chemistry of our cells, as determined by the genome, what is the chemical difference between what I will and others don't? Why do I will to pursue this issue to a conclusion, whilst others will to stop short? I contend that these questions are unanswerable under the religion of materialism, and we ought, if we are to be intellectually honest, seek the answers elsewhere. If, in this intellectual endeavour, I ask: Who is willing to come with me, do we accept that nobody has a free choice, and I only undertake this challenge because I am compelled to by my genes? Of note is that none of my family is, or ever has been, similarly inclined. With no apologies at all, I contend that philosophers like Arthur Schopenhauer are delusional, for they do not fully comprehend the significance of what they pontificate. He has chosen (willed) to make a statement, but at the same time asserts that he had no choice over willing so to do. The question becomes: how can he know the truth of what he has stated, and why did he bother in the first place, if he understood what he was saying? It seems obvious to me that he did not.

I earlier explained my Primary Axioms, #2 being that *all knowledge is built upon prior knowledge*. Whatever we are conscious of knowing is built upon a chain of data ancestors. Knowledge, both true and false, is of two conceptual types: data and information. Data is the conceptualisation of entities, and can be instantiated symbolically in material mediums, but information is a transient derivation of data correlated in context. This partly explains why people can reach different conclusions from the same data. Another part of the explanation concerns the breadth and depth of data acquired and retained by an individual. This in turn is influenced by an individual's worldview, which can be traced to culture, nurture, education, training, religion, experience, and other factors. In short, I am accepting that what we do is determined, but that we are able to resolve the relative weights of the deterministic factors leading to the result, which is our free will. Much like scientists modelling neural networks, the structure of our deterministic network is variable given the circumstances, or as in the case of network modelling, what we want the answer to be. Curiously, this accords with my Primary Axiom #1, that *nothing can explain itself*. In this case, the mind cannot explain itself, for to whom would it do the explaining? The mind can explain some of the factors, and can explain some of the processes in a general way, but it cannot delve into itself to the extent of apprehending the first cause of any particular mental process. Thus, in some ways, our free will is hidden from us.

I am a determinist, believing that every process or action, mental or physical, has one or more primary causes precipitating a chain of secondary causes and effects. You see, we all have backgrounds, cultures, education, and other factors which influence our worldview. Though all of these can influence our decision making, where our free will comes into effect is the weights we apply, sometimes unconsciously, to those individual influences. We can act or speak spontaneously, but the spontaneity is a function of an already established hierarchy of influences. On the other hand, we can deeply ponder an issue before taking action, and in this case, we are deliberately weighing up the relativity of factors, clearly evidencing free will as we can still decide to not act, or act contrary to our measured conclusions. Spontaneity might appear as us having no free will, and indeed the behaviour of children and the immature might suggest this, but the factors influencing actions are sometimes just instinct, and at other times a lack of experience to develop a more robust process of decision making. I reject the more common understanding of determinism which leads to the conclusion that our free will is an illusion. Similar to the mindset that accepts randomness, a failure to properly apprehend a material cause is not reason to conclude that there cannot be an immaterial one.

A conclusion is where one stops thinking, something I refuse to do.

However, there must always be a process for any conclusion that we reach, which may or may not result in an act of commission or omission. Just as in neural network modelling, each conceptual node in our minds has a weight determined by innumerable factors, that combined, resulted in an outcome, even if that outcome is tentative or null. In computing, irrespective of the architecture, an input event starts the process rolling: I would expect that a similar initiator precipitates all cognitive activity. What started it, and where the process goes from there is anyone's guess, especially as I know from experience that in mental processes, *non sequiturs*, both tangential and otherwise, are more common than not. Some people can express themselves succinctly, or otherwise tell a tale in summary form, whilst others will begin with a theme and be distracted onto other themes by every fact recounted, even to the point of forgetting where they started. "Now, where was I?"

Whilst I know little on the subject, I would assume that the task of psychiatrists and the like is to identify, and understand the priorities of, individual issues that they can extract from the minds of their patients, whether the patient is aware of them or not. In a sense, they then "re-weight" the conceptual priorities of their patients to achieve a different outcome. It is interesting that sometimes this re-weighting appears permanent, and at other times not, but the process is clearly deterministic. As to how this process could rewire the neural network in the patient's brain, is a question I will leave others to ponder, for I cannot imagine how a vocalisation by a doctor could result in chemical changes in neurons.

Agency, Determinism, and Free Will (again)

This brings us back to the subject of dualism, whereby an agency separate to the brain is the determining factor of decision making. In their book, *The Grand Design*[2], the authors assert that free will is an illusion. My first reaction to the book's title was: Why did the authors use the word, *design*? Design and purpose are antithetical to the evolution hypothesis, yet so often, scientists cannot avoid using that term. Before reviewing their assertions, consider this. When Albert Einstein presented his theories on relativity, did he derive them from deliberate, conscious thought, freely willed, or did they just come to him as he sat relaxing in the sun? When Stephen Hawking, renowned for his accomplishments as a theoretical physicist, chose that profession and expounded as he did, was that of his own free will, or was he a helpless participant in the events? As noted earlier, in his book, "*Science on Trial – The Case For Evolution*", Professor

Douglas Futuyma stated that Darwin drew "his evidence from comparative anatomy, embryology, behaviour, geographic variation, the geographic distribution of species, the study of rudimentary organs, atavistic variations (throwbacks), and the geological record to show how all of biology provides testimony that species have descended with modification from common ancestors."[3] How could the good Professor know that, if he did not freely will to study Darwin's works? More to the point, how could Darwin draw conclusions from so many factors if he lacked the free will to investigate as he did, freely choosing to evaluate the evidence before him, using deliberate cognitive processes rather than just accepting whatever his brain chose to deliver? Is his theory of evolution a concept that simply spilled from his brain as he sailed around on HMS Beagle?

Now, back to *The Grand Design*:

"The molecular basis of biology shows that biological processes are governed by the laws of physics and chemistry and therefore are as determined as the orbits of the planets. Recent experiments in neuroscience support the view that it is our physical brain, following the known laws of science, that determines our actions and not some agency that exists outside those laws ... so it seems that we are no more than biological machines and that free will is just and illusion."

I cannot know whether Hawking and Mlodinow believed what they wrote, or whether they were so beholden to philosophical materialism that they were not bothered to investigate further, but one truth that I will assert: *Recent experiments in neuroscience* DO NOT *support the view that it is our physical brain ... that determines our actions*. Certainly, it is our physical brain that initiates the motor neurons that through our nervous systems, cause our physical actions, but there is no evidence of what determines our brain to initiate such actions. I have found no articles that attempt to describe cognitive intentionality as being a function of the physical brain. Their assertion: "*the known laws of science ... determines our actions*" is **demonstrably false**, for there is no science that is able to account for the conceptual. Hawking, Mlodinow, and others of a similar belief are duplicitous in claiming that they know of scientific laws, whether in chemistry, physics, or quantum mechanics, that can derive the conceptual from the physical, and I defy any scientist, of any discipline, to prove otherwise. To put it bluntly, we are being lied to. A neuroscientist, known for his steadfast atheism, offered:

"Free will *is* an illusion. Our wills are simply not of our own making. Our thoughts and intentions emerge from background causes of which we are unaware and over which we have no conscious control. We do not have the freedom we think we have."[4] (italics in original)

Note here that contrary to Hawking *et al*, this neuroscientist admits that we are unaware of the background causes for our thoughts and intentions, although obviously, he still believes that such causes are entirely material in nature. I would offer that all books written by authors who subscribe to this "no free will" theory ought to start their books with a caveat:

"This book has not been written of my own free will – I am not responsible for its contents – evolution made me do it."

If you, the reader, also subscribe to this theory, then you have no reason to criticise any mistakes herein - they are not my fault - and it is only an accident of evolution that I am writing it, and not you. Now, back to something more plausible (I hope).

In more recent times, Dr Hannah Critchlow, a neuroscientist at the University of Cambridge, published her book, *"The Science of Fate"*[5], writing of "how much of our life is predetermined at birth and to what extent we are in control of our destiny". A review can be found online here[6]. Quoting from that review, in response to a question concerning the role that genes play in our political views, she responded:

"There have definitely been studies that have looked at different brain profiles associated with ideology. People who are very conservative seem to have a much larger volume and a much more sensitive amygdala – the area of the brain that is involved in perceptions of fear. People who are more liberal seem to have a greater weighting on the region of the brain that is engaged in future planning and more collaborative partnerships. They don't seem sensitive to immediate threats; instead, they are looking to the future. What we see in propaganda through the centuries is that if you heighten someone's fear response using environmental manipulation, you are more likely to make them vote in a right-wing way."

I cannot comment on the research, but in modern Western democracies, there is ample evidence that fear *has* become a common socialist tactic to convince voters to swing to the left. In the USA in particular, the Democrats have used fear of the future under a second term by Donald Trump, as President, to convince people to vote against him. What does that say about Dr. Critchlow's conclusions? Is it not the Left who champion *environmental manipulation*, especially in relation to climate change? Another quote, in response to a question concerning how to change the minds of people:

"It's very difficult. Once you have built up a perception of the world, you will ignore any information to the contrary. Your brain is already taking up about 20% of your energy, so changing the way that you think is going to be quite cognitively costly. And it might be quite socially costly too."

I wonder what Critchlow is implying by *cognitively costly*? Is she disagreeing with Hawking *et al*, accepting that we do have free will, and are able to perform deliberate cognitive activity? How can we change the way we think other than by willing to do so? My perception of the world, ingrained from childhood by the Catholic Church, did change over time as I researched that religion and related matters, refuting Critchlow's assertion that *"you will ignore any information to the contrary"*. I have earlier noted others who have changed their minds on a variety of subjects to which they were culturally conditioned. As for being *"quite socially costly too"*, that is a phenomenon that we observe concerning academics and scientists who do not accept the evolution narrative, but are afraid to publicly express their views. Finally, from that review:

"As we learn more about how our brains give rise to the staggering breadth of different behaviours, we see how each of us has a unique cartography of the mind, like a roadway that maps our choices and our strengths. The more we can appreciate that we are each different and that actually that's a good thing for the species as a whole, we should understand that this neurodiversity shouldn't be wiped out."

From a biological perspective, how is it that people of very close similarity in their DNA, could end up with such a variety of *brain mappings*? If the genome of parents is replicated in their offspring, how is it that in lower order animals, the brain mapping is the same, but in humans, the same genetic rules do not apply? There is something seriously illogical about Dr Critchlow's assertions, and her conclusions from what is learned of the structure of the brain. I would suggest that contrary to common wisdom, she has confused correlation with causation, no doubt because believing in philosophical materialism and evolution, she has no other answer. Perhaps she has not accepted that her own warning: *"Once you have built up a perception of the world, you will ignore any information to the contrary"*, applies at least as strongly to herself as any other. There is an old expression for that: *hoist on your own petard*!

Another commentator noted, although I have misplaced the source: "The irony that these dimwits don't seem to realise is that if the way we think is genetically pre-programmed, then no-one, including themselves, can know whether what they think and believe is true. It could be that their minds are "diseased" which causes them to think and believe as they do, whereas us with healthy minds think correctly and discern truth". [... and the audience was heard to exclaim with one voice: *Amen*!]

As an aside, what continues to befuddle me is why people such as Sam Harris and Richard Dawkins are so adamant that there is no God, when they are also adamant that our thoughts, and theirs, are not of our/their own making. I do not know whether God exists or not, but no-one who believes that they have no control over their own thoughts, can make truth statements about anything whatsoever. If they have no control over their own thoughts, then neither can they claim to be capable of logical and rational thinking. Stephen Hawking has been lauded as one of the greatest intellects of our time, yet according to him, he has no control over his intellect, and thus has no reason to believe in what he thinks (or thought).

And this is the greatest irony: such scientists cannot know whether they are correct in their beliefs, when they believe that they have no control over their beliefs. Of *paramount significance*, and this bears repeating over and over again, is why they believe in their own beliefs. Let me state emphatically (and therefore in bold font): **It takes a peculiar form of *cognitive dissonance* to claim that you have no basis for believing what you believe to be true, yet at the same time assert a truth statement that you have acknowledged that you cannot know to be true.** That is the paradoxical proposition of the evolutionists.

What arrogance to make such pronouncements, and write so many books, when by their own admission, they CANNOT know of what they speak or write! Do they not understand that we have no reason to agree with them? In my considered opinion, it takes a special kind of stupidity to contend as they do, but then again, who knows, perhaps that is not "my" opinion. It is an unresolvable paradox that people who demonstrably believe in their own beliefs, at the same time assert that they have no reason so to do.

Well, I do have a solution: they know not of what they speak. Either that or they speak falsely. Can anyone truly believe that the entire history of scientific discovery *"emerged from background causes of which we are unaware and over which we exert no conscious control"*? Do we really have any reason to accept the evolution narrative, as earlier published in Charles Darwin's *"The Origin of Species by Means of Natural Selection"*[5] and *"The Descent of Man"*[6], if what Hawking *et al* assert is actually true?

Now, think about that for a moment, as a purely biological process behaving in accordance with the known laws of physics and chemistry. If the material monists and evolutionists are right, then **the evolution process itself, even before there was life, had the intelligence to develop a neural system capable of "knowing" the cause of its own existence**. Let me restate that: evolutionists argue that we, and supposedly they, have no free will, and no ability to state truths other than the truths that evolution has imprinted in their minds. The question becomes: how can evolution know of itself, and its processes and achievements to imprint them in our minds? Evolutionists may claim that they believe in evolution because evolution itself taught them that it was true, but why did it not teach others such as myself that it was true? I must protest: that is grossly unfair! If the evolutionists are right that we have no free will, why did a Professor of Evolutionary Biology assert as he did that Darwin "drew his evidence from comparative anatomy, embryology, behaviour, geographic variation, the geographic distribution of species, the study of rudimentary organs, atavistic variations (throwbacks), etc."[3] Why did Darwin need to do so? I will insist on my Primary Axiom #1, that *nothing can explain itself*. The evolutionists believe to the contrary: that evolution can explain itself, a process beyond explanation.

Let me confess that whilst I have no idea of the cause of human existence, my logic, faulty as it may be, convinces me that *"to know"* requires intelligence, and if evolution as a process is incapable of *"knowing"*, then for evolution to be true at all, there had to be an external intelligent agency behind it. It is this line of thinking which allows me to accept *directed evolution*, rather than the *undirected* as espoused by the evolutionists.

Implications for Justice

Just briefly, if it is true that we cannot trust our own beliefs, and those of others, what is the point of social justice issues, and the court system? Why would we believe the prosecution's case, the prosecution witnesses, the defence case, and the defence witnesses, if whatever they profess to believe that they witnessed, could well be untrue? If you were a member of the jury, why would you accept the direction of the Presiding Judge, and by what moral or intellectual right could you decide the case, when your own potentially faulty beliefs are added to the mix?

Societies can only operate in harmony when the people believe that their beliefs are justified. Again, the irony is that all scientific endeavour is predicated on knowing, and believing, in truth, despite many of them asserting that they have no reason to do so.

Simply madness masquerading as scholarship.

References:

1. Tallis, Raymond, *Aping Mankind*, Routledge Classics, New York, NY, 2016, pp. 51-52

2. Hawking, Stephen, and Mlodinow, Leonard, *The Grand Design*, Bantam Books, New York, NY, 2010

3. Futuyma, Douglas J., *Science on Trial – The Case for Evolution*, Pantheon Books, New York, 1982, p. 36

4. Harris, Sam, *Free Will*, Free Press, New York, NY, 2012, p. 5

5. Critchlow, Hannah, *The Science of Fate: Why Your Future is More Predictable Than You Think*, Hodder & Stoughton, London, UK, 2019

6. https://www.theguardian.com/science/2019/may/11/neuroscientist-dr-hannah-critchlow-science-of-fate-interview?

7. Darwin, Charles, *The Origin of Species by Means of Natural Selection*, Penguin Books, London, 1985

8. Darwin, Charles, *The Descent of Man*, Wordsworth Editions, London, UK, 2013

Chapter 5-6: Self-Organization

"Organizing is what you do before you do something, so that when you do it, it is not all mixed up."

~ A.A. Milne (1882-1965), English author ~

If the organiser both participates in, and is the subject of, the re-organising, it seems unlikely that the organiser could keep control whilst being shuffled around. As I have learned from my experiences as a project manager, effective organisation requires a plan, with external monitoring of the execution of the plan, and timely corrective intervention when inevitably, something goes awry. In a later chapter, we will discuss the concept of neural plasticity, and what that would mean for the stability of the organiser – the mind or brain.

If the mind is a property of the brain, and not a separate entity, or is an activity of the brain distributed across mental organs of computation, as some would claim, then we must begin with the ontology of the brain to determine its capabilities and limitations. If we fail to definitively identy the brain as a biological organ, we run the risk of attributing properties and activities of which it is incapable. We do know that the brain does perform some autonomous functions, but it is the volitional functions which are the most problematic. One of the more perplexing autonomous functions is *self-organization*. There can be no doubt that it does occur in biology, because from the moment of conception, cells self-organise over a period of roughly nine months into a functioning human baby. Cells continue to self-replicate during our lifetime, and thus self-organization is an intrinsic behaviour of biological organisms.

In biological systems, self-organisation is described thus:

"Self-organization in biology can be observed in spontaneous folding of proteins and other biomacromolecules, formation of lipid bilayer membranes, pattern formation and morphogenesis in developmental biology, the coordination of human movement, social behaviour in insects (bees, ants, termites) and mammals, and flocking behaviour in birds and fish.

The mathematical biologist Stuart Kauffman and other structuralists have suggested that self-organization may play roles alongside natural selection in three areas of evolutionary biology, namely population dynamics, molecular evolution, and morphogenesis. However, this does not take into account the essential role of energy in driving biochemical reactions in cells. The systems of reactions in any cell are self-catalyzing but not simply self-organizing as they are thermodynamically open systems relying on a continuous input of energy."[1]

As you will see from the above description, there is no sense of *conceptual self-organisation*, which would be necessary for the neuronal network of the brain to structure itself around its understanding of an external reality. Biological self-organisation deals only with the biological self, and is a closed-system other than it can be influenced by its environment, relying as it does on a continuous input of energy. There is undisputed evidence of reorganisation of the brain's networks, from the earliest days from conception, but the questions to be asked are: (1) the causes of that reorganisation; and (2), whether there are any internal causes apart from the biological? Again, I would point out that data and information are conceptual, but morphogenesis can only operate on the physical.

Cell Differentiation

In his interesting, but logically flawed treatise, "*The Origins of Order: Self-Organization and Selection in Evolution*"[2], Stuart Kauffman, a theoretical biologist, and complex systems researcher, offered an alternative to natural selection as a driver of evolution. As Stephen Meyer introduces this subject,

"Kauffman advanced a comprehensive alternative theory to account for the emergence of new form. In addition, he advanced a specific proposal for explaining the Cambrian explosion. Kauffman notes the development of animal body plans involves two phases: cell differentiation and body-plan morphogenesis (cell organization). He explores the possibility that self-organizational processes at work today – specifically in cell-differentiation and body-plan formation – might help explain how new animal forms originated in the past. Kauffman proposes, first, that gene regulatory networks in animal cells – genes that regulate other genes – influence cell differentiation. They do this by generating predictable 'pathways of differentiation', patterns by which one type of cell will emerge from another over the course of embryological development as cells divide."[3]

It is at this point that I will justify my words, "logically flawed". I have confessed before my interest in origins, and so when I read hypotheses which rely on *what is*, rather than what supposedly *was*, my antennae begin to tingle. When one uses *gene regulatory networks* to explain the initial development of the genome, it should be obvious that the argument begs the question. Yes, the theory *might help explain how new animal forms originated*, but only from an already established form with a developed genome. The danger is in extrapolating the theory further back than biology allows. Continuing Meyer's commentary,

"He [Kauffman] invokes an idea sketched out in the 1940s by the famous English mathematician Alan Turing. Turing proposed that specific arrangement of cells in animal development might ultimately derive from the diffusion and specific arrangement of crucial molecules – presumably something like the morphogen proteins present in embryonic cells ... Turing postulated that the distribution of these molecules might have originated in the first place independently of such information as the result of simple chemical reactions. He imagined one molecule producing both a copy of itself ('autocatalyzing') *and*, in addition, producing a different molecule as well. Then he envisioned one of these molecules inhibiting the production of the other, thus allowing, through repeated cycles, the production of more and more of one molecule and less and less of the other."[4]

The obvious question is: If cell replication results in two competing molecules, one inhibiting the other, what is the probability of either surviving? The possibilities are endless. Mathematics is one thing, chemistry another. To support this hypothesis, a scientist should identify the chemicals and/or proteins which would enable this process. As Meyer argues,

"Kauffman does not mention any specific chemical or proteins that would behave in the way he envisions. Instead, he describes the behaviour of hypothetical molecules ... More important, Kauffman offers no evidence that chemical interacting in the way he envisions could create specific *biologically relevant* configurations or distributions or morphogen proteins – apart, that is, from the process that generate specifically arranged distributions of these proteins in *pre-existing information-rich* embryonic cells today ... Kauffman himself seems tacitly to acknowledge the difficulty of generating biological specificity from the reactions of chemicals alone." (p. 296) [italics in original]

The problem, from the point of view of origins, is that Kauffman's theory cannot explain the origin of the genetic regulatory networks that are necessary for cell differentiation. Nevertheless, Kauffman "acknowledges that the predictable pathways of differentiation that characterize this process *derive from pre-existing* gene regulatory networks ... thus, the self-organizational process that Kauffman cites cannot *explain* the origin of genetic information, because it derives from it, as Kauffman's own description reveals." (Meyer p. 297). Returning to the context of the mind-brain conundrum, I would reiterate that genetic information is different to conceptual data, and self-organisation at the genetic level is in no way comparable to the prerequisite conditions for self-organisation at the conceptual level. Even if a plausible explanation is found at the genetic level, it would tell us nothing about how the mind self-organises around the concepts of data and information.

I applaud scientists who make the effort to seek understanding of the mysteries of life, but it does seem that eventually, one must come to the conclusion that they are looking in the wrong place. It is rather pointless continually looking in the same place muttering over and over again, "It must be here, somewhere", when all the evidence uncovered so far would suggest that it is not. To date, scientists have proven that life comes from life, which doesn't answer the question: where or how did life come from non-life in the first place? If the genome is a product of evolution, then the genome itself cannot be part of the answer.

Search Space

A "search space" does not necessarily imply the area to be *intelligently* searched; rather, it is the domain of a function to be optimised by whatever process. In abiogenesis (chemical evolution), where a porridge of primordial goop is said to give rise to amino acids, which in turn manage to self-organise into just the right teams which can combine to form proteins, the search space is beyond our knowing. How big, or small, was the bowl of porridge, and how many potential players were there? More importantly, remembering that to be suitable for selection in the team, only 21 out of 400 types are eligible, and they must all be left-handed. Clearly, biology is not politically correct. Not knowing the composition of the primordial porridge, we cannot know the distribution of suitable candidates in amongst the many more unsuitable. Whichever way you think about it, the likelihood of forming just one team, let alone a whole Olympic village necessary to create a cell, is impossibly small.

In genetic mutation, we have a similar problem. Evolution theory relies on optimisation of genetic mutation and natural selection effecting fitness. Predicting how quickly a population will adapt, and the type of beneficial mutations that will fuel that adaptation, requires estimates of the additive genetic variance in fitness and of the beneficial mutation rate, and the distribution of beneficial effects. In my research, of admittedly mostly online sources, I have found studies that estimate beneficial fitness-altering mutations being in the range, 6 - 50%. On the other hand, most mutation-accumulation experiments appear to identify few, if any, beneficial mutations. There is a suggestion in the literature that analyses of mutation-accumulation data typically assume that selection is directional, but ignore that whilst mutations increase fitness in regard to one component, they may also lower fitness in regard to lifetime fitness, or another fitness component (*i.e.*, antagonistic pleiotropy). Genes are both pleiotropic and polygenic, and thus predicting the outcome of a point mutation in such a complex organism (search space) is problematic.

In population genetics, we find the term, "mutation–selection balance", which is defined as equilibrium in the number of deleterious alleles in a population. This occurs when the rate at which deleterious alleles are created by mutation equals the rate at which deleterious alleles are eliminated by selection. An allele is simply a variant of a gene. Allele frequency, or gene frequency, is the relative frequency of an allele at a particular locus in a population. Microevolution, a term not favoured by evolutionists, is the change in allele frequencies that occurs over time within a population. Now, the majority of genetic mutations are neutral or deleterious, with a consensus that beneficial mutations are relatively rare. We know that the unit of inheritance is not a single gene or nucleotide pair, but a block of nucleotides. Thus, deleterious alleles must be inherited far more frequently than beneficial alleles. This brings us to the question: Is there such a phenomenon as a "normal" mutation rate, and what would be the effect?

Ontology

Ontology is defined as the philosophical study of being. Part of the goal is to identify the categories of being, thereby leading us to comprehend the nature of things, their properties, activities, and their relationship to other things. When we identify an entity and categorise it as a fish, we assume certain properties, behaviours, and the environments in which it can sustain itself. We can also identify a sub-category, such as salt water or fresh water, and thus know other things.

In this chapter, I wish to discuss the ontology of the mind-brain complex in relation to its ability for self-organization, defined as:

"a process by which systems that are in general composed of many parts spontaneously acquire their structure or function without specific interference from an agent that is not part of the system. Another term for this is spontaneous order, the process whereby some form of overall order arises from local interactions between parts of an initially disordered system. The process can be spontaneous when sufficient energy is available, not needing control by any external agent. Self-organization is said to occur in many physical, chemical, biological, robotic, and cognitive systems, with examples found in crystallization, thermal convection of fluids, chemical oscillation, and animal swarming."[5]

This phenomenon is also claimed to occur in neuronal circuits as found in the brain, but to what extent, and for what purpose, is what we need to examine. The brain is a biological organ, so let us review what is known about biology.

Biology is the scientific study of life and living organisms. It is concerned with physical structure, chemical processes, molecular interactions, physiological mechanisms, development, and evolution. In biology, it is accepted that the cell is the basic unit of life, that genes are the basic units of heredity, and that evolution is the process whereby species change over time. Living organisms are open systems that survive by transforming energy and decreasing their local entropy to maintain a stable and vital condition defined as homeostasis. Thus, without a continuing supply of the appropriate nutrients, an organism will deteriorate and die.

Digging ever deeper, we must now consider the science of *synergetics*. Synergetics deals with both material and immaterial systems comprised of component parts. The focus is on spontaneous activities, i.e. the emergence of new properties, structures, processes, and/or functions through self-organization. The question we must ask ourselves is whether there are general principles of self-organization irrespective of the nature of the individual parts of a system? In particular, whereas humans can self-organize based on cognitive rules, can the component parts of a human, i.e., atoms, molecules, and cells, organize themselves in a cognitive manner? More specifically, can neural cells, as found in the brain, self-organize in ways which the cells in other biological organs cannot, even though they are comprised of precisely the same basic chemicals, albeit arranged in a different fashion? Put another way, are the organs of cognition themselves capable of cognitive behaviour? If the answer is "yes", what are the properties, or structure, which allow neural cells to behave in such a manner, whilst others cannot?

Our next step is to understand what is meant by cognitive behaviour. *Cognition*, as defined in my trusty Cassell dictionary, is "the faculty of perceiving, conceiving, and knowing, as distinct from feelings and the will". Another definition states that "cognition is the mental action or process of acquiring knowledge and understanding through thought, experience, and the senses. It encompasses many aspects of intellectual functions and processes such as attention, the formation of knowledge, memory and working memory, judgment and evaluation, reasoning and computation, problem solving and decision making, comprehension and production of language. Cognitive processes use existing knowledge and generate new knowledge."

One question that arises, similar to earlier, is whether neural cells or neural networks, which form the architecture of our brains, are themselves capable of thinking their own thoughts, and making their own decisions, independent of the thoughts which their collective activities are said to represent? If brain activity is the evidence of thinking, what is the evidence of a neural network having thought processes independent of the thought processes that can be externally observed to be occurring? Let me try that another way. Other than processes which occur naturally in accordance with our understanding of the laws of science, all processes require a supervisory process to initiate and direct them, much as an orchestra requires a conductor. Given that the brain's neural network, being biological, can only do what it can and *must* do in accordance with the laws of science, what is the supervisory process that enables it to do other than what such laws would initiate? I am not confident that I expressed that in a clear, logical, or convincing manner, but I hope that you can glimpse, dimly even, what I am driving at.

Cognitive (conceptual) organisation is not the same as physical organisation, and especially not the same as the organisation of symbolic representations of the conceptual. Whilst it could be said, hypothetically, that physical matter can self-organise around its intrinsic properties, that is not the same as self-organising around the concepts that the material represents. These words that you are reading have not been self-organised as they persisted in an electronic file: they were organised by a (relatively) intelligent external agent and fortunately, remained so organised.

Summary

The referenced Wikipedia entry[1] also states: "Examples of self-organization include crystallization, thermal convection of fluids, chemical oscillation, animal swarming, neural circuits, and artificial neural networks." We also find, "A neural circuit is a population of neurons interconnected by synapses to carry out a specific function when activated. Neural circuits interconnect to one another to form large scale brain networks. Biological neural networks have inspired the design of artificial neural networks, but artificial neural networks are usually not strict copies of their biological counterparts" – of course not.

There is an element of the salesman's technique of "bait-and-switch" in such descriptions, by introducing examples which are scientifically proven (crystallization), to examples which are observed and the causes unknown (swarming), to examples which have no scientific foundation whatsoever. As we will discuss in a later chapter regarding marble statues, the intrinsic properties of marble render it incapable of self-organisation, a fact that sculptors would no doubt celebrate. We can't have marble statues bending down, straightening their arms, turning their heads, or otherwise changing their stance. Neural networks, as in the brain, cannot be asserted as examples of self-organisation without scientific evidence that such is even possible. There is a failure of logic in the statement: *interconnected ... to carry out a specific function*. It is antithetical to the evolution narrative to claim that biological entities evolve, or otherwise behave, in a purposeful manner to achieve a specific function previously "unknown" to them. Purpose is synonymous with intentionality, and intentionality is a function of intelligence, not bio-chemistry.

Biological entities can, and must, self-organise, if at all, based on their biological properties, but can do no more than that. Despite repeated assertions and assumptions by scientists, there is no scientific evidence that neuronal circuits can self-organise around the semantic layer representing the concepts of the source as detected and transmitted through sensory inputs, and a great deal of scientific evidence that they cannot.

References:

1. https://en.wikipedia.org/wiki/Self-organization

2. Kaufmann, Stuart A., *The Origins of Order: Self-Organization and Selection in Evolution*, Oxford University Press, Oxford, UK, 1992

3. Meyer, Stephen C., *Darwin's Doubts: The Explosive Origin of Animal Life and The Case for Intelligent Design*, HarperCollins, New York, 2013, p. 294

4. *Ibid*, pp. 295-296

5. https://en.wikipedia.org/wiki/Ontology

Chapter 5-7: Emergence

"In philosophy, systems theory, science, and art, emergence occurs when an entity is observed to have properties its parts do not have on their own. These properties or behaviours emerge only when the parts interact in a wider whole."

~ Wikipedia ~

Wikipedia offers this: "For example, smooth forward motion emerges when a bicycle and its rider interoperate, but neither part can produce the behaviour on their own."

In this example, what is seen is not an *emergent* property or behaviour, but an effect resulting from a cause. When stationary, do the bicycle and rider evidence emergence? If not, then when in motion, the motion is an activity enabled by the combined properties of the bicycle and rider, not an emergent property. A stone may gather moss, but a rolling stone not so. Is the combination of a stone, a slope, and gravity evidence of an emergent property of this system? As I teeter when standing on the edge of a high cliff, are we witnessing an emergent property of the circumstances? For a property to be emergent, it must provide the entity with the capability of autonomous activity, which is entirely consistent with the laws of physics and chemistry. The property must not only enable and regulate the activities of the entity, but the entity can only do what it must do, given its own material properties and those of other entities with which it interacts in a specified environment. It is a category mistake to conflate mechanical and biological entities in this context, but not unusual as we earlier discussed concerning transferred epithets. In the example of the bicycle, we have two entities: (1) the rider who is capable of moving his/her legs in a predefined pattern, and (2), a mechanical device designed to take advantage of that movement. Compare the specifications of this system with the evolutionists' claim of the mind being an emergent property of the brain. The activities of the mind owe nothing to the properties of the brain, and according the evolution narrative, the mind is not designed by an intelligent agent. It cannot do so itself.

Examined closely, definitions of emergence are usually nothing but literary sleight of hand, because in those terms, any and every activity or effect could be termed, *emergent*. This is another example of scientists co-opting common terminology for their own purposes, and being disingenuous whilst doing so. To put it bluntly, we are being conned.

Let us go back to basics to understand this issue. The word, *emerge*, means to "move out of, or away from, something and become visible", or "to become apparent or prominent". Revisiting some fundamentals, the world consists of entities to which we apply conceptual labels, and which have properties which allow, constrain, and otherwise regulate the activities of the entity. Activities of a composite entity are dependent on the intrinsic properties of the components, and the interactions of those components in an environmental setting. When chemicals are mixed in a beaker such that the fluid changes colour, or a filament glows when conducting an electric current, it is deceptive to claim that such effects are *emergent behaviours*. When water changes from a fluid to a solid, with different properties, it is deceptive to claim that such properties are emergent.

There are many wonderful phenomena said to "emerge" in nature by a combination of law and chance, the snowflake being a common example. The binding properties of the water molecule dictate what angles are permissible, and the chance atmospheric conditions determine the unique structure of the ice crystal. It is said that no two ice crystals, or snowflakes, ever look alike, but whilst possibly true, I do not believe that we can know that. The important point to note, in the context of "the mind is an emergent property of the brain", is that no crystal formation or similar process can result in a conceptual statement. If hail falling to ground ever spelled a word or phrase, no-one would think that the hail self-organised that way, or that the words were an emergent property of water and environmental circumstance.

I can understand why evolutionists, and scientists holding to philosophical materialism have so readily accepted *emergence* as the answer to the mystery of life – what else can they do? Here, presented with so much academic authority, is the silver bullet explaining the unexplainable. Restating the definition: *emergence* is a process whereby larger entities, patterns, and regularities arise through interactions among smaller or simpler entities that themselves do not exhibit such properties. Unstated is the assumption that the smaller, simpler entities have the requisite properties that allow autonomous interactions, resulting in new patterns and regularities. For example, sand can be arranged into an endless variety of patterns, including sculptures of other entities, but sand is incapable of doing so itself, and cannot, for example, in combination with lime, water, and other chemicals, autonomously create cement. Yes, the interactions will, if stimulated, result in larger entities with additional properties, but the result is still nothing more than a mixture of chemicals. If these chemicals were just dumped one on top of the other, they would not form cement – an external agency is required to do the mixing.

Now this is where the claims for emergence are disingenuous, *the simpler entities do not exhibit the properties of the larger entities*, for there is an untold story. The truth is that for emergence to occur, the smaller entities must have intrinsic properties such that they can autonomously interact with one another. It is not enough to make bald, unsubstantiated assertions as in the following examples, because following the rule of causality, these intrinsic properties must be demonstrated. Consider these articles:

"Each level of biological organization builds upon the previous level, and is more complex. Moving up the hierarchy, each level acquires new emergent properties that are determined by the interactions between the individual parts. When cells are broken down into bits of membrane and liquids, these parts themselves cannot carry out the business of living. For example, you can take apart a lump of coal, rearrange the pieces in any order, and still have a lump of coal with the same function as the original one. But, if you slice apart a living plant and rearrange the pieces, the plant is no longer functional as a complete plant, because it depends on the exact order of those pieces. In the living world, the whole is indeed more than the sum of its parts. The emergent properties created by the interactions between levels of biological organization are new, unique characteristics. **These properties are governed by the laws of chemistry and physics.**"[1] [emphasis mine]

Curiously, the author is agreeing with my contention, for if the properties, and thus the interactions, *are governed by the laws of chemistry and physics*, then the smaller entities must be comprised of just the right chemicals, in just the right proportions, in just the right purity, under just the right environmental conditions, for the interactions to autonomously occur. Looking back, perhaps this is why my chemistry experiments at school so often failed. Despite my physically mixing the right ingredients, emergence chose to not occur.

"These emergent properties are due to the arrangement and interactions of parts as complexity increases. For example, although photosynthesis occurs in an intact chloroplast, it will not take place in a disorganized test-tube mixture of chlorophyll and other chloroplast molecules. The coordinated processes of photosynthesis require a specific organization of these molecules in the chloroplast. Isolated components of living systems, serving as the objects of study in a reductionist approach to biology, lack a number of significant properties that emerge at higher levels of organization. Emergent properties are not unique to life. A box of bicycle parts won't transport you anywhere, but if they are arranged in a certain way, you can pedal to your chosen destination. Compared with such nonliving examples, however, biological systems are far more complex, making the emergent properties of life especially challenging to study."[2]

Here, the author confirms what I have been saying. Before emergence can be offered as the scientific salvation of evolution theory, it must be scientifically proven. As has been demonstrated with attempts at synthesising chemical evolution (abiogenesis), the task is more than just challenging, it is proving impossible.

There is an interesting, but misleading article, online here[3], *"Emergence: The Remarkable Simplicity of Complexity"* attributed to Andy Martin and Kristian Helmerson of Monash University. They state: "Emergence describes the ability of individual components of a large system to work together to give rise to dramatic and diverse behaviour". By that definition, the motion of a motor vehicle caused by mechanical components working together can be called *emergence*. Similarly, the sailing of a ship or the flight of an aeroplane. As a scientific definition, it is appallingly bad. The article states:

"Recent work by Enkeleida Lushi and colleagues from Brown University showed how bacteria in a drop of water spontaneously form a bi-directional vortex, with the bacteria near the centre of the droplet circulating in the opposite direction to those near the edge. Since the bacteria do not consciously decide to create the bi-directional vortex, such behaviour is said to be "emergent".

Unlike music from an orchestra led by the conductor, emergent behaviour arises spontaneously due to (often simple) interactions of the constituent parts with each other and the surrounding environment. Here, there is no "leader" deciding on the behaviour of the system."

White phosphorus is pyrophoric (self-ignites on contact with air), burns fiercely, and can ignite cloth, fuel, ammunition, and other combustibles including human flesh; salt and other compounds are hygroscopic: do we declare those reactions to be *emergent*? That there is no "leader" in these processes is a straw man argument, deflecting attention from the realities of naturally occurring chemical reactions. Once again, we encounter this "bait and switch" tactic so beloved of salespeople, and now apparently, scientists. They want us to believe that if organic entities can exhibit emergent properties or activities, then the mind can be an emergent property of the brain. The intellectual teasing continues:

"Many biological systems commonly exhibit emergent behaviour. The complex behaviour of flocks of birds, colonies of ants, swarms of bees and schools of fish emerges from the interactions of the constituent parts of the respective systems.

Consider an ant colony. In the absence of centralised decision making, ant colonies exhibit complex, problem solving behaviour. This behaviour emerges from the reaction of individual ants to simple chemical stimuli – from larvae, other ants, intruders, food and waste. In turn, each ant produces chemical signals, providing a stimulus that other ants respond to. From simple interactions leading to self-organisation, ant colonies have demonstrated the ability to collectively solve geometric problems, such as optimising their foraging route to and from food resources.

The idea of emergence, though, isn't confined to biological systems. It pervades all areas of science and is a manifestation of other complex interacting systems in our daily lives, such as stock markets, the connectivity of the internet, and traffic flow."

I would offer that "complex, problem solving behaviour" is a cognitive function, not a chemical one. The claim that "ant colonies have demonstrated the ability to collectively solve geometric problems" is in the realm of fantasy, not science. Having lived in areas plagued by ants, my observation is that there is nothing optimal about their foraging route. Further "bait" is offered by referencing stock markets and other activities of intelligent agents. There are three levels of activities that these physicists are attempting to conflate: chemical, biological, and cognitive. Chemicals can only do what their properties allow; biological organisms do what their cellular structures determine they must do, based on their genetic make-up; and intelligent agents *can do whatever they like*, such is the freedom of the intellect. As I read in another commentary,

"Martin and Helmerson present examples of emergent phenomena that, due to complex interactions, are "greater than the sum of their parts." What they fail to point out is that the "cellular automata" of flagellated bacteria in a vortex, or honeybees in a hive, or patterns made by a school of fish, are dependent on information coded in genes -- not just the objects and the environment. A school of dead fish may self-organize in a whirlpool, but they will never show the responsive patterns of living fish. Yet these two authors fail to make any distinction between life and non-life ... There needs to be a differentiation made between the emergent properties of information-coded things (like cells, robots, or ants) and the emergent properties of dead things."[4]

I must agree, as we can see from these statements:

"In physics, magnetism of everyday materials emerges from the spontaneous alignment of the magnetic moment of billions of electrons. Similarly, phenomena such as superconductivity and superfluidity emerge from the cooperative flow of electrons and atoms, respectively, at temperatures close to absolute zero (-273C). On a much larger scale, the structure of the universe emerges from the gravitational attraction of stars.

In chemistry, many atoms combine to form macromolecules with structures that emerge from the secondary interaction of the atoms, which determines their function in molecular biology. In turn, cells emerge from the interaction of many of these macromolecules – resulting in cell biology."

Notice how the authors imply a smooth transition from physics to biology, as if between the two was a smooth super highway, instead of what is truly there – an unbridgeable chasm. In passing, I continue to be bemused by statements such as "the structure of the universe emerges from the gravitational attraction of stars", for whilst I have no knowledge to disagree regarding the structure, I am well aware that so far, cosmologists have not been able to explain the formation of first-generation stars. In his book, "*A Brief History of Time*"[5], Stephen Hawking admits that we do not really know how the initial stars were formed, and Professor Abraham Loeb of Harvard University's Centre for Astrophysics has said, "*We don't understand star formation at a fundamental level*"[6]. That being so, one can only conclude that the structure of the universe initially emerged from another process altogether, but back to the subject.

I was curious about their use of an observation by German scientist and engineer Jochen Fromm:

"one water molecule is not fluid

one gold atom is not metallic

one neuron is not conscious

one amino acid is not alive"

I pondered their source, as it is found in David Etkin's book, "*Disaster Theory: An Interdisciplinary Approach to Concepts and Causes*"[7], where perhaps he quoted from Jochen Fromm on the emergence of complexity.

The issue is that there is no useful correspondence between these entities. Water is unique, often described as the strangest liquid on the planet, and cannot be used to segue into a discussion on emergence. Gold remains gold, and is inert, so I have no idea of how that relates to emergence. A neuron is not conscious, but whether consciousness emerges from a collection of neurons is what we are here to discuss. And as for amino acids, only specified arrangements can exhibit new properties, and there is a great chasm between amino acids and living organisms. I do hope that the authors were not attempting to suggest that living cells, or even proteins, emerge autonomously by chance from a random collection of amino acids. The authors, to their credit, admitted to the enormity of the task of bridging this chasm in their conclusion headed, "Bridging the levels of complexity":

"Despite the ubiquity of emergent behaviour there remains no deep understanding of emergence. At each level of complexity, new laws, properties and phenomena arise and herein lies the problem. Properties describing one level of a complex system do not necessarily explain another level, despite how intrinsically connected the two may be. Understanding the emergence of the structure of molecules does not necessarily allow one to predict the emergence of cellular biology.

While controlled experiments such as those of Enkeleida Lushi and colleagues may help to identify the essential ingredients of emergent behaviour, **new ways of thinking** about emergence that go beyond conventional modelling of specific systems are required. These would allow us to determine unifying principles of emergent behaviour – bridging all levels of complexity.

As such, understanding and harnessing the fundamental organising principles of emergence remains one of the **grand challenges** of science." [emphasis mine, underline in original]

"There remains no deep understanding of emergence" ... in which case, can we be confident that emergence is autonomous, and does not require an external agent? Of course, scientists can define their way out of problematic examples, and define their way into examples which appear suitable, but this is a function of philosophy, not science.

Design and Specified Information

New (old) ways of thinking have already been offered not only by Intelligent Design proponents, but even as far back as the 4th century BCE. In Book XII of *Metaphysics*, Aristotle observed, "[Design] is a principle of movement in something other than the thing moved; nature is a principle in the thing itself"[8]; and in Book II of *Physics*, Aristotle opined that design is what brings to completion that which nature itself cannot bring to a finish[9]. That notwithstanding, it is notable how often evolutionists use the concepts of design and purpose in their own narratives, almost as *Freudian slips*. Of course, such concepts are antithetical to philosophical materialism, but when the evidence points that way, it does seem unscientific to not at least consider the validity of such proposals. When one ponders the improbability of just the right twenty out of an available hundred or so, left-handed amino acids accidentally meeting in the one place with no right-handed amino acids there present, and queuing in just the right sequence to form just one protein, none of which are stable in solution, there does seem a real issue of specified information.

The underlying question is whether nature is complete in and of itself, needing no external influence. The materialists would argue, yes, just as Richard Dawkins does when he would have us believe that even human intelligence, which implements design, is part of our *extended phenotype*[10]. In other words, evolution, which is not teleological, is responsible for the development of teleological functions in the human brain. There is no evidence for this, other than the commonly observed phenomenon of human intelligence. There can be no scientific proof, one way or the other, but the rational approach is to rely on first principles to understand what is possible, and what is not. I agree with Aristotle, that *nature is a principle in the thing itself*, which I would paraphrase as - nature can only do what is intrinsic to itself. If "nature" is as defined by the materialists, consisting entirely of the physical, then nature cannot be responsible for anything which we would define as *conceptual*. Purpose and design are conceptual, not physical, and cannot be intrinsic to nature. Design is teleological, and always expresses purpose. A natural corollary to my primary axiom that *nothing can explain itself*, is that *nothing can design itself*.

You may have encountered the terms, *specified complexity* or *complex specified information* attributed to William Dembski, but as these involve complex mathematics and concepts which I only partly comprehend, they contribute nothing to my thinking. I will leave those for readers who do understand them.

References:

1. Sylvia S. Mader, *Biology* (10th ed., 2010), Ch. 1. *A View of Life*

2. Jane B. Reece, Lisa A. Urry, et al. *Campbell Biology* (10th ed., 2014), Ch. 1. *Evolution, the Themes of Biology, and Scientific Inquiry*

3. https://theconversation.com/emergence-the-remarkable-simplicity-of-complexity-30973

4. http://www.evolutionnews.org/2014/10/emergence_is_re090261.html

5. Hawking, Stephen W., *A Brief History of Time*, Bantam Press, Great Britain, 1988, p. 130

6. Marcus Chown, *Let There Be Light*, New Scientist 157 (2120):26-30, 1998

7. Etkin, David, *Disaster Theory: An Interdisciplinary Approach to Concepts and Causes*, Butterworth-Heinemann, Oxford, UK, 2016

8. Aristotle, *Metaphysics*, trans. W.D. Ross, XII.3 (1070a, 5-10), in Richard McKeon, ed. *The Basic Works of Aristotle* (New York: Random House, 1941), 874 cited in Dembski, *Being as Communion*,

p. 54

9. Aristotle, *Physics*, trans. R.P. Hardie and R.K. Gaye, II.8 (199a, 15-20), in Richard McKeon, ed. *The Basic Works of Aristotle* (New York: Random House, 1941), 250 cited in Dembski, *Being as Communion*, p. 54

10. Dawkins, Richard, *The Extended Phenotype: The Long Reach of the Gene*, Oxford University Press, Oxford, UK, 1982

Chapter 5-8: Doubting the Science of Evolution

"There is no refutation of Darwinian evolution in existence. If a refutation ever were to come about, it would come from a scientist, and not an idiot."

~ Richard Dawkins, evolutionary biologist, author, and renowned anti-theist ~

Let me emphasize that it is not my intention to dogmatically deny that evolution ever happened, for such is beyond my remit and knowledge. In this chapter, I simply want to evidence the doubt that exists, even amongst evolutionists, as to the mechanisms of evolution. Being a process-oriented analyst, I do question whether a process could have existed if thousands of highly qualified scientists, working countless hours, have so far yet to agree on what it is. If one studies the literature, it seems that the whole argument is based on bones and comparative anatomy, with some DNA commonality thrown in. There is no substantive, plausible, or demonstrated evidence of process, and it is process which is the argument of evolution. If one has no clue as to how it happened, then one cannot be dogmatic that it did happen.

Firstly, what is "evolution"? In truth, it is the overarching, philosophical term for unknown processes, which philosophical materialists use in an attempt to explain the existence of varying life forms. There is far less science behind it than the public is given to believe. Stephen Meyer laments, "Evolution can refer to anything from trivial cyclical change within the limits of a preexisting gene pool to the creation of entirely novel genetic information and structure as the result of natural selection and random mutations"[1]. In one clumsy example, Richard Dawkins claimed that the ability of guppies to change colour was evidence of evolution[2], but in truth, it was evidence of adaption based on existing genes. When threatened, guppies can resort to camouflage, but when the threat is removed, the normal rainbow colour returns. Whilst adaptation is a process within the evolutionary paradigm, it offers no evidence in support of the *microbes to man* hypothesis.

The tragedy of the behaviour of so many evolutionists is that they resort to ridiculing and mocking those who dare to question what I have concluded, is their religion masquerading as science. In this chapter, I will introduce evidence from well-credentialled scientists, not idiots, that all is not well with the Darwinian evolution hypothesis. I do not know the truth of evolution, and given the degree of debate concerning the mechanisms of *descent with modification*, I contend that no-one does, especially not Richard Dawkins, as I have previously written about.[3] It is fair to say that many religious people reject evolution entirely, for no reason other than that it conflicts with their interpretation of the Book of Genesis, but others seem to find a compromise. Books by Christian authors, Denis Alexander[4], Francis S. Collins (Head of Human Genome Project)[5], and John H. Walton[6] are prime examples. Other scientists, who accepted the evolution hypothesis during the days of their formal scientific training, subsequently changed their minds as their later research led them to doubt what they had earlier been taught. One of the best reads in that regard is *"Genetic Entropy"*[7] by Dr. John C. Sanford, co-developer of the gene gun technology and holder of some thirty patents in his chosen field of research. To Richard Dawkins, these men may be idiots rather than scientists, but the scientific world would disagree.

What is the evolution hypothesis? Quoting Professor Andrews, whose description, I would offer, is as accurate as any:

"Briefly stated, the theory claims that organisms evolve by a dual process consisting of (1) random genetic mutations (changes in the organism's DNA produced by a variety of causes) followed by (2) 'natural selection' of those members of a population to which mutations have imparted superior reproductive capacity. Although it is admitted that genetic mutations are overwhelmingly damaging or neutral in their effect, it is held that favourable mutations (that is, those that improve reproductive success) do occasionally take place. These beneficial mutations then spread through the population because their owners reproduce more successfully than others."[8]

Already I can hear the objection: What about survival success? Simply put, if the organism does not reproduce, then there is no survival success. In a practical sense, the two concepts are synonymous. A challenging point made by Andrews is that humans are "seriously overevolved". There is a great deal about the human condition for which evolution has no explanatory power, in either survival or reproductive success. On the contrary, there are human behaviours that argue against evolution, such as the free will to not reproduce. Evolutionists do attempt to explain these oddities, but for the most part, they wander off the reservation, as we shall see. In my research, I have encountered innumerable questions by highly qualified scientists, including evolutionary biologists. Let us hear what they have to say.

Douglas J. Futuyma is an American evolutionary biologist, and Distinguished Professor in the Department of Ecology and Evolution at Stony Brook University. His textbook, "Evolutionary Biology"[9], is (or was) in wide use in American universities. Gently tiptoeing around the probabilities of favourable genetic mutation in the available time span, and perhaps with good reason, he noted, "If the population size is, say, 10^6, about 10 progeny in every generation will carry a new mutation at a given locus. More realistically, if the population size is about 1000 to 10,000, a mutant offspring will appear only once in a hundred or so generations and is thus effectively a unique event." (p. 242) On mutation rates, an online paper offers,

"We will assume that 90 percent of the DNA is non-coding and that mutations there are almost always neutral. For the remaining 10 percent, we will assume that half of the mutations are neutral, and the rest, 9/10 are harmful and one is beneficial. We will assume that about half of the harmful mutations are fatal. So for a typical set of 200 mutations, 190 will be neutral, 5 will be fatal, 4 will be harmful but not fatal, and 1 will be beneficial."[10]

I am in no position to argue this to a conclusion, but I believe that science has moved on from the 90% assumption. However, combining these two quotations, one has to wonder. As animal gestation periods increase with animal size, and here I am thinking of dinosaurs and other critters far larger than their modern counterparts, how could evolution have occurred as claimed. For example, the gestation period for a cow averages 286 days, whilst for an African elephant it is 645 days; how much longer for much larger animals in the past? Reportedly, most female African elephants give birth for the first time between 14 and 15 years old, and slightly later for Asian elephants. Thus, the generation turn rate (a term I coined) is say, 14 years, and according to Futuyma, "a mutant offspring will appear only once in a hundred or so generations", i.e. once every 1400 years. If only 0.5% of mutations are beneficial, progress would be exceedingly slow. I acknowledge the hypothesis of *convergent evolution*, the independent evolution of similar features in species of different periods or epochs in time, but even so, that is a philosophical proposition, not a scientific fact. Writing in a similar vein, on the improbability of whales evolving from land dwelling quadrupeds in a mere 10 million years, Robert Wesson observed:

"Genetic considerations also point up the difficulty of the whale's rapid evolution. By Mayr's calculation, in a rapidly evolving line an organ may enlarge about 1 to 10 percent per million years, but organs of the whale-in-becoming must have grown about ten times more rapidly over 10 million years. Perhaps 300 generations are required for a gene substitution. Moreover, mutations need to occur many times, even with considerable selective advantage, in order to have a good chance of becoming fixed. Considering the length of whale generations, the rarity with which needed mutations are likely to appear, and the multitude of mutations needed to convert a land animal into a whale, it is easy to conclude that gradualist natural selection of random variations cannot account for this animal. After their perplexing rapid development, both whales and bats have for many millions of years evolved slowly, supposedly because their populations mingle widely, with no territoriality and much dispersal. Perhaps we should not expect to understand major evolutionary innovations. None has ever been observed; indeed, no one has ever observed a mutation's making even the beginnings of a new organ. Innovation is the central problem that has troubled evolutionists ever since Darwin, and it is no less mysterious today than when he published his great book."[11]

In reading evolutionist's arguments concerning genetic mutations, their narratives give the impression that individual mutations are inherited individually, which lends credence to their theories. However, genetic inheritance is not at all like that. In his book, *"The Greatest Show on Earth"*[12], Richard Dawkins employs literary *sleight of hand* when he asks the reader to believe that a single point mutation in a nucleotide is selectable. It is not. Accepting for the moment that a beneficial mutation, one that adds useful genetic information, is even possible, scientific evidence is that deleterious mutations would far outnumber beneficial ones. Human nucleotides exist in large clusters or blocks, ranging in size from ten thousand to one million. These linked blocks are inherited as single units and never break apart. The human genome has approximately 100,000 to 200,000 linkage blocks.[13] In any linkage block, the ratio of deleterious to beneficial mutations is likely to be high, effectively masking any benefit the beneficial mutation might confer. As Dr. Sanford puts it, "Since the large majority of mutations are deleterious, each mutation cluster will have an increasingly negative affect on fitness each generation."[14] This point should not be missed, as it is supported in other academic papers such as this one, found online here[15]. Entitled "The frailty of adaptive hypotheses for the origins of organismal complexity", the author argues that natural selection is given far too much credence with far too little evidence. I should emphasise that the author does not dispute evolution, but proposes other mechanisms as the primary causes. He notes,

"There is no evidence at any level of biological organization that natural selection is a directional force encouraging complexity. In contrast, substantial evidence exists that a reduction in the efficiency of selection drives the evolution of genomic complexity ... By reducing the efficiency of selection, random genetic drift imposes a high degree of directionality on evolution by increasing the likelihood of fixation of deleterious mutations and decreasing that of beneficial mutations."

In his closing comments, the author takes his colleagues to task:

"Four of the major buzzwords in biology today are complexity, modularity, evolvability, and robustness, and it is often claimed that ill-defined mechanisms not previously appreciated by evolutionary biologists must be invoked to explain the existence of emergent properties that putatively enhance the long-term success of extant taxa. This stance is not very different from the intelligent-design philosophy of invoking unknown mechanisms to explain biodiversity. Although those who promote the concept of the adaptive evolution of the above features are by no means intelligent-design advocates, the burden of evidence for invoking an all-powerful guiding hand of natural selection should be no less stringent than one would demand of a creationist. If evolutionary science is to move forward, the standards of the field should be set no lower than in any other area of inquiry."

Mutations are inherited in linked blocks, not as single point mutations as Dawkins and others would have us believe. Every linked block will have a far greater number of deleterious mutations than the hypothetical beneficial ones. There is no way for natural selection to eradicate the bad mutations without eradicating the hypothetical good ones, so even if there was a beneficial mutation, every generation must have a corresponding increase in deleterious mutations. Darwin claimed, and Dawkins[16] agreed, that natural selection is daily or even hourly scrutinising every variation, even the slightest, rejecting that which is bad and preserving and adding up all that is good. This is nothing but romantic nonsense and wishful thinking. Natural selection cannot reject the bad without rejecting the good, nor can it preserve and add up all that is good without doing the same for all that is bad. Modern science effectively refutes Dawkins' notion that in relation to enzyme performance, "even tenfold or twofold would be enough for natural selection to get an adequate grip"[17].

Dr Sanford adds further commentary that refutes Dawkins (and Darwin):

"The gap between molecules and the whole organism is profound. Part of this gap involves size ... standing between a nucleotide and an individual organism are many different *levels of organization*. For example, a single nucleotide may affect a specific gene's transcription, which may then affect mRNA processing, which may then affect the abundance of a given enzyme, which may then affect a given metabolic pathway, which may then affect the division of a cell, which may then affect a certain tissue, which may then affect the whole organism, which may then affect the probability of reproduction, which may then affect the chance that the specific mutation gets passed onto the next generation. Massive amounts of uncertainty and dilution are added at each organization level, resulting in massive increase in "noise", and a loss of resolution. There must be a *vanishingly small* correlation between any given nucleotide (a single molecule), and a whole organism's probability of reproductive success." (*Genetic Entropy*, pp. 48-49) [italics in original]

The issue that Darwin could not have been aware, and Dawkins seems to choose to be unaware, is the *pleiotropic* and *polygenic* nature of proteins, which make it impossible to predict the outcome of single point mutations. Another interesting issue, raised in the academic paper mentioned above, is the efficiency and survival rates of unicellular versus multicellular organisms.

"Complex, multicellularity has only arisen twice, once in animals and once in vascular plants ... given the massive global dominance of unicellular species over multicellular eukaryotes, both in terms of species richness and numbers of individuals, if there is an advantage of organismal complexity, one can only marvel at the inability of natural selection to promote it. Multicellular species experience reduced population sizes, reduced recombination rates, and increased deleterious mutation rates, all of which diminish the efficiency of selection. It may be no coincidence that such species also have substantially higher extinction rates than do unicellular taxa."

Evolution-doubters, including many scientists, differentiate between micro-evolution and macro-evolution, although quite predictably, evolutionists choose to deny that there is any difference. Reviewing an earlier study, Professor Andrews puts it this way:

" ... the conclusion was that although Darwinian processes can and do produce minor changes in the characteristics of populations (microevolution), it is incapable of creating the major changes required to transform one kind of creature into another (macroevolution). Evidence from centuries of artificial selection by human intervention, as practiced by plant and animal breeders, supports this conclusion. While many new varieties and breeds (of, say, cats or dogs) have been generated, artificial selection never produces new kinds of organisms ... There are natural barriers to macroevolution that no amount of human ingenuity can overcome. Some of these barriers may well be surmounted using 'genetic engineering' in which scientists deliberately 'edit' the DNA of an organism ... but genetic engineering requires the skilled and purposeful manipulation of organisms by intelligent agents; it doesn't happen by chance or accident."[18]

No doubt, evolutionists will still respond, "yes, but it could happen." However, evolutionary biologist Professor Futuyma, with a deeper understanding of genetic complexity, expressed his own doubt: "It is important to ask, though, whether true evolutionary novelties actually arise by mutation. For example, can both a new enzyme and the regulatory system that modulates its production arise by mutation?"[19] Despite his doubts, Futuyma asserts, "just because we don't know *how* evolution occurred, does not justify doubt about *whether* it occurred."[20] I would offer that here, this very competent scientist has succumbed to interpreting evidence within the paradigm of evolution, even though the science does not point him in that direction. In the scientific method, hypotheses need to be supported by the science, and if they are not, then there is reason to not accept them.

This is the question with which so many evolutionary biologists continue to grapple. So little was known of genetics and microbiology in the days of Darwin, and those who formulated the Neo-Darwinian Synthesis, that to misquote The White Queen in *Alice's Adventures in*

Wonderland, they *believed as many as six impossible things before breakfast*. An interesting study is "*The Altenberg 16*"[21] by Suzan Mazur. The book opens the Introduction with the opinion of cytogeneticist, Antonio Lima-de-Faria, in his book, "*Evolution without Selection*"[22]: "There has never been a theory of evolution." I will not here review Mazur's book, for to attempt a synopsis would do it an injustice, such is the amount of material therein. I will content myself with quoting from the Introduction to give you a sense of the study:

"This book ... looks at the rivalry in science today surrounding attempts to discover the elusive process of evolution, as rethinking evolution is pushed to the political front burner in hopes that 'survival of the fittest' ideology can be replaced with a more humane explanation for our existence ... Evolutionary science is as much about the posturing, salesmanship, stonewalling and bullying that goes on as it is about actual scientific theory. It is a social discourse involving hypotheses of staggering complexity with scientists, recipients of the biggest grants of any intellectuals, assuming the power of politicians while engaged in *Animal House* pie-throwing and name calling ... Perhaps the most egregious display of commercial dishonesty in this year's [Ed. 2009] celebration of Charles Darwin's *On the Origin of Species* – the so-called theory of evolution by natural selection, i.e., survival of the fittest, a brand foisted on us 150 years ago.

Scientists agree that natural selection can occur. But the scientific community also know that natural selection has little to do with long-term changes in populations." (p. v)

I can highly recommend this book if you would like an exposé on the multitude of scientific opinions on just how evolution could have worked. Every hypothesis has failed so badly that at least one evolutionary biologist has posited that the genome must have preceded all evolution, rather than being a product of evolution. As for abiogenesis, chemical evolution, the belief that life arose from non-life, a solution to that quandary is as far away as ever. Again, quoting Professor Andrews,

"Furthermore, the emergence of the hypothetic first living organism from nonliving starting materials (a process often called 'chemical evolution') is today commonly attributed to fortuitous but entirely undirected physical and chemical processes which are as yet unknown. Such undirected processes have never been observed in the laboratory and are never likely to be observed, in spite of decades of effort by origin-of-life researchers. It is true that artificial life of a kind has been created by chemists such as Craig Ventnor using sophisticated techniques to imitate the DNA found in nature. But this has only been achieved under the most precise control and direction of skilled scientists."[23]

For a description both informative and entertaining, in layman's terms, of the complexity, and dare I say, impossibility of chemical evolution, I can highly recommend Professor Andrews' chapter entitled, *"Life in a cake mixer"*[24]. Yes, I know, the title of the book can be a little off-putting, but even theists can be amongst the best of scientists, as Francis Collins and John Sanford have demonstrated. I am reminded of a remark by evolutionist, Stephen Jay Gould, expressing his curiosity that so many of his fellow scientists actually believed in God. In passing, Gould was an advocate of the hypothesis that science and religion were non-overlapping magisteria, and in his book, "Rocks of Ages"[25], explained his view of the proper relationship between the two. In his review of this book, H. Allen Orr commented on "a peculiar property of the topic":

"Talk of the relationship between science and religion routinely reduces normally sensible people, as if by magic, to idiots. Or if not idiots, charlatans. The problem is that you are, in your bones, either for religion or against it and that's the end of the matter. If you're for the Pope, certain arguments, often ludicrous, get trotted out, predictably and automatically. And if you're for Science, another set, often disingenuous, gets trotted out, just as predictably and just as automatically. This doesn't mean the two sides are evenly matched. It just means the discussion suffers from an unusual amount of intellectual dishonesty.

The good news is that Gould avoids most of the usual dishonesties in *Rocks of Ages.*"[26]

Putting that contentious issue aside, and now turning our attention to natural selection, I earlier referred to *"Beyond Natural Selection"*[27] by Robert Wesson. As noted in the front inside flap:

"Humans are not simply the result of a mechanistic process ... [Wesson] emphasizes the importance for evolution of inner direction and the self-organizing capacities of life. This view is better able to account for the chaotic nature of the evolutionary process and the inherent propensity of complex dynamic systems to grow more complex with time."

The fallacy should be obvious. If life started from non-life and evolved from simple systems, *the inherent propensity of complex dynamic systems to grow more complex with time* cannot account for it. I read the book with interest, hoping to find resolution of this conundrum, but alas, none was to be found. Nevertheless, Robert Wesson was able to expose many of the scientific and logical failures of Neo-Darwinism, whilst falling into similar traps himself. I do not intend to further review his book here, as I suspect that enough has been said concerning the doubtful nature of evolutionary "science", insofar as it is said to explain the *microbes to man* idea. I would stress that if evolutionary biology is ever to be proven, or even taken seriously as an explanation for the human condition, then it must address and come to grips with the concept of *de novo* - starting from the beginning. Starting in the middle is nowhere good enough, at least not for me.

Where ARE the Fossils?

Charles Darwin wrote: "As on the theory of natural selection an interminable number of intermediate forms must have existed, linking together all the species in each group by gradations as fine as our present varieties, it may be asked, Why do we not see these linking forms all around us?"[28] He continues with a somewhat plausible explanation, ending with "I can answer these questions and grave objections only on the supposition that the geological record is far more imperfect than most geologists believe", but I sense from his writing that he was not entirely convinced. In an earlier chapter, I discussed the issue of *handicapped fossils*, and the probability that there should have been more of these than examples of fully functional creatures. Without going into detail, we have the conundrum of the Cambrian Explosion. From the evidence, the Cambrian period saw an incredible increase in the number of fully-formed phenotypes, with no fossil evidence of development. A typical apologetic is "an artefact of our incomplete sampling of an incomplete fossil record". One is caused to wonder why ardent fossil hunters would have been satisfied with an *incomplete sampling*?

If fossils provide evidence of former types of organisms, why would not an absence of fossils be evidence that evolution did *not* occur as proposed? Absence of evidence is not necessarily evidence of absence, but it some cases, it most likely is. "Many palaeontologists and evolutionary biologists now concede that the long-sought-after Precambrian fossils, those necessary to document a Darwinian account of the origin of animal life, are missing. Scientists are especially candid about this when addressing each other in the technical peer-reviewed literature."[29] Evolutionists have proposed the concept of *deep divergence* to explain the problem, resorting to comparative genetic analysis as the preferred approach. This requires a relatively constant biological clock with beneficial mutations occurring at a regular rate. However, as was cautioned in one paper in the journal of *Molecular Biology and Evolution*, "The rate of molecular evolution can vary considerably among different organisms, challenging the concept of a 'molecular clock'."[30] Another comment, "the idea that there is a universal molecular clock ticking away has long since been discredited."[31] I am indebted to Stephen Meyer for these quotes.

I am not here arguing that evolution did not occur over eons. I am merely seeking to demonstrate that the *science of evolution* continues to evolve with new hypotheses being posited, and just as quickly debated and sometimes discredited. It is difficult to accept that any scientist, in any discipline, knows the story, which entitles one to be sceptical of the overarching narrative, especially as some issues would appear to be beyond resolution irrespective of the path taken.

Mathematics, Probabilities, and Time

When evolutionists argue that give enough time, any biological evolution could occur, one must ask: Has there been enough time? Current thinking is that life arose on Earth somewhere between 3.7 and 4.4 billion years ago, not long after the oceans formed about 4.41 billion years back. The birth of the planet itself is dated at 4.54 billion years ago. Scientists propose that oceans may have appeared first in the Hadean Eon, as soon as two hundred million years after the Earth formed. Abiogenesis (chemical evolution) is accepted as fact, primarily because it was the necessary prerequisite for life to exist, but there is as yet, no accepted hypothesis to describe the chemical process. The data that I am about to present has likely been superseded by later research, so I am not asserting its accuracy. However, it is inaccurate only in degree, not magnitude.

I found reference to a project that was attempting to define the minimal set of genes that could sustain life[32]. The genomic DNA of the smallest known bacterium contains about 160,000 base pairs (of nucleotides), but this bacterium is incapable of reproduction on its own. The smallest known free-living organism has a genome of 582,970 base pairs corresponding to about 480 proteins[33], so you can see that if this is claimed as the starting point for evolutionary biology, the study of living organisms, an enormous amount of prior evolution still needs to be explained. We should now look at the probability of proteins forming by chance from some imagined primordial soup. In passing, we often find the Miller-Urey experiment of 1952-3 still quoted as proving abiogenesis, for it demonstrated that several organic compounds could be formed spontaneously, in what they suggested was a simulation of Earth's early environment. It can be demonstrated that their results in no way proved the possibility of chemical evolution, and whilst subsequent experiments have made some progress, it has been insufficient.

There are about five hundred amino acids found to occur naturally, being both left and right-handed. Proteins use only twenty-one, specific left-handed amino acids in their formation. A protein molecule consists of one or more long chains of these amino acids, typically around 400 in length. Proteins differ from one another primarily in their sequence of amino acids, which in living organisms, is dictated by the nucleotide sequence of their genes, and which usually results in protein folding into a specific three-dimensional structure that determines its activity. The question to be asked is this: what is the probability of just the right twenty-one, left-handed amino acids coming together in a primordial soup, sequenced in just the right way to form a useful protein, even before there was DNA to guide the assembly? Neither amino acids nor proteins are stable in solution, yet even to produce small bacteria, some 50,000 to 1 million proteins are required. In recent research, under clinical conditions, this minimal number has been reduced to 182, but that is still a very large number to achieve by chance.

Mathematics is not my strong point, but if it is yours, you can do the calculations. I have read that the probability of creating a single protein of 150 amino acids, by chance, is 1 in 10^{77}. The probability of that improbable protein associating with another improbable protein in amongst all that junk stuff seems rather unlikely. Would there be enough time? Could this happen in 4 billion years, a mere 4×10^8 (or 10^9 depending on your definition of a billion)?

Despite these difficulties, some researchers are diligently attempting to produce synthetic bacteria for useful purposes, such as generating hydrogen. If you are interested, there is a brief article online here[34].

The Mystery of the Origin of Life

James M. Tour, Ph.D., a professor of Chemistry, Computer Science, Materials Science, and NanoEngineering, has researched this issue extensively, and would appear to be well qualified to do so. In his descriptions of the complexity of a single cell, and the biological mechanisms necessary for its production, he explains the functions of enzymes, and asserts that there were no enzymes in a pre-biotic world, which evidences a substantial obstacle for the theory of abiogenesis. In a presentation on this mystery, he provides an update on the status of modern experimentation, noting that in truth, no progress toward substantive proof has been demonstrated since the failure of the Miller-Urey experiments.

Research in this area continues, with encouraging results according to some. As seen in the Miller-Urey experiment, and as explained by James Tour, such activities are conducted in tightly controlled conditions, which may or may not replicate conditions on the early Earth. In this particular experiment to which I refer below, reagents derived from living organisms were used to "coax" the process along, which is a debatable practice in the context of abiogenesis, but I accept that these scientists were not attempting to prove that hypothesis. Nevertheless, the results were of interest. The full report can be found online here[35], concluding "The importance of this work – it's hard to describe," said Lane. "You see these genome sequences and try and reconstruct what the cell might look like, but you can't do that with any real power. Finally, you see what the cell looks like and it's not what anyone expected."

Now, back to James Tour. The following has been extracted from a longish presentation available online here[36].

"Almost every chemical synthesis experiment in Origin of Life (OoL) research can be summed up by a protocol analogous to this:

- Purchase some chemicals, generally in high purity, from a chemical company.

- Mix those chemicals together in water in high concentrations or in a specific order under some set of carefully devised conditions in a modern laboratory.

- Obtain a mixture of compounds that have a resemblance to one or more of the basic four classes of chemicals needed for life: carbohydrates, nucleic acids, amino acids, or lipids.

- Publish a paper making bold assertions about OoL from these functionless crude mixtures of sterochemically scrambled intermediates, much like Miller did in 1952.

- Engage with the ever-gullible press to dial-up the knob of unjustified extrapolation.

- Watch the mesmerized layperson exclaim, "You see, scientists understand how life was formed!"

- Encourage a generation of science textbook writers to make colourful deceptive cartoons of raw chemicals assembling into cells, which then emerge as slithering creatures from a prehistoric pond."

Ok, the style of this explanation is facetious, and perhaps unworthy of a scientist, but the truth is there. With all that is known about genetics and biology today, there is still no known pathway for reverse-engineering life to non-life, to demonstrate how chemical evolution (abiogenesis) could be possible. Tour continues with the synthesis problem, but before we start, a definition: *chirality* is a property of asymmetry, the word derived from the Greek word for *hand*, a familiar chiral object. An object or a system is chiral if it is distinguishable from its mirror image; that is, it cannot be superimposed onto it. This was mentioned earlier in reference to left-handed amino acids.

- Molecules that compose living systems almost always show homochirality.

- When building molecular systems, constant redesigns are needed which take the synthesis back to step 1. It is often impossible to remove a moiety (part of something which can be divided) once it has been added to a molecule.

- The synthetic reactions do not know how to stop their current course of progression, or why to stop. There is no targeted goal.

- Time, although claimed to be the great saviour of abiogenesis, can actually be the enemy. For example, carbohydrates are kinetic products: 'caramelization' and Cannizzaro reactions ($CH_2O + RCHO \rightarrow HCO_2H + RCH_2OH$) ensue.

- A prebiotic system does not have the ability to easily purify the structures.

- Reagent addition-order is essential.

- The parameters of temperature, pressure, solvent, light or no light, pH, atmospheric gases or no gases, have to be carefully controlled in order to build complex molecular structures.

- The characterization at each step is essential for the chemist, but hard in a prebiotic system to consider because it knows nothing of molecular structure.

- The mass transfer problem will be the killer of all routes.

- Nature keeps no laboratory notebook so it cannot go back for more material.

My apologies, but another definition to help explain this last part: *Interactomics* is a discipline at the intersection of bioinformatics and biology that deals with studying both the interactions and the consequences of those interactions between and among proteins, and other molecules within a cell. Regarding

"the non-covalent interactive connectivity within a functioning cell, nobody knows how a viable cell emerges from the massive combinatorial complexity of its molecular components. And of course, nobody has ever synthetically mimicked it. An interactome is the whole set of molecular interactions in a particular cell. If one merely considers all protein-protein interactome combinations in just a single yeast cell, the result is an estimated $10^{79,000,000,000}$ combinations.[37] (10^{90} is the estimated number of elemental particles in the universe.)" (Wikipedia)

Summary

My primary purpose in this possibly overlong chapter, is to highlight that far too much faith is put in evolution by scientists working in a variety of fields. It is as if evolution can explain *everything* regarding what I have termed, *The Exceptional Human Condition*, but when examined closely, as guided by the writings of numerous scientists, it in fact appears to explain *nothing* at all. Earlier I opined that *nothing can explain itself*, and this is particularly true of evolution. Even the best modern researchers cannot explain evolution, and it is my contention that unless there is a rethink in this area, the mind-brain conundrum will never be properly understood, let alone be solved.

If you would like to further research the works of scientists who, like those in Suzan Mazur's book, believe that a rethink of evolution is well overdue, you could start with this list of over sixty provided on this website[38].

References:

1. Meyer, Stephen C., *Darwin's Doubts: The Explosive Origin of Animal Life and The Case for Intelligent Design*, HarperCollins, New York, 2013, p. x

2. Dawkins, Richard, *The Greatest Show on Earth – the Evidence for Evolution*, Bantam Press, London, 2009, pp. 133–138

3. Talbot, Wayne, *The Dawkins Deficiency – Why Evolution is Not the Greatest Show on Earth*, Deep River Books, Sisters, OR, 2011

4. Alexander, Denis, Creation or Evolution: Do We Have To Choose? Monarch Books, Oxford, UK, 2008

5. Collins, Francis S., The Language of God, Free Press, Simon & Schuster, New York, 2006

6. Walton, John H., *The Lost World of Genesis One: Ancient Cosmology and the Origins Debate*, IVP Academic, Downers Grove, Illinois, 2017

7. Sanford, Dr John C., *Genetic Entropy & The Mystery of the Genome*, FMS Publications, Waterloo, New York, 2008

8. Andrews, Professor E.H., *What is Man? Adam, Alien, or Ape?* Thomas Nelson Publishers, Nashville, TN, 2018, pp. 7-8

9. Futuyma, Douglas J., *Evolutionary Biology*, Sinauer Associates Inc., Sunderland, Massachusetts, 1979

10. http://www.cs.unc.edu/~plaisted/ce/rates.html

11. Wesson, Robert, *Beyond Natural Selection*, A Bradford Book, The MIT Press, Cambridge Massachusetts, 1991, pp. 52-53

12. Dawkins, *Ibid*

13. Sanford, *Ibid*, p. 82

14. *Ibid*, p. 87

15. www.pnas.org/cgi/doi/10.1073/pnas.0702207104

16. Dawkins, *Ibid*, p. 64

17. *Ibid*, p. 214

18. Andrews, *Ibid*, p. 9

19. Futuyma, Ibid, p.

20. Futuyma, Douglas J., *Evolution as Fact and Theory*, BIOS 56, 1985, p. 8

21. Mazur, Suzan, *The Altenberg 16: An Expose of the Evolution Industry*, North Atlantic Books, Berkeley, CA, 2010

22. Lima-de-Faria, Antonio, *Evolution without Selection. Form and Function by Autoevolution*, Elsevier, New York, NY, 1988

23. Andrews, *Ibid*, pp. 9-10

24. Andrews, Professor E.H., *Who Made God? Searching for a Theory of Everything*, EP Books, Darlington, England, 2009, pp. 194-210

25. Gould, Stephen Jay, *Rocks of Ages: Science and Religion in the Fullness of Life*, Ballantine Books (Random House), New York, NY, 1999

26. https://bostonreview.net/archives/BR24.5/orr.html

27. Wesson, Robert, *Beyond Natural Selection*, A Bradford Book, The MIT Press, Cambridge Massachusetts, 1991

28. Darwin, Charles, *The Origin of Species by Means of Natural Selection*, Penguin Books, London, 1985, p. 438

29. Meyer, *Ibid*, p. 101 quoting Richard Monastersky, "Ancient Animal Sheds False Identity", *Science News* 152 (1997):32

30. Ho, Simon, Matthew J. Phillips, Alexei J. Drummond, and Alan

Cooper, "Accuracy of Rate Estimation Using Relaxed-Clock Models with a Critical Focus on the Early Metazoan Radiation", *Molecular Biology and Evolution* 22, no. 5 (2005): 1355-63

31. Smith, Andrew B., and Kevin J. Peterson, "Dating the Time and Origin of Major Clades", *Annual Review of Earth and Planetary Sciences* 30 (2002):65-88

32. Glass, J.I., Assad-Garcia, M., Alperovich, et al., *Essential Genes of a Minimal Bacterium*, Proceedings of the National Academy of Sciences 103 (2006), pp. 425-430

33. Walton, John, *The Origin of Life, Should Christians Embrace Evolution*, Inter-Varsity Press, Nottingham, England, 2009, p. 189

34. https://www.nature.com/news/2006/061009/full/news061009-10.html

35. https://www.theguardian.com/science/2020/jan/15/breakthrough-gives-insight-into-early-complex-life-on-earth

36. https://www.youtube.com/watch?v=zU7Lww-sBPg

37. Peter Tompa and George D. Rose, *Protein Science* **2011**, 20, 2074-2079. Department o Structural Biology, Vrije Universiteit, Belgium, and the Department of Biophysics, John Hopkins University, Baltimore, Maryland.

38. https://www.thethirdwayofevolution.com/people

Chapter 5-9: Is Darwinism Broken?

"ID proponents point to a chasm that divides how evolution and its evidence are presented to the public, and how scientists themselves discuss it behind closed doors and in technical publications. This chasm has been well hidden from laypeople, yet it was clear to anyone who attended the Royal Society conference."

~ Paul Nelson and David Klinghoffer ~

The quote is taken from an article, accessible here[1], on the conference mentioned below. I was delighted at the use of the term, *chasm*, finding someone agreeing with my perception of the problem, leading to my entitling this book as I have. I have included this chapter to provide more substantiation of my contention, that consensus on evolution seems to stretch ever further out of reach as more is learned of the human condition. As I have elsewhere opined, even Darwin would not believe in his own theories, if he knew what science has subsequently discovered in genetics and microbiology. He may have reflected, "In retrospect, it just seemed like a good idea at the time".

In November 2016, the Royal Society, arguably the world's most distinguished and historic scientific organization, organised a three-day conference in London, the Overview later stating:

"Developments in evolutionary biology and adjacent fields have produced calls for revision of the standard theory of evolution, although the issues involved remain hotly contested. This meeting presented these developments and arguments and encouraged cross-disciplinary discussion, which involved the humanities and social sciences in order to provide further analytical perspectives and explore the social and philosophical implications."[2]

In the opinion of many attending, Darwinian theory is broken, and may not be fixable. Of note is that this historic meeting went mostly unreported by the media, unsurprisingly. I would encourage readers to review the credentials of those organising and attending the conference, for they are not at all whom Dawkins would describe as idiots. Then, perhaps, you will understand how significant this conference was. Quoting from the report:

"What's really notable, however, is that such a thoroughly mainstream body should so openly acknowledge problems with orthodox neo-Darwinian theory. Indeed, though presenters ignored, dismissed, or mocked the theory of intelligent design, the proceedings perfectly illustrated a point made by our colleague Stephen Meyer, author of the New York Times bestseller "Darwin's Doubt: The Explosive Origin of Animal Life and the Case for Intelligent Design." Dr. Meyer, a Cambridge University-trained philosopher of science, writes provocatively in the book's Prologue:

The technical literature in biology is now replete with world-class biologists routinely expressing doubts about various aspects of neo-Darwinian theory, and especially about its central tenet, namely the alleged creative power of the natural selection and mutation mechanism. Nevertheless, popular defenses of the theory continue apace, rarely if ever acknowledging the growing body of critical scientific opinion about the standing of the theory. Rarely has there been such a great disparity between the popular perception of a theory and its actual standing in the relevant peer-reviewed science literature."

The tragedy, and dare I say, dishonesty, of some sections of the scientific community is that they prefer some knowledge to be kept secret. The opening presentation at the Royal Society by one of those world-class biologists, Austrian evolutionary theorist Gerd Müller, underscored exactly Meyer's contention. Dr. Müller opened the meeting by discussing several of the fundamental "explanatory deficits" of "the modern synthesis," that is, textbook neo-Darwinian theory. According to Müller, the as yet unsolved problems include those of explaining:

- Phenotypic complexity (the origin of eyes, ears, body plans, i.e., the anatomical and structural features of living creatures);

- Phenotypic novelty, i.e., the origin of new forms throughout the history of life (for example, the mammalian radiation some 66 million years ago, in which the major orders of mammals, such as cetaceans, bats, carnivores, enter the fossil record, or even more dramatically, the Cambrian explosion, with most animal body plans appearing more or less without antecedents); and finally

- Non-gradual forms or modes of transition, where you see abrupt discontinuities in the fossil record between different types.

As Müller has explained in a 2003 work ("On the Origin of Organismal Form," with Stuart Newman), although "the neo-Darwinian paradigm still represents the central explanatory framework of evolution, as represented by recent textbooks" it "has no theory of the generative." In other words, the neo-Darwinian mechanism of mutation and natural selection lacks the creative power to generate the novel anatomical traits and forms of life that have arisen during the history of life. Yet, as Müller noted, neo-Darwinian theory continues to be presented to the public via textbooks as the canonical understanding of how new living forms arose – reflecting precisely the tension between the perceived and actual status of the theory that Meyer described in "Darwin's Doubt."

Searching the literature, I have found commentary by numerous scientists, including evolutionists, on this *perceived and actual status of the theory*. When people, especially scientists are secretive, deceptive, disingenuous, and downright dishonest, I must ask: What are they hiding? The authors of "*What Darwin Got Wrong*" were disposed to comment on this issue, as earlier noted[3]. There is sufficient evidence from within the scientific community itself, to doubt the overarching narrative of evolution: that all life on Earth arose from a single common ancestor which itself arose from an inorganic form. Introducing the concept of convergent evolution does not solve the problem – it just alters the degree, but not to any significant extent. Evolution *may be true*, but unless and until scientists can identify the mechanisms and processes which could give rise to the human condition as we understand it, we have no reason to accept Darwinism in any of its iterations.

In reviewing amongst hundreds of other resources, the studies and conclusions of Cambrian palaeontologists, and the commentaries of the sixteen influential evolutionary biologists who attended the 2008 Altenburg conference[4], Stephen Meyer discusses the alternative theories which have been proposed over the past few decades, concluding:

"Each of these new theories attempts to answer the increasingly urgent question: After Darwin – or neo-Darwinism – *what*?"[5]

What, indeed.

References:

1. https://www.cnsnews.com/commentary/david-klinghoffer/scientists-confirm-darwinism-broken

2. https://royalsociety.org/science-events-and-lectures/2016/11/evolutionary-biology/

3. Fodor, Jerry and Piatelli-Palmarini, Massimo, *What Darwin Got Wrong*, Picador, New York, 2011, p. xxii

4. Mazur, Suzan, *The Altenberg 16: An Expose of the Evolution Industry*, North Atlantic Books, Berkeley, CA, 2010

5. Meyer, Stephen C., *Darwin's Doubts: The Explosive Origin of Animal Life and The Case for Intelligent Design*, HarperCollins, New York, 2013, p. 292

Chapter 5-10: An Ancient View

"Real knowledge is to know the extent of one's ignorance."

~ Confucius ~

Snobs, Bumpkins, and Dinosaurs

Forgive this brief aside to overcome what C.S. Lewis described as: *chronological snobbery*, defined thus: "an argument that the thinking, art, or science of an earlier time is inherently inferior to that of the present, simply by virtue of its temporal priority or the belief that since civilization has advanced in certain areas, people of earlier time periods were less intelligent"[1].

This article provides an insight:

"It is the sort of belief that sits very comfortably in the subconscious, giving one the warm glow of knowing that we are faster, better, wiser, more advanced, and more knowledgeable than our parents and forebears. Yet one of the problems Lewis noticed in the myth was that such superiority tends to produce not wisdom but ignorance ... Of course, such "chronological snobbery" does not like to admit its own existence. No snob likes to be thought of as an ignorant bumpkin."[2]

My point, which I am about to evidence, is that much of what perplexes us today, perplexed people of the past, to such an extent that they thought logically about the fundamentals of existence. Some scientific principles credited to scholars of later centuries, were known of much earlier, as the title of this chapter seeks to imply. Ok, perhaps ancient is an exaggeration, but what follows dates from centuries ago. I encountered this work when I was researching matters theological, and was surprised to learn that what had been going through my mind, was similar to what had occupied greater minds than mine centuries before. The surprise was not so much that people had been thinking, but that our thoughts across the centuries ran parallel, even though our understanding of science was so very different. I had concluded certain fundamentals, which we have discussed in earlier chapters, but here I will briefly outline what a great man once concluded.

Maimonides

This man is Moses ben Maimon (1135-1204), usually known simply as *Maimonides*. In Jewish commentaries, he is referred to as *Rambam*. Maimonides was a medieval Sephardic Jewish philosopher who became one of the most prolific and influential Torah scholars of the Middle Ages. I should also mention that he was a preeminent astronomer[3] of his time, and achieved fame as a physician, writing medical treatises on a number of diseases and their cures.

His most enduring writing, and most controversial in some rabbinic circles, being originally banned, is *"The Guide of the Perplexed"*. It is no longer available in its original language, but we do have access to a translation dating from 1885[4]. It is from this that I shall freely quote, noting that the copyright has long expired. If you are interested, there is a useful review online here[5]. Maimonides offered twenty-six propositions regarding the fundamentals of existence. Before we get to them, it should be understood that some of his words may not have the same meaning with which the modern reader may be familiar, and that the book from which I am quoting contains numerous annotations and comments to explicate the propositions. I shall list them as given, but before doing so, I will provide these definitions to disambiguate some of his propositions:

- *Essential* – an attribute without which, an entity cannot be what it is said to be.

- *Accidental* - an attribute that may or may not belong to an entity, without affecting its essence.

The problem here, from the outset, is that these are Aristotelian philosophical definitions over which there is much debate, and generally not accepted in modern philosophy. However, rather than engage in the debate over *first-level properties* and the like, I will attempt to provide a simple explanation of how Maimonides used these terms (students of philosophy can suspend their own understanding).

Later we shall refer back to some of them to show their relevance, and how at times, I have arrived at a similar conclusion via a different train of thought. As one comic quipped: *"There is nothing new in this world, just lots of things we had forgotten"*. In my case, I confess that I had never been previously informed to have forgotten, but now to Maimonides' propositions, only some of which are relevant to our current study:

1. The existence of an infinite magnitude is impossible.

2. The co-existence of an infinite number of finite magnitudes is impossible.

3. The existence of an infinite number of causes and effects is impossible, even if these were not magnitudes.

4. Four categories are subject to change: substance, quantity, quality, and place.

5. Motion implies change and transition from potentiality to actuality.

6. The motion of a thing is either essential or accidental; or it is due to an external force, or to the participation of the thing in the motion of another thing.

7. Things which are changeable are, at the same time, divisible. Hence everything that moves is divisible and consequently corporeal.

8. A thing that moves accidentally must come to rest, because it does not move of its own accord; hence accidental movement cannot continue forever.

9. A corporeal thing that sets another corporeal thing in motion can only effect this by setting itself in motion at the time it causes the other thing to move.

10. A thing which is said to be contained in a corporeal object must satisfy either one of the two following conditions: it either exists through that object, as is the case with accidents, or is the cause of the existence of that object; such as, e.g., its essential property.

11. Among the things which exist through a material object, there are some things which participate in the division of that object, and are therefore accidentally divisible.

12. A force which occupies all parts of a corporeal object is finite, that object itself being finite.

13. None of the several kinds of change can be continuous, except motion from place to place, provided it be circular.

14. Locomotion is in the natural order of the several kinds of motion the first and foremost.

15. Time is an accident that is related and joined to motion in such a manner that one is never found without the other. Motion is only possible in time, and the idea of time cannot be conceived otherwise than in connection with motion; things which do not move have no relation to time.

16. Incorporeal bodies can only be numbered when they are forces situated in a body ... hence purely spiritual beings, which are neither corporeal nor forces situated in corporeal objects, cannot be counted, except when considered as causes and effects.

17. When an object moves, there must be some agent that moves it, either without that object, as, e.g., in the case of a stone set in motion by the hand; or within, e.g., when the body of a living creature moves.

18. Everything that passes over from a state of potentiality to that of actuality, is caused to do so by some external agent; because if that agent existed in the thing itself, and no obstacle prevented the transition, the thing would never be in a state of potentiality, but always in that state of actuality.

19. A thing that owes its existence to certain causes, has in itself merely the possibility of existence; for only if these causes exist, the thing likewise exists. It does not exist if the causes do not exist at all, or if they have ceased to exist, or if there has

been a change in the relation which implies the existence of that thing as a necessary consequence of those causes.

20. A thing which has in itself the necessity of existence cannot have for its existence any cause whatever.

21. A thing composed of two elements has necessarily their composition as the cause of its present existence. Its existence is therefore not necessitated by its own essence; it depends on the existence of its component parts and their combination.

22. Material objects are always composed of two elements [at least], and are without exception due to accidents. The two compound elements of all bodies are substance and form. The accidents attributed to material objects are quantity, geometrical form, and position.

23. Everything that exists potentially, and whose essence includes a certain state of possibility, may at some time be without actual existence.

24. That which is potentially a certain thing is necessarily material, for the state of possibility is always connected to matter.

25. Each compound substance consists of matter and form, and requires an agent for its existence, viz., a force which sets the substance in motion, and thereby enables it to receive a certain form. The force which thus prepares the substance of a certain individual being, is called the immediate motor.

26. Time and motion are eternal, constant, and in actual existence.

Observations

Note that Propositions 17 & 18 predate by several centuries, the scientific concept of inertia, and Newton's First Law of Motion; i.e., an object will remain at rest or in uniform motion in a straight line unless acted upon by an external force. Proposition 18 also relates to what is now termed, *potential energy*. Proposition 15 is interesting in how it relates time and motion, affirming what I and others believe, that without motion, there is no time because essentially, time is a measurement of change. In that sense, "time" is an *accidental property* of existence. Proposition 4 states: "Four categories are subject to change: substance, quantity, quality, and place". If none of these properties of an object, e.g., a piece of granite, experience change, then time is not a property of that piece of rock.

There is one more nuance found here, of interest perhaps only to spiritualists: the meaning of "substance", which in other contexts, is rendered as "form". *Form* is an essential property, but it need not relate to physical substances. As mentioned elsewhere, there is a meta-language overlaying the writings of scientists and philosophers which I contend, is not explicable in evolutionary or biological terms.

On a lighter note, laypersons such as myself, look for ways to explain the scientific in non-scientific, everyday laypersons' terminology, which in this case is *stuff*.

Understanding "Stuff"

"I'm not materialistic, I just love stuff."

~ with apologies for misquoting Kylie Jenner ~

Stuff (noun):

1.   Matter, material, or articles of a specified or indeterminate kind that are being referred to, indicated, or implied.

2.   The basic constituents or characteristics of the stuff in (1).

Yes, I know, "stuff" is not a technical term, but I would offer that it does an admirable job of conveying reality, especially when one is referring to the *indeterminate* kind. The other advantage of using this term is that no-one, especially those in the scientific community, is likely to challenge me on my definition, or usage, of the term. Carl Sagan famously said: "The Cosmos is all that is or was or ever will be", implying that everything is stuff that can be detected, either by our senses, or by scientific instruments. The first lesson here is that not only does it takes stuff to detect stuff, but that stuff can only detect stuff – nothing else. In more formal terms, this can be described as *philosophical materialism*. As you read this, you may suspect that I am attempting to *convey* information, and whilst that indeed is my desired outcome, it is not what I am accomplishing – it is you, the reader, who is *deriving* information, from what you see written on this page. Confused? Let me explain, or rather, reiterate an earlier explanation. As an aside, this happens when one writes individual chapters in isolation, and then attempts to organise them in a logical sequence without repeating what is in then, preceding chapters. I trust that you will accept my apology.

The letters, words, and punctuation on this page are stuff: your sense of sight being derived from stuff, it is able to detect this stuff. But what is it that your eyes are detecting? To understand this, we need to take a pace to the rear, or numerous paces if I can put it like that. What is written text, or more precisely, written language? Written language is an invention that is used to represent a spoken or gestural language by means of a writing system. All writing systems utilise a language (there about 7,000 of those), and a predefined set of symbols known as an alphabet, which are used in predefined groups. In the general sense, this is a form of coding. Whether the grouping of the letters of the alphabet are secret, revealed on a "need to know" basis, or whether it is public, and can be used to search a dictionary (a form of code book), the process is the same. When writing, we encode the message, and when we read, we decode the message. Albert Einstein is said to have remarked that reading and writing are the most cognitively intense processes we humans undertake (not his exact words), and whilst I have no idea of what the great man had in mind at the time, my purpose here is to explain what is in the mind of a rather lesser man.

It is my contention that our human existence entails more than just "stuff".

References:

1. https://en.wikipedia.org/wiki/Chronological_snobbery

2. https://www.crossway.org/articles/how-to-fight-chronological-snobbery/

3. Twersky, Isadore, *A Maimonides Reader*, Behrman House, Springfield, NJ, 1972, pp. 464–73 (abridged account)

4. Maimonides, *The Guide of the Perplexed*, A translation from the original text with annotations by M. Friedlander, Trubner & Co., London, UK, 1885

5. https://plato.stanford.edu/entries/maimonides/

Part 6: The Building Blocks

"To put it bluntly, I seem to have a whole superstructure with no foundation. But I'm working on the foundation."

~ Marilyn Monroe ~

This quotation resonated with me because in a sense, it describes our understanding of the subject of this study. We can see the superstructure, but we do not understand its foundations. Too often, scientists attempt to disguise their ignorance by using language which on first reading, seems to convey an apt description, but on reflection, tells us nothing at all.

The primary theme of this study is the mind-brain conundrum: whether the mind-brain is a single physical entity, or whether the mind is a separate immaterial entity that interacts with the material brain. Earlier I submitted that *entities* have *properties* which determine the *activities* (or behaviours) of the entity: in other words, entities can only do what their properties allow. This is the basis of *Materials Science*, defined thus: "an interdisciplinary field involving the properties of matter and its applications to various areas of science and engineering. It includes elements of applied physics and chemistry, as well as chemical, mechanical, civil and electrical engineering"[1]. From Wikipedia, "Materials science is also an important part of forensic engineering and failure analysis – investigating materials, products, structures or components which fail or do not function as intended, causing personal injury or damage to property. Such investigations are key to understanding, for example, the causes of various aviation accidents and incidents."[2] It is materials science which gives engineers the confidence to build everything from bridges to skyscrapers, and gliders to spacecraft, because they understand the behaviours of materials under a range of conditions, and trust that such behaviours are reliable.

Sadly, it would appear that the disciplines of materials science have escaped the attention of biologist and neuroscientists, especially those offering materialistic explications on the mind-brain conundrum.

In this section, we are going to review relevant issues in biology to ascertain whether the behaviours of the mind can be explained by the materials of the brain. The human brain and nervous system are extremely complex, with some factors having been proven to be true, and others mere speculation driven by a commitment to philosophical materialism. A refusal to even consider the existence of the immaterial has scientists offering quite ridiculous explanations for the behaviour of the human mind - we shall refute these as we encounter them.

The following example, although in a different context, illustrates the nonsense that can arise if materials science is not taken into account. Writing on the subject of miracles in *The blind watchmaker*, Richard Dawkins spoke of how a marble statue might wave a hand by sheer coincidence; here is what he had to say:

"A miracle is something that happens, but which is exceedingly surprising. If a marble statue of the Virgin Mary suddenly waved its hand at us we should treat it as a miracle, because all our experience and knowledge tells us that marble doesn't behave like that."

He continued:

"In the case of the marble statue, molecules in solid marble are continually jostling against one another in random directions. The jostlings of the different molecules cancel one another out, so that the whole hand of the statue stays still. But if, by sheer coincidence, all the molecules just happened to move in the same direction at the same moment the hand would move back. In this way it is *possible* for a marble statue to wave at us. It could happen. The odds against such a coincidence are unimaginably great but they are not incalculably great. A physicist colleague has kindly calculated them for me. The number is so large that the entire age of the universe so far is too short a time to write out all the noughts! It is theoretically possible for a cow to jump over the moon with something like the same improbability. The conclusion to this part of the argument is that we can calculate our way into regions of miraculous improbability far greater than we can imagine as plausible."[3]

This all sounds wonderful stuff if one knows nothing of materials science, but fortunately for us, an acknowledged expert in this discipline, Professor Edgar Andrews, is on hand to expose Dawkins' error (and ignorance). He points out that while molecules in gases may exhibit random movement, the same is not true of crystalline solids like marble. You can read his full explanation in his referenced book, but my point is that many scientists appear to believe that you can posit the probability of an event without first verifying that the event is even possible within the laws of science as we know them. The irony of Dawkins' example is that if such an event were observed to occur, the best explanation would be that it truly was a miracle! Professor Andrews concluded: *"The idea that the internal motion of atoms in a lump of crystalline rock could somehow cause that lump to move from here to there is scientifically ridiculous. Harry Potter it may be, but science it is not."*[4]

I contend that the same is true when scientists attribute behaviours to the brain when materials science would argue against such being possible. If these phenomena cannot be attributed to a material brain, then logically, the existence of an immaterial mind is worthy of investigation.

References:

1. https://www.sciencedaily.com/terms/materials_science.htm

2. https://en.wikipedia.org/wiki/Materials_science

3. Dawkins, Richard, *The blind watchmaker*, Norton Press, New York NY, 1996, pp. 159-160

4. Andrews, Professor E.H., *Who Made God*?, EP Books, Darlington, England, 2009, p. 157-159

Chapter 6-1: The Genome

"The human genome contains so much data that, it has been calculated, it would fill 43 volumes of Webster's International Dictionary."

~ Iain McGilchrist, psychiatrist, writer, and former Oxford literary scholar ~

I have written this chapter for two reasons: firstly, to refresh my own memory, and secondly, for any reader who may have a similar need. Let me state unequivocally that what follows is my understanding derived from numerous academic sources, and thus is possibly subject to cognitive mutation (excuse the pun). However, I am hopeful that it does not wander too far from scientific truth. Quoting from Dr. John Sanford's book, *"Genetic Entropy"*[1],

"There is simply no human technology that serves as an adequate analogy for the complexity of a human life. The genome is the instruction manual encoding all the information needed for life!

We have thus far only discovered the first dimension of this book of life: a linear sequence of four types of extremely small molecules called nucleotides. These small molecules make up the individual steps of the spiral-staircase structure of DNA. These molecules are the letters of the genetic code, and are shown symbolically as A, T, C, and G. These letters are strung together like a linear text. They are not just symbolically shown as letters, they are very literally the *letters* of our instruction manual. Small clusters or motifs of these four molecular letters make up the *words* of our manual, which combine to form genes (the *chapters* of our manual), which combine to form chromosomes (the *volumes* of our manual), which combine to form the whole genome (the entire *library*).

A complete human genome consists of two sets of 3 billion individual letters each. Only a very small fraction of this genetic library is required to directly encode the roughly 100,000 different human proteins and the uncounted number of functional RNA molecules found within our cells. Each of these protein and RNA molecules are essentially miniature machine, each with hundreds of component parts, and with its own exquisite complexity, design, and function.

But the genome's *linear* information, equivalent to many complete sets of a large encyclopedia, is not enough to explain the complexity of life." (pp. 2-3)

I shall state this another way, again primarily for my own benefit, because the terminology is often confusing and I like to build a picture in my mind. Please note that the numbers given here are approximate, as in my research, I have found variances depending on the source.

The genetic code is the set of rules by which chemical instructions encoded in genetic material (DNA or RNA sequences) is translated into proteins (amino acid sequences) by living cells. Specifically, the code defines a mapping between tri-nucleotide sequences called codons and amino acids; every triplet of nucleotides in a nucleic acid sequence specifies a single amino acid. Because the vast majority of genes are encoded with exactly the same code, this particular code is often referred to as the canonical or standard genetic code, or simply the genetic code, though in fact there are many variant codes; thus, the canonical genetic code is not universal. For example, in humans, protein synthesis in mitochondria relies on a genetic code that varies from the canonical code. The genome of an organism is inscribed in DNA, or in some viruses, RNA. The portion of the genome that codes for a protein or an RNA is referred to as a gene.

The substance of what follows is taken from here[2]; the full article may be of interest.

For decades, it was recognised that only about 1% of DNA coded for proteins, and it was thought that the other 99% was "junk DNA" left over from eons of evolution. Apparently, the genome did not purge itself of DNA past its "best before" date. It is now understood that some (or most, if not all) of such DNA is functional, in that it controls gene activity. Identified are promoters, enhancers, silencers, and insulators. The referenced source provides explanations for the functions of these regulatory genes, but they are of no specific interest to my arguments, other than to reiterate the musings of evolutionary biologist, Professor Futuyma: "It is important to ask, though, whether true evolutionary novelties actually arise by mutation. For example, can both a new enzyme and the regulatory system that modulates its production arise by mutation?"[3] Even more intriguing is how genetic mutation managed to evolve its own regulatory mechanisms. If these non-coding DNA sequences are required to act as regulatory elements, determining when and where genes are turned on and off, it suggests that these were needed to have evolved either prior to, or contemporaneously with, the protein-coding DNA sequences which they controlled. That is a very big ask.

The complexity as described above should give those believing in undirected evolution pause for thought – the evidence of intelligent design is everywhere. Whilst at a simplistic level, it is easy enough to imagine mutations altering the genome with a consequent change in the phenotype, it is another matter entirely to propose how such biological machinery simply evolved over time. We must apply *systems thinking* to such hypotheses. What would be a plausible sequence for the evolution of this machinery in step with the one percent coding DNA such that changes in the genotype would result in phenotypes offering improved survival and/or reproduction success? Which structural elements could be left out?

On average, a human body contains about 37 trillion cells, each cell containing 46 chromosomes. A chromosome is made of bundles of looping coils which when unravelled, would result in about a six-foot long double strand of DNA (deoxyribonucleic acid). A DNA molecule is composed of two chains of nucleotides, that coil around each other to form a double helix structure carrying genetic instructions for the development, functioning, growth and reproduction. There are four types of DNA nucleotides: adenine (A), cytosine (C), guanine (G), and thymine (T). Each side of the double helix links to the other forming a ladder-like structure, with the restriction that an A can only pair with a T, and a C can only pair with a G. There are about six billion of these pairs of nucleotides in each cell. Humans have roughly 23,000 genes, a gene being a distinct stretch of DNA. Genes vary in size, from just a few thousand pairs of nucleotides (or "base pairs") to over two million base pairs. Diagrammatically,

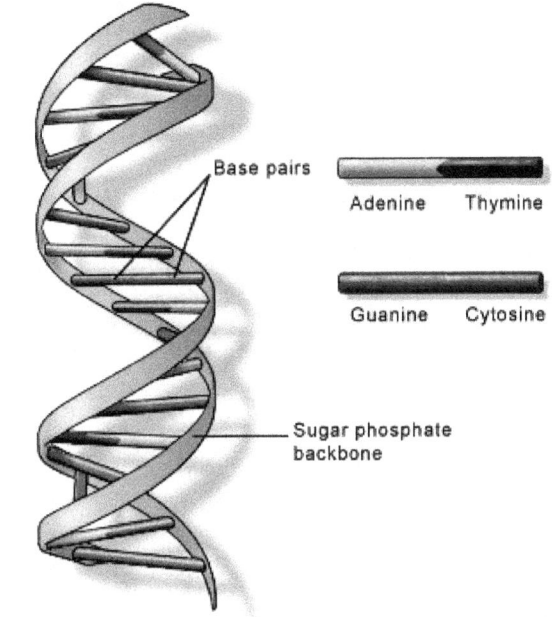

U.S. National Library of Medicine

I do not intend to further review that complexity, as it would serve no purpose here, but I would highly recommend Dr. Sanford's book – it is not overly long, and written in terms that a layman (me) can readily understand. The issue that I wish to emphasise, as before, is *genetic mutation*. This is claimed by evolutionists to be the primary mechanism of evolution, somehow capable of building the genome, and introducing variations which give rise to newer and more complex body forms, the human being the highest and most complex level achieved so far. But the question is: can random mutations achieve that?

Modern medical research is focused on identifying mutant genes which cause disease, including aging. It is believed that disease can be entirely eradicated if such deleterious genes are identified, and eliminated before reproduction occurs. One difficulty is that deleterious mutations may be occurring faster than they can be identified, because our genes being pleiotropic and polygenic, the effects may not be discernible for generations. A pleiotropic gene is one that influences two or more seemingly unrelated phenotypic traits. Mutation in a pleiotropic gene may have an effect on several traits simultaneously, due to the gene coding for a product used by a myriad of cells or different targets that have the same signalling function. Polygenic traits are controlled by two or more than two genes (usually by many different genes) at different loci on different chromosomes. Thus it can be seen that whilst a gene may be identified as the cause of a disease, care needs to be taken that varying (designing a mutation of) a gene does not have unforeseen consequences, which is why medical trials take so long. Scientists have the ability to design genetic variations, but evolution had to have relied entirely on random variations.

It is worthwhile reviewing the process and units of inheritance, especially as it has taken time for me to wrap my mind around it. As this is not in my knowledge bank, I shall quote from here[4]:

"Chromosomes are passed from parents to offspring via sperm and eggs. The specific kind of chromosome that contains a gene determines how that gene is inherited. There are three major kinds of chromosomes: autosomes, sex chromosomes and mitochondrial.

Most chromosomes are present in two copies in both men and women and are called autosomes. These chromosomes are numbered from 1 to 22. While most cells in our body have two copies of each autosome, sperm and eggs carry only one copy of each autosome. When a sperm fertilizes an egg, the embryo now contains two copies of each autosome, one from the father and one from the mother. Consequently, each person has two copies (alleles) of every gene carried on an autosome: one inherited from their father and one from their mother.

The X and Y chromosomes are the "sex chromosomes". Women have two copies of the X chromosome, one from their father and one from their mother. Men have one X chromosome, from their mother, and one Y chromosome, from their father. Because of this, a man will pass his copy of the X chromosome—and the genes it contains—only to his daughters and his copy of the Y chromosome only to his sons. Women have two alleles of every gene on the X chromosome, one inherited from their father and one from their mother. Men have a single allele of each gene on the X chromosome, inherited from their mother, and a single allele of each gene on the Y chromosome, from their father.

Mitochondrial chromosomes are inherited solely from the mother. Men inherit their mother's mitochondrial genes but do not pass them to their offspring."

My point here is that the unit of inheritance is not a single nucleotide, and not a single gene: it is the chromosomes of both parents, depending on the sex of the offspring. An interesting aspect of the genome is the proportion that is considered functional, and the rest once referred to as "junk DNA", but now more scientifically as *non-functional*. If mutations occurred uniformly across the entire genome, then the lower the percentage of functional genes, the lower the probability that mutations would occur in that portion. Selection can only occur based on mutations in the functional zone, and if most mutations are neutral or deleterious, then there could be little opportunity for selection to occur on the few beneficial mutations that occurred in that functional zone. In his book, Dr. Sanford provides graphs based on analyses by Motoo Kimura, a Japanese biologist best known for introducing the neutral theory of molecular evolution in 1968. He became one of the most influential theoretical population geneticists. It is important to pause here to consider the implications of this neutral theory. If genetic mutation is essentially neutral in its effect, then its ability to power evolution must be suspect. Attempting to precis Dr. Sanford's commentary, he stated that Kimura believed that most mutations, although deleterious, are effectively neutral, meaning that they are not subject to selection. If that be true, then the same must be said for the much lower occurrence of beneficial mutations. Estimation of the ratio of beneficial to deleterious mutations vary from one in a thousand to one in a million, but according to some studies (Bataillon, 2000; Elena et al., 1998), "the actual rate of beneficial mutations is so low as to thwart any actual measurement" (p. 24). He adds,

"For your possible interest, geneticists agree that the frequency of highly deleterious mutations is almost zero, while 'minor' mutations are intermediate in frequency. Minor mutations are believed to outnumber major mutations by about 10-50 fold (Crow, 1997), but near-neutrals vastly outnumber them both." (p. 31)

Again, if that be so regarding deleterious mutations, what does that tell us about beneficial mutations?

Mutations can occur as *point mutations*, being a change in a single nucleotide, or *chromosomal* mutations. Point mutations are of three types, and occur during replication; *substitution* where one base is incorrectly substituted; *insertion* where one or more nucleotides are inserted into replicating DNA; and *deletion* where one or more nucleotides are skipped or otherwise excised. DNA and RNA nucleotides are in codons (groups of 3). Where insertion or deletion occurs, this causes a *frame shift* (or framing error) because the number of nucleotides in a sequence is no longer divisible by three. These are inevitably deleterious. By analogy, consider a code of three letters; you send this message: "THE CAT SAT MAT", but a *deletion* (mutation) of the nucleotide "H" occurred in the first codon resulting in the message, "TEC ATS ATM AT". Simply gibberish. Note that adding, subtracting, or even substituting a single letter from a pre-arranged set does not add new information, it only corrupts the message.

Chromosomal mutations occur where fragments of chromosomes are deleted, duplicated, inverted, translocated to different chromosomes, or otherwise rearranged, resulting in changes such as modification of gene dosage, the complete absence of genes, or the alteration of gene sequence. The type of variation that occurs when entire areas of chromosomes are duplicated or lost, called copy number variation (CNV), has especially important implications for human disease and evolution. It is more difficult to generalize about the phenotypic effects of these chromosomal mutations. If the breakpoints of the mutation divide a protein, that protein will be lost in the mutant organism. But if the break is between proteins, any effect will depend on whether the expression of a gene depends on its position in the genome.

Irrespective of whether the majority of inherited genes are beneficial or deleterious, the question must be asked: Can any mutation add *genuinely new* genetic information? My Primary Axiom #4 states that *no coding system can devise or maintain itself*. The molecules referred to as A, T, C, and G are the basic units of our genetic code, which according to evolution, was devised by genetic mutation, a claim which to my mind, is illogical. One cannot claim that a coding system was devised by a mechanism which relied on that coding system for its devising. In the human genome, the search space (the domain to be optimised) for genetic variation is some 3 billion letters. Even though that is not a linear string, that it is not further lessens the probability of adding new information. Dr. Sanford uses the analogy of "misspellings", which I believe to be apt for helping us understand the issue. In a book of say, 200 thousand words, a single misspelling would be unlikely to alter the meaning of the book, or even a sentence or phrase. Two adjacent misspellings might do so, but only in a deleterious way, i.e., they would make comprehension more difficult. What they would not do is add new information or context to what the author was intending to convey. Well, they might, however improbably, but would it be useful for the purposes of the narrative?

This is the perspective we need to take of genetic mutations. They have to be evaluated in the context of the overall narrative conveyed by the relevant chapter of the book. The most significant single issue related to genetic mutation is that each is a "misspelling". The search space for misspellings is in the order of 3 billion letters, with only about 100,000 being used for encoding. The probability of misspellings being additive in the one direction, in adjacent loci, over generations, is beyond my imagining, although clearly not beyond the imagining of the evolutionists.

There is no answer as to how the genome came to be, but the evidence *against* genetic mutation does seem to be compelling. Mutation can clearly make the "genetic book" incomprehensible, but I cannot perceive how it could write a comprehensible book, or make a book more comprehensible.

Cell Division (Mitosis)

Why is a kidney *kidney-shaped*, and the heart, *heart-shaped*? It would appear logical that as cells divide, daughter cells could not be exact replicas of the parent, for there must be what I would term, *structural information* that causes the final shape. The same must hold true for bone cells, else our skeletal structure would resemble that made from LEGO® blocks. In terms of DNA content, daughter cells are identical to the parent. Despite the DNA equivalence however, daughter cells may each become completely different cell types.

"This is because certain molecules are unequally distributed between the daughter cells during mitosis ... Another reason is that after dividing, they are connected to different neighboring cells. These neighbors can touch the daughter cells through cell-to-cell communication by proteins. The neighbor of one daughter cell may activate certain protein receptors on that daughter cell, while the neighbor of the second daughter cell may activate a different set of protein receptors on the second daughter cell."[5]

Sometimes, academic descriptions raise more questions than they answer. The use of the word, "may", suggests to me that biologists "may" have an understanding of *what* is happening, but do not know the *why*. If such connections to neighboring cells were simply random, there could be no predictability of result – sometimes a baby would be formed, sometimes not. The article mentions *"different maturation programs"*, but how are they determined? Logically, they must exist, else randomness would be the order of the day. As best scientists have determined so far, the answer must be found in *epigenetics*. We briefly discussed this in an earlier chapter – *briefly* because little is known, but the issue is critical to our understanding of the neural structure of the brain.

Asymmetric Cell Division

I have included this subject because it has piqued my interest, especially as it might have implications for how neural cells self-organise, although at present I cannot imagine just how. Nevertheless, it introduces another complexity which evolutionists would need to explain, adding further to their burden of scientific proof. This seems to get ever further out of reach. Like me, you may also find the following of interest.

Rupert Sheldrake, PhD, is a biologist and author, initially working in developmental biology at Cambridge University. In a later appointment, he was Principal Plant Physiologist at the International Crops Research Institute for the Semi-Arid Tropics in Hyderabad, India. He made numerous discoveries in plant development, one of which he mentions here:

"In the course of my PhD project, I made an original discovery: dying cells play a major part in the regulation of plant growth, releasing the plant hormone auxin as they break down in the process of 'programmed cell death'. Inside growing plants, new wood cells dissolve themselves as they die, leaving their cellulose walls as microscopic tubes through which water is conducted in stems, roots, and veins of leaves. I discovered that auxin is produced as cells die, that dying cells stimulate more growth; more growth leads to more death, and hence to more growth.

When I returned to Cambridge, I developed a new hypothesis of ageing in plants and animals, including humans. All cells age. When they stop growing, they eventually die. My hypothesis is about rejuvenation, and proposes that harmful waste products build up in all cells, causing them to age, but they can produce rejuvenated daughter cells by asymmetric cell divisions in which one cell receives most of these waste products and is doomed, while the other is wiped clean. The most rejuvenated of all cells are eggs. In both plants and animals, two successive cell divisions (meiosis) produce an egg cell and three sister cells, which quickly die. My hypothesis was published in Nature in 1974[6]. 'Programmed cell death', or 'apoptosis', has since become a major field of research, important for our understanding of diseases such as cancer and HIV, as well as tissue regeneration through stem cells. Many stem cells divide asymmetrically, producing a new, rejuvenated stem cell and a cell that differentiates, ages and dies. My hypothesis is that the rejuvenation of stem cells through cell division depends on their sisters paying the price of mortality."[7]

My fascination here is that if such complexity exists in organisms as simple as plants, relatively speaking, how much more is there to be learned about the complexity in humans, and how does that influence our thinking about the plausibility of the *microbes to man* evolution hypothesis?

References:

1. Sanford, Dr John C., *Genetic Entropy & The Mystery of the Genome*, FMS Publications, Waterloo, New York, 2008

2. https://medlineplus.gov/genetics/understanding/basics/noncodingdna/

3. Futuyma, Douglas J., *Evolutionary Biology*, Sinauer Associates Inc., Sunderland, Massachusetts, 1979, p. 252

4. https://genos.co/resources/inherited.html, citing Alberts, B. et al (2008), *Molecular biology of the cell*, New York, NY: Garland Science; Hartl, D.L. and Jones E.W. (2005), *Genetics: analysis of genes and genomes*, Sudbury, MA: Jones and Bartlett Publishers.

5. https://education.seattlepi.com/daughter-cells-mitosis-identical-parent-4720.html

6. Sheldrake, Rupert, 'The ageing, growth and death of cells' (1974), *Nature*, 250, 381-50

7. Sheldrake, Rupert, *The Science Delusion*, Coronet, London, UK, 2013, pp. 1-2

Chapter 6-2: Neuron Communications

"The fact that three-fifths of an octopus' neurons are not in their brain, but in their arms, suggests that each arm has a mind of its own."

~ Unknown ~

A thought to ponder: Is the arm of an octopus capable of cognitive behaviour, and if not, why would a human brain, consisting of the same cell types, be capable of cognitive behaviour? If it is the structure of the neural cells, then in the context of the brain's functioning, what structured the network in any predetermined manner, let alone to be a symbolic representation of conceptual data?

Because I am interested in origins, and the implications for evolution theory, I am interested to know how the nervous system got started. If genetic mutation is the cause of development of different types of cell, what was the precursor to the neuron? What part of the DNA mutated, and from what to what? All cells have some commonality in their structure, but the neuron has some unique features like axons and dendrites. Whilst some neurons have direct electrical connection between their membranes, others rely on these two cytoplasmic extensions attached to the cell (bipolar neurons) for communications. If axons evolved before dendrites, did axons connect to other axons, or did they have nothing with which to connect? If there were no connections, then proto-neurons just floated about with no obvious purpose. A nerve fibre is comprised of adjacent neurons which through the interaction at the synapse, ripple an electric pulse (action potential) all the way down the line. At each end of nerve fibres serving the senses, there is a chemical mix that represents the transduction of energy by the sensory neuron. As asserted in Primary Axion #8, a system is irreducibly complex, and where the basic components are missing, you don't have a system.

Revisiting an earlier topic, the human nervous system comprises two primary arrangements: The Central Nervous System (CNS) and the Peripheral Nervous System (PNS). The CNS comprises the brain subtended by the spinal cord, with the PNS branching from there. The PNS is of two divisions. The sensory division transmits stimuli from outside and within the body to the CNS, whilst the motor division transmits stimuli from the CNS to organs of the body. The brain is composed of neural circuits, defined as populations of neurons interconnected by synapses to carry out a specific function when activated. Neural circuits interconnect to one another to form large scale brain networks. The immediate questions arise: what defines the function; what orchestrates the organisation of the network for specific functions; and what informs a neural circuit of its function?

We ended the previous chapter discussing the logical conclusion that during mitosis, daughter cells could not be exact replicas of parents in all circumstances, because otherwise, all daughter cells would exhibit the same behaviours. Deep within the system, there must be structural information which causes cells to associate with other cells in specific ways. My thoughts turn immediately to evolution: that would have been a huge problem to solve. This differentiation in daughter cells, and how they connect with one another, leads us to consider the structuring of neural networks in developing babies and animals. Observation tells us that there must be specific, predefined, neural structures for each animal type. We know that replication in mitosis is not one-hundred percent perfect, so some allowance must be made for variation, but in the main, the brains within a given species, human or otherwise, must be exactly the same as one another.

According to the scientific literature, at birth, a baby's brain contains approximately 100 billion neurons. These are almost all the neurons the brain will ever have.

"In the brain, the neurons are there at birth, as well as some synapses. As the neurons mature, more and more synapses are made. At birth, the number of synapses per neuron is 2,500, but by age two or three, it's about 15,000 per neuron. The brain eliminates connections that are seldom or never used, which is a normal part of brain development."[1]

Consider the complexity of 100 billion neurons each with 2,500 synapses, and wonder at the orchestration of the network structuring. The instructions for the patterns are somehow contained in the genome, and are expressed during development of the foetus. Maturation of the neurons is said to be, partially at least, a function of age and experience, with neural patterns developing in response to occurrences within the development of the baby, such as sight, walking, language, and so on. I intend going no further down this path other than to ponder two issues: (1), what pre-programmed, biological mechanisms are in charge of these processes; and (2), how could evolution have accounted for such complex logic, structures and behaviours? Make that three: what would Darwin, being the intelligent chap that he was, have thought about his evolution hypothesis had he known then, what we know now?

Structure of Neurons

Neurons are considered the most polymorphic cells in the body, as they defy formal classification on the basis of shape, location, function, fine structure or transmitter substance.

Unless otherwise stated, the following descriptions have been copied, with permission, from this website[2]. If the reader is already familiar with the biochemistry, you will likely skip through, but for others (and myself), my purpose here is to provide an overview. A neuron, (nerve cell) is an electrically excitable cell that communicates with other cells either via specialized connections called synapses, or by direct connection through adjacent membranes. It is the main component of nervous tissue. The colour diagram of this copied picture, with explanation, is available here[3]:

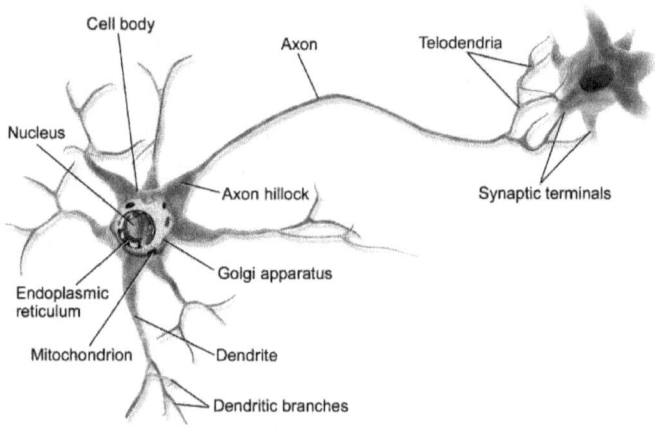

Permission not required

The *soma*, or cell body of a neuron, is the non-process portion of a neuron or other brain cell type, containing the cell nucleus. In other words, the soma plays no part in communications, but exists to support the communicating structures. Energy producing metabolism and the synthesis of the macromolecules used by the cell to maintain its structure and execute its function are the principal activities of the neuronal soma. The nucleus is the source of most of the RNA that is produced in neurons. In general, most proteins are produced from mRNAs that do not travel far from the cell nucleus. This creates a challenge for supplying new proteins to axon endings that can be a meter or more away from the soma.

A group of connected neurons is called a neural circuit. Neurons are classified in a variety of ways, depending on where they are found and the functions they perform.

By Structure:

- Unipolar neurons are typically sensory neurons with receptors located within the skin, joints, muscles, and internal organs. The axons of such neurons are usually long, terminating in the spinal cord.

- Bipolar neurons are neurons that have two cytoplasmic extensions attached to their cell body. These extensions are an axon and a dendrite, both of which can have many smaller branches.

- Multipolar neurons are defined as a type of neuron that possesses a single axon and many dendrites (and dendritic branches), allowing for the integration of a great deal of information from other neurons.

The anatomical terminology above can be a little confusing (I found it so), but we shall go with what we have.

By Connection:

There are three classes of neurons: afferent neurons, efferent neurons, and interneurons.

- Afferent neurons carry information from tissues and organs into the central nervous system.

- Efferent neurons transport signals from the central nervous system to the effector cells.

- Interneurons connect neurons within the central nervous system.

By Function:

- Sensory neurons carry signals from sense organs to the spinal cord and brain.

- Relay neurons carry messages between sensory or motor neurons and the central nervous system.

- Motor neurons carry signals from the CNS to muscles, motor neurons are connected to the relay neurons. The signal passes between the neurons via synapses. Synapses are microscopic voids between cells where chemicals are released from the axon terminal of one cell to specialized chemical receptors on the dendrite of the receiving cell.

A typical neuron consists of a cell body (soma), dendrites, and a single axon. The soma is usually compact. The axon and dendrites are filaments that extrude from it. Dendrites typically branch profusely and extend a few hundred micrometres from the soma. The axon leaves the soma at a swelling called the axon hillock, and travels for as far as 1 meter in humans or more in other species. It branches but usually maintains a constant diameter. At the farthest tip of the axon's branches are axon terminals, where the neuron can transmit a signal across the synapse to another cell. Neurons may lack dendrites or have no axon. In other words, some neurons do not have receptors and cannot receive signals, and some may lack the ability to send signals to other neurons.

Genes are described as *polygenic* and *pleiotropic*, but I cannot find that terminology relating to neurons, although such terminology is apt. As the text is not explicit on this issue, I will insert my own observation and terminology. It is the branching structure of the axon that identifies a neuron as *polyfunctional*, i.e., an action potential in a single, sending (presynaptic) neuron can result in signalling to multiple receiving (postsynaptic) neurons, influencing the behaviour of neurons downstream in each neural circuit in the network. Because a single neuron can receive signals from multiple sending neurons, it can be described as *polydependent*; i.e., its behaviour is subject to multiple influences. According to one study, "The receiving neuron integrates all the information it receives from all its afferents and may produce output in turn, releasing neurotransmitters from its own axon."[4] Depending on the circuitry of the network, not every afferent may be active at the same time, so not every synapsed dendrite would necessarily result in a change in its membrane potential. The neuron diagrammed below is described as *multipolar*: the type of neuron that possesses a single axon and many dendrites (and dendritic branches).

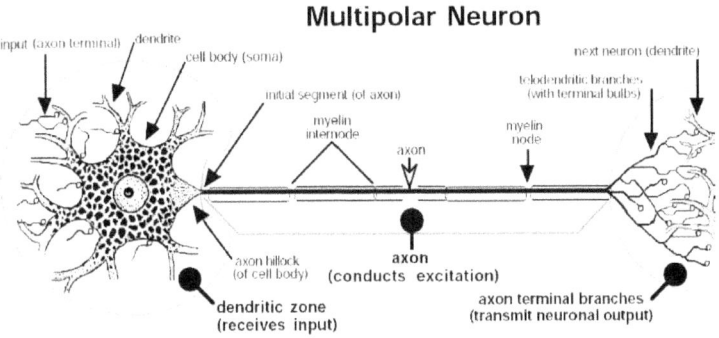

Courtesy of the free online resource, *Neurohistology.pdf*

My reason for going into such detail is to evidence the complexity of the human nervous system, and question how such elegant and functional complexity could have arisen through the millions (or billions) of small accidental mutations of undirected evolution. Again, using *systems thinking*, some types of neurons are specific to the type of biological function, which raises the question of the sequence of evolutionary small steps to ensure that the requisite support functions were available as and when necessary for the *system* to work at all. Whilst mutations neutral in effect could persist in an organism, biological organs not connected to a system were unlikely to have continued to evolve without any contribution to survival or reproduction.

How Neurons Communicate

Neural circuits do indeed carry signals, but there the similarity with computers ends. A digital signal is one that represents data as a sequence of discrete values, which can only take on one of a finite number of values in any given instance. Signalling between neurons is entirely different, because the determination of whether a neurotransmitter is released across a synapse is related to electronic gradients across the membrane of the soma (cell body) of the presynaptic neuron. The composition of the neurotransmitter, which determines the response in the postsynaptic neuron, varies significantly. Neurotransmitters are chemical messengers comprising small molecules with amine functional groups or amino acid derivatives. Synapses can be chemical, communicating through chemical messengers or electrical, where there is a flow of ions between the cells. Whilst the chemicals composing neurotransmitters have been identified, what might be termed its *logical* coding system is unknown.

Activation of synapses involves electrical potential variations as little as 15mV (millivolts). If you have studied electrical engineering or related subjects, you will understand the engineering precision required to manage such fine voltages. I may be deemed incredulous, but I struggle to understand how evolutionary processes could arrive at such solutions, purely through whatever mechanisms are now favoured by evolutionists.

Repeating an earlier observation, "The receiving neuron integrates all the information it receives from all its afferents and may produce output in turn, releasing neurotransmitters from its own axon." Referring again to this text[5],

"Neurons are essentially electrical devices ... Action potentials are the fundamental units of communication between neurons and occur when the sum total of all of the excitatory and inhibitory inputs makes the neuron's membrane potential reach around -50 mV, a value called the action potential threshold. Neurons talk to each other across synapses. When an action potential reaches the presynaptic terminal, it causes neurotransmitter to be released from the neuron into the synaptic cleft, a 20–40nm gap between the *pre*synaptic axon terminal and the *post*synaptic dendrite."

An interesting aspect of this description is whether or not a neuron will synapse with another is dependent upon the chemical compatibility of the pre and postsynaptic neurons. In earlier chapter, we learned that during the early development of the brain of babies, "The growing tips of axons (called growth cones) apparently recognize and respond to various molecular signals, which guide axons and nerve branches to their appropriate targets and eliminate those that try to synapse with inappropriate targets."[6] From this we can see that not every neuron can synapse with any other: there is specificity of the presynaptic neurotransmitters and the neurotransmitters of the dendrites of the postsynaptic neuron. Apart from complicating our understanding, this raises further questions for evolution. These processes suggest that for some period, there were limited numbers of the types of neurons, and thus limited numbers of connections. When a mutation caused a new type of neuron, there needed to be further mutations to develop even more types that were compatible. But then the ever-present stumbling block: neurons fall under the category of somatic cells, but only mutations of gamete cells (germ line) can be inherited.

For further details, there are some useful explanatory diagrams on the referenced website. At this point, we need to disambiguate the confusing terminology we find in many descriptions.

Information, Messages, & Signals

Many scientific descriptions use these words interchangeably, but the reader needs to be aware that in scientific literature at least, each should only be used in specific contexts. We have dealt at length with the concept of *information*, so we can put that aside as an inappropriate term to be used in neural communications. As described above, what passes along a neural pathway is an electrical pulse, with no semantic layer whatsoever. Each neuron has its own action potential which never varies, and according to the literature, the responses are always **all** or **none**. If none, then the signalling stops. Conceptually, it is little different to an electric current, except that as each stimulated neuron synapses with another, it returns to what may be termed an inactive (off) state in that its membrane returns to its resting state. This impulse could be termed a *signal*, provided that one is aware that it only ever has one message: GO! *Messages* are never conveyed between neurons: chemical messages are only conveyed to cells at the end of a neural pathway.

To reiterate, neurons communicate *with one another* via the synapse. The firing of an action potential in a presynaptic (sending) neuron results in the release of neurotransmitters across the synaptic cleft, causing the postsynaptic (receiving) neuron to be either more or less likely to fire its own action potential. And so it goes, down the neural pathway – no information, no messages, no semantic layer in the signal, just an electrical impulse - until it reaches a terminus, for example in a muscle or other type of cell. At this point, the released neurotransmitters can be described as chemical messages specific for the type of cell into which they are released. The question that arises is what determines the starting and ending neuron in the brain's neural circuits, and with what does the terminus neuron communicate? I will attempt to answer that question in a later chapter on the anatomy of the brain.

Notwithstanding the detail covered above, these explanations of the physical mechanisms are of limited value in promoting an understanding of how neural pathways work in the brain, and how such signalling relates to the *conceptual*, the domain of the mind. By analogy, perhaps poor, telling a student how to form the letter "a" informs him/her of nothing at all about how to read. One cannot teach Morse Code by telling a budding telegraphist nothing more than that he/she presses down on a mechanical key to make an electrical connection. Of interest is the initiating impulse for brain activity, where sensory inputs are not involved. Reading the literature is like catching a word or two in the middle of a conversation: words are being exchanged, but what started the conversation, its purpose, and where it is heading, are all unspecified. Neurons exchange signals, but to what end, and what do they mean? One text stated that axons conduct information encoded in the form of action potentials, but I believe this to be incorrect: action potentials are simply signals – there is no encoding. However, that text is encouraging in that *encoding* is recognised as an essential process in communications, but discouraging in the use of the term, *information*. We will come back to this.

The other important issue here is that explaining the *"what is"* of neuron signalling takes us no closer to understanding how it *"came to be"*. The specification of millivolts and specific chemicals, positive and negative ions, and other complexities, must raise serious questions about origins. The detailed specificity of the interactions would argue against undirected evolution.

Again, I would ask the reader to ponder the complexity not of just the neural network in the brain, but how the central and peripheral nervous systems are dependent upon the very same types of processes. The issues to be considered are the *evolution sequence* and *timeline*. The evolution of earliest organisms said to have any type of sensory or nervous system, must have been preceded by the evolution of the nervous system itself. How long that would take is anyone's guess, but eons would appear far too short. The other issue is functionality. What imperative could there have been for partially functional neurons to have continued to have evolved, when logically, partially functional must be considered equivalent to dysfunctional. I would repeat my Primary Axiom #8: *a system, of whatever simplicity, is irreducibly complex.*

A *single neuron* is itself, a complex system, as the foregoing discussion has evidenced. In fact, all biological cells are complex systems in their own right. To prove otherwise, evolutionists must propose an evolutionary sequence where each component serves sufficient function to be considered for selection.

What Do Neurons Communicate?

There are two types of synapse. A *chemical* synapse transforms an electrical signal (the action potential in the presynaptic neuron) into a chemical signal (which is the neurotransmitter) and back to an electrical signal (the postsynaptic potential) in the postsynaptic neuron. In *electrical* synapses, the membrane of one cell is directly adjacent to the membrane of a second cell via a small series of pores known as gap junctions. Gap junctions allow direct electrical contact between two cells, so if the membrane potential of one changes, the membrane potential of the other one changes instantaneously. The chemical signal (neurotransmitter) is of particular interest, as its composition and that of dendrites on other neurons determine whether the two will synapse, but again, this only happens where there is a terminus in another cell type. Neurotransmitters regulate such functions as heart rate, breathing, digestion, appetite, and so on, but how they function related to cognitive activity is not understood, even if some literature adds them into lists without explanation.

As before, neurons are classified as *unipolar, bipolar,* or *multipolar,* and may be *afferent, efferent,* or *interneurons*. They are further classified as *sensory, relay,* or *motor*. I shall try to deal with these complications in as simple a fashion as possible, with limited technical terminology, my goal being to differentiate where true messaging occurs, and where it does not. It was earlier stated that "relay neurons carry messages between sensory or motor neurons and the central nervous system", which would seem to make them synonymous with interneurons. Communication along a neural pathway (nerve fibre) is limited to signalling caused by the relay of stimulations of action potentials across them – there is no messaging along the pathway. There is a form of messaging across the synaptic cleft, via chemical neurotransmitters, but the function here is simply to modulate the polarity of the postsynaptic membrane. This could be interpreted simply as the command: "carry the message", without specifying the content of the message. True messaging, that which contains a semantic (meaning) layer, occurs only at the beginning and end of neural pathways. The following description seems to suggest that I am correct in my understanding:

"A major role of sensory receptors is to help us learn about the environment around us, or about the state of our internal environment. Stimuli from varying sources, and of different types, are received and changed into the electrochemical signals of the nervous system. This occurs when a stimulus changes the cell membrane potential of a sensory neuron. The stimulus causes the sensory cell to produce an action potential that is relayed into the central nervous system (CNS), where it is integrated with other sensory information—or sometimes higher cognitive functions—to become a **conscious perception** of that stimulus. The central integration may then lead to a motor response."[7] (emphasis mine)

Ah! A *conscious perception*! Consciousness, the state of being aware of and responsive to one's surroundings is a much-debated concept, as to whether it is immaterial as I and others contend, or simply a function of material interactions as evolutionists contend. That aside, and continuing with an understanding of neural communications, in the surface of the skin are specialised sensory neurons called *exteroceptors*, which respond to sensations such as pressure, temperature change, touch, or pain. These cells that interpret information about our external reality are of three types:

1. A neuron that has a free nerve ending, with dendrites embedded in tissue that would receive a sensation.

2. A neuron that has an encapsulated ending in which the sensory nerve endings are encapsulated in connective tissue that enhances their sensitivity; or

3. A specialized receptor cell, which has distinct structural components that interpret a specific type of stimulus.

The pain and temperature receptors in the dermis of the skin are neurons with free nerve endings. Neurons with encapsulated nerve endings that respond to pressure and touch are also located in the dermis. It is logical to assume that there is some chemical differentiation between how these sensations affect the neuron. From there, relay neurons convey not the chemical message itself, but the fact that there is a message to be conveyed. Recall earlier that neurons communicate via an action potential in the presynaptic neuron causing a neurotransmitter to be released into the synaptic cleft, resulting in a spike of electrical activity (pulse) in the postsynaptic neuron that travels down the length of a nerve fibre, neuron to neuron, via these synapses. I have studied some technical literature on what happens next, but I see no point in quoting descriptions which I do not properly understand. Suffice to say that the individual pulses travel via nerve bundles terminating in the somatosensory cortex, on the surface of which it is claimed, is *mapped the body* allowing information from specific regions of the body to be analyzed. I am curious about this concept of the body being mapped in the cortex. In communications, there are two methods of identifying a source: either (1), the message contains data identifying the source, or (2), there is a dedicated connection. I wonder whether this mapping has been scientifically determined, or whether based on what is known about communications, it is an inference to best explanation? Moving on, again, I am not comfortable with the term, *analyzed*, meaning to examine methodically and in detail, typically in order to explain and interpret. Analysis is my stock-in-trade, and I am very much aware of the cognitive nature of the activity. The term arose in this text:

"The human nervous system comprises the brain, spinal cord, and peripheral nerves and is composed of several billion nerve cells or neurons. Specific populations of neurons, together with associated fiber tracts, are organized anatomically into pathways involved in sensory, motor, and cognitive function. A significant proportion of the brain carries out a sensory analysis of the world using information supplied from specialized receptors in skin, joints, ear, eye, nose, and tongue. This information is analyzed in primary sensory areas before being integrated into an image of the "state of the world." This image is used in cognitive processing resulting in motor action and a behavioral response to incoming sensory data. This chapter provides a guide to the parts of the nervous system, the structure of sensory systems, and describes how this information is used in more complex brain functions."[8]

I would not for a moment dispute the scientific texts in this area, apart from perhaps the terminology which would suggest something more than is really occurring. We KNOW that the system works, and I am confident in the scientific research that has resulted in a correct understanding of the architecture and physiology of the human nervous system. However, whilst the "what" is undoubtedly correct, the "how" may not be. As my primary focus is the mind-brain conundrum, any inaccuracies in the scientific understanding of the PNS and CNS, and to an extent, the cortex and its components such as the somatosensory cortex, are not necessarily relevant. In short, it is the ability to "visualise" from purely physical and chemical stimuli which has me intrigued. My journey into the sensory systems is necessary to disambiguate the terminology generally encountered. Is the term, *messaging*, accurate in the context of transmissions along a neural pathway, or as I have come to believe, *signalling* would be more accurate? As I mention elsewhere, there are two methods for identifying the source of a communication: by a dedicated connection, or by the message content. As neural pathways do not convey messages *pe se*, then the first method must be the case. The scientific literature would seem to agree, as we have seen, it mentions *mapping of the body*: somewhere in the cortex region, such mapping allows the source of the signal to be "understood", and thus knows what the message would be. This explanation works for some sensory perceptions, but not all: the messaging through the visual and auditory systems must be entirely different, as no pre-set internal mapping could exist for external realities not yet experienced.

Whilst one source states "the 'language' of the nervous system is electric signals", we can only understand this as described above – there is no semantic layer in the signal. As an aside, I recall my days in air traffic control over a half-century or more ago, when the technology of radios was primitive by today's standards. It was possible for the radio in an aircraft to transmit a carrier wave at a particular frequency, but without modulation, the latter being the semantic layer. You can hear this carrier wave when clicking the press-to-talk button on and off. This gave rise to a condition we termed, a no-radio aircraft. We would hear a longish carrier wave, but without modulation, and respond by asking a series of questions to which the pilot would respond with one click for yes, two for no, and three for say again. Our first question was always: Are you declaring an emergency? If the answer was "yes", this got very serious, and we had a sequence of questions where we attempted to ascertain the type of emergency, whether the pilot could recover his aircraft to our base, the identity of the aircraft, and so on. If the initial response was "no", not an emergency, we would first confirm not an emergency, and then the fun began. Our base served the Aircraft Research & Development Unit, the pilots all being highly qualified test pilots, most of whom were accustomed to testing our imagination in trying to ascertain their status. As could be expected, they were very creative and we often lost the contest. Whoever lost paid for the beers that evening in the Officers' Mess. However, I digress, but wanted to emphasise that the signal transmitted between neurons could be compared to this issue: a carrier wave but with no modulation (semantic layer).

Neurotransmitters, described as chemical messengers that transmit a message from a nerve cell across the synapse to a target cell, have been identified as amino acids, peptides, monoamines, etc., and whilst there is some understanding of the function of these messengers in managing physiological functions, there is none in relation to cognitive functions. Whilst the semantic layer in sensory systems is understood to some extent, the same is not true of the neural circuits in the brain. Disappointingly for my study, I can find no literature which assists in understanding how the transmission of electrical impulses along neural pathways in the brain convey a semantic layer, or how meaning is derived from the architecture of the network. If the signal transmitted via the nervous system contains no semantic layer, how are the sources differentiated? If the signalling along a nerve is as I have described it above, with no semantic content transmitted across neurons, this suggests that the source is recognised in the cortex by the particular nerve fibre. This is not scientific fact – it is supposition on my part, based on my own experience with matters technical. I have been unable to source scientific literature which disambiguates this issue.

As earlier emphasised, at best such signalling systems in the PNS and CNS communicate neither data nor information. In the body, both interoceptors and exteroceptors receive a chemical message appropriate to the location of the sensor. At the other end of the neural pathway, that chemical message is conveyed to whatever organ is structured to process it. In the evolution of the neural architecture within the body, there needed to have been a process whereby whatever "message" is chemically encoded in the sensory neuron is understood by the target organ. In a sense, this can be considered and encode/decode protocol, but as asserted by Primary Axiom #5, such protocols must be devised and implemented simultaneously for communications of any form to occur.

Just how evolution achieved such an amazing system is beyond my imagination, but that said, the context of my study is the mind-brain conundrum, and how what the mind conceptualises can result in a physical action, such as what I am doing here in writing these sentences.

Within the Brain

What I am attempting to understand, and as yet failing to do so, is to identify what would *initiate* a communication in the brain's neural network to have my fingers clumsily finding their way across the keyboard. Remember that sensory neurons respond to stimuli that affect the cells of the sensory organs, sending signals to the spinal cord or brain. Motor neurons receive signals from the brain and spinal cord to control everything within our bodies. Sensory neurons can trigger activity in the brain further triggering motor neurons to respond. Ignoring functions entirely internal to our biological workings, what is the mechanism that stimulates the brain to both initiate and respond to *cognitive activity*? We explore this further in later chapters, but for now, I just want to have a clearer understanding of the processes within neural networks such as the brain.

Connections in the brain's network are unlike the connections in a PNS nerve fibre. In the PNS, and this is a generalisation, relay neurons connect to the next one in the circuit, each connection stimulated by the same electro-chemical potential. If there were variations in the behaviour of the relay neurons, the integrity of the signal would be compromised. The target neuron, I presume, is always predetermined.

In a neural network, there must be a method of determining to which neuron(s) the source neuron should connect. From a physical standpoint, this would require that the target neurons have properties representative of the logical connections to be made by the source neuron. For example, in English at least, most words have a semantic range which is *contrast* and *context* dependent. My thesaurus lists 45 synonyms for the word, "hot". If my brain remembered them all, that would require at least that many connections, plus many more to differentiate both the contrast and context chains. The contrast domain provides the qualitative determination from, for example, tepid to scorching, whilst the context domain identifies the appropriateness of the adjective to the subject. We might describe the day as "boiling hot", which is an acceptable metaphor derived from the more appropriate usage, as in "boiling water". Whilst "fiery" is a synonym for "hot", it would be unusual to use that adjective in relation to the temperature of water, as fiery has the connotation of burning.

All along, we must not forget that in the brain, in the context of cognitive activity, neural patterns are symbolic physical representations of conceptual entities. We must question whether any part of the brain, being of the same biological construction throughout, could be capable of performing both encode and decode functions required for the storage and retrieval of such concepts. Using the computer analogy, it seems unlikely that the same region of the brain could perform both computing and storage functions. Neuro-specialists have identified regions of the brain responsible for specific processing, but have not differentiated the processing regions from the storage regions. I am reminded of an explanation that I read in an article on Alzheimer's disease. The text stated that whilst people 60 years and older will often complain of loss of memory: *"The information is always in the brain, it is the "processor" that is lacking."* The obvious question arises: What is the processor? From my understanding of neuroanatomy, admittedly limited, it would seem that the data processes itself, which I find most curious.

Here are other examples of this mystery. When reading cells in a spreadsheet, we may at times not comprehend the precise meaning. For example, consider a column showing financial values displaying a cell as "$ 3.6". If the format allows for only one decimal place, the ".6" might be specifically ".60", rounded up from ".58", or down from ".64". To correctly interpret the precise meaning, not only must the brain process the visual representation, it must also know that formatting rules may apply, and must have neural connections that apply rules to what is seen. A more complex example is shown in this string of letters:

- Tihs artlice amis to porvdie cliarty reagdring why it is poslisbe taht we are albe to raed wrods, eevn wehn the letrtes are julbmed;

which we interpret as:

- This article aims to provide clarity regarding why it is possible that we are able to read words, even when the letters are jumbled.

This link[9] offers some brief observations, but being one with considerable problem-solving experience, coupled with competency in both computer programming and the English language, I would struggle to know where to begin programming the necessary interpretation / translation logic. Curiously, evolution is claimed to have performed this momentous accomplishment for us, and stored it in our neural network, although apparently not every brain has been so gifted. In this example, we have the issue of the eye encoding the letters as they appear, into an electro-chemical transmission to the brain via the visual cortex, whereby some process is invoked to both decode the transmission and transform it from a superficially incomprehensible physical message, to a comprehensible conceptual one. The bridge between the *physical* and *conceptual* remains as elusive and incomprehensible as ever, but it must exist in what we loosely refer to as the *mind*.

Without exception, the resources that I have studied make reference to "information flow" between neurons. This could be due to familiarity on the part of the authors, or it could be that the authors truly believe that information does flow, but quite obviously, they have not thought very deeply about the issue. Earlier I asserted that information is *conceptual*, not *physical*, and we know that communications across neurons is entirely physical. But putting that aside, when neuroscientists "listen in" to neuron communications, they know neither the language being spoken, nor the topic of conversation. Further, other than in defined neural conversations, neither the initiator of the conversation, nor the reason for initiation of the conversation, are known. By analogy, the scenarios offered by scientists are like my stumbling across a conversation between Russian and Polish diplomats speaking in the native language of one of them. I would have no idea of who started the conversation, and whether they were discussing a national revolution, or where to place their bets in the next race at the local track. In truth, other than by logical assumption, I could not be sure whether they were both discussing the same topic at all. All I could know was that a conversation was in progress.

In both the PNS and CNS, messaging is generally initiated by sensors reporting events, both internal and external, such as related to the condition of an organ, or perception by eyes, ears, or touch. Internally initiated messaging is autonomous – a function of the biology. The message is technical in nature, analogous to telemetry within an aeroplane, in that it automatically conveys the state or health of an organ, or convey commands regarding the workings of the system, utilising a coding system consistent throughout the body. Medical research has made great strides in interpreting these messages in terms of what chemicals, such as hormones, are to be released depending on the circumstance. It is important to note that the context of the process is the biology of the organs and nervous system. In a sense, it can be said that the internal system "knows" everything there is to know about itself. Put another way, the system understands the semantic content (meaning) of any internally initiated message.

Messaging initiated by the sensory system in response to external events is different, in that there may be implications for the internal system, such as when injury or discomfort occurs, in addition to what the message conveys to the brain about the external environment. For example, extremely bright light causes discomfort to the eye, but what the eye sees is still conveyed to the visual cortex via the optic nerve, although perhaps with less clarity.

The eye transduces the light energy into an electro-chemical message, containing a semantic layer (meaning) for both the discomfort condition, and the visual scene. The internal biological system obviously "understands" the first, but conceptually, not the second. The discomfort message may result in an autonomous response by the brain, such as blinking or turning away, but the "comprehension" of the message is purely biological. The semantic layer of the message, conveying the meaning of the visual is incomprehensible to the internal neural system, and may be irrelevant to the biological workings of the body. Comprehension, if it is achieved, is believed to be a function of the visual cortex, which we will discuss in a later chapter.

Neural Plasticity

It was thought for many years that once the brain matured, the organisation of the brain was fixed. Whilst it was recognised that new neural patterns must arise with the learning of both cognitive and manual skills, without the necessary investigative tools, anything more was just speculation. Modern methods of mapping the brain has led to a clearer understanding of which parts of the brain are involved in specific tasks, and have revealed some of the secrets of memory. Thus arose the term, *neural plasticity*, in recognition of the fact that the brain is not *entirely* hard-wired. Raymond Tallis notes:

"The immense literature on brain plasticity has examined changes at both the microscopic and macroscopic levels occurring in response to experience. Microscopic studies have shown that repeated activity in a particular neural pathway that includes a synapse may result in increased ease with which the synapse is crossed: less is lost in translation from the presynaptic to the postsynaptic neuron. Also, where pathways are active together, they are more likely to make contact and their activity is more likely to be coherent: "nerves that fire together wire together", to use the famous aphorism attributed to neuropsychologist Donald Hebb. At the macroscopic level, it is possible to see changes in different parts of the brain in the size of the areas dedicated to different parts of the body in response to recurrent stimulated or their involvement in repetitive activities. The quantity of brain active in tasks involving fine touch discrimination, for example, expands and contracts in response to the amount and kind of workload carried by different hands."[10]

Neural-physiology explains muscle memory: the phenomenon that repetitive activity of a particular type leads to better performance, and why, for example, once you learn to ride a bicycle, you never forget. The latter is not entirely true, as I can testify: remembering depends on how well one learned to ride in the first place! That aside, this increasing knowledge of the plasticity of the brain is significantly contributing to the rehabilitation of people with damage to the nervous system. As Tallis comments: "Those involved ... have been excited by the increasing evidence of recovery based on *reorganisation* of the brain circuitry, with new connections being formed and dormant areas waking up."[11] I am reminded of an occurrence in my local region where a young man was told by his doctors that he would never walk again, yet by courage and persistence with the aid of physiotherapists, he did eventually walk again and even returned to his sporting love, rugby football.

Neural plasticity must have its down side, when, for example, we repeatedly do something the wrong way, repeatedly learn the wrong lesson, or repeatedly accept a specific untruth as truth. Although indoctrinators and marketing people (they are synonymous in practice) know that their techniques work, it is only since the discovery of neural plasticity that we have some understanding as to why. In the evolution context, I ponder the scenarios where neural plasticity may have been more deleterious than beneficial.

I can understand neural plasticity in relation to manual activities, for the mechanisms and processes are within a defined environment (e.g., the human body), as stimulated by the sensory system. However, plasticity in relation to cognitive activity is somewhat more problematic. No doubt, the reorganisation processes within the brain are the same, but it is the stimulation that needs explaining: how does the conceptual translate into the physical?

Summary

Let me confess that whilst I understand all this at one level, the deeper I go, the less that I comprehend. My purpose in this longish description is to demonstrate that fundamentally, all nerve cells behave in accordance with their cellular properties, which are dependent upon numerous factors including the properties of chemicals involved in synaptic signalling. The presynaptic neuron(s) stimulates the postsynaptic neuron(s) depending on variations in the signals it receives from its own presynaptic neuron(s). Whilst we have some understanding of "what" happens in these processes, we have no understanding of how multiple individual signals in a complex, interrelated neural network together combine to convey a message (semantic layer). Crucially, there is nothing in the scientific descriptions that have any sense of cognition. For the most part, they speak of signals passing from one or more (polyfunctional) nodes in a network, to one or more (polydependent) nodes. On a few occasions, the text wanders in its terminology from *signals* to *messages* and *information*, but I suspect that this is more a slip of the pen than a deliberate attempt to differentiate the type of communication (at least I hope so). To be clear, as the text generally gets right, what passes **between** neurons is an electrical or chemical signal – *no data, and no information*. What passes **along** neural pathways is an electrical pulse.

Brain cells are nerve cells in the brain region, not too different to nerve cells elsewhere, other than that they are said to be *specialized* to carry "messages" through an electro-chemical process. Specialization infers purpose, which in evolutionary terms, means adaptation, but the more we learn of neural complexity, the more difficult it becomes to find a process. Now, to return to our primary question: What is it about the cells in the brain that allow them to self-organise around concepts, such as the outside world, of which they cannot be aware? We do have evidence of organisation, but no scientific underpinnings for self-organisation of the type that is claimed for neural networks. If we wish to understand the "how" of the organisation, we will need to look beyond the biochemical properties of the structure itself.

There can be no doubt that the mind and brain are interconnected, and interdependent. Without the mind, the brain is limited to autonomous responses, and without the brain, the mind can have no relationship with the body. I have earlier asserted that the brain cannot store data, which is conceptual, but only symbolic representations of physical data in patterns, although just how that is done remains a mystery. I have also asserted that information is the correlation of data, and it seems logical to accept that neural networks provide that correlation, but again, only at the symbolic representational level. Computer databases do the same, with relationships established via their structures, but such structures are designed by intelligent agencies for specific purposes. In passing, sometimes even these intelligent agencies have got it wrong, as anyone who has volunteered to be a software *Beta* tester can attest. Science has demonstrated that memory loss, especially with aging, is associated with deterioration of the brain itself, so the relationship between mind and brain has been substantiated. However, there remains a twist: sometimes memory returns whilst there is no change in the organic structure of the brain.

I must admit that I continue to struggle to visualise the biological structure of the brain. When I read that each neuron can be connected to as many as 15,000 synapses, my imagination fails me. The following images, one simple, the other less so, are the best that I have found, but they don't really help to visualise the large complexity.

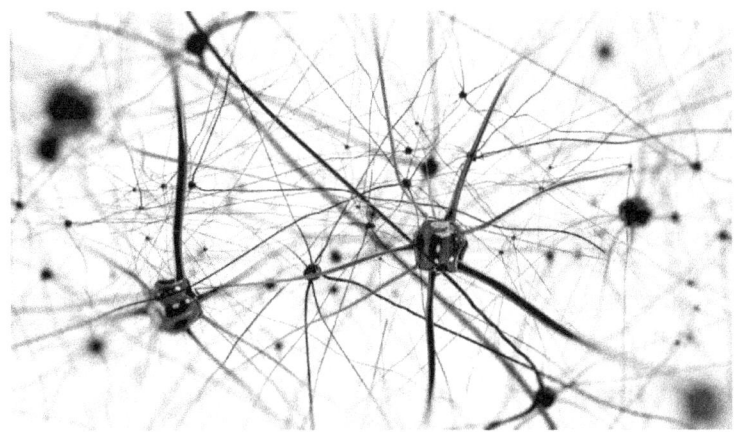

Reproduced under licence from Shutterstock

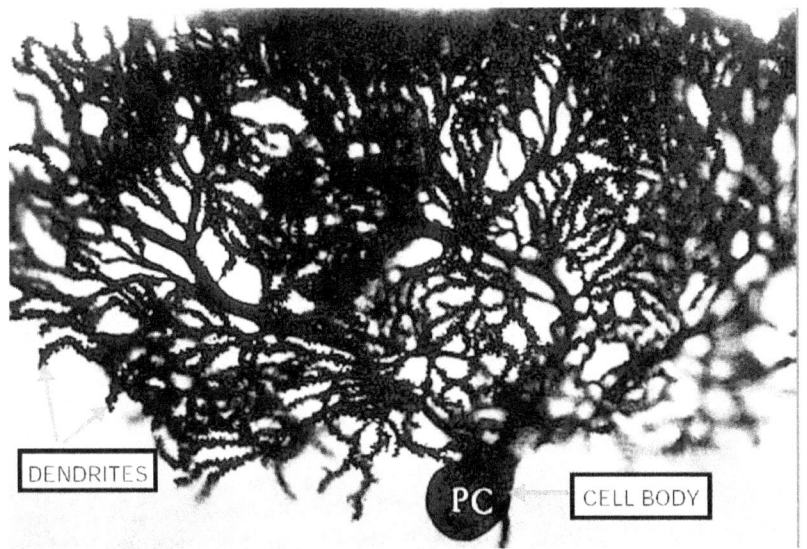

Courtesy of the free online resource, *Neurohistology.pdf*

References:

1. https://extension.umaine.edu/publications/4356e/

2. https://en.wikipedia.org/wiki/Neuron

3. https://commons.wikimedia.org/w/index.php?curid=28761830

4. Gluck, Mark A., and Myers, Catherine E., *Gateway to Memory: An Introduction to Neural Network Modeling*, The MOT press, Cambridge, MA, 2001, p. 45

5. https://qbi.uq.edu.au/brain-basics/brain/brain-physiology/action-potentials-and-synapses

6. https://www.britannica.com/science/human-nervous-system

7. https://opentextbc.ca/anatomyandphysiology/chapter/14-1-sensory-perception/

8. Preedy, Victor R., *The Neuroscience of Cocaine: Mechanisms and Treatment*, Academic Press, Cambridge, MA, 2017, Ch. 1

9. https://www.lifehack.org/423422/it-doesnt-seem-to-make-sense-that-we-can-read-this-but-science-explains-why

10. Tallis, Raymond, *Aping Mankind*, Routledge Classics, New York, NY, 2016, p. 26

11. *Ibid*, p. 27

Chapter 6-3: The Nervous System

"There is no fixed physical reality, no single perception of the world, just numerous ways of interpreting world views as dictated by one's nervous system and the specific environment of our planetary existence."

~ Deepak Chopra ~

There is not a lot more to be said here in terms of the biological workings of the nervous system, most of the detail being covered in the previous chapter on neurons and how they work together. The important point to note, in the context of evolution, is that it is not just a single system, but a complex arrangement of multiple systems with both individual and coordinated functions. The bio-technology of these systems is essentially the same, but there is a degree of *calibration* and *tuning*, if I can use those terms, for each to perform the functions as we know them today.

A most significant consideration is that way back in the distant past, some prototype nervous system must have been the very first achievement of evolution. Even insects have a brain and a nervous system. According to this online source[1],

"The origin of brains goes back to 750 million years ago. Ediacaran Biota is considered the grand ancestor of all species. Since then many species had been branched out and took their own development by adjusting the environment they faced. So even if we take an example of worms as a big category, their structure of neuron system varies depends on their genetic groups. In other words, some of their brains are close to related insects, and some are related to the group we human beings belong."

If there is one component, or sub-system, at a level of evolutionary development dating back 750 million years that is indispensable to life, it is the nervous system. All sensory systems, all internal biological workings, and all movement of extremities, rely entirely on a nervous system for communications. The question becomes: what is the minimal biological structure of a neuron, and how neurons connect, that would enable the workings of the brain, and whatever connects to the brain? If all brains are composed of neural networks, then evolution of the neuron must have preceded evolution of the brain.

We earlier had a brief look at the *soma* of a neuron, but it is worth reconsidering now to ponder the sequence of evolution from the perspective of survival advantage. The membrane of the neuron functions as a receptive surface over its entire extent. Specific inputs, referred to as *afferents*, from firstly, as electrical gradients from adjacent cell membranes, and secondly, as chemical neurotransmitters from presynaptic neurons, are received primarily on the surface of the cell body and on the surface of dendrites. In the latter case, the soma is where the signals from multiple dendrites are joined and passed on. The soma and the nucleus do not play an active role in the transmission of the neural signal. Instead, these two structures serve to maintain the cell and keep the neuron functional. The soma contains numerous specialised structures (organelles) involved in a variety of cell functions, including production of RNA that directs the synthesis of proteins, and production of mitochondria which provides energy for the cell. So far, not a lot different to other cells. But back to the dendrites, those tree-like extensions at the input end of a neuron that help increase the surface area of the cell body. These tiny protrusions receive signals from other neurons and transmit electrical stimulation to the soma. Other than where neurons communicate membrane to membrane, without dendrites, the soma has no function, and there can be no nervous system. The question arises: did dendrites evolve at the same time as the particular cell type, the soma, or did somas evolve and replicate, and only later join together in chains whilst having neither a particular starting nor ending point? When did the axon evolve in the sequence? What evolutionary advantage could they have offered when they had no function?

Recent research has shown that whilst neurons are unique in their structure and function, the soma is not much different to other cells, and that other cell types also have a membrane potential.

"Whilst the phenomenon of an electrical resting membrane potential (RMP) is a central tenet of biology, it is nearly always discussed as a phenomenon that facilitates the propagation of action potentials in excitable tissue, muscle, and nerve. However, as ion channel research shifts beyond these tissues, it became clear that the RMP is a feature of virtually all cells studied. The RMP is maintained by the cell's compliment of ion channels. Transcriptome sequencing is increasingly revealing that equally rich compliments of ion channels exist in both excitable and non-excitable tissue. In this review, we discuss a range of critical roles that the RMP has in a variety of cell types beyond the action potential. Whereas most biologists would perceive that the RMP is *primarily* about excitability, the data show that in fact excitability is only a small part of it. Emerging evidence show that a dynamic membrane potential is critical for many other processes including cell cycle, cell-volume control, proliferation, muscle contraction (even in the absence of an action potential), and wound healing. Modulation of the RMP is therefore a potential target for many new drugs targeting a range of diseases and biological functions from cancer through to wound healing and is likely to be key to the development of successful stem cell therapies."[2]

Fascinating! I am a fan of medical research, for apart from the advantages it offers for the healing and prevention of disease, it further illustrates the complexity of biology and the consequent difficulties such complexity poses for the evolution hypothesis. From the above quotation, we can accept that the soma of the neuron may well have evolved from an existing complex cell type which preceded other development, but it still tells us nothing about the evolution of dendrites and axons.

Earlier, for the sake of simplicity of the model, I misrepresented dendrites in postsynaptic neurons as being only *receivers* of neurotransmitters from presynaptic neurons. In fact, the dendrites of many postsynaptic neural populations *transmit* information back to their synaptic inputs by releasing neuroactive substances. Modulation of neuronal function by dendritic transmitter release is a widespread phenomenon and is specific neither to a localized part of the brain nor to a particular subtype of signalling molecule. I add this for the sake of completeness, and to evidence that signalling across neurons and neural pathways is far more complex than was earlier thought. To understand how this works in any given neural network must be beyond our comprehension, for the number of possible or even actual connections is likewise incomprehensible.

It is important to consider that in the evolutionary sequence, the nervous system must have preceded all development of all organisms with multiple organs, and all organisms capable of movement in relation to their external environment. The nervous system functions as the internal communications channel, as the telemetry system for monitoring both its internal and external sensors, and for two-way communications with faculties capable of movement, such as fins, legs, and so on. In short, all movement, whether as simple as that of a sea slug, or as complex as flying, requires both a communications channel for movement instruction, and a monitoring system for the movement itself. People lacking the faculty of sight, nevertheless "know" the positioning of fingers, arms, legs, and so on. Even a fish cannot "know" that its mouth is properly open without internal sensors providing the status feedback.

I shall leave that subject there, for anything more would be pure conjecture on my part. The issue that I wanted to highlight is that a form of nervous system must have preceded development of the brain and other organs, and itself preceded by the evolution of neurons and their connectivity. No organ could communicate with another, nor could there be two-way communication between an organ and the brain without a nervous system. No fin, tail, muscle, tendon, or other extremity could move in a regulated or deliberate manner without a nervous system providing the necessary communications, and a feedback system to confirm that the instructions had been carried out as intended. Communications systems require specificity of protocols and encoding systems, plus a method of calibration and verification (Primary Axioms 4 - 8).

I am asked to believe that evolution, via genetic mutation or whatever mechanisms are currently in vogue amongst evolutionists, was capable of such achievement. Perhaps it was, but given the specified complexity of the nervous system as discussed in this study, it does seem unlikely.

References:

1. http://web.colby.edu/st132origins/2017/10/31/human-brains-vs-insects-brains/

2. https://www.frontiersin.org/articles/10.3389/fphys.2018.01661/full

Chapter 6-4: Neural Network Modelling

"Your brain does not manufacture thoughts; your thoughts shape neural networks."

~ Deepak Chopra ~

In the context of evolution, the value of understanding modern *Neural Network Modelling* is to appreciate the enormity of the difficulty that this subject presents to the evolution hypothesis. Some truly amazing research has been performed, with very laudable results, such as in developing speech-to-text software. Progress in this area will provide solutions to many problems, but in contrast, will likely ring the death-knell of any ideas developed to support Neo-Darwinism. Adaptation, a form of learning, can be *taught*, but whether it can be *learned* autonomously is subject to serious debate. That is not to deny the natural process of phenotypic adaptation as an evolutionary phenomenon, but to question whether cognitive knowledge can arise through the same process.

"The power of nodes in a neural network (and neurons in the brain) comes not so much from their individual power, but from the collective power that emerges when many such devices are interconnected in a network."[1] I would be interested to know the nature of this "power", and into what it does it emerge – what does this power power? The question becomes, of course: What determines the structure of the interconnections, and what is the organising mechanism? The referenced study provides an absorbing discussion on the development of neural network modelling, and the degree of trial and error undergone to have the model provide the right answers. Of utmost importance is this: the right answer was **predetermined**, unlike in the evolution case (*if you don't know where you are going ...*). Earlier I mentioned the complexity associated with *polyfunctional* neurons (nodes) connected to *polydependent* neurons (nodes), with the outcome of any set of connections being dependent on resolving the contribution of the afferents. To reiterate, *afferent* neurons are sensory neurons that carry nerve impulses from sensory stimuli towards the central nervous system and brain, while *efferent* neurons are motor neurons that carry neural impulses away from the central nervous system and towards muscles to cause movement. The brain, or perhaps just the cortex, as a form of neural network manager, must contain efferent neurons for the purposes of bodily management, but these are unnecessary for cognitive functions other than that cognitive functions may stimulate them.

In the development of functional neural networks:

"The key to learning in neural networks is changing the weights between nodes ... intended to roughly capture what happens during learning in real neurons, in which synaptic strengths are altered, thereby altering the ability of some neurons to cause firing in other neurons. The earliest approaches to the problem of how to change the weights in a neural network came neither from psychology nor from neuroscience, but rather from engineering."[2]

Let us put this in perspective, but before doing so, let me comment on *altering synaptic strengths*. Without delving into the chemistry, a synaptic strength would depend on the chemical potency of the neurotransmitters from the presynaptic neuron, and the receptiveness of the postsynaptic neuron, usually a dendrite. It has been found that the synaptic strength can also be modulated by the release of neurotransmitters from the dendrites into other molecules. However, these are biological functions, not cognitive. It has been demonstrated that such chemical activity can affect moods and other aspects of behaviour, the domain of psychiatry and psychologists, which many claim relate to cognition. Such activity strongly supports the view of those rejecting the notion of dualism, of any form, because many aspects of what appear to be cognitive behaviours are demonstrably of a chemical in nature. We also know that behaviours can be altered or modified by specific chemicals, so those beholden to philosophical materialism do have a point. There is no argument that repetition can cause a strengthening of synaptic connections, but that has no relevance to the mechanism of the physical transforming into the conceptual.

The development of neural models is aimed at AI (Artificial Intelligence), the goal being adaptive computer models that in effect, set them free to teach themselves as new data is acquired. It is believed by some that in emulating the human mind, one could emulate evolution. The issue of how such a neural model could begin development of itself without an external, intelligent agent to bootstrap the process, is of no concern in network modelling.

However, it should be of concern to evolutionary theory. Weighting of contributing factors is not a new approach – it has long been a feature of logical evaluations; predictive modelling as occurs in commercial applications such as sales forecasting; and in recent years, climate modelling. When you do not get the answer which seems right, for whatever reason, you continue to manipulate the factors (termed "forcing" factors in climate modelling) until you do. By way of a simple mathematical model, if you had four factors to consider in evaluating three proposals, you could weight the factors by relative importance such as A=1, B=0.8, C=0.6, and D=0.3, and scoring each proposal as meeting the criteria either yes (1) or no (0). Thus, the evaluations might be:

Proposal 1: $1*1 + 0.8*0 + 0.6*0 + 0.3*1 = 1.3$

Proposal 2: $1*0 + 0.8*1 + 0.6*1 + 0.4*1 = 1.8$

Proposal 3: $1*0 + 0.8*1 + 0.6*1 + 0.3*0 = 1.4$

Note here that Proposals 2 & 3 do not satisfy the criteria considered to be of the highest importance, "A" (value 1), yet both score higher than the one proposal that does. This is more often the case than not, because the weighting factors are determined subjectively, and despite what should be obvious, you cannot get an objective result from subjective premises. In climate modelling, there is no prior scientific determination of the relative weights of the forcing factors. The best that can be done is to vary the weights until a model predicts a known reality. This, to my knowledge, has never occurred in any of the 20 models currently being investigated by the IPCC.

In neural network modelling, this is achieved by varying the weights between nodes until the predicted result is achieved.

"In real brains, of course, synaptic strengths are not set 'by hand' **but learned through repeated exposures** to regularities in the environment. Moreover, in very complex problems such as speech transcription, the network might need hundreds or thousands of weights to encode highly complex relationships between the input patterns and the output patterns. In such a case, it is difficult or impossible to choose all the appropriate weights by hand. Thus, Widrow and Hoff[3] needed a learning rule: an algorithm that would allow the network to adapt its weights to solve an arbitrary problem."[4] [emphasis mine]

"Learned through repeated exposures" might sound like a scientific statement, but it is lacking. In many scenarios, it is just as easy to learn the wrong lesson than the right, and how would the neural network know without a goal to be achieved? Repeated exposure could lead to a strengthening of a synaptic response long before that response had the feedback from wherever to confirm that it was evolving in a useful direction. The other question is whether synaptic strengths could be simultaneously manipulated in a complete mini-network need to excite motor neurons to the right target, or whether they could only occur on a single neuron or neural circuit at a time. In neural network modelling, it is advisable to alter the weight of only one node at a time to ensure that the response is accurately related to the change. If more than one change is performed at a time, the precise contribution of each change to the overall result is not easily determined. They could in fact cancel one another out. If hundreds of thousands, or even just hundreds of individual neurons and/or synapses are required to construct a network capable of resulting in a behavioural change based on repeated exposure to regularities in the environment, one must question the time scale for this to be achieved, and the probability of the requisite changes being maintained through the *required number of generations* of the organism. I would offer that expression of this process is more the product of wishful thinking than science, most especially because according to genetic research, neurons in the brain are of the somatic line, not gamete (germ) line, and thus irrespective of the *number of generations*, few if any neuronal changes are inherited.

Consider these requirements in the context of the evolution of human speech. The "mind", however defined, chooses to articulate one or more sounds. The thought must be transcribed in the neural network such that the correct motor neurons are transmitted through the nervous system to activate the appropriate organs to produce the sounds. In modern language, speech is created with pulmonary pressure provided by the lungs that generates sound by phonation through the glottis in the larynx, that then is modified by the vocal tract into different vowels and consonants. In earlier times, grunts and the like were less complex, but the same *systems approach* must be considered: coordination of lungs and whatever organs were involved, in making any type of sound whatsoever. We could posit that the first sounds were involuntary, and that the organism then experimented, but this changes the problem only in degree. Up in the far reaches of the mind, the organism formulated a thought of what it wanted to achieve, and the neural network set about structuring itself to produce an output which it had never before produced, although it could be said that it had a template from the original involuntary sound.

Sounds have a pattern, which can be detected by the ear of the emitting organism (bio-feedback), but if the initiation was a conceptual pattern in the mind, how was it conveyed to the brain as to what was wanted? Again, I could accept trial and error, but if thousands of synaptic connections were required to be established, with just the right voltage and chemical settings, how long would that take? We know that animals, and people, are born with the intrinsic ability to make vocalisations, but given the complexity of the process, I nevertheless struggle to conceive of evolution being able to accomplish it. It is reasonable to conclude that we are born with pre-set synaptic patterns on which new patterns could be modelled. As scientists working in the field of neural network modelling have discovered, getting started is the hard part, but even more, validating that the correct answer has been achieved is pivotal to success. Noting that in neural network modelling, the right answer is pre-specified to validate the model, the question becomes: Without the oversight of human intelligence, how could an evolving brain "know" that it had the right answer?

"To train a network to predict or classify inputs correctly, Widrow and Hoff noted that there needs to be an external signal, distinct from the network's actual response, that specifies the desired output. This signal [is] often called a **teaching input**. Given such a teaching input, the network can compare its own output against the desired output, and then adjust the weights so that, in future, errors are less likely."[4] [emphasis in original]

In evolution theory, there could be no *teaching input*, without which, how could it "learn" that its interpretation was correct? Trial and error is an improbable process for evolution, for the only mechanism that could answer the question of truth or error, is natural selection. Selection cannot operate at the microbiological level of a single neuron, and it is problematic that it could operate on any small segment of a neural network.

A fascinating revelation from research into neural network modelling and AI, is the improbability of evolution being able to accomplish a level of "intelligence" that even the best evolved intelligences cannot match.

The other question which puzzles me is this. Consider that as a communications channel, a telephone relies on two mediums: sound waves and electronic signalling, each of which has its own standardised protocol. When considering sounds both sent and received by humans, two separate communication channels are involved: motor neurons to the organs of speech, and sensory neurons from the sensor (ears). In between, the *Central Nervous System* acts as a signal processor. When a singer experiments with new vocalisations, such as in a cappella, the new sound is imagined, vocalised, and then the listener hears what was imagined. This means that the neural connections that enabled the vocalisation, were mirrored by the connections in the brain of another person as they processed the sensory input. Fascinating, most especially how that could happen via evolution.

As earlier noted, the receiving neuron integrates all the information it receives from all its afferents (inputs) to determine its output. According to Gluck, "the rule for computing output response is a weighted sum."[5] With but a limited understanding of the science, I would assume that the neural response is similar. In explaining how neural networks are *trained*, the author notes that with repeated trials, errors should gradually reduce to zero. Referring to individual connections:

"... the blame for an incorrect response or the credit for a correct response lies in the weights. But there may be many weights in the network, and not all of them may deserve blame or credit equally. This is a fundamental problem for training neural networks and is often referred to as a credit assignment problem because the issue is how to assign credit (or blame) for a network's output performance to the weights in the network ... the approach is called *error-correction learning* because it specifies that each weight should be adjusted according to its particular contribution to the total network error."[5] (italics in original)

There are about 86 billion neurons in the adult human brain, with anywhere from 100 to 1,000 trillion connections, depending on the opinion of the experts. Each connection has a "weight" determined by chemical and physical properties. The trials described above were on networks of just a few thousand connections, and often less to be able to focus on specific problems. Evolutionists would have us believe that, absent of deleterious mutations, evolution managed to resolve the weighting issue on a grand scale, even without a "teaching input". Even starting from a lower level of complexity, the issue of validation of results remains unsolved. There was no *desired* outcome, and in the sensory system, no way of validating that the signal processor (the brain) correctly interpreted the inputs (Primary Axiom #6).

Returning to the proposition that *synaptic strengths are learned through repeated exposures to regularities in the environment*, studies by Pavlov, Kamin, and others on animals have attempted to understand the relationship between Unconditioned and Conditioned stimuli (US and CS). Most are probably familiar with some of Pavlov's experiments with dogs. Further studies have concluded that "Apparently, co-occurrence alone is not sufficient for learning. Instead, blocking suggests that *learning only occurs when the CS is a useful predictor of the US*"[6] (italics in original). Consider that such conclusions relate to what I might describe as macro-behaviour, contrasted with micro-behaviour that might occur when individual neurons, or small multiples of neurons, undergo changes in synaptic strength. If these small changes do not result in useful predictions, then they could not contribute to learning. Just where the critical mass is achieved is anyone's guess. Further on the subject of conditioning, a form of learning, and without detailing the experiments on the blocking effect,

"The blocking effect exposed a challenge for simple theories of classical conditioning. It suggested that cues do not merely acquire strength on the basis of their individual relationships with the US; rather, *cues appear to compete with one another for associative strength* ... This study and others like it led to a growing view among psychologists that to produce effective conditioning, a stimulus cue (CS) must impart reliable information about the expected occurrence of the outcome (US). Moreover, even if a given cue is predictive of a US, it might not become conditioned if its usefulness has been pre-empted by another co-occurring cue that is a better predictor of the US or that has a longer history of successfully predicting the US. The result of all these findings was to show that 'simple" Pavlovian conditioning was not nearly as simple as had once been thought!"[7] (italics in original)

Taking those conclusions back to the premise that *synaptic strengths are learned through repeated exposures to regularities in the environment*, and accepting that conditioning and learning are synonymous, it becomes ever more difficult to accept that this proposition has any validity. The complications for evolutionary development are beyond comprehension.

Pattern Recognition

Another approach to neural network modelling, which may better represent how neural networks in the brain might be formed, is given here[8]. Quoting from that article:

"Neural networks are a set of algorithms, modelled loosely after the human brain, that are designed to recognize patterns. They interpret sensory data through a kind of machine perception, labelling or clustering raw input. The patterns they recognize are numerical, contained in vectors, into which all real-world data, be it images, sound, text or time series, must be translated."

Whilst the article refers to "cluster and classify", I would think that logically, one must classify before clustering. It also states that this is a layer "on top of the data you store and manage", which is true in the context as given, but how could that be instantiated in the human brain? The article is very informative, and I would recommend it, but my issue is not how scientists are able to do this with computers, but how that is relevant in the context of the human brain. We must ever be wary of transferred epithets.

If you do read the article, note the mention of *labelled datasets* and *the correlation between labels and data*. These statements correspond to my earlier discussion where I stated: "entities are identified by names assigned to them by intelligent agencies – the entity cares not a whit whether it has a name or not. Whilst "a name" is a physical entity instantiated symbolically, "name of" is not a physical entity that exists in reality – it is just a conceptual label that we apply to an entity." The author of the discussion above has it right: Deep learning does not require labels to detect similarities, because labels are arbitrarily assigned by humans, and are not the entities involved in the processing. The same must be said for the processing in the human brain: the brain cares not whether the encoded symbolic representations of data have names or not. However, deep learning still requires algorithms for detecting similarities, for as curious as it may seem, not all similarities are logically similar. I have experience with this in my work on Business Intelligence applications, where context still remains critical. What is not explicitly stated in the quoted article is that you cannot just feed the model with data without expressing the context of that data – no point in analysing digital voice data if the intention was to recognise faces.

In deep learning, the outcomes are eventually abstracted to data names for human interpretation, based on the data dictionary compiled in the first place – humans are not expected to interpret the learnings at the level of processing: symbolic representations. This poses a problem for our understanding of deep learning in the brain, as it works at the symbolic physical level, not at the abstracted level of data names which humans understand. An obvious problem from evolutionists and material monists is: How does the mind-brain complex "explain" the workings of the brain to the owner, if the mind is of the same substance, structure, and type of processing as the material brain? Further, if the intended output is *visualisation*, this being the case for our sense of sight, what is the process for converting electro-chemical signals into pictures as viewed by the "mind's eye"?

The single biggest issue regarding pattern recognition in the brain is the assertion that neural networks are organised in patterns. Patterns cannot explain themselves (Primary Axiom #1), most especially in conceptual terms that the patterns are meant to represent. If the entire neural network complex of the brain is organised in patterns, what is left to interpret the patterns? Not only can patterns not explain themselves, they cannot explain other patterns structured similarly to themselves. It is all very well to scoff at the concept of the *Ghost in the Machine*, but on the evidence, it most likely exists.

Speech Decoding Technology

Fascinating progress has been made in artificially translating brain activity into speech, leading to technologies that would enable the mute to "speak out loud", as it were, far more comprehensibly than in earlier synthesizers as used by the late Stephen Hawking. Before getting too excited however, the technology does not involve interpreting the *conceptual* from the *physical* neural activity in the brain. Rather, "They then trained a machine learning algorithm to be able to match patterns of electrical activity in the brain with the vocal movements this would produce, such as pressing the lips together, tightening vocal cords and shifting the tip of the tongue to the roof of the mouth. They describe the technology as a "virtual vocal tract" that can be controlled directly by the brain to produce a synthetic approximation of a person's voice."[9] In other words, the researchers were still operating within the biology of the central nervous system, a field of ever-increasing understanding.

This gets us no closer to understanding the mind-brain conundrum, as the thoughts behind the intended speech still remain a mystery. In some ways, the technology is similar to that used in developing bionic limbs and hands. The researchers have been able to identify the behaviour of motor neurons in the nerves serving the particular body part, and mimic them synthetically. This medical research is proving of immense benefit to the human condition. The issue for me in this study, however, is how the cognitive activity as the root cause of decision making in both speech and movement, is translated into the neural activity that then actions those thoughts through the central and peripheral nervous systems.

That connection remains a mystery.

Summary

Note the significant statement above: "All classification tasks depend upon labelled datasets; that is, **humans must transfer their knowledge to the dataset** in order for a neural network to learn the correlation between labels and data." [emphasis mine] This is the key problem for evolution: How could "knowledge" be transferred to the brain's neural network when knowledge is conceptual, yet the network is only a biological symbolic representation? As per Primary Axiom #2, *all knowledge is built upon prior knowledge,* but the embryonic brain of the evolution narrative had none to start with. Note also the statement in relation to the first method of neural network modelling we discussed: "The key to learning in neural networks is changing the weights between nodes." This is a task that only human intelligence can perform, yet evolutionists would have us believe that effectively, human intelligence evolved from human intelligence! Perhaps that is a misstatement, but what would be the biological mechanism in neural networks that could achieve the equivalent of *changing the weights between nodes* in order to achieve a predetermined output? In the context of evolution, there is never a predetermined or correct result!

Developing computing techniques based on how it is believed that the brain's neural network functions is paying great dividends. Many useful applications are being derived to serve human interests, but one must be wary of *transferred epithets*. Neural network modelling is based on how it is *believed* that neural networks operate, but then additional functionality is provided to make the system work as intended. What seems to be happening, as it so often does, is that anthropomorphic conceptions are applied to the mechanistic, but then the enhancements to the mechanistic are read back into the anthropomorphic without substantiation. The complexity of the research and experimentation should raise serious doubts that the model upon which this activity is based could have arisen via undirected evolution.

It simply does not sound plausible.

References:

1. Gluck, Mark A., and Myers, Catherine E., *Gateway to Memory: An Introduction to Neural Network Modeling*, The MOT press, Cambridge, MA, 2001, p. 49

2. *Ibid*, p. 51

3. Widrow, B., and Hoff, M., *Adaptive Switching Circuits*, Institute of radio Engineers, Western Electronic Show and Convention Record (1960), 4, pp. 96-104

4. Gluck, *Ibid*, p. 52

5. *Ibid*, p. 55

6. *Ibid*, p. 63, citing, Kamin L. (1969). Predictability, surprise, attention and conditioning. In B. Campbell & R. Church (Eds.), *Punishment and Aversive Behaviour*, pp. 279-296), New York, Appleton-Century-Crofts.

7. *Ibid*, p. 65

8. https://pathmind.com/wiki/neural-network

9. https://www.theguardian.com/science/2019/apr/24/scientists-create-decoder-to-turn-brain-activity-into-speech-parkinsons-als-throat-cancer

Part 7: The Sensory Messaging System

"The strong man is the one who is able to intercept at will, the communication between the senses and the mind."

~ Napoleon Bonaparte ~

Curious, is it not, that a military dictator understood this relationship? Even more, he confirmed his belief in free will, and the ability of a "strong man" to not respond automatically to sensations detected by our sensory system.

I recall reading in one of Richard Dawkins' books that some sight is better than no sight. As written, this is demonstrably false. One might offer that some hearing is better than no hearing, or some sense of smell is better than no sense of smell, but if you have ever visited a large dairy farm on a hot day, you would question the latter. Similarly, if you suffer from severe tinnitus, or have your car radio turned up too loud as you get to the extreme range of your favourite radio station, and all you experience is static, you would realise that in some cases, no hearing is better than some. As for sight, we cannot know how or when the sensory signal processor "understood" than an image through a lens is inverted, but before that happened, how did the critters not fall down holes or off cliffs, thinking that they were going up not down? The issue is the *quality* of the sensory experience, as perceived by the mind-brain complex. The question for evolution is: How did the process of verification and validation occur biologically (see Chapter 4-5)? I cannot know, but in all probability, early sensory development would have resulted in "noise" rather than a validated, quality sensory experience, raising the question of why any such faculty was continued. Let me repeat that for emphasis: any early sensory experience, especially in the visual and auditory systems, could only have been **noise**, not a faithful reproduction of what was experienced. Richard Dawkins' proposition is demonstrably false.

We should continue to be aware that between reality, and our sense of it, is a complex sensory communications system which to be reliable, needs to be verified and validated by an agency external to the system itself (Primary Axiom #6). Our interpretation and identification of physical realities is conceptual, not physical. Consider the concept of colour. Colour does not exist as a physical reality – there is no such entity as red, or blue, or green, other than as words. Technically speaking, our sight is constrained to the visible light spectrum - that portion of the electromagnetic spectrum (approximately 380 to 740 nanometres) to which the human eye is sensitive. What we perceive as deep blue is around 400 nanometres; red is 700. Just who decided that around 400 is "blue" and not "yum", and 700 is "red" and not "yuk", is beyond my knowing. But the point should be clear. As I explained earlier concerning entities, properties, and activities – the latter represent reality, but how we perceive them is conceptual. Philosophically, one can ask: Well, how do we know that our perception matches reality? A good question, one which I choose to not pursue here, other than to respond with my own question: What evolutionary processes could have resulted in such marvellously reliable, sensory information systems whereby our perception *could* match reality?

In this section, I will discuss some technical aspects of our *auditory* and *visual* systems. The reason that these two systems are our special focus here is that it is through these two systems that we learn. I cannot conceive of how you could teach someone who is both totally blind and totally deaf. It is through our ears and eyes that we learn of concepts, and how these relate to the real world. Therefore, the messaging in these two systems is very different to the somatosensory systems, and internal biological communications. You may recall that in general, nerves in the PNS and CNS do not physically carry messages, neuron to neuron. Neurons talk to the next in line simply to say, "relay the message", without "knowing" or relaying the message itself. The message is predefined by the channel dedicated to that specific message. As the primary theme of this book concerns the mind-brain conundrum, then we must deal with how the physical is relayed to the brain to be conceptualised by the mind.

My objective is not to offer a treatise on these subjects, but simply to reveal the complexity of such systems, and as ever, question their origins. Even in their astonishingly perfected state as we know them, all senses dependent upon the Peripheral and Central Nervous Systems occasionally fail due to "line faults", and other errors in signal processing. Early last year, I developed a debilitating bout of *brachial neuritis*[1], a type of peripheral neuropathy that affected both my shoulders and my lower neck. This disease was characterized by pain, very often extreme in the early onset, and a loss of function in the nerves carrying signals to and from my brain and spinal cord (the central nervous system). This nerve fault is of no known origin, nor unfortunately, cure. The best that one could hope for was that it would go away in its own good time, but that could take months or even years. In my case, it has continued for eighteen months and in the early stages, I was functionally incompetent both physically and mentally. But this got me thinking about diseases during evolution. I cannot recall ever reading of diseases in the context of evolution. It seems logical to me that without the benefits of modern medicine, earlier species would have been susceptible to diseases, particularly as viruses can mutate rapidly and overcome natural resistance. Witness the current COVID-19 pandemic. Of course, animals have access to natural remedies in foliage and plants, which could have ameliorated the problem, but still. It is generally taught that dinosaurs were wiped out by the effects of volcanic activity or an asteroid striking the planet – nobody truly knows – but could their demise have been caused by a disease, especially one affecting the nervous system? If mutations beneficial to survival and reproduction were the foundation for evolution, would that not have been the case for viruses and bacteria which whilst beneficial for them, were deleterious for other species? A thought to ponder.

To reiterate, what this neuritis suggested to me, from an evolutionary perspective, is why should we think that in the early development of the nervous system, it all went well? In the previous section, we reviewed what is known about the biological construction of our nervous system, so that we could understand not just the engineering difficulties, but those associated with developing a "noise free" communications channel. I can personally attest that my nervous system was anything but "noise free"!

In that context, we should consider the difficulties faced by Guglielmo Giovanni Maria Marconi, inventor and electrical engineer, known for his pioneering work on long-distance radio transmission and the radio telegraph system. A brief review is revealing, but I will leave that to you should you be so interested. Telegraphs were notoriously unreliable during not infrequent storms and atmospheric electrical activity. We might consider the implications of whether early nervous systems were likewise susceptible to electrical interference, especially if myelin sheathing was a much later development. Now that would be challenging.

Telemetry Systems

You probably haven't thought of the human body in this context, but we are more "wired" with sensors than the Lockheed Martin F-35 Lightning II, or the Boeing 787 Dreamliner, during their development and acceptance testing. Most modern vehicles of all types contain sensors to monitor and log critical aspects of their performance, from engines to frame to instrumentation. In some cases, they are capable of limited "repair" when redundancy is built in, but otherwise, the vehicle or aeroplane is plugged into another device to download the logged data from which an analysis can be performed, and maintenance indicated. The technology behind this is astounding, and has added greatly to the reliability of modern transport.

Calibration

Anybody who has worked in instrumentation is well aware of the need for accurate calibration of instruments. I remember assisting in compass swings on Douglas DC-3 aircraft, a process of checking the magnetic compass in the aircraft against accurate markings on a hard stand. The metal in the aircraft could affect compass accuracy, and with magnetic variation taken into account, a compass correction card was affixed next to the instrument. The same was true of airspeed indicators. I recall learning the difference between indicated, rectified, and true airspeed. If you used the wrong one in air navigation, you would likely be late for dinner. In more modern times, hand-held radar guns used by the police to check vehicle speeds have been called into question due to their questionable accuracy. It would appear that regular calibration had not been performed.

Now, the interesting aspect of instrument calibration is that one needs a reliably accurate method for performing the calibration – instruments cannot calibrate themselves (Primary Axiom #6). Sensors developed for modern technological applications are designed to monitor specific conditions, such as distortion or heat stress, for which there must be standards and allowable variations established. Few instruments will perform to maximum efficiency in all conditions, and thus error limitations are evaluated and included in the design brief. Even earlier determinations of "standards" included specifications of environmental conditions, particularly temperature and humidity. For example, this book[2] describes the struggles and eventual success of designing a time-keeping device that maintained accuracy during a sea voyage from England to America, despite the varying climatic conditions encountered. Whilst latitude could be determined from the angle of the sun and other celestial bodies, longitude required an accurate measurement of the time since leaving a known location, without which reliable navigation could not be accomplished. Another example: simple rulers of wood and metal would be of slightly different lengths if cut under substantially varying temperatures. As an aside, that is my excuse for being inept at carpentry. But still the question: how does one calibrate the calibration equipment? I will leave that subject there for you to think about in the context of evolution: what was the calibration process for the myriad sensor applications in the human body?

I will contend that modern advanced technology is overshadowed by what evolution is claimed to have accomplished with its biological telemetry system. Such a system makes modern technology primitive by comparison. The next chapters discuss the detail which has me doubting that undirected evolution is capable of such achievements.

References:

1. https://www.hopkinsmedicine.org/health/conditions-and-diseases/brachial-neuritis

2. Sobel, Dava, *Longitude*, Fourth Estate, Harper Collins, UK, 1998

Chapter 7-1: Human Telemetry System

"The thing I'm most interested in is the nervous system. How do brains grow? How do genes build complicated nervous systems?"

~ Sydney Brenner, South African biologist, Nobel Prize in Physiology 2002 ~

We are wired with a nerve structure fed from thousands of receptors throughout our bodies. Take a moment to think of the implications of that in the context of undirected evolution underpinned by random genetic mutations. I cannot begin to imagine the sequence of development, let alone how the integrity of the system could have been verified. So, let us have a look at some biological details. Before beginning, the terms *receptor* and *sensor* are synonymous in this context. A sensor, in this case biological, is a device that detects and responds to inputs from the physical environment, both internal and external. Internally, the sensor primarily detects chemical changes, but externally, the specific input could be light, heat, motion, moisture, pressure, friction, or any one of a great number of other environmental phenomena. Sensors are also termed, transducers, because they convert (transduce) energy from one form to another.

In physiology, sensory transduction is the conversion of a sensory stimulus from one energy form to another. Transduction in the nervous system typically refers to stimulus-alerting events wherein a physical stimulus is converted into an action potential, which is transmitted along axons towards the central nervous system for integration. It is a step in the larger process of sensory processing. Somatosensory cells convert the energy in a stimulus into an electrical signal.

Receptor Types

Receptors in the body are cells that interpret both our internal and external environments. They are neurons of three types: (1) that have a free nerve ending, with dendrites embedded in tissue that would receive a sensation, e.g., pain and temperature receptors in the dermis of the skin; (2) neurons that have an encapsulated ending in which the sensory nerve endings are encapsulated in connective tissue that enhances their sensitivity, e.g., lamellated corpuscles also in the dermis which respond to pressure and touch; and (3), a specialized receptor cell which has distinct structural components that interpret a specific type of stimulus, e.g., cells in the retina that respond to light stimuli (photoreceptors). The latter are of particular interest in this study as they release neurotransmitters onto bipolar cells, which then synapse with the neurons in biologically predetermined nerve fibres.

A second way we can study receptors is based on their location relative to the stimuli. For example, an *exteroceptor* is a receptor that is located near a stimulus in the external environment, such as the somatosensory receptors that are located in the skin. An *interoceptor* is one that interprets stimuli from internal organs and tissues, such as the receptors that sense the increase in blood pressure in the aorta or carotid sinus. A *proprioceptor* is a receptor located near a moving part of the body, such as a muscle, that interprets the positions of the tissues as they move. In summary, we have sensors (receptors) in all parts of our bodies providing telemetry on our internal functioning and how external factors are influencing our well-being.

During research for this subject, I encountered a remarkable organ called the *Golgi Tendon Organ*, and as you may find it as fascinating as I have, we shall take a brief diversion to examine it. "The golgi tendon organ is a proprioceptor, sense organ that receives information from the tendon, that senses TENSION. When you lift weights, the golgi tendon organ is the sense organ that tells you how much tension the muscle is exerting. If there is too much muscle tension the golgi tendon organ will inhibit the muscle from creating any force (via a reflex arc), thus protecting the you from injuring itself."[1] That may be all very well, but why is it that sometimes (ok, too often) I overstretch and do injure myself. That aside, I find the images on this website having me again question how evolution could have accomplished such fine specificity as in these organs (see also the Muscle Spindle).

Golgi Tendon Organ

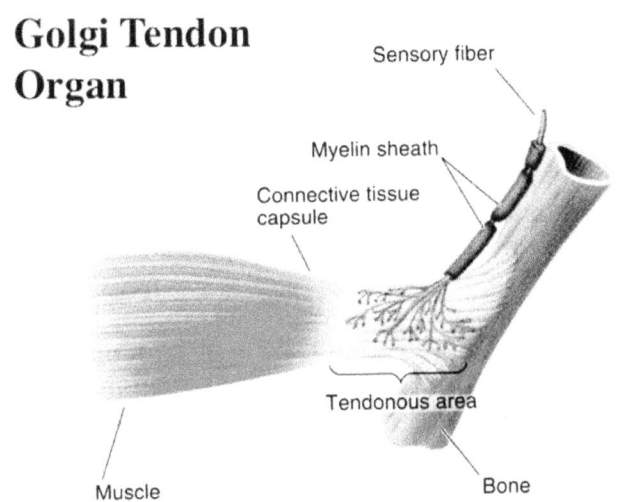

Yet another way of looking at receptors, one which is of particular interest to me from a technical perspective, is how they function as *transducers*. A transducer is a device that converts energy, usually a signal, from one form to another. For example, in the application of sonar, a transducer is the device that converts electrical energy into a sonic pulse that is "pinged" to detect the presence of other objects. In biology, the receptor transduces stimuli into membrane potential changes. Now to get just a tad technical, stimuli are of three general types. Some stimuli are ions and macromolecules that affect transmembrane receptor proteins when these chemicals diffuse across the cell membrane. Some stimuli are physical variations in the environment that affect receptor cell membrane potentials. Other stimuli include the electromagnetic radiation from visible light. For humans, the only electromagnetic energy that is perceived by our eyes is visible light. Some other organisms have receptors that humans lack, such as the heat sensors of snakes, the ultraviolet light sensors of bees, or magnetic receptors in migratory birds. I am not entirely confident of the latter, primarily because the magnetic north pole wanders around a bit, and the poles have been shown to have reversed. Would that have northern hemisphere birds flying *north* for the winter? (just joking) In my brief experience of air navigation, magnetic variation was an important consideration lest I guided the pilot to the wrong destination. I do wonder how birds can be aware of these variations and make the necessary allowances, generation to generation. For example, using the much valued 1:60 rule from air navigation, a 1° error over 60 miles would have you off-course by 1 mile. Therefore, a 1° change in magnetic variation over 3,600 miles would result in the birds landing 60 miles from their intended destination! There must be other factors at play.

In modern technological applications, transducers are designed for different types of energy transformation, based on the input stimulus. Thus, we have detectors for carbon dioxide, carbon monoxide, nuclear radiation, sound waves, light waves, heat, etc. Our biological transducers follow the same approach. The aptly named *chemoreceptor* interprets chemical stimuli, such as an object's taste or smell. *Osmoreceptors* respond to solute concentrations of body fluids. Interestingly, although pain is perceived as a nerve issue, it is primarily a chemical sense that interprets the presence of chemicals from tissue damage, or similar intense stimuli, through a *nociceptor*. Physical stimuli, such as pressure and vibration, as well as the sensation of sound and body position (balance), are interpreted through a *mechanoreceptor*. Another physical stimulus that has its own type of receptor is temperature, which is sensed, not surprisingly, by a *thermoreceptor* that is sensitive to either temperatures above or below normal body temperature.

Interpretation of Encoded Output

This is where the evolution hypothesis becomes extremely problematic. Receptors (biological transducers) convert energy from one form to another. For example, heat is the transfer of thermal energy from a body (object, air, water) of higher temperature to some part of our body at a lower temperature, whereas cold is the reverse. The temperature variation energises the relevant neurons to transmit a signal to the brain, which in turn interprets the signal in terms of any necessary response, such as sending motor neurons to the appropriate muscles or tendons. I am aware that many do not accept the term, *encoded*, to refer to electro-chemical signals travelling throughout our nervous system, but this online entry might help to convince you:

"Each of the senses is referred to as a *sensory modality*. Modality refers to the way that information is **encoded**, which is similar to the idea of transduction. The main sensory modalities can be described on the basis of how each is transduced. The chemical senses are taste and smell. The general sense that is usually referred to as touch includes chemical sensation in the form of nociception, or pain. Pressure, vibration, muscle stretch, and the movement of hair by an external stimulus, are all sensed by mechanoreceptors. Hearing and balance are also sensed by mechanoreceptors. Finally, vision involves the activation of photoreceptors."[2] [emphasis mine]

At the detailed functional level, transduction involves two processes. At each stage, the transducer must be capable of "understanding" the nature of the energy with which it is intended to deal. For example, a diaphragm may be designed to transduce sound waves, but it will not transduce light waves. Additionally, its efficiency may be subject to variations in heat, light, moisture, a range of electromagnetic radiation or even sound waves which are outside its design brief. Firstly, a transducer must *decode* the input energy according to its characteristics, and *re-encode* that energy based on its own physical characteristics, before it can further encode the signal into another energy form for downstream transmission of the message. It is true that these decode-encode processes occur simultaneously, but I mention this to emphasise that there are three forms of energy that are encoded in a manner appropriate to their form. A sound wave has its own encoded pattern of frequency, wavelength, amplitude, etc., and when it strikes the transducer such as a diaphragm, the diaphragm will vibrate in a particular way, interpreting (decoding) the input signal in terms of its encoded pattern. That is why some diaphragms are better than others in handling a range of frequencies and amplitudes. The energy of the sound wave is converted into mechanical energy in the diaphragm itself, again in a regulated (encoded) manner, depending on how it was interpreted (decoded). If the shape of the diaphragm is not designed correctly, or is deformed, the decoding will not be true to the input signal, resulting in poor encoding of the output to the third stage – conversion to electrical energy. This is done by the voice coil and magnet.

Remember this principle for all that follows. The technical specifications of transducers need to be very precise for the intended purpose, if energy conversions are to be faithful to the semantic content of the input message, and subsequently faithfully retransmitted with the encoding required of the next medium. In modern technologies, the transducing mechanism is specifically designed for the energy modes to be transduced. The biological sensors (transducers) in the human body operate at superfine levels of electrical and chemical specificity. In a laboratory in earlier days, these levels would have been very difficult to obtain, and then maintained for whatever experiments needed to be performed. Think about that in the context of evolution and random genetic mutation.

The purpose of this chapter is to make you aware of the complexity, and superfine specificity, of the thousands of receptors (transducers) throughout the human body. The technical specifications for the transduction and subsequent faithful transmission through the nervous system, of the many types of messages, would fill volumes if documented. But then, of course, we have the problem of how the mind-brain complex managed to learn what those messages meant.

Earlier I mentioned Werner Gitt's five-layer model of communication, but the one of most relevance here is *Semantics* – a common understanding of the idea / meaning of the communication. Whilst transducers (sensors) transform energy from one form to another, they must also faithfully reproduce the semantic layer. This is why the specificity of the decoding / encoding is so critical to the performance of sensors, lest the wrong message is sent. Again, I question how evolution could have continued with a less than accurate transmission of the semantic layer, unless of course, it just managed by chance to get it right from the outset, a highly improbable occurrence.

References:

1. https://www.unm.edu/~lkravitz/Exercise%20Phys/spindleGTO.html

2. https://opentextbc.ca/anatomyandphysiology/chapter/14-1-sensory-perception/

Chapter 7-2: The Auditory System

"With the sense of sight, the idea communicates the emotion, whereas, with sound, the emotion communicates the idea, which is more direct and therefore more powerful."

~ Alfred North Whitehead, English philosopher, definer of process philosophy ~"

William Sanford Nye, popularly known as Bill Nye the Science Guy, an American science communicator, television presenter, and mechanical engineer, observed: *"One of the drawbacks of English is you can't spell things by hearing them."* What I like about these observations is the acceptance of sight and hearing as being primarily about ideas, which are conceptual not physical. I have quoted them because they are so pertinent to our discussion here, evidencing as they do, that what we sense physically through our auditory system does not directly equate to the concepts behind the sounds. A great deal of intelligent signal processing must be performed, and even then, it is so easy to get it wrong. As I will continue to stress, the physical brain is incapable of such activities because no arrangement of chemicals is capable. As we proceed through the complexity of the auditory system, keep in mind that for undirected evolution to be true, each of the components must have evolved, either concurrently or in an appropriate sequence, with sufficient capability to warrant selection. Sounds which approximated the noise of a poorly tuned radio, or a bad case of tinnitus, would hardly provide survival advantage.

For the following discussion, I wish to acknowledge the Open Textbook[1] project which has made so much material available for people to use without prior permission, without which I would be required to obtain numerous permissions for other copyright material. My purpose in referring to this technical text, here much paraphrased by me, is no more than to expose the complexity of the auditory system, leaving the reader to contemplate how undirected evolution could be responsible. I would ask the reader to keep in mind that some terminology correctly *identifies* a process, without explaining the how of it. The "how" is the most important aspect of the process, as discussed in the previous chapter on receptors and transducers.

Hearing, or audition, is the transduction of sound waves into a neural signal that is made possible by the structures of the ear, as shown below.

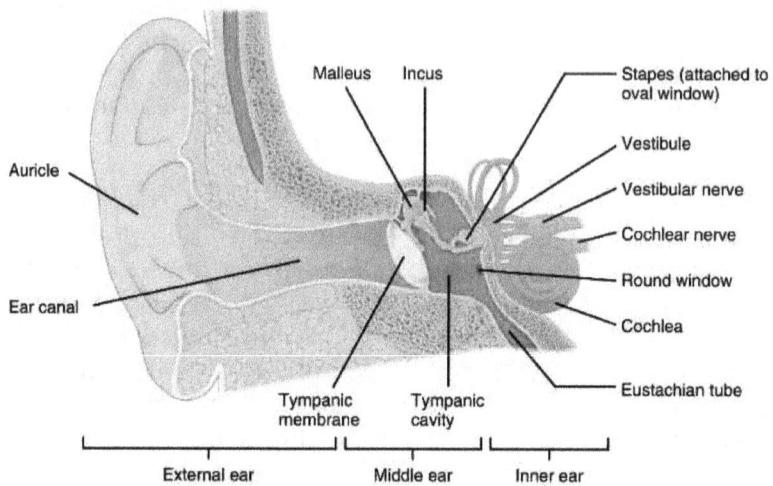

The semi-rigid protuberance that sticks out from the side of our head is known as the *auricle*. The concave C-shaped curves of the auricle direct sound waves toward the auditory canal, much as the dish of a radio telescope focuses and amplifies radio waves from space onto the antenna horn. Quoting the text,

"The canal enters the skull through the external auditory meatus of the temporal bone. At the end of the auditory canal is the *tympanic membrane*, or ear drum, which vibrates after it is struck by sound waves. The auricle, ear canal, and tympanic membrane are often referred to as the *external ear*. The *middle ear* consists of a space spanned by three small bones called the *ossicles*. The three ossicles are the *malleus, incus,* and *stapes*, which are Latin names that roughly translate to hammer, anvil, and stirrup. The malleus is attached to the tympanic membrane and articulates with the incus. The incus, in turn, articulates with the stapes. The stapes is then attached to the *inner ear*, where the sound waves will be transduced into a neural signal. The middle ear is connected to the pharynx through the Eustachian tube, which helps equilibrate air pressure across the tympanic membrane."

The text is an example of the "what" in mechanistic terms, omitting the "how" in conceptual terms, which is the theme of this chapter. Technically, this quoted statement omits the essential detail: *"where the sound waves will be transduced into a neural signal"*. In the middle ear, the energy of the sound waves has already been transduced into mechanical energy in the tympanic membrane, and subsequently transferred into the mechanical energy that moves the ossicles. Digging ever deeper, there is a further transfer of mechanical energy from one ossicle to another. Here, the shape and geometry of the ossicles is critical to the faithful transmission of the message content of the energy. I am not a mechanical engineer, but I can imagine the work required to design such a system of levers. The following diagram further evidences the complexity of the design:

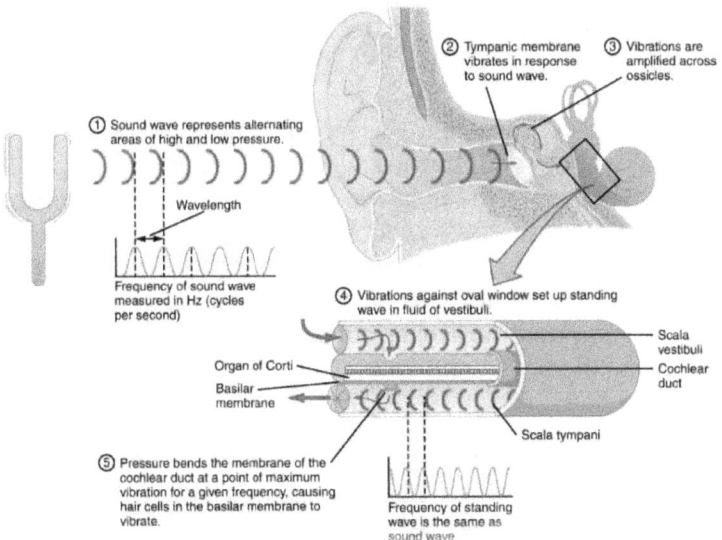

I would direct your attention to note (5), and the comment further to the right: "Frequency of standing wave is the same as the sound wave". There is energy transduction of the sound wave through the tympanic membrane, then through the ossicles, and again through the oval window sending pressure waves through the membrane of the cochlear duct, resulting in a faithful replication of the frequency of the original sound waves. As an engineering exercise, this would require painstaking experimentation to achieve the requisite level of accuracy. Let me again quote from the references article lest I have misrepresented the process:

"The inner ear is often described as a bony labyrinth, as it is composed of a series of canals embedded within the temporal bone. It has two separate regions, the *cochlea* and the *vestibule*, which are responsible for hearing and balance, respectively. The neural signals from these two regions are relayed to the brain stem through separate fiber bundles. However, these two distinct bundles travel together from the inner ear to the brain stem as the vestibulocochlear nerve. Sound is transduced into neural signals within the cochlear region of the inner ear, which contains the sensory neurons of the *spiral ganglia*. These ganglia are located within the spiral-shaped cochlea of the inner ear. The cochlea is attached to the stapes through the *oval window*.

The oval window is located at the beginning of a fluid-filled tube within the cochlea called the *scala vestibuli*. The scala vestibuli extends from the oval window, travelling above the *cochlear duct*, which is the central cavity of the cochlea that contains the sound-transducing neurons. At the uppermost tip of the cochlea, the scala vestibuli curves over the top of the cochlear duct. The fluid-filled tube, now called the *scala tympani*, returns to the base of the cochlea, this time travelling under the cochlear duct. The scala tympani ends at the *round window*, which is covered by a membrane that contains the fluid within the scala. As vibrations of the ossicles travel through the oval window, the fluid of the scala vestibuli and scala tympani moves in a wave-like motion. The frequencies of the fluid waves match the frequencies of the sound waves. The membrane covering the round window will bulge out or pucker in with the movement of the fluid within the scala tympani."

Intriguing is it not, how these several levels of energy transduction became so finely tuned that *the frequencies of the fluid waves match the frequencies of the sound waves* - exactly. From my limited understanding of electro-mechanical engineering, the most intricate part of the system is the transduction of the mechanical energy from the moving hair cells to neural signals, as described here:

"The organs of Corti contain *hair cells*, which are named for the hair-like *stereocilia* extending from the cell's apical surfaces. The stereocilia are an array of microvilli-like structures arranged from tallest to shortest. Protein fibers tether adjacent hairs together within each array, such that the array will bend in response to movements of the basilar membrane. The stereocilia extend up from the hair cells to the overlying *tectorial membrane*, which is attached medially to the organ of Corti. When the pressure waves from the scala move the basilar membrane, the tectorial membrane slides across the stereocilia. This bends the stereocilia either toward or away from the tallest member of each array. When the stereocilia bend toward the tallest member of their array, tension in the protein tethers opens ion channels in the hair cell membrane. This will depolarize the hair cell membrane, triggering nerve impulses that travel down the afferent nerve fibers attached to the hair cells. When the stereocilia bend toward the shortest member of their array, the tension on the tethers slackens and the ion channels close. When no sound is present, and the stereocilia are standing straight, a small amount of tension still exists on the tethers, keeping the membrane potential of the hair cell slightly depolarized."

Incredibly, to my mind at least, there are approximately 30,000 nerve fibres in the human auditory nerve, and about 50,000 in that of a cat. These fibres are attached to the hair cells, transmitting a signal of the corresponding frequency to the auditory cortex via the cochlea nerve. The neurons in the nerve fibres are a little different to neurons in other nerve systems, but they function roughly the same in that an action potential ripples from neuron to neuron, so again, there is no messaging *per se*. The "message" of each nerve fibre is a function of the identity of the nerve fibre itself, which only relates to a specific frequency. This is not precisely correct, but conveys the general sense of how the auditory signalling occurs. To correct the explanations given in some online sources, the auditory nerves in each ear do NOT send "sound vibrations" to the cochlear nucleus; nerves do not vibrate. The cochlear nuclei do NOT process the "sound information" received and organize it according to pitch, duration, and intensity — I am unaware of how this process occurs, but it is based on signals received via individual nerve fibres: information is only derived by correlation of the meaning of those signals. In some unexplained manner, these signals are sent to different parts of the brain for further processing and interpretation. It is believed that the auditory cortex is what enables us to understand a conversation. The technology of the auditory cortex is quite amazing, for it is said to analyze time-varying signals in what the ears pick up, allowing us to perceive individual syllables and words, and thus process speech. All along, we must never forget that sending and receiving speech involves the conversion / interpretation of physical signalling in nerve fibres to the conceptual aspects of language. No plausible explanation has been offered for how this works.

I cannot help but marvel at the scientific research which has taken us so far into the biology and physiology of hearing. On the other hand, I must remark on the technical inexactitudes of the text, and the hollowness of some statements. For example, "the auditory nerve in each ear sends sound vibrations to the cochlear nucleus": the auditory nerve does not send *sound vibrations*, but only electro-chemical representations of sound vibrations. You may not find this important, but it is this laziness in texts, effectively transferred epithets, that amount to misinformation, misleading the reader and I suspect, the author. Another example that you will find in the full text is: "to help give sound meaning". You see, in the context of speech, "meaning" is entirely conceptual, requiring pre-knowledge of language and words, and the processing of these data concepts in context to derive meaning. The entire description of *how we hear* entails physical organs of reception and transduction, physical lines of transmission, and physical organs of processing. Any discussion on how that converts to the conceptual is omitted – the hardest part is yet to be covered.

In passing, I do wonder about "nerve bundles", i.e., bundles of nerve fibres in close proximity to one another. Before myelin sheathing evolved, what prevented cross contamination of signals in fibres running closely parallel with one another? I have no understanding of the distances between fibres, and the thresholds in millivolts where cross contamination would occur, but nevertheless, that was one more technical problem for evolution to solve.

I will leave that subject there, for I believe that I have presented more than adequate evidence for the reader to ponder. How many genetic mutations, and how long would it have taken, for evolution to achieve this highly complex and beautifully constructed, yet finely tuned auditory system? I am not one given to accepting the maxim, "that all things are possible given enough time". The problem is, you see, that we can have no idea of what *enough time* would be in this instance.

Still unresolved, let me add, is how the mind-brain complex came to understand the semantic layer of the neural signals. Understanding is conceptual, not physical – we cannot allow ourselves to overlook this issue at any stage, for it remains central to the question of whether the mind can be material, or must it be mysteriously immaterial.

References:

1. https://open.bccampus.ca/open-textbook-101/

Chapter 7-3: Echo Location & Ranging

"Only echoes answer me."

~ Anton Chekhov, Swan Song ~

How poignant, and so true.

With practical experience in the use of RADAR (Radio Detection and Ranging) as a military air traffic controller, and a lesser knowledge of SONAR (Sound Navigation & Ranging) through AJASS (Australian Joint Anti-Submarine School) training, I have a particular curiosity as to how echo location and ranging could have evolved in organisms, absent of a reliable method of equipment calibration and verification of results. I am all too aware of the technical limitations of equipment, and the environmental circumstances hindering accurate interpretation of reflected signals. Although none of this can disprove evolutionary theory, it should give readers serious concern over the possibility of undirected evolution being responsible for the development of these amazing capabilities. We will start with some basics.

If I understand evolution theory correctly, organisms are unlikely to develop facilities for which they have no use. For example, fish evolving in the deep, dark depths of the ocean, or in deep, dark subterranean caverns, would not develop a sense of sight, because they would have no use for it. It would appear that some deep-sea creatures provide their own light source, although just how useful it is, and for what purpose, remains a matter of conjecture. In a book defending evolution, although I can no longer find the entry, and anyway, it matters not whether it is authoritative, I read that an owl's hearing is so acute that it can discriminate direction to one thousandth of one degree. Apart from finding no practical use for such fine accuracy, I considered this to be technically and biologically improbable, and thus sought to test the proposition using geometry and the properties of sound. The point is not worth arguing, and I do not intend to pursue it further here, but I found that the principles are useful as a segue into genuine phenomena that we will discuss below.

Relevant Properties of Sound

If you are already familiar with these, I apologise for the presumption on my part, but they are worthy of a brief review.

Sound waves are characterised by intensity, amplitude, wavelength, time-period, frequency and speed. Intensity, measured in decibels (dB), is the amount of energy in a sound wave, and is proportionate to its amplitude. Amplitude is the fluctuation or displacement of a wave from its mean value, and is the extent to which the medium's particles are displaced. If a sound is free to spread out in all directions, its intensity will decline with the inverse square law. A noise that is 100 dB at one metre will have an intensity of only 1/100 as much at ten metres, i.e., 20 decibels less (80 dB), and only 60dB at 100m. According to the inverse square law, it can be shown that for each doubling of distance from a point source, the sound pressure level decreases by approximately 6 dB.

To experience the full characteristics of a sound wave, the surface area of the receiver must be at least equal to the amplitude of the sound wave. In the production of sound, this is why speakers are of different sizes to best reproduce complex sounds dependent on their amplitude. But back to our auditory systems. In terms of direction, the asymmetry of time lag, wave length and tone detected by each of two ears helps us to determine the general direction of a sound source, but no more than that. Determining the precise direction, measured in degrees or less, as in the unlikely example of an owl's hearing mentioned earlier, is far more problematic. The ears of some owls are oriented asymmetrically in the vertical plane, the left ear being positioned lower than the right, which gives them the ability to perceive both the azimuth and elevation of a sound source simultaneously, reportedly with high precision, although I am unsure of the *quantification* of the *qualitative* adjective, "high". In contrast, owls with symmetrical ears most likely must listen from at least two different head positions to locate a sound in both axes, which is why we humans have a tendency to cock our heads when attempting to locate the source of a sound. The question we must ask is: What cognition is required to understand the concepts of azimuth and elevation that would lead evolution down the required path?

Using the aforementioned 1:60 rule, a 1° angle subtends 1 mile at a distance of 60 miles. Converted to metres (easier math), and assuming that owls only care about prey within swooping distance, say 25 metres, a targeted mouse of a width of 10cm would require an azimuth discrimination of approximately 0.25° (proper mathematicians can correct my error). Despite the latter, you can get some sense of the target discrimination requirements that evolution would have needed to develop.

Sound Localization

Putting aside a discussion on the emission of sound pulses, let us consider the issue of determining the source of a reflected signal, and here I will quote from an article on how it is said to work for owls. The principles must be the same for all auditory systems, irrespective of their sensitivity. Owls have "bilateral ear asymmetry", meaning that the positioning of each ear on the head is not symmetrical, as it is for humans and most creatures. This phenomenon is said to have been achieved by morphological adaptations, but I am not confident. As per earlier discussions, if the positioning of ears moves around, and the positioning is critical to auditory accuracy, why wouldn't any change be deselected? Again, we must apply systems thinking: any repositioning must be accompanied by complementary changes in how the auditory cortex "understands" the new variations in signalling via the auditory nerve. Given that such "intelligence" must follow the anatomical change, I would expect such a change would be deleterious rather than beneficial in the first instance.

"Owls must be able to determine the necessary angle of descent, i.e. the elevation, in addition to azimuth (horizontal angle to the sound). This bi-coordinate sound localization is accomplished through two binaural cues: the interaural time difference (ITD) and the interaural level difference (ILD), also known as the interaural intensity difference (IID). The ability in owls is unusual; in ground-bound mammals such as mice, ITD and ILD are not utilized in the same manner. In these mammals, ITDs tend to be utilized for localization of lower frequency sounds, while ILDs tend to be used for higher frequency sounds. ITD occurs whenever the distance from the source of sound to the two ears is different, resulting in differences in the arrival times of the sound at the two ears. When the sound source is directly in front of the owl, there is no ITD, i.e. the ITD is zero. In sound localization, ITDs are used as cues for location in the azimuth. ITD changes systematically with azimuth. Sounds to the right arrive first at the right ear; sounds to the left arrive first at the left ear.

In mammals there is a level difference in sounds at the two ears caused by the sound-shadowing effect of the head. But in many species of owls, level differences arise primarily for sounds that are shifted above or below the elevation of the horizontal plane. This is due to the asymmetry in placement of the ear openings in the owl's head, such that sounds from below the owl are louder in the left and sounds from above are louder in the right. IID is a measure of the difference in the level of the sound as it reaches each ear. In many owls, IIDs for high-frequency sounds (higher than 4 or 5 kHz) are the principal cues for locating sound elevation."[1]

This is an opportune moment to reflect on the complexity discussed above, in the context of evolutionary processes. Again, we have the twin issues of computation of differentiation in neural signals, and the comprehension of what such differentiation tells the owl about its external environment. If such sound localising ability is essential to the survival of the owl, how could any equivalent prototype owl have survived during the hundreds or thousands of years that would have been required to achieve the level of precision that we now understand exists? The evolutionist's answer that owls weren't truly owls until they had developed a useful capacity for sound localisation is somewhat acceptable, I would grant that. However, the most problematic aspect relates to cognition, not simply the physical structure of the sensory organ. Let us continue with the earlier quoted article. It is not my intention to blind the reader with science, nor to assume any level of scientific authority. Scientists have achieved an incredible level of understanding of the subject, and we must commend them for that, but in doing so, they have revealed complexities which evolutionists cannot explain.

"The *axons* of the *auditory* nerve originate from the hair cells of the cochlea in the inner ear. Different sound frequencies are encoded by different fibers of the auditory nerve, arranged along the length of the auditory nerve, but codes for the timing and level of the sound are not segregated within the auditory nerve. Instead, the ITD is encoded by *phase locking*, i.e. firing at or near a particular phase angle of the sinusoidal stimulus sound wave, and the IID is encoded by spike rate. Both parameters are carried by each fiber of the auditory nerve."

I am curious as to how "Both parameters are carried by each fiber of the auditory nerve", i.e., sound frequency and ITD (interaural time difference). A nerve fibre carries nothing but an electrical pulse, said to be of the same strength for any given neuron, so I would have thought that only one parameter could be carried by a nerve fibre. Nonetheless, a minor point. We know that the technical research and development which gave rise to the Cochlear Implant[2] had to be correct, otherwise the device would not work as it does, so I will not contend with the science of this process.

I cannot even begin to imagine the evolutionary mechanisms, or sequence, for the development of such an intricate and finely tuned sound localization system. Yes, I am incredulous, as Richard Dawkins would surely accuse me. I cannot dogmatically assert that undirected evolution is incapable of such a magnificent feat of biological engineering, but at the same time, scientists are still a long way from proving that undirected evolution is so capable. Consider the technological advances that have enabled scientists to understand the intricacies of the auditory system, including such technical aspects as *the phase angle of the sinusoidal stimulus sound wave*, and *encoding by phase locking*. Is any evolutionist, or any evolutionary biologist, able to begin to suggest how such a technical solution was achieved by evolution?

It is said that "Owls can use a hundredth of a millisecond difference in the time a sound reaches one ear or the other to fix the direction of a mouse. One ear of the barn owl is tuned to lower frequencies to locate a sound in the horizontal plane, the other to higher frequencies for the vertical plane."[1] *A hundredth of a millisecond difference*, i.e. 0.00001 of a second! There is still much to be learned about the speed of neuronal transmissions, but consider that most signals are passed via neurotransmitter molecules that travel across the small spaces between the nerve cells called synapses. This process takes more time (at least 0.5 ms per synapse) than if the signal was continually passed within the single neuron. Just how the auditory system can determine a 0.01 ms difference when the general transmission speed is much slower is beyond my understanding, but the point to note is that for the system to recognise a difference, there must be a reference point. In other words, there must be some specific encoding in the neural signal that allows the signal processing system to recognise impulses received almost simultaneously through each ear as having precisely the same source in space-time. As far as I can understand, this is a function as previously described:

"This bends the stereocilia either toward or away from the tallest member of each array. When the stereocilia bend toward the tallest member of their array, tension in the protein tethers opens ion channels in the hair cell membrane. This will depolarize the hair cell membrane, triggering nerve impulses that travel down the afferent nerve fibers attached to the hair cells. When the stereocilia bend toward the shortest member of their array, the tension on the tethers slackens and the ion channels close."

I am still amazed at how such technical precision could be accomplished through the processes of undirected evolution. I am even more amazed at the process whereby the owl comprehended the significance of a one millisecond difference in the arrival time of a sound. Why were sounds arriving at each ear at different times believed to have had the same source? The intelligence of the signal processing is outstanding, and hardly credible as arising through the purported processes of undirected evolution.

Another issue is the speed of neural signals. A great deal of research has been undertaken to determine the conduction velocity in nerve fibres. A nerve impulse (action potential) is the signal that is transmitted along an axon that enables nerve cells to communicate and to activate many different systems in an organism. Neurons, as we have seen, generate action potentials caused by a change in the neuron membrane permeability, resulting in a change in distribution of ions across the membrane, which in turn leads to a change in electrical charge (potential) across the membrane. Changes in electrical potential have been experimentally detected as the action potential passes along the axon of the neuron. The question becomes: Do all neural pathways conduct signals at precisely the same speeds? The answer is: No. Depending on their thickness and construction, and especially *myelination* (insulation), different types of nerve fibres in the PNS and CNS conduct signals at different speeds. The myelin geometry, diameter and internodal length, is therefore a critical parameter of the nerve conduction velocity and changes in this geometry underlie numerous functional properties of the brain and nerves. In the auditory nerves, this myelin contributes to axonal protection and allows for efficient action potential transmission, and thus speed. Thus, to achieve a hundredth of a millisecond difference in the time a sound reaches one ear or the other, the biological construction of each auditory nerve fibre must be incredibly, and precisely the same. Is such biological construction accuracy possible? I cannot know, but given that the precision in the symmetry of other biological features is less than optimal, I would be surprised that there could be no variations in the symmetry of neuronal pathways.

Again, I mention this only to highlight the incredible technology involved.

How many mutations would it take for that to develop, and how did the owl's brain learn what the frequency differences meant? Why wasn't it just understood to be noise, an irritation to be shut out? Why wasn't it like being in a crowded room and overhearing one conversation in one ear, and a different conversation in the other ear, especially if it took evolutionary periods for the synchronisation to be achieved? Scientists are able to tell us how the eyes, ears, and other sensory receptors work today, and where the signals are processed in the brain, but that tells us nothing about how an organic system could have learned that, for example, electrical impulses travelling along the individual fibres in the optic nerve from the eye carry data that represents visual information that can be interpreted in terms of colour, shape, texture, depth, distance, and movement, or that a frequency differential in sound represents movement and direction.

The *Doppler Effect*—the change in frequency of a wave for an observer when there is relative movement between the observer and the sound source—was only understood in the mid-nineteenth century, but evolution theory asks us to believe that this understanding evolved biologically in animals.

We have not touched on two associated issues: (1) echo *ranging*; and (2), far more importantly perhaps, how a bat, for instance, could recognise that a reflected signal represented food versus non-food. I would think the latter to be crucial for survival in the dark. I will leave you to ponder these for yourself, as I have long pondered them without anything approaching understanding. Similarly, we have not discussed the bio-acoustical modelling of dolphins, but adding more issues of complexity is not going to help us understand.

Echo location and ranging involve the transmission of sonic pulses, and the interpretation of the reflected pulses. "Trial and error" is a deliberate experimental, cognitive process, not a matter of mere chance, for the "correct" interpretation is not a single event. Developing a system requires the identification and retention of the successful steps, which provide the foundation for subsequent experimental steps. Minimal functionality requires a series of proven steps, in the case of evolution, instantiated in the genome for inheritance by subsequent generations. This is a matter of complete wonderment to me, let alone the earlier mentioned subjects of telemetry and calibration. I wonder what Charles Darwin would have thought had he known this level of complex detail?

References:

1. https://en.wikipedia.org/wiki/Sound_localization_in_owls

2. www.cochlear.com

Chapter 7-4: Evolution of the Eye

"The eye is the most refined of our senses, the one which communicates most directly with our mind, our consciousness."

~ Robert Delaunay, French artist, co-founder of the Orphism art movement ~

I suspect that scientists pay far too little attention to the experiences of artists, people who look beyond the material to the conceptual, striving to portray their own perceptions in different ways. That such perceptual variations exist is, to my mind, compelling evidence that all cannot be entirely material, for the material can only be what it must be.

That aside, I have a particular interest in the hypothetical evolution of the eye, because it contradicts everything that I know about information processing. Whilst evolutionists pay attention to the biology, and conjure up just-so stories in an attempt to explain incremental functionality during the many evolutionary steps, they seem to entirely ignore the fundamental role of the eye – it is a sensor that conveys, via the optic nerve, symbolic representations of its external reality for visualisation by the "mind's eye". This is a key issue that evolutionists ignore when they propose their theories of how the eye developed, one which underpins much of my doubts about undirected evolution.

Key fact: Only the eye "sees" - the brain can but "visualise".

Some scientific texts allude to this reality, but fail to discuss the implications for evolution, other than by proposing "just so" stories. None of the texts that I have read deal with sight from a truly systems perspective, with no acknowledgment of the irreducible complexity of the functions of the system; we will come back to that in the next chapter. I can find no explanations of how the sensory input could be processed via the visual cortex to convert electro-chemical signals into a symbolic representation of three-dimensional data, stored as a neuronal pattern in the brain, which can then be correlated into conceptual information.

The question is simply this: Could the eye, from its earliest embryonic state, evolve without the presence of both a communications channel and a back-end processor to enable functionality? By analogy, could the heart evolve without arteries, and destination for the blood? The eye is just one complex component, the receptor, of an even more complex information processing system, which has no functionality until all components and linkages exist. Before reviewing evolutionists' explanations for how the eye could have evolved in such a short timescale, or even at all, we should spend a little time on the anatomy of the eye to be able to keep these so-called "explanations in perspective.

Anatomy of the Eye

To keep this short, I will limit the discussion to the retina where the transduction of light occurs. Sensors in the body transduce certain types of energy into electro-chemical signals for transmission to the cortex, where it is said to become a conscious perception of the external reality. In our system of sight, the lens is not the transducer, and by itself would provide no survival advantage. The lens not only captures the light energy, but importantly focuses and transmits it to photoreceptors termed rods and cones which do the real work. Following is a brief discussion on the complexity of this system, to highlight the paucity of explanations by evolutionists. If evolutionists are to attempt an explanation of the eye, they must address the complexity as revealed on this online site[1], which I highly recommend for your attention, as the coloured diagrams are very informative. I have skipped any discussion on the nature of light here, and how it interreacts with matter, the reflected light being altered by the molecular structure of the object which is how our visual system can discriminate a scene before it. If you are interested in learning more about light, this website[2] is useful.

That said, it is noteworthy that the retina of the eye is comprised of 10 to 18 distinct layers, depending on which source you access. Consider this complexity:

"The retina is a layered structure with ten distinct layers of neurons interconnected by synapses. The cells subdivide into three basic cell types: photoreceptor cells, neuronal cells, and glial cells ... Within these layers of the retina, we find multiple different types of cells with specific jobs that help transmit incoming photons into action potentials that the brain's cortices process into three-dimensional vision. The six different cell types in the retina include: Rods, Cones, Retinal Ganglion cells, Bipolar cells, Horizontal cells, and Amacrine cells."

It is believed that a large degree of processing of visual information occurs in the retina itself, before signals are sent to the visual cortex. Quoting extracts from the above referenced site[1]:

"The rods and cones are the site of transduction of light to a neural signal. Both rods and cones contain photopigments [Ed. unstable pigments that undergo a chemical change when they absorb light] … Visual signals leave the cones and rods, travel to the bipolar cells, and then to ganglion cells."

The following detail is given for the reader to, once again, understand the exquisite detail of the eye, asking you to ponder the likelihood of undirected evolution being capable of such achievement. Ganglion cells, which we haven't discussed so far, are the final neurons connecting the retinal to the nerve fibres in the optic nerve. The human optic nerve contains anywhere up to one and a half million nerve fibres, which are axons of the ganglion cells. Again, I must wonder at the opportunity for signal cross-contamination of so many nerve fibres in one such tiny bundle, but likely the weakness of the signals is the answer. Quoting from this educational website[3]:

"Ganglion cells collect information about the visual world from bipolar cells and amacrine cells (retinal interneurons). This information is in the form of chemical messages sensed by receptors on the ganglion cell membrane. Transmembrane receptors, in turn, transform the chemical messages into intracellular electrical signals. These are integrated within ganglion-cell dendrites and cell body, and 'digitized', probably in the initial segment of the ganglion-cell axon, into nerve spikes. Nerve spikes are a time-coded digital form of electrical signalling used to transmit nervous system information over long distances, in this case through the optic nerve and into brain visual centers.

Ganglion cells are also the most complex information processing systems in the vertebrate retina. It is a general experimental truth that an organism as a whole cannot behaviorally respond to visual stimuli that are not also detectable by individual ganglion cells. Different cells become selectively tuned to detect surprisingly subtle 'features' of the visual scene, including color, size, and direction and speed of motion. These are called 'trigger features'. Even so signals detected by ganglion cells may not have a unique interpretation. Equivalent signals might result from an object changing brightness, changing shape, or moving. It is up to the brain to determine the most likely interpretation of detected events and, in the context of events detected by other ganglion cells, take appropriate action.

Ganglion cell axons terminate in brain visual centers, principally the lateral geniculate nucleus and the superior colliculus. Ganglion cell axons are directed to specific visual centers depending of the visual 'trigger features' they encode. The optic nerve collects all the axons of the ganglion cells. In man this optic nerve bundle contains more than a million axons."

Researching further on the issue of *Ganglion cells are also the most complex information processing systems in the vertebrate retina*, described as "ganglion cell electrophysiology", I have been unable to find literature describing just how these cells perform "information processing", and can only conclude that as far as scientific research has taken us, this is an assumption based on the capabilities of other components. Certainly, there is a transformation from chemical to electrical messaging, but as discussed in earlier chapters, the analogy with *complex information processing* is false. You may recall an earlier discussion on action potentials in neurons, which are described as "all or nothing" - there is no electrical gradient variation in any given neuronal response. I was interested to learn that it was Edgar D. Adrian, winner of the 1932 Nobel prize in physiology and medicine, who pioneered the measurement of impulse discharges from individual peripheral nerve axons. This led to the discovery that nerve fibers transmit signals in the form of a temporal sequence of 'all or none' impulses."

Re-quoting: "It is up to the brain to determine the most likely interpretation of detected events and, in the context of events detected by other ganglion cells, take appropriate action", I wonder whether this is another inference to best explanation? I can find no explanation as to how the brain *interprets* detected events, especially events involving fast moving objects, the processing of which would equate to gigabytes, or even terabytes of data in real time. We know that the system works wonderfully well, I just wonder how it could have evolved, and whether purely physical processes are able to account for its full range of interpretation and internal visualisation. Remember, only the eye "sees", the mind-brain complex must visualise based on activity in neural circuits.

So now, let us hear from the evolutionists, but before we begin, I wish to refute the straw man argument often advanced, that Creationists claim that our eyes are "organs of perfection". I have read creationist literature extensively, and apart from some silliness found in Christian apologetics, creation scientists make no such claims. Let me demonstrate the depths of irrationality to which evolutionists often descend. One opponent of creationism opined: "The eyes are definitely not organs of perfection as creationists claim – we get short-sighted, develop glaucoma, cataracts and go blind." Indeed, we do – as all of our organs are subject to disease and decay. However, more than one scientist has opined, "Our eyes are perfectly adapted". So, it seems that evolution can achieve perfection, but God (?) cannot, even though the organ suffers the same deficiencies irrespective of how it came to be. Very curious, but here note that I am not defending creationism, nor advocating that "God did it", although he may have done. I just wanted to highlight some of the silliness that I have encountered in the literature.

Robin McKie, the Observer's science editor, offered his opinion that "In fact, the evolution of the human eye was a basic business. It evolved from simpler versions that in turn evolved from even simpler eyes that in turn evolved from basic light sensors. That is how natural selection operates. It acts on existing features of animals' bodies and slowly induces change that can eventually result in new species."[4] I might suggest to the Observer that they need a new science editor, for there is nothing simple about even the most primitive visual sensory system. The biological architecture is one thing – the informational content and comprehension entirely another. This website[5] provides another "explanation", if that is the right word. It begins by recounting Darwin's acknowledgement that his concept of evolution would have difficulty explaining the development of the eye. It then asserts: "Difficult, but not impossible. Scientists have come up with scenarios through which the first eye-like structure, a light-sensitive pigmented spot on the skin, could have gone through changes and complexities to form the human eye, with its many parts and astounding abilities."

"*Difficult, but not impossible*"? Only someone who chooses to limit understanding to the biological architecture could make such an assertion. Architecture is the enabler of the sensory system, not the system itself. The *system*, and I insist on that word, is about messaging: conveying semantic representations from one energy form to another, where a physical reality is perceived conceptually. Until evolutionists can address this primary issue, they have not made their case. For the suggested sequence of eye evolution, refer to the website article.

As explanations go, the above is very light on explanatory power. Hypotheses, scenarios and beliefs are not scientific evidence, but an impetus (hopefully) for further research. Starting with the idea of a "simple light-sensitive spot on the skin", we must first understand what this means in technical terms. It can be said that our entire skin is sensitive to electro-magnetic radiation, from the infra-red to the ultra-violet. We experience this as heat and sunburn, even resulting in melanomas. Within this range is the visible light spectrum, typically in the range 380 – 700 nanometres. The evolutionists' claim is that one or more cells mutated such that rather than, for example, the UV radiation causing concentration of melanin in the skin, it resulted in significant sensitivity to this narrow spectrum, with reduced or insignificant sensitivity to the remainder of the electro-magnetic radiation spectrum. To be *light-sensitive* means that the cell operates as a transducer converting the energy of electro-magnetic radiation within a specified range to a neural signal that is comprehensible to the organism's brain as light, a concept of which it is unaware. In colloquial terms, one might suggest that the organism responded: "What was that?" Each and every evolutionary explanation that I have read entirely ignores the crucial issue of cognitive comprehension. One can understand that sensors operating to monitor internal bodily functions communicate using electro-chemical protocols and coding systems that are comprehensible to other organic elements, but the same cannot be said of the sight and hearing sensory systems that attempt to comprehend an external reality.

Consider this "just-so" story: "The simple light-sensitive spot on the skin of some ancestral creature gave it some tiny survival advantage, perhaps allowing it to evade a predator." The pre-conditions for this story are numerous, starting with the spectrum of radiation to which the spot was sensitive, the visualisation in the primitive brain, and recognition that what was perceived was in fact a predator. If it was a predator, and this ancestral creature did not recognise it as such, it would likely have ended its life as lunch (end of evolutionary development). There are far more variations on how this supposed survival advantage would have failed, compared to the very few on how it could have succeeded.

As for a scientific calculation that only 364,000 years were needed, we will come back to that. The most important issues are those discussed in earlier chapters concerning the human *nervous*, *telemetry*, and *calibration* systems. No matter what may have happened on the surface of the skin, the mutation must have included the correct transducing properties, converting light energy to electrochemical energy. There needed to have been the identification of a prespecified and calibrated nerve fibre to the visual cortex; which in turn was capable of converting that signal to a conceptual perception of reality. It should be seen that any genetic mutation effecting the surface of the skin is just the start – so much more needs to be engineered downstream, without which, the mutation has no relevance whatsoever.

Here is an extract from an alternate explanation which offers a little more promise:

"Researchers ... have discovered that the light-sensitive cells of our eyes, the rods and cones, are of unexpected evolutionary origin – they come from an ancient population of light-sensitive cells that were initially located in the brain ... By studying a "living fossil", I was immediately intrigued by the idea that both of these light-sensitive cells may have the same evolutionary origin."

The article goes on to describe the concept of "molecular fingerprints". Such fingerprints represent a combination of molecules which are unique for specific cells, and based on the paradigm of evolution, the researchers concluded that such cells shared a common ancestor cell. Investigating further, it was found that the light-sensitive molecule, opsin, in the living fossil, a worm, "strongly resembled" the opsin in more recent vertebrates.

"When I saw this vertebrate-type molecule active in the cells of the Playtnereis brain – it was clear that these cells and the vertebrate rods and cones shared a molecular fingerprint. This was concrete evidence of common evolutionary origin. **We had finally solved one of the big mysteries in human eye evolution.**"[6] [emphasis mine]

Well, not really. At best it may solve the mystery of the origin of light-sensitive cells, but there is far more to our sense of sight than the optical sensor. These scientists have no more "elucidated the evolutionary origin of the human eye" than one could solve the mystery of the genome by seeing the birth of a baby. This is akin to a novice chess player moving his first pawn without it being taken, and declaring, "I won, I won!". Intrigued, I wanted to know that nature of an "opsin", but contented myself with "Opsins are a group of proteins, made light-sensitive, via the chromophore *retinal* (or a variant) found in photoreceptor cells of the retina". Now, even if *opsins* are of a truly ancient origin, it tells us nothing of how they evolved to become light sensitive, so I asked: How do proteins evolve? I thought that I had found the answer, only to once again have my hopes dashed by this article entitled, *"At Last, the Details of How Proteins Evolve?"*[7] The article offered much, beginning:

"How did proteins evolve? It is a difficult question because, setting aside many other problems, the very starting point — the protein-coding gene — is highly complex. A large number of random mutations would seem to be required before you have a functional protein that helps the organism. Too often such problems are solved with vague accounts of "adaptations" and "selection pressure" doing the job."

Apparently, according to the article, "ground-breaking research" managed to provide "a step-by-step, detailed, description of the evolution of a new protein-coding gene and associated regulatory DNA sequences." The researchers claimed that their new research "provides the specific sequence of mutations, leading to the new gene." Excuse me for being sceptical, but goal-focused experiments in laboratories cannot replicate conditions from way back in time. The Urey-Miller experiments of the 1950s made similar claims, since having been debunked. Assumptions can be made, but they do not necessarily represent truth.

Reading further,

"This is a monumental finding. Having the scientific details, down to the level of specific mutations, of how a new protein-coding gene evolved — not from a related gene but from non-coding DNA — is something evolutionists could only dream of only a few short years ago. **There's only one problem: it is all junk science.**" (emphasis mine)

I will leave you to read the rest of the article for yourself, and decide for yourself which narrative is likely authoritative and true. For myself, it is not that important as I have more important fish to fry (again, excuse the pun). The important question to ask is: What survival advantage would an opsin cell on the surface offer, unless there was a pathway to convey that sensation to the brain, as primitive as that may have been? More importantly, how would the brain recognise the sensation of light, never before having the *concept* of light? Even if an ancient population of light-sensitive cells were initially located in the brain, that still doesn't explain how the brain "understood" their function, both within the brain and when they migrated to the surface. Whilst in the brain, the relationship between the opsin cell and the neuronal network was entirely chemical, eliciting a chemical response to other parts of the body, as in the circadian rhythm. A sense of sight is not like that, because the response to a visual signal is in the end, *conceptual* not physical.

Evolution Timeline - Only 364,000 years?

What of the time scale for the evolutionary development? It is said that eyes appear in the fossil record about 540 million years ago, at the beginning of the Cambrian period. According to an entry in Wikipedia:

"The first fossils that might represent animals appear towards the end of the Precambrian, around 610 million years ago, and are known as the Ediacaran or Vendean biota. These are difficult to relate to later fossils, however. Some may represent precursors of modern phyla, but they may be separate groups, and it is possible they are not really animals at all. Aside from them, most known animal phyla make a more or less simultaneous appearance during the Cambrian period, about 542 million years ago. It is still disputed whether this event, called the Cambrian explosion, represents a rapid divergence between different groups or a change in conditions that made fossilization possible. However, some paleontologists and geologists would suggest that animals appeared much earlier than previously thought, possibly even as early as 1 billion years ago."[8]

All of which is rather inconclusive, but by some experts' reckoning, there may have been as little as 60 million years for the visual system to have evolved. Given the complexity of sight, and the significant amount of brain and mind development needed to complete the system, it is difficult to understand why animal brain and mind development went into semi-hibernation for the next 540+ million years: there is a logical disconnect here.

Consider also that "according to one scientist's calculations, only 364,000 years would have been needed for a camera-like eye to evolve from a light-sensitive patch,"[9] but for that to be true, the whole visual sensory system as will be later described must have evolved in the same timescale, or shorter. This example demonstrates how some evolution proponents fail to understand the *systems nature* of organisms: they argue against irreducible complexity by concentrating on some small (but significant) component, but ignore the fact that if would provide no survival benefit without complementary components within an appropriate timescale, lest the mutation be later discarded as offering no benefit.

The eye has been around for over 540 million years; thus, the essential components of cognitive ability in relation to sight have been around for the same period. Now, unless the evolutionists want to claim that the eye developed anatomically without being connected to the visual system, or that the cognitive abilities needed to implement the visual system and the other senses is so limited as to not require much evolution at all, there is this very strange hiatus in the development of the mental faculties of animals which is, so far, without explanation.

Even if one is able to surmount the irreducible complexity issue of the sensory systems: sight, sound, smell, taste, and, touch, one is then left to explain why evolution took time out to the tune of 500+ million years with the brain. An explanation along the lines of the size of the brain cavity, for example, would then have to be justified in comparisons of all animals that have senses, and the variations in mental capability based on brain sizes from the smallest to the largest.

Continuing, the detail for this can be found on the website of The Royal Society Proceedings – Biological Sciences[10]. The 1994 paper by Dan-Eric Nilsson and Susanne Pelger, from which I shall quote, is entitled, "A pessimistic estimate of the time required for an eye to evolve". The Summary stated:

"Theoretical considerations of eye design allow us to find routes along which the optical structures of eyes may have evolved. If selection constantly favours an increase in the amount of detectable spatial information, a light-sensitive patch will gradually turn into a focused lens eye through continuous small improvements of design. An upper limit for the number of generations required for the complete transformation can be calculated with a minimum of assumptions. Even with a consistently pessimistic approach the time required becomes amazingly short: only a few hundred thousand years."

It did not take long to understand that the title of the paper was deceptive, and dare I say disingenuous:

"The evolution of complex structures, however, involves modifications of a large number of separate quantitative characters, and in addition there may be discrete innovations and an unknown number of hidden but necessary phenotypic changes. These complications seem effectively to prevent evolution rate estimates for entire organs and other complex structures. An eye is unique in this respect because the structures necessary for image formation, although there may be several, are all typically quantitative in their nature, and can be treated as local modifications of pre-existing tissues.

Taking a patch of pigmented light-sensitive epithelium as the starting point, we avoid the more inaccessible problem of photoreceptor cell evolution (Goldsmith 1990; Land & Fernald 1992). Thus, if the objective is limited to finding the number of generations required for the evolution of an eye's optical geometry, then the problem becomes solvable."

Ah! *The more inaccessible problem of photoreceptor cell evolution*! So, the objective was NOT to develop "a pessimistic estimate of the time required for an eye to evolve" as stated, but simply **limited** to *the evolution of an eye's optical geometry*, i.e. the lens. The eye is much more than a lens, as already discussed. And as for the "*unknown number of hidden but necessary phenotypic changes*", the less said the better. The paper continues,

"We have made such calculations by outlining a plausible sequence of alterations leading from a light sensitive spot all the way to a fully developed lens eye. The model sequence is made such that every part of it, no matter how small, results in an increase of the spatial information the eye can detect. The amount of morphological change required for the whole sequence is then used to calculate the number of generations required."

The further I read this paper, the more it condemns itself:

"Ideally we would like selection to work on a single function throughout the entire sequence. Fortunately, spatial resolution, i.e. visual acuity, is just such a fundamental aspect and it provides the sole reason for an eye's optical design (Snyder *et al.* 1977; Nilsson 1990; Warrant & McIntyre 1993)."

I read through the rest of the paper with little interest, for it explained in exquisite detail how the lens of the eye might have evolved. The lens is, without question, the least complex component of the eye, just as the eye, from a systems perspective, is the least complex component of the sensory system we call sight.

Summary

No matter how biologists approach the evolution of the eye, their explanations always contain "just-so stories", defined in Wikipedia as:

"In science and philosophy, a just-so story is an unverifiable narrative explanation for a cultural practice, a biological trait, or behaviour of humans or other animals. The pejorative nature of the expression is an implicit criticism that reminds the hearer of the essentially fictional and unprovable nature of such an explanation. Such tales are common in folklore and mythology."

I have no problem with such stories when offered as hypotheses, for what else can we do with mysterious happenings of the past. However, I do object when scientific explanations, claiming to offer scientific evidence or proof, embed these just-so stories as part of that evidence or proof. This practice is far too common in the evolution narrative. Again, let me emphasise that evolution narratives of the eye have no explanatory power for our sense of sight. However, this explanation[8] in Wikipedia is a worthwhile read, as it narrates a well-attested description of how the eye likely evolved. I am sympathetic to some scientific accounts as I respect the researchers and the amazing discoveries that they have made. As I have said before, I do not entirely reject the possibility, just *undirected* evolution, but just don't ask me the identity of the *director*.

Next, we will consider the complexity of our sense of sight from a systems perspective, the purpose being to allow the reader to ponder the likelihood of such arising through genetic mutation. If evolution proceeded through small progressive steps, with steps in a particular non-intentional direction being retained with or without function, why did this occur before the system was complete in terms of reception, transduction, encoded transmission, decoding in the visual cortex or similar, and eventual cognitive perception of the external reality?

References:

1. https://courses.lumenlearning.com/wm-biology2/chapter/transduction-of-light/

2. https://andor.oxinst.com/learning/view/article/what-is-light

3. https://webvision.med.utah.edu/book/part-ii-anatomy-and-physiology-of-the-retina/

4. https://www.theguardian.com/commentisfree/2018/aug/19/intelligent-design-how-come-he-made-so-many-blunders

5. Nilsson, *Evolution of the Eye*, 2001,

    https://www.pbs.org/wgbh/evolution/library/01/1/l_011_01.html

6. https://www.sciencedaily.com/releases/2004/10/041030215105.htm

7. https://evolutionnews.org/2019/02/at-last-the-details-of-how-proteins-evolve/

8. https://en.wikipedia.org/wiki/Evolution_of_the_eye

9. https://www.pbs.org/wgbh/evolution/library/01/1/l_011_01.html

10. https://royalsocietypublishing.org/doi/10.1098/rspb.1994.0048

Chapter 7-5: Irreducibility of Sight

"If it could be demonstrated that any complex organ existed which could not possibly have been formed by numerous, successive, slight modifications, my theory would absolutely break down."

~ Charles Darwin, Origin of Species ~

Here again, Charles Darwin missed the point. Our bodies are composed of organs, that is true, but the organs work as essential components of a biological system, and as injury, disease and old age have demonstrated, deterioration or failure of a single organ can have disastrous consequences for the system. This is relatively easy to understand in purely internal biological systems, but far more difficult when dealing with biological systems that source external realities to be comprehended cognitively. We must think, not just in terms of individual components which arguably, may not be irreducibly complex in isolation, but in *systems* which *are* irreducibly complex if they are to achieve any useful functionality whatsoever (Primary Axiom #8). If survival and reproduction success are key to evolution of species, we must ask what advantage could be offered by an *incomplete system*.

As an aside, whilst we talk of *seeing*, that is truly a metaphor for what we really do: *visualise*. The dictionary defines "to see" as *perceive with the eyes; discern visually*. This is very close to the truth, but I doubt whether many people see this (alternate meaning). The eye deals with light, the brain deals with perception. No light reaches the visual cortex or the brain's neural circuitry: these organs need to make sense of electro-chemical signals, and it is the "mind's eye" that does the visualising, just as it does with memory recall. In a genuine technical sense, the visual cortex does complex signal processing of messages received via the optic nerve, before passing results to the brain for further processing, resulting in visualisation. The question becomes: how does the material brain project immaterial visualisations, and onto what canvas? The difficulty for evolution is to show how the eye and backend signal processes could have evolved in a complementary, contemporaneous manner. *Sight* is a "system", perhaps the most complex in the human body, and it will not do to argue that individual components are not necessarily irreducible - we must apply *systems thinking*.

Now, to begin:

"Try describing an elephant to a blind man"

The saying above evidences our understanding that it is practically impossible to explain something to someone who has no prior knowledge of the concept, the entity, or any of its components. We might as well try to explain a *tretwopary* or a *melopricant*: where do we start? With that in mind, let us dig a little deeper. This chapter repeats some essential explanations from earlier chapters in an endeavour to make it somewhat self-contained. I suspect that I am the only one who would attempt to read this book in a single sitting, as has been necessitated in editing the draft manuscript!

In this chapter, I contend that despite the claim that, for example, the human eye may be irreducibly complex, it is the *least complex* component of our sensory sight system, and though explanations of the evolution of the eye are suppositional, the eye (or at least, the lens) is the *most easily explained* component in terms of evolution theory. Sensory systems: sight, hearing, taste, smell, and touch are essentially information systems and need to be understood as such before trying to explain their development from an evolution perspective. The architecture and chemistry alone are insufficient.

My premise is that in the case of sight, the data symbols received from a primitive photosensitive cell were beyond comprehension by a primitive brain, and that the only source of understanding about that message stream was the very symbol set itself that the brain did not understand. Thus, no acquisition of information was possible (Primary Axiom #2).

Biological Information Processing

It may be possible, in evolutionary terms, to explain the biological components of our senses, the sensory receptors such as the eye, for example, but it is another matter entirely to posit that starting from a zero knowledge base, evolution is capable of developing the encode/decode mechanisms needed for the transmission of symbolic representations of an external reality, the conceptualisation of that reality, the correlation in context to derive information, and the formulation of an appropriate response. Our five senses provide a symbolic representation of our external environment, one that the earliest organisms *knew* nothing about, yet their very survival was dependent upon a correct interpretation of its sensory inputs.

This is the very essence of the argument.

At some stage of evolution, the earliest brains were empty of information about the external environment in which the organism existed. According to the evolution narrative, the only source of information about that environment was the data stream provided by one or more of the five senses. The only source of information about what that data meant was that very same data stream. As I have contended in Primary Axiom #1, that *nothing can explain itself*, then the organism could have learned nothing from that sensor. Our knowledge and experience of data, information systems, and language tells us that even the best analysis by the best human intellects is incapable of decoding data symbols without at external frame of reference. Absent of an external frame of reference for the brain to decipher the meaning of the sense data stream, I contend that it is impossible for the sensory systems to biologically evolve.

Think of the primitive brain of the earliest organism as empty of any knowledge, experience, or instinct; How did it acquire any? For the sense of *touch*, how did the brain know what that sense data meant, without any previous experience of sensory systems? The sense of touch differentiates temperature, pressure, friction, and texture – how did the brain learn to differentiate the sematic layer when it had no knowledge of semantics? In the context of the *human telemetry system*, previously discussed, what biological mechanisms allowed such differentiation? How could it evolve an appropriate response to the level of pressure, or to the relativity of the external temperature and its own temperature? We know that pain represents a distress signal from some part of the body that something has gone wrong, and is interpreted in the mind-brain complex in both a physical and emotional context (it hurts!) We also know that the signal processing in the nervous system sometimes gets it wrong (referred pain). Although the encoded pattern in the neural transmission would have been different, why wasn't the first light signal, for example, interpreted as a different degree of heat, pressure, or pain? The brain receives a signal stream in the form of electrochemical impulses via the nervous system: How could a system evolve whereby the brain understands that any particular signal represents information about sight, sound, smell, taste, or touch? Now, one might argue that it is the characteristics of the sensors (transducers) that determine the *semantics* (see Chapter 3-6) of the message, and that is true, but the organism still needs to "understand" those characteristics in terms of the external reality. Additionally, communications require collaboration between sender and receiver to encode and decode the semantics of the message. It is difficult to conceive of evolution working from one end at a time, or both ends simultaneously.

As the eye evolved from, for example, a light sensitive cell on the brain, the brain did not understand that this new cell type was "light sensitive", this being a concept of which the organism could not have been aware. It was aware of a new sensation, but could not have known its significance. Tightly close your eyes, and you will have some idea of what that early organism would have experienced – not particularly useful. Irrespective of where in the hypothesised sequence of eye evolution you choose to focus, the brain having no previous knowledge or experience of eyes or sight, did not know that it was an eye (Primary Axiom #9), it did not know that the nerve impulses represented light, and it had no framework for evaluating and understanding what those light signals meant. This might seem strange at first, but if understanding evolved via genetic mutation, there is no reason why colour was not interpreted as texture, or shape as distance, as all of these concepts are simply represented as a data stream of electro-chemical impulses along a nervous system pathway. Remember too that light received by the eye as we know it today, is deconstructed into anything up to 1.5 million individual electrical impulses for onward transmission by the same number of nerve fibres in the optic nerve. Somewhere downstream, likely partially in the visual cortex, and partially in the brain, these individual signals are recombined into the picture that we visualise. Evolution would have started with a smaller number of nerve fibres, and thus a lower level of visual acuity, but the question to be asked is at what stage of development would the embryonic eye have offered survival advantage? Scientists can speculate on the sequence of development, because that sequence leads to the end goal, but that is begging the question – we cannot know that such was the sequence as evolution is not goal directed.

At the level being described here, what is transferred from the retina is simply a set of symbols representing elements of the picture captured by the eye (think pixels in the digital world). Within this picture are the conceptual elements of colour, shape, texture, distance, and so on, none of which the primitive brain could have been aware. Information is an intelligent interpretation of the conceptual data within a referential framework. A suggestion of Intelligent Design is hard to avoid. Consider the Search for Extra-terrestrial Intelligence (SETI) researchers scanning the cosmos for radio waves: when they find a pattern, they assume intelligence, but without knowing the coding of the signal and the language that the coding represents, they cannot understand the message and would not know how to respond. They only know that it may be a message by the existence of a pattern within a defined range of the electromagnetic spectrum; outside that range they would assume a different source, and even recognising a pattern requires intelligence. What would take code breakers with powerful computers possibly years to decode and interpret, is said to have evolved biologically at the same time as the eye and ear sensory systems evolved, in just a few million years or less.

The Inter-Relationship of Senses

I contend that our sense of touch, by itself, cannot convey cognitive information. Imagine an unfortunate man who is both totally blind and totally deaf (not just legally so): How could you teach him to "read" Braille? How could you teach him to identify and name fruits such as strawberries or raspberries, or to identify and name certain odours and develop the "nose" of a wine taster? Touch, taste, and smell work to the extent of differentiating between sensations, but without other inputs, cannot contribute to communicable information or knowledge. Our auditory and visual systems are essential to the interpretation of the other senses, and are the only systems capable of confirming common conceptualisations.

We can also question the origin of concepts such as desirable versus undesirable. Was it a genetic mutation that inspired an organism to remark to itself: "That smells nice" or "that tastes delicious"?

I accept that chemistry plays the major role in the identification of odours and tastes, and perhaps again it is chemistry that determines an individual's preferences, but chemistry itself cannot give rise to the notion of conceptual preferences, even though the chemical gradients represented by those preferences is understood. The link between the sensations of touch, taste, and smell, and the cognitive recognition of what such sensations represent, falls into the domain of the conceptual, not the physical, especially when we take the next step to communicate our responses to these sensations to other sentient beings.

Now let us look more closely at just one sense: that of sight.

What is Light?

One explanation offers: "Light transmits spatial and temporal information."[1] I hope by now that you understand that light does no such thing. For the layman, such as myself, light is perhaps the most mysterious phenomenon of our experience. Continuing from that source,

"The properties of light can be derived from the theory of classical electromagnetism, in which light is described as ... a traveling wave. However, this wave theory ... is not sufficient to explain the properties of light at very low intensities. At that level a quantum theory is needed to explain the characteristics of light and to explain the interactions of light with atoms and molecules. In its simplest form, quantum theory describes light as consisting of discrete packets of energy, called photons."

The article goes on to explain that neither the wave or particle model is adequate to explain light, and that a new theory termed quantum electrodynamics (QED) is now regarded as correct by physicists. This combines the concepts of electromagnetism, quantum mechanics, and the special theory of relativity. Is it any wonder that I do not understand light? Nevertheless, I contend that as with any electromagnetic transmission, only encoded symbolic representations of data can be conveyed. Evolution is not a sufficient explanation for how organisms learned what those transmissions meant in terms of their external reality. Remember that what we experience as heat (thermal radiation) is electromagnetic radiation also, but we do not contend that heat conveys cognitive information, other than in a comparative sense. When we shine a torch (flashlight), we do not contend that it is conveying information. Curiously, when we see that light reflected from an object, it is claimed that the reflection conveys information about that object. How can information be added by that object? When reflected from an object, the characteristics of the light are changed depending on the intrinsic properties of the object, and it is these changed characteristics that allow us to interpret and differentiate different realities. But the problem remains: What agent "explains" these signals to us? It is claimed that this occurs in the visual cortex, partially at least, but the cortex is just a particular form of neural network: how did it learn the signal encoding system used by the eye?

One article claims: "It is no accident that humans can 'see' light"[2], though just what prompted the author to make such a ridiculous statement is quite beyond me.

When light strikes an object, it interacts with that object and its properties become altered. By studying light that has originated or interacted with matter, scientists can determine many of the properties of that matter. This an issue of cognition, supported by many years of scientific research, yet evolutionists contend that this "science" was acquired by whatever evolutionary mechanisms are now in vogue, absent of any external, explanatory agent. It is the "language" of light that evolutionists are unable to explain.

The Concepts of Light and Sight

Imagine that some thousands of years ago, a mountain tribe suffered a disease or mutation such that all members became blind. Generation after generation were born blind and eventually even the legends of the elders telling of being able to see were lost. Everyone knew that they had these soft spots in their heads which hurt when poked, but nobody knew if they had some intended function. Over time, the very concepts of light and sight no longer existed within their tribal knowledge. As a doctor specialising in diseases and abnormalities of the eye, you realise that with particular treatment you can restore the sight of these people. Assuming that you are fluent in the local language, how would you describe what you can do for them? How would you convey the concept of sight to people to whom such an idea was beyond their understanding? You could hardly just walk up to them and exclaim: "I can restore your sight!"

My point is that this is the very same problem that primitive organisms would have faced if sight did in fact evolve organically in an undirected fashion. When the first light sensitive cell hypothetically evolved, the organism had no way of understanding that the sensation it experienced represented light signals which might help it to understand its external environment: light and sight were concepts unknown to it. As earlier, *"Try describing an elephant to a blind man"*; well, even more problematic, try describing sight to those who have no knowledge or experience of sight, which was the situation of these early organisms.

The Training of Sight

Those familiar with the settlement of Australia by Europeans in the 19th century, and the even earlier settlements in the Americas and Africa, would have heard of the uncanny ability of the indigenous population to track people and animals. It was not so much that their visual acuity was better, but that they had learned to understand what they were seeing. It was found that this tracking ability could be learned and taught to others. In military field craft, soldiers are taught to actively look for particular visual clues and features. In my school army cadet days, we undertook night "lantern stalks" (creeping up on enemy headquarters) and later in life, the lessons learned regarding discrimination of objects in low light were put to good use in orienteering at night. All of this experience demonstrates that while many people simply "see" passively, it is possible to engage the intellect and "look" actively, thus apprehending much more.

With the advent of the aeroplane came aerial photography and its application during wartime as a method of intelligence gathering. Photographic analysis was a difficult skill to acquire - many people could look at the same picture but offer differing opinions as to what they were seeing, or rather *thought* they were seeing. The underlying lesson is that sight is more a function of intellect than passively receiving light signals through the eyes. Put another way, it is intelligent data processing.

Understanding Messaging

As discussed in an earlier chapter, data itself is *conceptual*, conveyed in any communications channel via symbolic representations appropriate to the communications medium. *Cognitive information* is derived by intelligently processing this input against a pre-loaded referential framework of conceptual and contextual data. Using a computer analogy, master files represented *conceptual* data, application files provided *context*, and transaction filed provide the *input* data. If a transaction file was input to a computer that did not already have access to the corresponding master and application files, no computing could result. This an apt analogy for our sense of sight: the biological "knowledge" equivalent to these reference files needs to be preloaded before sense can be made of input light signals. The question becomes: where could these have come from when their only source was the light source that as yet they could not comprehend?

The point to grasp is that message transmissions through any medium consist of symbols, not data or information: human sensory input is of that nature.

With apologies to Claude Shannon, Werner Gitt and other notables who have contributed so much to our understanding on this subject, I would contend that in the context of this discussion, none of these files contain *information* in the true sense: each contains *data* which only becomes usable information when intelligently correlated. I would further contend that no single transmission in any form can stand alone as information - absent of a preloaded conceptual and contextual framework in the recipient, it can only ever be a collection of meaningless symbols. This is easily demonstrated by simply setting down everything you have to know before you can read and understand these words written here, just as I earlier related concerning *menus*, or what you would have to know before reading a medical journal in a foreign language in an unfamiliar script such as Hebrew or Chinese.

Consider that you have four documents before you: one in English, one in French, one in Chinese, and one in Russian. You are told that each contains the same narrative. Consider that your only language is English (I speak Australian, purportedly a form of English). You can recognise some, or possibly all of the elements in the English version, can recognise some of the symbols in French, but have no idea about the other two. The issue is the multiple levels of coding, starting with the symbol layer. One document has symbols (letters of the alphabet) which you recognise, one has symbols which you mostly recognise, and the other two could be anything. At the next level, vocabulary and grammar, you may be conversant with all of the words in the English version, and some of the words in the French version due to their similarity. But then we have context and the difficulty associated with words having a semantic range, sometimes culturally conditioned. As earlier, in the English tradition, a "rubber" is used to erase mistakes, whereas in America it is used to prevent a mistake of an entirely different kind. In Australia, a "thong" is a form of casual footwear, but in America it is an item of clothing used to (partially) cover parts of the body unmentionable in polite company. Even colloquialisms play their part. An acquaintance of mine has a different understanding of the word "next" in the context of days of the week. If today is Monday, next Friday is just four days away, but for him, the day that I perceive as *next* is *this*, and the next is eleven days away. For me, next and this are synonymous in this context. Unsurprisingly, we sometimes struggle with appointments.

Punctuation is also significant in conveying data in the correct sense. Consider these words from the song, *I Don't Know How to Love Him*: "For I have had so many men before in so many ways he's just one more." Does the comma go after the word "before", or after the word "ways"? The difference in the information conveyed is significant. Finally, speech communication also depends on pronunciation, which in many cases does not follow the rules and has to be taught. I am still bemused by the American pronunciation of *Kansas* versus *Arkansas*, and *buoy* verses *buoyant*, but lest you think me racist, we have even more words in Australia which still confuse me. I am criticised when I fail to use the local pronunciation, especially of place names in different Australian States.

The point to understand is that the pathway from symbols stored on physical media, whether synthetic or organic, to information, is tortuous and multi-layered; a system far beyond the capabilities of undirected organic mutation and selection. Organic material contains none of the intrinsic properties which can account for the complex organisation of the storage and communication of the elements of cognitive information.

Understanding Meaning & Intentionality

Earlier I mentioned Werner Gitt's five-layer model of communications: Apobetics, Pragmatics, Semantics, Syntax, and Statistics. In this section, I wish to discuss just the first two: *Apobetics* relates to the intended result of a communication; *Pragmatics* to the expected action. Consider your intentionality when reading this or any written word, or of observing a picture, diagram, or scene. Sometimes you just browse, with no intention of memorising or paying particular attention. Other times, you observe for the purpose of responding to an immediate situation, but have no conscious need to remember beyond that occasion, even though you may learn a lesson for the future. At the most deliberate end is study, where your intention is to retain what you have read to a predetermined level of recall. During my training as an air traffic controller, we had three levels of knowledge: immediate, intermediate, and reference. Obviously, some situations required an immediate response using standard terminology (patter) in conformance with separation and other rules. There was little or any time to think (or panic). At a less urgent level, there was the opportunity to confer or with coordinate with others, where even some discussion was permissible, such as determining an airways clearance. The third level related to non-urgent activities, such as how to paint an obstruction. I would expect that most professions have similar protocols.

My point here is that any communication or message, whether perceived by our auditory or visual sensory systems, is insufficient in and of itself to effectively address the Apobetics and Pragmatic layers. The mind-brain complex must complete the task, and often gets it wrong. Primary Axiom #2 states that *all knowledge is built upon prior knowledge*, and it is this prior knowledge which forms the foundation for interpreting these two layers. However, in most cases, the intentionality arises within the individual, irrespective of any previous experience. From an evolutionary perspective, how did that arise? It is one thing to receive a communication via light or sound, but entirely another to know, or decide, what to do about it. Purpose is antithetical to evolution theory, yet purpose and intentionality are inherent in all communications. Evolutionists would have us believe that purpose, which is conceptual, arose through biological processes, implemented via neural networks. But before the neural network could be structured, the concept must have been determined, including its relevance to other concepts, and its relative importance.

For that to happen, there must be a controlling processor, that decodes existing neural structures into conceptual forms, before processing them in context with the new message to form information, and then re-encode the result back into a new neural network structure. Remember that neural networks are nothing but encoded, symbolic representations of external realities and internal cognitive concepts, and according to Primary Axiom #1, are incapable of explaining themselves.

Understanding Input Devices

We have five physical senses: sight, hearing, smell, taste, and touch, and each requires unique processes, whether physical and/or chemical. What may not be obvious from an evolution standpoint is that for the brain to process these inputs, it must first know about them (*concept*) and how to differentiate amongst them (*context*) (Primary Axiom #9). Early computers used card readers as input devices, printers for output, and magnetic tape for storage (input & output). When new devices such as disk drives, bar code readers, plotters, etc were invented, new programs were needed to "teach" the central processor about these new senses. Even today, if you attach a new type of reader or printer to your computer, you will get the message "device not recognised" or similar, if you do not preload the appropriate software. It is axiomatic that an unknown input device cannot autonomously teach the central processor about its presence, its function, the type of data it wishes to transmit, the context of that data, or the protocol to be used. Note that I said *autonomously* - an intelligent external agent may well have preloaded the software necessary to achieve those tasks.

The same lessons apply to our five senses. If we were to hypothesise that *touch* was the first sense, how would a primitive brain come to understand that a new sensation from say *light* was not just a variation of *touch*? I would take your mind back to the discussion on *Knowing That You Can*.

Signal Processing

All communications can be studied from the perspective of signal processing, and without delving too deeply, we should consider just a few aspects of the protocols. Before we start, let us lighten the mood with a little humour, although I apologise to those on whom the humour will be lost - it is a *nerdy* joke, but my hope is that it will stimulate thinking in the minds of those familiar with the technology of TCP (Transmission Control Protocol).

"Hi, I'd like to hear a TCP joke."

"Hello, would you like to hear a TCP joke?"

"Yes, I'd like to hear a TCP joke."

"OK, I'll tell you a TCP joke."

"OK, I will hear a TCP joke."

"Are you ready to hear a TCP joke?"

"Yes, I am ready to hear a TCP joke."

"OK, I am about to send the TCP joke. It will last 10 seconds, it has two characters, it does not have a setting, it ends with a punchline."

"OK, I am ready to get your TCP joke that will last 10 seconds, has two characters, does not have an explicit setting, and ends with a punchline."

"I'm sorry, your connection has timed out."

… "Hello, would you like to hear a TCP joke?"

The joke evidences the handshake protocol in place to ensure data integrity when two devices seek to communicate with one another. Without such safeguards, data integrity cannot be guaranteed. My point in quoting this joke is to consider this issue in relation to the supposed evolution of nerve pathways in the human body, especially those that convey "information" to the brain. Remember that the term "communicate" is conceptual: what biological entities do is to "interact" dependent upon their respective chemical properties. In the earliest organisms, how could they have known that their function was to *communicate*, what protocols and encoding were to be used, and how reliable could those pathways have been? Essentially, each nerve fibre conveys specific content, not by "messaging" *per se*, but by the chemical release at the destination being prespecified for the fibre. The nerve fibre only relays that the sensory neuron at the beginning of the nerve fibre experienced something to which it was susceptible. In some ways, this is similar to the TCP protocol where the sender tells the receiver that there is a message to be received, without saying what that message is; or alternatively, like your device telling you: "You've got mail".

Keeping that in mind, let us continue.

As previously discussed, (perhaps more than once), all transmissions, be they electronic, visual, or audible are encoded using a method appropriate to the medium. Paper transmissions are encoded in language using the symbol set appropriate for that language; sound makes use of wavelength, frequency and amplitude; and light makes use of waves and particles in a way that I cannot even begin to understand. No matter, it still seems to work. The issue is that for communication to occur, both the sender and receiver must have a common understanding, not just that the signal represents a communication, but of the communication protocol, the symbols and their arrangement, and both must have equipment capable of encoding and decoding the signals. That is the principle, and it can be implemented in various ways (Primary Axiom #5). Consider nerve fibres which do not convey messages, but nevertheless manage to communicate. At one end of the nerve is a sensory receptor which chemically, stimulates the adjacent relay neuron into an action potential which is relayed up the nerve fibre to a terminus neuron which responds with the appropriate chemicals. The question becomes: how were the beginning and end chemically synchronised?

Now think of the eye. It receives light signals containing data about size, shape, colour, texture, brightness, contrast, distance, movement, etc. The eye must decode these signals and re-encode them using a protocol suitable for transmission to the brain via the optic nerve and visual cortex. In this instance, the encoding is achieved by having separate nerve fibres for each attribute of the light, which are reconciled by the mapping in the cortex as we have earlier discussed. Upon receipt, the brain must store and correlate the individual pieces of data represented by the signals to form a mental picture, but even then, how does it know that its perception, the visualisation in the mind's eye, matches the scene captured by the lens? Movement may be interpreted by comparing signals in a manner equivalent to frame-by-frame analysis, but that is a very intensive data processing activity: did the early eyes not comprehend movement? How did the brain learn of the concepts conveyed in the signal such as colour, shape, intensity, and texture? How did the visual system, like the auditory, understand such concepts as phase locking? Evolutionists like to claim that some sight is better than no sight, but I would contend that this can only be true provided that the perceived image matches reality: what if objects approaching were perceived as receding? Ouch! What if a shadow was perceived as a solid object? Strangely, I have noticed both people and animals walking around shadows, as if ... what exactly?

As we discussed earlier in relation to *search space*, an empty or embryonic brain contains nothing which can be used to compare the inputs signals, and thus has no cognitive sense of reality to begin with.

When the telephone was invented, the physical encode/decode mechanisms were simply the reverse of one another, allowing sound to be converted to electrical signals then reconverted back to sound. Sight has an entirely different problem because the decode mechanism in the brain is entirely different organically to the encode mechanism in the eye as the brain or cortex do not deal with light, and the conversion is to yet another format for storage and interpretation. These two encoding mechanisms must have developed independently, yet had to be coherent and comprehensible with no methodology for verifying system integrity and data validation. Again, I am not a mathematician, but the odds against two coding systems developing independently, yet coherently, in any timespan, must argue against it ever happening.

Data Storage and Retrieval

Continuing our computer analogy, our brain is said to be the central processing unit and just as importantly, our data storage unit, reportedly with an equivalent capacity of 256 billion gigabytes (or thereabouts). In data structuring analysis, there is always a compromise to be made between storage and retrieval efficiency. The primary difference from an analysis perspective is whether to establish the correlations in the storage structure itself, thus extending the storage time but optimising the retrieval process, or whether to optimise the storage process and later mine the data looking for the correlations. In other words, should the data be indexed for retrieval rather than just being sequentially or randomly distributed across the storage device, in the case of the brain, via neuronal networks. It is said that the brain is made up of more than 100 billion nerve cells that communicate with each other via trillions of connections (synapses), so storage capacity and connections would not seem to be a problem. From our experience in data analysis, data structuring, and data mining, we know that it requires intelligence to structure data and indices for retrieval, and even greater intelligence to make sense of unstructured data.

Either way, considerable understanding of the data is required. Is it possible that undirected evolutionary processes could account for the efficient storage and processing of the *conceptual* dimension of the data, when it only deals in a *physical*, symbolic representation of that data? Without some method of intelligent structuring, the brain is left with an incredibly large search space, but then, where does it begin?

Now let us apply that to the storage, retrieval, and processing of visual data. Does the brain store the data then analyse, or analyse and then store, all in real time? Going back to the supposed beginnings of sight, on what basis did the primitive brain decide where to store and how to correlate data which was at that time, just a meaningless stream of symbols? One could posit that the neural network patterns were based on similarities in the input signals, but that gets us no closer to the relationships of conceptual data, and the visualisation of it. What was the source of knowledge and intelligence that provided the logic of data processing?

Correlation and Pattern Recognition

In the papers that I have read on the subject, scientists discuss the correlation of data stored at various locations in the brain. As best as I understand it, no-one has any idea of how or why that occurs. Imagine a hard drive with terabytes of data and those little bits autonomously arranging themselves into comprehensible patterns.

Quite apart from aspects of materials science, you would assert such to be impossible, but that is what evolution claims, for an organic medium at least. It is possible for chemicals to self-organise based on their physical properties, but what physical properties are expressed in the brain's neural networks such that self-organisation based on conceptual abstractions would be possible? The brain consists of neurons, synapses, axons, etc., and in each class, there is no differentiation: every neuron is fundamentally like every other neuron, and so forth. Now, even if there are differences such as in electrical potential or electro-chemical gradients, the differences must occur based on physical properties in a regulated manner for there to be consistency. Even then, the matter itself can have no "understanding" of what those material differences mean in terms of its external reality. Remember that what are stored are *symbols*: encoded representations of abstractions at a very primitive level.

In the case of chemical self-organisation, the conditions are preloaded in the chemical properties, and thus the manner of organisation is pre-specified. When it comes to data patterns and correlation however, there are no pre-specified properties of the storage material that are relevant to the data which is represented, whether the medium be paper, silicon, or an organic equivalent. Being immaterial, conceptual patterns cannot be autonomously manipulated in a regulated manner by the storage material itself, although changes to the material can corrupt the data being represented. I do that regularly whilst typing, but sometimes Microsoft, an intelligently devised external agent, automatically corrects my mistakes for me, and sometimes it makes them worse. I still cannot get Microsoft Word to permanently retain my language selection as English (Australian): quite autonomously, it chooses to revert to English (US) and reports spelling mistakes which are not. It also has strange quirks regarding grammar, wanting to change, for example, "Pattern … " to " Pattern …". Why the space preceding the initial letter, P?

Pattern recognition and data correlation must be learned, and that requires an intelligent agent that itself is preloaded with conceptual and contextual data, and has a method of verification.

Facial Recognition

Facial recognition has become an important tool for security and it is easy for us to think, "Wow! Aren't computers smart!" The "intelligence" of facial recognition is actually an illusion: it is an algorithmic application of comparing data points, just as in fingerprint matching, and it does that very well, but what the technology cannot do is identify what *type* of face is being scanned, at least not so far as my research has revealed. In 2012, Google fed 10 million images of cat faces into a very powerful computer system, designed specifically for one purpose: that an algorithm could learn from a sufficient number of examples to identify what was being seen. The experiment was partially successful but struggled with variations in size, positioning, setting and complexity. Once expanded to encompass 20,000 potential categories of object, the identification process managed just 15.8% accuracy: a huge improvement on previous efforts, but nowhere near approaching the accuracy of the human mind.

The question raised here concerns the likelihood of evolution being able to explain how facial recognition by humans is so superior to efforts to date using the best intelligence and technology, particularly when an essential component would have been absent: a method of verification that what the animal thought it was seeing matched the reality.

In an interesting study by Raymond Tallis, where he cogently argues that "human beings cannot be understood purely in biological terms"[3], he makes the point that brains do not store images. I believe that we have discussed the structure and limitations of brain matter to accept that assertion.

Irreducible Complexity

Our sense of sight has many more components than described here. The eye is a complex organ which would have taken a considerable time to evolve, the hypothesis made even more problematic by the claim that it happened numerous times in separate species (convergent evolution). Considering the eye as the input device, the system requires a reliable communications channel (the optic nerve) to convey the symbolic representations of an external reality to the central processing unit (the brain) via the visual cortex, itself providing a level of distributed processing. We have previously discussed the complexity of communications protocols, revealing that very demanding criteria are required to ensure reliability and minimise data corruption.

A certain amount of signal processing occurs in the eye itself; particular receptor cells have been identified in terms of function: M cells are sensitive to depth and indifferent to color; P cells are sensitive to color and shape; K cells are sensitive to color and indifferent to shape or depth. As an aside, my personal experience has me doubting that an individual cell can be sensitive to depth. I recall being taught in high school that it was our bifocal vision that allowed depth perception: somehow the vision system made a calculation based on the delta of the eye positions. Wondering how pirates with an eye patch could make their way around so well, I used to close one eye, and then the other, looking at various objects to determine if they seemed closer or further away, and concluded that what I was being taught was likely not true. Sensibly, I did not seek to refute our rather stern teacher. On a visual approach to landing, pilots judge height above the ground by the perceived width of the runway, which gets tricky if you switch between a Tiger Moth and a Boeing 747. As an Air Traffic Controller, I could spot aircraft at varying distances in a clear sky, without any surrounding structures to act as guides, and sometimes I could judge their distances reasonably well. I have no idea of how that worked, I just know that it did. Incidentally, there is a very interesting discussion here[4] regarding how bees land without crashing, which further illustrates how sight is a cognitive activity more than just the passive reception of light. Whilst judging a landing on smooth horizontal surfaces is relatively easy for the suitably skilled, landing on irregular surfaces of any vertical or horizontal surface orientation requires different guidance strategies, as carrier pilots know only too well. For the bees, however:

"To the amazement of engineers who had unsuccessfully tried lasers, radars, sonars and GPS technology in striving to design autonomous landing systems for flying robots, the bees' guidance strategy is "surprisingly simple".[5] Experiments show that bees land safely by simply ensuring that the surface they are approaching expands at a constant rate within their field of vision.[6] This is a form of *optic flow* monitoring,[7] which we have noted before.[8]"

As was known from aviation, when approaching a runway at a constant speed, the image of the runway widening gets faster the closer you get. According to the scientific research, bees have solved the problem by keeping the rate of expansion of the image constant using a biological autopilot which automatically slowed the speed of approach. The conclusion of the researcher was, as would be expected from an evolutionist, "It's surely self-evident that no 'guidance strategy' came about by itself." Indeed, but how? Something similar must happen when driving or riding at high speed. We judge our closing rate with a cause to slow down, like a corner, a slower vehicle, or a speed restricted area. I have experience with that, and fortunately have never failed to judge correctly. But I digress, now back to the discussion.

The question we must ask is how an undirected process could inform the brain about these different signal types and how they are identified. The signals are transmitted to different parts of the brain for parallel processing, a very efficient process but one that brings with it a whole lot of complexity. The point to note is that not only does the brain have the problem of decoding different types of messages (from the M, P, and K cells), but it has to recombine these signals into a single image - a complex task of co-ordinated parallel processing.

I have been aware for some time that the image received by the eye is inverted in transmission to the brain, just as trying to use an astronomical telescope for terrestrial purposes results in inverted images. I have wondered how the primitive organism that first developed an embryonic lens ever understood that the image was inverted, and then by evolutionary processes managed a corrective. Just recently I encountered another curiosity. My wife underwent surgery to correct a tear (as in *laceration*) in the back of the macular of her left eye. The procedure involved draining the fluid and inserting a gas bubble. As the gas bubble was absorbed, the eye would refill with fluid and the level could be seen as a line. Over the days that this process occurred, my wife perceived that the line was moving from top to bottom, whereas in truth it was bottom up – the fluid replacing the lighter gas. Here is evidence of the inverted image, yet curiously in this instance, the brain did not understand that it was inverted, even though it knew that all other images were. I have no explanation, and I doubt that anyone does, but it does suggest that the correction of the inverted image is not an autonomous response by the brain.

But back to the main theme, finally we have the processor itself which if the evolution narrative is true, progressively evolved from practically nothing to something hugely complex. If we examine each of the components of the *sight system*, it is difficult to identify a useful function for any one of them operating independently except perhaps the brain. However, absent of any preloaded data to interpret input signals from wherever, it is no more useful than a computer without an operating system. It can be argued that the brain could have evolved independently for other functions, but the same argument could not be made for those functions pertaining to the sense of sight.

I contend, with absolute confidence, that a system of sight is irreducibly complex. That is not to say that just our system of sight as we have it today, is irreducibly complex, but that even a more primitive and less discerning system is irreducibly complex, especially the processing in the brain that converts the multi-threaded signalling from the eye into an accurate visualisation of the external reality.

Inheriting Knowledge

Let us suppose, contrary to all reason, and everything that we know about how knowledge is acquired, that a primitive organism somehow began developing a *sense* of sight. Maybe it wandered from sunlight into shadow and after doing that several times, came to "understand" these variations in sensation as representative of its external environment. Just what it understood and how it got it right is anyone's guess, but let us assume that it happened. How is this knowledge then inherited by its offspring for further development?

If the genome is the vehicle of inheritance, then sensory experience must somehow be stored therein and be progressively built upon for further cognitive development to occur. I have researched this question, not finding the answers that I was seeking, but reconsider this, already quoted elsewhere:

"Between conception and age three, a child's brain undergoes an impressive amount of change. At birth, it already has about all of the neurons it will ever have. It doubles in size in the first year, and by age three it has reached 80 percent of its adult volume. Even more importantly, synapses are formed at a faster rate during these years than at any other time. In fact, the brain creates many more of them than it needs: at age two or three, the brain has up to twice as many synapses as it will have in adulthood. These surplus connections are gradually eliminated throughout childhood and adolescence, a process sometimes referred to as blooming and pruning."[9]

Fascinating, as it shows that the development of the neural connections that manage the autonomous processing by the brain, are orchestrated by specific biological cells as determined by the genome. As discussed earlier regarding genetic inheritance, although all cells contain a copy of the DNA of parents, only mutations occurring in a germ-line cell (i.e., egg or sperm cells) of the parents can be passed on to their offspring. Other mutations, i.e., somatic (non-reproductive) mutations occurring in cells found elsewhere in bodies of the parents, are not inherited – there is no known feedback mechanism to the germ-line, although there is encouraging research in this area. Thus, the hypothesis that mutations acquired by parents during their lifetimes, from whatever cause, can be inherited by offspring, is largely *false*. This was the concept of *inherited characteristics*, beyond those received by the parents from their parents. On this evidence, the "knowledge" acquired during the life of an organism cannot be passed on to subsequent generations. This poses a particular problem for evolution where it is claimed that certain abilities, e.g., flying, were acquired over generations through successive learning through those generations.

I have no answer as to how birds "learned" to fly before they were complete birds, but I do wonder.

Putting it all together

I could continue to introduce even greater complexities that are known to exist, but I believe that we have enough to draw some logical conclusions. Over the past sixty years, we have come to understand a great deal about the nature of information and how it is processed. Scientists have been working on artificial intelligence with limited success, but it would seem probable that intelligence and information can only be the offspring of a higher intelligence. Even where nature evidences patterns, such as in the human brain, we have reason to question how such patterns could be the result of progressively acquired, inherited physical properties. But the stumbling block for all such evolutionary hypotheses is that patterns cannot explain themselves (Primary Axiom #1), nor can they be recognised other than in chemical complementarity, because "pattern" is a concept which must be learned. A pattern is a form of information, but without an understanding of what is regular and irregular, it is nothing more than a series of data points. Most importantly, any pattern that arises based on the laws of physics and chemistry cannot represent a conceptual abstraction of which it is unaware.

Modern technology has shown that the path to pattern matching is not about pictures or images, but about matching selected data points in a digital image, which are unique for each person. If the computational theory of mind is in any way correct, then the human ability to recognise faces and shapes must implement a similar strategy.

We often hear the term, *emergent properties of the brain*, to account for intelligence and knowledge, but just briefly, what is really meant is emergent properties of the **mind**. You may believe that the mind is nothing more than a description of brain processes but even so, *emergence* requires something from which to emerge, and that something must have properties which are foundational to the properties of that which emerges. Emergence cannot explain its own origins, as we have noted before. To borrow the words of Joseph Keating, emergence "is no more than a 'pseudo-explanation', and may deceive us into believing we have explained some aspect of biology when in fact we have only labelled our ignorance."[10]

Our system of sight is a process by which external light signals are converted to an electro-chemical data stream which is fed to the brain for processing and storage. The data points of light, which are themselves symbolic representations of a physical reality, must be decoded and then re-encoded in a regulated manner using a protocol that is comprehensible by the recipient. The brain then stores these symbolic representations in a manner that allows correlation and future processing. Evolutionists would have us believe that this highly complex system arose through undirected processes with continual improvement through generations of mutation and selection. However, there is nothing in these processes which can begin to explain how raw data received as symbolic representations of reality, through a light sensitive organ, could be processed without the pre-loading of the meta-data that allows the processor to make sense of the raw data. In short, the only source of data was the very channel that the organism neither recognised nor understood.

Without the meta-data that establishes the relationship between the physical symbols and the abstractions they represent, no meaningful storage or processing can occur. Without the back-end storage, retrieval, and processing of these symbolic representations, the input device has no useful function. Without an input device, the storage and retrieval mechanisms have no function.

Just like a computer system, our sensory sight system is irreducibly complex.

References:

1. https://www.britannica.com/science/light

2. https://andor.oxinst.com/learning/view/article/what-is-light

3. Tallis, Raymond, *Aping Mankind*, Routledge Classics, New York, NY, 2016

4. Ross, J., *Bees no drones when it comes to landing*, theaustralian.com, 29 October 2013

5. Baird, E., Boeddeker, N., Ibbotson, M., and Srinivasan, M., *A universal strategy for visually guided landing*, Proceedings of the National Academy of Sciences (USA) **110**(46):18686–18691, 2013

6. Esch, H., Zhang, S., Srinivasan, M.V. and Tautz, J., Honeybee dances communicate distances measured by optic flow, *Nature* **411**(6837):581–583, 31 May 2001

7. Sarfati, J., *Can it bee? Creation* **25**(2):44–45, 2003; creation.com/bee

8. Ross, *Ibid*

9. http://www.urbanchildinstitute.org/why-0-3/baby-and-brain

10. Keating, Joseph C. Jnr., *The Meaning of Innate*, Journal of the Canadian Chiropractic Association, **46(1)**, 4-10

Chapter 7-6: The BOLT

"Complexity is not just a phenomenon, it can be perceived as a property of a system"

~ Pearl Zhu, Digital Maturity ~

I dithered over including this chapter in an already overly long manuscript, but finally concluded that it might add value for some readers struggling with the concepts offered. To that end, it might well serve to confuse rather than elucidate, and if so, I apologise, but understand that in all likelihood, my mind works very differently to yours. I have been convinced of this after three decades working with other IT professionals, who often wondered what I was on about.

The purpose of this chapter is that by demonstrating the incredible complexity of the messaging accomplished by our system of sight, we have good reason to question how such a system could derive from the claimed processes of evolution. For this, I will deconstruct the visual components of a common article: a bolt. If you have ever worked with inventory systems concerned with fasteners, worked in mechanical trades, or as I have done, messed around with restoring vintage motorcycles, you will likely understand that bolts come in a frustrating variety of specifications. Putting aside the functional purpose of different bolt types, I shall simply refer to the most common purpose: joining two pieces of material together.

We start with fastener standards such as metric, imperial, DIN, ISO, SAE, etc. There are different grades of material and finish: brass, hardened steel, stainless steel, and hot-dipped galvanised for example. In most cases, these can be visually identified by their finish, texture and colour. Lengths are dependent upon standards, as are diameters. Some bolts are threaded for their entire lengths, others not so. The bolt head, which determines the tool to be used, can be square, hexagonal, or socket, and frustratingly, the tool shape can be slotted, Phillips, square, or Torx (star shaped). My tool kit has grown progressively over the years due to manufacturers' preferences changing for no apparent reason. The pitch of the thread is determined by the fastener standard, which leads to people such as myself having a container full of mismatched nuts and bolts. You might wonder where I am going with all this, but we are getting there.

The point is that during the period when I was heavily involved in restoring vintage motorcycles, I could visually identify bolts of a variety of specifications. Not only that, but I could look at a task and visualise what type of bolt was required. Looking at any bolt on the bench, the detailed specifications would be encoded in the light signal reflected off the bolt. My eyes would transduce the light energy into electro-chemical signals for messaging via the optic nerve, re-encoding the specifications of the bolt using a method currently beyond our comprehension. This is the key point: the semantic layer encoded in the light signal was faithfully reproduced in a different coding system in the brain, retaining all of the fine detail of the bolt specifications: length, diameter, type, thread, colour, finish, etc.

The optic nerve conveyed this detail by rippling electrical signals from one neuron to the next, as earlier discussed. If you remember, our understanding of this messaging process is limited to the chemical and electrical interactions, but most importantly, that was all that was happening – just physical activities: there was no "understanding" as such, of the semantic layer being encoded. The transduction of the light signal results in a multiplexed message containing encoded details from each of numerous receptor neuron types, which we can only assume gets sorted in the visual cortex before being stored in the brain proper. The optic nerve, made up of neurons, consists of over one million nerve fibres, which goes some way to explaining how such complex signalling can occur simultaneously, especially when watching events encompassing multiple activities.

The issue is the granularity of the message. No doubt, the million plus nerve fibres are sufficient to handle the multiple aspects of the detail to be conveyed, but apart from what we have already discussed in earlier chapters, let us see what clarity I can bring by approaching it from a data processing perspective.

In inventory systems, each type of bolt is given a discrete part number. The file record for that part contains fields for each of the individual characteristics – length, type, etc. To improve access time for searches, each field may be indexed, or new fields created that group characteristics that are common. As an aside, similar functionality must exist in the mind-brain complex for me to identify what type of bolt I need for a specific task. During processing, these fields are represented in a digital format (logically, ones and zeros), with the fields predefined so that programmers know what they are dealing with. Messaging via the optic nerve is similar in that the semantic layer is encoded across multiple nerve fibres, the firing of which is compared to digital processing – some fibres are "on", others "off". The question I ask is: how does the visual cortex recombine the signals conveyed via individual nerve fibres, and where is, or what is, the process that provides the equivalent of the "file definition", that allows the brain to correlate the signals with the characteristics of the object seen? The view of the bolt is encoded by the eye for transmission via the optic nerve, and then stored in memory, although whether any subsequent decode / encode process occurs is unknown. Just how much memory capacity is required to store the details of the bolt is unknown, because the method of encoding and storage is unknown, other than it is believed to be via neural patterns. However, having seen the bolt, I can close my eyes and visualise it, recalling from memory, the various characteristics. I could also articulate those characteristics, or make a rough drawing.

To achieve this, some process in the mind-brain complex reconstructs a visual image from the physically encoded representations in the neural network. The brain does not store images: it can only store encoded representations based on the intrinsic properties of the neural connections. We use the term, the *mind's eye*, to locate such visualisations, and are certain of the reality of this phenomenon because we can describe or draw what we are mentally "seeing".

To summarise, light signals are transduced by the eye into electro-chemical signals for transmission via multiple fibres in the optic nerve to the visual cortex, where complex signal processing occurs. Encoded representations of the characteristics of semantic layer of the light signal are stored in neural networks. These characteristics are the properties of the object, in this case a bolt, and from data processing experience, we know that the value of each must be correlated to provide identification of any unique occurrence. Some values are provided from the visual event, and others from memory previously acquired; these include:

    a.   Type: anchor, eye, carriage, shoulder.

    b.   Material: stainless steel, brass, galvanised.

    c.   Standard: metric, imperial, SAE.

    d.   Head: Hex, Allen, Phillips.

    e.   Length: 2 inch, 25 mm.

f.   Diameter: ¾, M4, 8 mm.

g.   Thread: UNC, UNF, UNEF

h.   Colour / Finish: natural, anodised, plated.

An object as simple as a bolt is obviously quite complex when the detailed specifications are considered. The brain is tasked with correctly correlating each of these characteristics to, in the first instance, visualise the object, and secondly, to understand and convey the specifications. My purpose in conveying this detail is not to discuss bolts *per se*, but to illustrate the complexity with which the visual signalling system must cope, and the correlations to be performed with previously stored data (memory).

We know at a physical level, how neurons communicate with one another, and we have some understanding of the coding protocol, in that each nerve fibre represents a specific chemical complex. With the triggered nerve fibre firing, the combination of active fibres is interpreted in some mysterious way. But now for the truly mysterious part – a stumbling block for evolution: *visualisation*. We use the term, "mind' eye" to locate where in our mind-brain complex this is achieved, but this metaphor is the best that we can do.

I want a scientific explanation to replace the metaphor.

Part 8: The Mind-Brain Complex – What Is It?

"If our brains were simple enough for us to understand them, we'd be so simple that we couldn't."

~ Ian Stewart, The Collapse of Chaos: Discovering Simplicity in a Complex World ~

Neurology and psychology are two overlapping sciences. The question is: Are these two different perspectives on the same physical behaviours, as some would claim, or as others would claim, sciences related to entirely different behaviours, one physical, the other not? Another question arises: How does one make a scientific argument over a scientific fact that is disallowed by a definition, based on philosophical materialism? Unsurprisingly, we encounter such unsubstantiated, nonsensical assertions as this by Daniel Dennett:

"There is only one sort of stuff, namely matter – the physical stuff of physics, chemistry, and physiology – and the mind is somehow nothing but a physical phenomenon. In short, the mind is the brain … we can (in principle) account for every mental phenomenon using the same physical principles, laws, and raw materials that suffice to explain radioactivity, continental drift, photosynthesis, reproduction, nutrition, and growth."[1]

It is all very well to assert that, *in principle*, we can account for every mental phenomenon, but an entirely different matter to actually do so. Part of my goal in this study is to convince you that Dennett is entirely wrong, no doubt seduced by his adherence to philosophical materialism. To do that we must consider Metaphysics: *the branch of philosophy that deals with the first principles of things.*

Before delving into the nature of the mind-brain complex, we must first deal with some issues that scientists commenting on the subject seem to overlook. A lack of precision at this point will allow dubious claims to be made, and in truth, this is what I have found in my research. It is logical, to my way of thinking, that before we can discuss, for example, what the mind does, or how it works, we ought first to ascertain just what the mind is. To do this, we must review the nature of things, both organic and inorganic.

Entities, Properties, and Activities

To review an earlier discussion, an *entity* is a thing with distinct and independent existence. In this discussion, organs of the body can rightly be termed, entities. *Properties* are defined as attributes, qualities, or characteristics of an entity. *Activities* are what entities do. Putting these together, the activities of an entity are allowed, constrained and otherwise determined by its properties; in this sense, we can consider properties to be *regulators*. In other words, entities can *only* do what their properties *allow*, and unless constrained by an external influence, *must* do what such properties prescribe. This is all to the good, for otherwise our hearts might stop beating for no reason at all. In the context of biological organs, the relevant properties are defined by the types of cells comprising the organ. We need delve no deeper than that (for now at least).

Let us discuss properties. Elasticity, for example, is a property found in both organic and inorganic matter, being the ability of an object or material to resume its normal shape after being stretched or compressed. If you are interested, you could research the constraints in linearly elastic materials, but that is of no interest to me here, other than to note that whilst elasticity is a property of some materials, it is constrained by other properties. Many organs exhibit a degree of elasticity based on their function, but malfunction when that degree is compromised. The heart is one example, and the bowel in cats another (my father died from symptoms associated with an enlarged heart, and my cat from megacolon, just in case you were curious about why I chose those examples).

Now, consider the activity in developing a crystalline structure such as a snowflake, or other naturally occurring crystalline materials, e.g., salt. Such structures always have geometric shapes such as triangles, rectangles, squares, and cubes. The shapes are a direct result of the type of molecules and atoms that make up the crystal, but the molecular structure of crystals disallows curves, circles, and spheres. A diamond, whilst considered a crystal, is of a unique form, being made from compressed carbon under intense heat and pressure, rather than growing as other crystalline structures do. My point is to emphasise that in both organic and inorganic materials, activity is regulated by the intrinsic properties of that material, and other materials with which it interacts in any given environment.

So, we have organs (entities) which have properties based on the biology of the relevant cell types. These properties determine and regulate the activities of the organ. Without attempting an explanation, especially as I only have the slightest familiarity with such things, I believe it logical to assert that all biological activity is consonant with what we understand to be the laws of chemistry and physics. I am confident in saying that all biological research is predicated on that being true. The question becomes: What activities can biological entities perform? The short answer is: many, but for a longer answer, I will need to consult an authoritative source, as here[2]:

"A definition of biological activity is proposed that is superficially analogous to the equation relating the thermodynamic activity of a solute to its concentration via an activity coefficient. The biological activity of a molecular entity is defined as $A=cf$, where A is the activity, c the amount-of-substance concentration, and f is a parameter designated as "inherent activity." Units and dimensions are determined by the type of activity, catalytic (katal) or binding (mol^{-1} L). The measurand is described by a chemical equation that identifies the entity for which an activity is being monitored. This definition of biological activity has the advantage of separating the chemical characterization of the entity in terms of structure and amount from the assessment of biological activity. Ideally, a homogeneous entity is used for the measurement of f. In instances where impure materials are used or the chemical equation defining the activity is unknown, the evaluated parameter should be designated as f' to denote its empirical nature. Any measurement of f or f' should be qualified with an appropriate estimate of measurement uncertainty."

Now, I have little comprehension of what that all means, but would offer that the authors, who work in the fields of Clinical Biochemistry, Molecular and Microbial Sciences, and Analytical Chemistry, are well qualified in their respective fields. It is apparent that according to this definition, biological activity is determined entirely by chemical reactions alone, and can do no more, nor less, than the chemical properties allow. As best as we can understand, chemicals are non-sentient, i.e., they lack the capacity to feel, perceive, or experience *subjectively*. All scientific research is conducted under the assumption that everything that happens in nature does so *objectively*: materials have no say in what they do – they just do what they *must*. Humans experience physical events both objectively and subjectively, but where subjectively, we can be deceived because at the cognitive level, we always experience reality from our own perspective. Chemicals, in that sense, always experience their reality objectively, just being and doing what they *must*.

What is obvious to me is that biological organs, of all types, can only do *objectively* what their chemical properties allow them to do, and no more. *Subjective* activities are not within the domain of biological organs – they cannot speculate, make decisions, or conceptualise their environment.

Architecture, Structure, and Networks

The human body would seem to evidence design, although evolutionists whilst often using the terminology of design, deny its influence because they are afraid of whom the designer might be. They deny design in favour of genetic mutation, natural selection, and other biological processes, giving rise to structures which have survival and reproduction benefits. Thus, microbes grew into man. Putting aside the origin of such structures, we need to review the nature of structures, identifying them as entities, properties, or activities as we did in an earlier chapter (*Foundational Propositions*). I believe it reasonable to conclude that they are complex entities composed of other entities, which give them their properties. For example, a building can be designed to insulate, let in light, or even facilitate wind currents, but it does none of these things by itself as a complex entity. Even insulation, whilst appearing to be an active agent, is actually passive, its properties disallowing heat transfer. It is the properties of an entity that allow, or disallow, activities by other entities such as heat, light, and wind.

In the human body, the skeletal structure prevents the internal organisms from collapsing into an untidy mess. The bones themselves perform few autonomous activities, other than growing and later deteriorating, all dependent upon the properties of their cells. In essence, structures are enablers of activities by other entities. A network is a type of structure, whether organic or inorganic. Digital computers can have internal networks, and be connected to external networks, but no network as an entity, nor its component entities, is capable of autonomous activity. Networks are in a sense, reactive enablers, which whilst allowing and regulating externally caused activities, dependent upon their individual properties, and their state of health, do not initiate activities themselves. Not for nothing do we describe machines as being sick, or computer programs as having bugs.

Now, this may all seem obvious and unworthy of comment, but as we will discover in the following chapters, these fundamentals are seemingly ignored by scientists when they attempt to explain the workings of the mind-brain complex. If you examined their wording carefully, it would be reasonable to conclude that at some level, *they know not of what they speak* - I hope to be able to convince you of that truth. On the other hand, they may well know, but are deliberately disingenuous, not wanting to run counter to established paradigms which have morphed into secular religions.

I encountered this interesting observation in an interview with Israeli historian Yuval Noah Harari, headlined: "Homo sapiens as we know them will disappear in a century or so". Harari commented:

"We know very little about the mind. We don't understand what it is, what are its functions and how it emerged. When billions of neurons in the brain fire electrical charges in a particular pattern, how does this create the mental experience, the subjective experience of love or anger or pain or pleasure? We have absolutely no idea. And because we understand so little about the mind, we also don't know how and why it emerged in the first place."[3]

Curiously, this author also asserts that we do not have free will. It seems logical to me that if you do not know the functions of the mind or how it operates, then you cannot be dogmatic on whether it exercises free will or not. The problem for this intellectual is that he operates within the paradigm of philosophical materialism, which forces him to believe that the brain and the mind are one and the same entity, despite the evidence to the contrary. He claims: "We have absolutely no idea", but incongruously asserts his ideas as truth. I won't be around in a century of so, but if I were to be, I would expect that we humans will still be around as we know them today, just as history evidences that Homo sapiens has not fundamentally changed for tens of thousands of years.

I find it curious (because I am curious about all manner of things, and this word looms large in my vocabulary) that someone could predict the end of Homo sapiens as we know it, to be taken over by artificial intelligence, whilst at the same time asserting that we don't know what intelligence really is. Without understanding human intelligence, any attempt to replicate it will always truly be, artificial, and not even beginning to approach true intelligence. Likewise, not knowing how it emerged, how can anyone be certain that it emerged through evolution? That is the subject of my earlier book, "*Information, Knowledge, Evolution and Self*", which argued that information and knowledge being conceptual, they cannot have a material or physical origin. No doubt, this chap is highly intelligent, just like, for example, Stephen Hawking, but not understanding his intelligence, he gives it free rein to wander off the reservation. If you objectively study his responses in this article, you will find that he contradicts himself on numerous points.

In her book, "*Touching A Nerve*", Patricia Churchland states:

"The human brain has been shaped by hundreds of millions of years of evolution. A powerful driver in the evolution of the brain was the importance of moving the body and making predictions so as to guide movement appropriately … complex brains evolved neuronal circuitry to model the body – its limbs, muscles, and innards – along with relevant aspects of the outside world."[4]

I am unsure of what Churchland means by "making predictions", but as a cognitive computational function, I would be interested to learn of how evolution could have been responsible. To suggest that the brain could shape its circuitry to model the body, let alone the outside world, is pure fantasy. The glaring flaw in Churchland's argument is that no part of the body can operate without the corresponding neuronal circuitry in the brain. No organ or limb could operate without a connection to the brain, and the brain "understanding" the purpose of the organ or limb (Primary Axiom #9). The neuronal circuitry could not have evolved in anticipation of each new feature, nor could any new feature have evolved to have contributed to survival or reproduction success, without the simultaneous corresponding modelling in the brain. I would emphasise the necessity for *systems thinking* in any attempt to argue the truth of undirected evolution. Now it is true that in a sense, the neuronal circuitry does represent a model of the body, in that the Central and Peripheral Nervous Systems, provide the necessary communications channels between the brain and other parts of the body, and the neuronal connections interpret sensory inputs and make the appropriate autonomous responses, but we can say no more than that. One needs to be beholden to evolutionary theory to make such statements, but there is no science behind that process, despite assertions to the contrary. We must understand that as with any other explanation, if evolution is said to explain everything, then in truth, it explains nothing – it is just a handwave.

As discussed earlier, it must be true that the cortex has a "map of the body", insofar as it determines the source of a nerve fibre by the terminal chemical communication of that fibre, for the fibre itself conveys no message other than an electrical pulse.

Whilst we can accept that the cortex maps the human body, we need to ponder whether the brain proper models its outside world. It clearly does so in some circumstances, as evidenced by the fact that a blind person can learn to navigate familiar surroundings, even without touching the obstacles in his/her path. I have tried walking around our largish house of fifteen years with my eyes closed, and I do it reasonably well. Clearly, something inside me knows its way around. Thus, I believe it to be true, based on practical experience, that the brain does indeed model "relevant aspects of the outside world" as stated.

I have some familiarity with 3D modelling in CAD/CAM applications, and know that the computer circuitry in some of those computers is no different to the circuitry in computers that processes Payroll or Inventory. In CAD applications, 3D modelling is the process of developing a mathematical representation of any surface of an object (either inanimate or living) in three dimensions using specialized software. Obviously, there is must be a software equivalent in biological organisms to accomplish modelling. What Churchland is suggesting, again by analogy, is that evolution developed "firmware", a form of persistent software, but physically in a dynamic structure capable of reorganising itself in response to external stimuli. I cannot but agree, but cannot begin to imagine how such a software equivalent could be represented in neural circuitry. If it is a function of the immaterial mind, as I would contend, then that is a different matter (pun intended).

There is an insoluble problem for how the neuronal circuitry could evolve through sensory inputs to model an external reality. In the chapter entitled, *The Sensory Conundrum*, I explain why I believe it impossible for this to occur purely by biological evolution. In earlier chapters, I explained the difficulties for evolution to account for neural networks.

"My" Mind - Brain

I have long been fascinated by how my mind and/or brain works, and often, *does not* work, most especially in its competence in processing symbols. In this context, alphabetic and numeric characters are classified as symbols of a particular type, meaning and purpose. Through experience, characterised by both satisfaction and frustration, my mind-brain complex competently processes the letters A-Z, punctuation marks, numerals 0-9, and simple operators such a +, -, /, and x. Beyond these, I tend to get muddled and no processing occurs.

The upshot of this is that I am reasonably literate in the English language, and can learn other Romance languages such as Italian, Spanish, and French. Due to my schooling, I also have some familiarity with Latin and Greek roots, which assists in English comprehension. Languages with alphabets based on other symbol sets, such as Greek, Russian, and Chinese, are quite simply beyond me. My daughter learned Japanese because we often hosted Japanese students on exchange, and to facilitate learning, she had magnets on the refrigerator with the various symbols. I would learn one or two on one day, and forget them the next, much to the disappointment of my precocious offspring. Because of other interests, I spent quite some time attempting to learn Hebrew, but alas, the symbols would never stick in my mind. Complicating the issue were variations in scripts, such variations in English seldom troubling me. At one stage in my working life, I had the task of summarising handwritten reports, quite short, from hundreds of different people whose expertise in handwriting varied from excellent to practically incomprehensible. Others in our group often consulted me over texts that they could not comprehend. Just why I have that skill is unknown to me – it has just been demonstrated that I have.

Mathematics is both a strong and weak point with me. I am very competent, although less so as old age creeps upon me, in simple and mental arithmetic. However, when it comes to what was known in my school days as Advanced Mathematics, requiring competence in logarithmic, exponential, trigonometric, and even more obtuse functions, I was always in danger of missing out on dinner, having accomplished nothing of value. As a consequence, I was never able to learn much in physics and chemistry, beyond the basics, but curiously, I am able to utilise basic physics is some quite innovative ways. Again, I do not know why – I just know that is it so.

The question becomes: What is behind such variations in the mental aptitude of people when it comes to processing specific tasks?

I have acquaintances who are experts in mathematics, physics, and chemistry, but whose literary skills have them struggling to comprise a text of 240 letters in Twitter. Very curious indeed!

What's Next

There is clearly a relationship between the mind and the brain. We have reviewed some fundamentals that we must use to identify the nature of the *mind*: is it a separate entity like the brain, a property of the brain, or a contingent activity of the brain? Our next step is to understand the nature of the brain itself, its properties, and the activities of which it could be capable.

It is worth noting this observation by Judith Grisel, a behavioural neuroscientist at Bucknell University, Pennsylvania, author of *"Never Enough: The Neuroscience and Experience of Addiction"*[5]:

"In more than 30 years as a neuroscientist, my most profound lesson has been that the brain and behaviour are products of multiple interacting influences, and the most powerful of these are located outside our heads, and therefore beyond the scope of any individual control. The brain acts as a conduit for such influences to shape who we are, but is not the source."

If the brain acts as a conduit, and our sensory organs are the inputs, to where does the conduit lead? Can the brain be both the conduit, and the processor of the inputs? In the computational theory of the mind, the brain is the processor and thus the source, just as in digital computing, the CPU is the source of transmissions to the output peripherals. What is it that shapes *who we are*? Of course, that is unanswerable without first identifying *who/what* we are, and *what* it is that identifies who we are: can it be the material brain, or must it be the immaterial mind?

References:

1. Dennett, Daniel C., *Consciousness Explained*, Penguin Press, London, England, 1991, p.33

2. https://link.springer.com/article/10.1007/s00769-006-0254-1

3. https://www.theguardian.com/culture/2017/mar/19/yuval-harari-sapiens-readers-questions-lucy-prebble-arianna-huffington-future-of-humanity

4. Churchland, Patricia S., *Touching A Nerve: Our Brains, Our Selves*, W.W. Norton & Company, New York, NY, 2013, pp. 33-34

5. Grisel, Judith, *Never Enough: The Neuroscience and Experience of Addiction*, Scribe Publications, London, UK, 2019

Chapter 8-1: What is the Brain?

"I was taught that the human brain was the crowning glory of evolution so far, but I think it's a very poor scheme for survival."

~ Kurt Vonnegut (1922-2007), American writer, satirist, and vocal critic of his society ~

No doubt, one can find numerous scientific descriptions of the brain, but for my purposes here, the following description, taken from a scientific study on language, is a sufficient place to begin:

"The brain itself is a very complex system ... The brain consists of gray matter and white matter. The gray matter is composed of about 100 billion neuronal cells that are interconnected via trillions of synapses. Each neuron has a number of connections via which it receives signals from other neurons (these are called dendrites), and it also has connections via which it forwards signals to other neurons (these are the axons). The axons come into contact with other neurons via synapses at which the transmission of the signals is realized by neurotransmitters. The white matter, in contrast, contains only few neuronal cells, being composed of fibre bundles connecting adjacent brain regions by short-range fiber bundles, or connecting more distant parts of the brain by long-range fibre bundles that guarantee communication between these. In their mature state, these fiber bundles are surrounded by myelin, which serves as insulation and enables rapid propagation of the signal. Both the gray and white matter are the basis for all cognitive abilities."[1]

My first point is to ask the reader to accept the accuracy of this passage, and the repeated use of the term, "signal". No mention of data, messages, or information, just the precise terminology of *signals*.

I will leave the reader to refer to whatever diagram best conveys this description pictorially to them. As an aside, I do wonder, from an evolutionary perspective, how and when the myelin-based insulation evolved? Of interest is a condition called *degenerative myelopathy*, a progressive disease of the spinal cord in older dogs. Demyelination is the destructive removal of myelin, the insulating and protective fatty protein that sheaths nerve cell axons. When axons become demyelinated, they transmit the nerve impulses 10 times slower than normal myelinated ones, and in some cases, they stop transmitting action potentials altogether. The question arises: could the nervous system have functioned effectively without the myelin sheath? This should remind us to always apply *systems thinking* to the process of evolution. If a lack of myelin sheathing represents a disease, it suggests that the earliest form of the nervous system, without myelin, was perhaps in the biological equivalent of a "diseased" state. Earlier we spoke of *communications integrity*, and the prerequisite quality control processes which must be implemented to verify the integrity of the communications channel. I have yet to encounter any discussions in evolution literature which addresses this issue. Continuing the quotation,

"The brain's functioning, however, is not yet completely understood. This holds for all different neural levels, from the single neurons and the communication between them up to the level of local circuits and the level of macrocircuits at which neuronal ensembles or even entire brain regions communicate."

Remember that neurons communicate with one another either by direct ion flow across their membranes, or through chemical neurotransmitters across the synaptic cleft. This creates an action potential which travels down the length of the neural circuit as an electrical pulse, analogous with the "ping" that you hear that accompanies the message, "you've got mail". There is no message content in the pulse, it is merely signalling that a chemical message is contained in the terminus of the circuit, telling the recipient to read the message in that particular mail box (nerve fibre).

In the context of my own study, the "how" of the brain's internal communications is less important than the "why", because the whole process occurs at a cellular level, and as such, the activity is electro-chemically based. If the study of the brain's internal communications is on the same basis as the study of cellular behaviour in other parts of the body, then no progress will be made in understanding language or any other form of cognition. Cells do what they do based on their intrinsic physical and chemical properties, and cannot do any more, or less, than such properties allow. Cognitive abilities have no such constraints, and thus I contend that we can never fully understand cognition by studying the physical structure of the brain. That is not to deny the science which has identified a relationship between cognitive functions and particular brain regions, but such relationships no more assist in understanding the fundamentals of cognition, than knowing where in a computer network particular processing is occurring.

Getting a little ahead of myself, if we hypothesise that the mind is the immaterial controller of the brain, then it is logical to expect to see biological activity in the brain, as it serves as the interface between the mind and the body. When a decision is made in the mind to initiate an action by the body, the brain is the organ which receives that thought and physically communicates the command. In more complex cognitive processes such as language, we should expect to see multiple brain regions involved, especially where language is to be articulated, or written as I am doing here. In a later chapter, we will have but a cursory look into the processing of language.

The study I have been referencing mostly covers grounds only partially relevant to my own purposes of understanding how the physical gives rise to the conceptual, but it is useful to review what is known about neuroanatomy. This we will do shortly. The author discusses "a functional neuroanatomical model of language comprehension" and provides a description of "the language networks connecting the different language-relevant brain regions, both structurally and functionally", later proposing a "neurocognitive model of the ontogeny of language"[2]. The author then admits that "the different language-relevant brain regions alone cannot explain language, but that the information exchange between these supported by white matter fiber tracts is crucial, both in language development and evolution". I continue to argue, obsessively perhaps, that there is no "information exchange" – merely an exchange of electrochemical signals; this is critical to understanding why we cannot understand how the mind works by examining physical structures. Whilst here, to save you looking up some scientific terms (which I had to do), "ontogeny" refers to the developmental history of an organism within its own lifetime, as distinct from "phylogeny", which refers to the evolutionary history of a species. In short, individual organisms develop (ontogeny), while species evolve (phylogeny).

Another attempt at explaining the brain is that by neuro-philosopher, Patricia Churchland in her book, "*Touching a Nerve*"[3]. The first chapter is entitled, "*Me, Myself, and My Brain*", where she contends that her brain and herself are inseparable. I cannot but agree, although ... Churchland continues,

"I think about my brain as *that* and about myself as *me*. I think about my brain as having neurons, but I think of me as having a memory. Still, I know that my memory is all about the neurons in my brain. Lately, I think about my brain in more intimate terms – as *me*." (p. 11) [italics in original]

In Churchland's terms, the biological entity known as the brain is capable of thinking about itself, but I contend that there is nothing in the structure of neurons, or in the neuronal structure of the brain, that would allow self-cognition in that way. Churchland is aware of the mind-brain conundrum, when she states:

"Some results are unnerving. Unconscious processes have been shown to play a major role in how we make decisions and solve problems. Even important decisions rely on unconscious brain activity. So you may wonder: How can I have control over a domain of brain activity I am not even aware of? Do I have control over brain activity I *am* aware of? And who is *I* here if the self is just one of the things my brain builds, with a lot of help, as it turns out, from the brain's unconscious activities?" (p. 12)

These are very important questions, which cause me to wonder why people are so dogmatic that the mind is a function of the brain, despite being unnerved by their belief. There is no physical part of the brain of which neuroscientists are unaware: it has been mapped in great detail. There *are* cognitive functions which have not been explained, but that is an entirely different matter. Churchland places great faith in promissory science, that one day all will be explained, which combined with her belief in evolution, goes partway to informing her worldview. This is foundational to what she terms, her unconscious brain activity.

Continuing Churchland's explanations, she offers:

"First, consider the brain circuitry organized to generate a neural model of the world outside the brain. Processes in this neural organization, model events in roughly the same way that the features of a map model the features of the environment." (p. 34)

That analogy is apt to an extent, but without an explanation of how a biological organism could self-organise to generate a neural model of an external reality, it is not overly useful. The next point to note that for any model to be observed, there must be an observer. What part of the brain is conscious of the model existing? Models do not, and cannot, interpret themselves, not even biological models.

We discussed modelling earlier, but the point to understand is that modelling involves conceptualisation and translation from one domain to another, e.g., from a *physical* reality to a *conceptual* map. This is more complex than perhaps Churchland might understand. Having studied Maps & Charts in an earlier life, I remember little other than the complexity of the different types of maps (Mercator projection, orthogonal projection, etc.), and that none were accurate representations other than in a specific application. The author continues with:

"Caution: Before getting too cozy with the map analogy, let me be clear about where it breaks down. When I consult a road map, there is the map in my hand and, quite separately, there is me. The map in my hand and I are not one. In the case of the brain, there is just the brain – there is no separate thing, me, apart from my brain."

Here, the author claims confirmation in Daniel Dennett's book, "Consciousness Explained"[4], but Edgar Andrews counters with impeccable logic:

"In the Analogy, this is like saying that the house (representing the physical brain) goes about its sophisticated functions but remains unoccupied. You, the owner, simply do not exist. She argues that external influences, imported by the physical senses, create physical maps of the external world in our physical brains, and then our brains then formulate appropriate responses which issue in suitable behaviour. In her materialistic scheme, no metaphysical "self" is need to create the brain maps, read them, or initiate responses. She continues: 'The "designer" of brain organization is not a human cartographer, but biological evolution … '."[5]

Andrews then reviews the concept of evolution *learning* from the consequences of an organism's experiences, whereby randomly generated "faulty maps" would be deselected, and only maps which provided survival benefit would be selected for reproduction. Whilst that sounds logical, it can only be so in scenarios that match the logic. Professor Andrews offers the example of an organism, not understanding the dangers of associating with venomous snakes, may get bitten and die. Sadly, the erroneous neural model (map), caused by a random genetic mutation, wasn't up to the task of making a wise choice about associating with venomous snakes. In truth, how would it know that a particular snake was venomous, other than by getting bitten and dying, unable to pass on that "knowledge" to subsequent generations? The difficulty for evolution is that mutations can be neutral in one environment, but deleterious in another. Neural circuits (brain maps) can be continually updated with false representations, becoming ever more detached from reality, until circumstances bring about a conflict with reality whereby that faulty representation is fatal. Too late for all the other generations, one must opine.

It is timely here to revisit the fallacy promoted by Richard Dawkins, as earlier explained: that in genetic inheritance, a single point mutation in a nucleotide is selectable. It is not. Whilst we cannot know the specifics of what may have been the case millions of years ago, we do know that human nucleotides exist in large clusters or blocks. These linked blocks are inherited as single units and never break apart. The human genome has approximately 100,000 to 200,000 linkage blocks. In any linkage block, the ratio of deleterious to beneficial mutations is likely to be high, effectively masking any benefit the beneficial mutation might confer. As Dr. Sanford puts it, "Since the large majority of mutations are deleterious, each mutation cluster will have an increasingly negative affect on fitness each generation." In one way, this argues for rapid deselection of organisms "infected" by deleterious mutations, but on the other hand, it similarly argues against the persistence of beneficial mutations. We should not forget that it is only from the germ-line, ova in women and sperm in men, that we inherit our DNA, and that there is no known feedback mechanism from other cells to update the germ-line. Thus, most mutations experienced in cell reproduction during the lifetimes of organisms are never passed on to offspring.

The *natural selection* model makes the unsubstantiated assumption that random deleterious mutations result in deselection within a generation or two, but the theory cannot have it both ways: because the unit of inheritance is a linked block, not single nucleotides, if deleterious mutations are deselected, then so too are any beneficial mutations, and even neutral mutations, within that block. One could consider the probability that even *deleterious* mutations are effectively *neutral* in all but specific circumstances. Such mutations could accumulate in a *potentially* adverse manner for untold generations, without the organism noticing. Consider a far too common problem in modern technology, whereby intelligently designed devices such as motor vehicles, can be on the road for years before a design fault is detected, and a vehicle recall initiated. By analogy, one could say that there was a deleterious mutation in the design process, which was effectively neutral until circumstances arose whereby the fault became apparent. If this happens in intelligent design processes by very experienced technologists, would it not also be possible in design processes subject to randomness? What the science of evolution cannot tell us is the relationship between mutations, and the environmental conditions prevalent at the time. Evolution would tell us that *it all worked out in the end*, but I suspect such is wishful thinking.

The issue of evolution "learning" from the experiences of individual organisms must be understood in the context of scenarios, and what it means *to learn*. Andrews notes:

"In other words, a learning process is involved, and learning requires *conscious* activity. No doubt someone will object to this last statement, pointing out that nonconscious robots and computers can learn. For example, that expensive voice-recognition software you bought for your computer had to learn the way you personally pronounce words before you could use it. Of course computers can learn, *but only if they have been preprogrammed to do so by a conscious agent*!"[6] (italics in original)

In the evolution scenarios, "learning" is the fortuitous outcome of chance encounters between mutations and environment, or put another way, it was all serendipitous - occurring or discovered by chance in a happy or beneficial way. Raymond Tallis provides another perspective on this issue of learning, in his discussion on perception of objects illuminated by a light source:

"This may seem to be a silly suggestion but let us stick with it for a moment and examine the actual things that are thought to trigger the nerve impulses that are in turn supposed to reveal the object. It is not the object that causes the perception of itself but its interaction with the light that results in my seeing it. This is a bit messy: the interaction is a fizz of events, not just a few neat straight lines connecting the object with the eye. [Ed. see Chapter 7-5 on Sight] The object has to be constructed from the interference with the light: a challenging task, to put it mildly. Indeed, it is so challenging that many neuropsychologists argue that the object that we experience is not really an object that is out there at all: it is a construct put together by the brain. This leads to the idea that the world we inhabit is *a mental model* that has only a tangential relation to what is "out there", an idea that has dominated cognitive psychology for many decades. Frith has gone further and argued that the contents of the mind are not real[7]."[8]

Is it just me, or do other not recognise that in this instance at least, Frith shows signs of desperation? You see, if the contents of the mind are not real, then the book that Frith thought that he wrote and published is not real, and we could not have read it. If we thought that we had read it, then it was just a mental model that we ourselves had constructed, and only by chance could it be similar to the mental model in Frith's mind. As a psychologist and Emeritus Professor of Neuroimaging, why does he imagine that his *mental model* of the reference works he studied was sufficiently reliable on which to base a successful career? I doubt that you could convince a watchmaker of old, a neurosurgeon, or a racing motorcyclist, that what they perceived was not real, merely a mental model that may or may not match reality. I wonder whether Frith has visited a dentist, or had his appendix removed? It is true that the contents of the mind are not real, in that they are mental constructs not physical entities. However, experience proves that such mental models do accurately reflect reality, otherwise nobody could learn anything, practice any skill, navigate across the world, or build such technological wonders as we have today. The models must be accurate, otherwise there could be no shared understanding or experience.

Tallis continues:

"If the objects we experience are actually constructed out of data that may mislead us, although they may be corrected by subsequent experience (otherwise we would not survive to be further deceived), then we have an interesting case of the pulled rug. The brain, which is supposed to be the passive recipient of energy from the outside world, now suddenly becomes something that actually constructs that outside world rather actively. Such activity seems to be at odds with the notion of the brain as a material object helplessly wired into the material world that surrounds it, via causal interactions guided by the laws of physical nature. One would like to know where, out of the electrochemical activity of the cortex and other bits of the nervous system, the ability to construct an illusory or approximate world arises. The brain, it seems, has the power to fight back and shape the world by which it is shaped. This, of course, relies on counter-causally directed intentionality."[9]

Sensory experiences arrive in the brain as electrochemical impulses, which then stimulate neurons already in the brain, apparently into some forms of networks. Given that *exteroceptors*, receptors that respond to stimuli from outside the body, transduce this external energy into encoded symbolic representations, on what basis does the cortex or upstream brain determine this new neural network model? If the model is determined by nothing other than the laws of chemistry and physics as we understand them, then the probability of the model accurately representing the external reality is from zero to none. Yet, we know from experience, that we instantly, and accurately, interpret realities that we have never encountered before. Try confidently climbing a high rock face being doubtful that what you perceive is not real! As a motorcyclist who has ridden new roads and tracks in many parts of the world, all without ending in tragedy, I can only conclude that my mental model of the road before me, disappearing under my wheels at speeds which the more cautious would baulk, must have been entirely accurate. As for potholes, tree branches, oil spills, and other road debris, or the local wildlife including kangaroos, wallabies, emus, goats, sheep, cattle, wombats and pigs, the less remembered the better.

The attempted explanations by cognitive psychologists are not supported by what we know to be true. Science cannot explain *intentionality* in scientific terms, for if the material brain alone determined intentions, then we would be unable to act, or not act, contrary to the predetermined intention. Chemicals, and biological constructs of chemicals, *cannot change their minds*: they have none. Whatever circuitous route is proposed for the workings of our neural networks, the fact remains that the outcome is inevitable – that's how chemistry works – there is just one possible outcome for any given set of circumstances.

Yet, we evidence many possible outcomes, often contrary to what our instincts would dictate. This is the curiosity of humankind – we can be what we ought not.

References:

1. Friederici, Angela D., *Language In Our Brain: The Origins of a Uniquely Human Capacity*, The MIT Press, Cambridge, MA, 2017, p. 5

2. *Ibid*, pp. 11-12

3. Churchland, Patricia S., *Touching A Nerve: Our Brains, Our*

Selves, W.W. Norton & Company, New York, NY, 2013

4. Dennett, Daniel C., *Consciousness Explained*, Penguin Press, London, England, 1991

5. Andrews, Professor E.H., *What is Man? Adam, Alien, or Ape?* Thomas Nelson Publishers, Nashville, TN, 2018, p. 187

6. *Ibid*, p. 188

7. Frith, Christopher D., *Making up the Mind: How the Brain Creates our Mental World*, Wiley-Blackwell, Oxford, UK, 2007

8. Tallis, Raymond, *Aping Mankind*, Routledge Classics, New York, NY, 2016, p. 110

9. *Ibid*, pp. 110-111

Chapter 8-2: Neuroanatomy

"When I look at the human brain, I'm still in awe of it."

~ Ben Carson, American neurosurgeon ~

I purchased this publication, the 4th Edition of *"Neuroanatomy Text and Atlas"*[1], hoping to find some answers on neural network structuring, but unfortunately my expectations were not met. That was not the author's fault – the book was intended for a different audience, but nevertheless, it did provide some useful commentary and diagrams. Having been published in 2012, I am confident that a later edition would offer more insights, but I doubt that any new discovery regarding the anatomy of the brain will resolve the physical/conceptual conundrum that so challenges scientists, or perhaps more correctly, is seemingly ignored by scientists. The text is replete with descriptions such as this: "... coverage of regional anatomy of the auditory system begins with the ear, where sounds are received and initially processed, and ends with the cerebral cortex, where our perceptions are formulated."[2] I can see students nodding in acceptance of this, but wait a minute – what do these words mean? What initial processing occurs in the ear? Actually, not much from a cognitive perspective: sounds travel through the outer and middle ear, being transduced from one energy form to another, and into the inner ear where the cochlea transduces the input signals into electrical nerve impulses. These are transmitted to the primary auditory cortex where initial processing occurs.

Not surprisingly, my most important issue has been glossed over: *"where our perceptions are formulated"*! The burning question is this: How are conceptual perceptions formulated by physical processes? The auditory cortex has a number of sub-divisions as discussed in an earlier chapter, but all this is still a matter of anatomy and physiology. I want to know how the physical becomes conceptual.

Structure of the Brain

The brain is not an independent organ like a heart or liver, but is part of the central nervous system. It interacts with other parts of the body through the Central and Peripheral Nervous systems, activated by sensor neurons, inputs from monitoring sensors, and responding with motor neurons to control bodily functions. It operates in an autonomous mode controlling involuntary actions such as breathing, and in a non-autonomous mode for the much-disputed voluntary activities such as composing music, writing poetry, or designing cars, houses, aeroplanes, or whatever.

The brain consists of two hemispheres, each hemisphere comprising four lobes with numerous folds. According to scientist who study such things, these folds do not all mature at the same time. The chemicals that foster brain development are released in waves; as a result, different areas of the brain evolve in a predictable sequence. From my background in manufacturing systems and processes, I continue to be fascinated by what might be described as the operational sequence of building the body and the brain. Somehow, those detailed process instructions come pre-packed in the genome, and only unpacked and activated according to some high-level master production schedule. Just how that could have developed is beyond my imaging. The timing of these developmental changes explains, in part, why there are "prime times" for certain kinds of learning and development. Different parts of the brain control different kinds of functions. Most of the activities that we think of as "brain work," like thinking, planning or remembering, are handled by the cerebral cortex, the uppermost, ridged portion of the brain. Other parts of the brain also play a role in memory and learning, including the thalamus, hippocampus, amygdala and basal forebrain. The hypothalamus and amygdala, as well as other parts of the brain, are also important in reacting to stress and controlling emotions. We have already discussed the fundamentals of neurons and their connections, so there is no need to repeat that detail here, other than to note that the neural network structure of the brain is very unlike the structure of the remainder of the CNS, and the PNS.

Interestingly, nerve cells proliferate before birth. I acknowledge that we have covered some of this in earlier chapters, but again, I think it worth repeating in a different context. I beg your indulgence – just skip the parts with which you are familiar. It is said that the embryonic brain of a foetus has approximately twice as many neurons as it will eventually need, although why evolution would result in redundancy, giving newborns a safety margin in developing a healthy brain is anyone's guess. Apparently, most of the excess neurons are shed in utero, and at birth, an infant has roughly 100 billion brain cells. Every neuron in the brain has an axon, and most have multiple dendrites, such that they are ready to begin making connections. As a child grows, the number of neurons remains relatively stable, but each cell grows, becoming bigger and heavier. The proliferation of dendrites accounts for some of this growth. The dendrites branch out, forming "dendrite trees" that can receive signals from many other neurons.

Quoting from here[3]:

"In the first decade of life, a child's brain forms trillions of connections or synapses. Axons connect to dendrites, and chemicals called neurotransmitters help send messages (called "impulses") across the resulting synapses. Each individual neuron may be connected to as many as 15,000 other neurons, forming a network of neural pathways that is immensely complex ... As the neurons mature, more and more synapses are made. At birth, the number of synapses per neuron is 2,500, but by age two or three, it's about 15,000 synapses per neuron."

At first reading, this assertion of up to 15,000 synapses per neuron seemed exaggerated, until further research revealed that in truth, it may be understated. This online article[4] makes interesting reading, but it offers nothing new in terms of understanding how a physical network of neurons, basically chemicals, electro-chemical reactions, and network patterns, could act in a cognitive fashion resulting in conceptualisations in the *mind's eye* (which is where?) This continues to be a problem for those researching neuronal networks and synaptic patterns.

Neural circuits, are said to be a population of neurons interconnected by synapses which carry out specific functions when activated. The brain is a construct of numerous neural circuits interconnected to one another. We have some understanding of the causes of activation both from the body to the brain, and the brain to the body, in autonomous functioning, but what of volitional activity: what is the activation? I suspect that it is this unanswerable question that has evolutionists arguing that we do not have free will, and everything that we ever think or do is a function of our evolutionary development. The other question is: by what imperatives are neural circuits caused to interconnect, other than as instructed by the genetic code during development of the embryo?

A simplistic explanation is given here:

"Basically, a neuron is just a node with many inputs and one output. A neural network consists of many interconnected neurons. In fact, it is a "simple" device that receives data at the input and provides a response. First, the neural network learns to correlate incoming and outcoming signals with each other — this is called learning. And then the neural network begins to work — it receives input data, generating output signals based on the accumulated knowledge."[5]

As an analogy this works reasonably well, but problems remain. A neural network does not receive *data*, but encoded symbolic signals which may or may not trigger an action potential in downstream neurons. Remember too that when an action potential causes a synapse to fire with neurotransmitters, the membrane returns to its rest state. Usage of the conceptual terms, "learning" and "knowledge" is disingenuous, for as the previous description made clear, neuronal networks operate purely at the physical level, whereas these terms relate to the conceptual. This online entry[6] giving a more accurate and detailed explanation, and generally avoids loose and deceptive terminology.

Thus, we have the architecture of the brain, much as we would have a description of, for example, the rooms in a house. In the latter, we designate areas as bedrooms, bathrooms, toilets, dining rooms, kitchens, and so on, appropriate to the activities which normally occur therein, but we do not expect the rooms themselves to perform those activities. The description of the brain takes a similar approach to designating rooms, but with one significant exception: it attempts to delineate not just areas of activity, but assigns autonomous activities to each "room". As Professor Andrews expressed it, "it is like saying that the house (representing the physical brain) goes about its sophisticated functions but remains unoccupied. You, the owner, simply do not exist." Alternatively, to put another way, it is like saying that the house is its own occupant – a highly illogical proposition. I would return your attention to the fundamentals of entities, properties, and activities. If the brain, as a biological organ, is an entity, it cannot also be its own activity. Its properties regulate its activities, but it must be understood what the properties allow, and disallow.

This brings us back to the differentiation of *autonomous* and *volitional* activities. Note the verbs often used about brain activity: *begin, controlled, coordinate, responsible, manage, involved,* and *contain*. The question we must ask is: which of these verbs, if any, represent activities of which biological entities are capable? Accepting that the brain is a biological organ which developed from embryonic cells, much as did the heart, lung, kidney, and liver, which of these verbs could be appropriate for describing the activities of these other organs, and in what manner? Can any of these other organs experience and respond *subjectively*, or only *objectively*, in accordance with the chemical properties of their constituent cells? If you answer, no, then you must ask: why would we think that the brain, of essentially the same composition, could do so?

Of course, there the house analogy ends, for surely, the house could not have designed, built, and equipped itself with the necessary appliances even whilst not anticipating occupants. That would be asking too much, yet that is what evolutionists are asking us to believe happened biologically, still without an occupant. Steven Pinker certainly believes so: "Computers are assembled according to a blueprint; brains must assemble themselves."[7]

There is something very mysterious about how the genome could contain all of the necessary instructions for assembling the brain, to make the multitude of neuronal connections that it does during gestation, and after. There can be no doubt that it does precisely that, but I question whether the process of undirected evolution is capable of achieving such an amazing set of instructions compacted into the structure of the genome. By analogy, the genome is primarily a zipped file containing a biological equivalent of WinZip©, which knows how to unzip itself to create thousands of functionality-specific files, which know how to cross-reference themselves to create a biological relational database. In an earlier life, I designed relational databases using techniques such as third-normal form to optimise and simplify the structures, to avoid duplication of data, and meet other performance criteria.

I am in absolute awe of the design of the genome. Were I tasked with designing a database to contain the overwhelming number of specifications for a human being, I would have no idea of where to begin.

Another Perspective

Let me end this chapter with a quote from a source that I promised I would not access. In my defence, it is worthy of consideration, despite the religious overtones. You may ignore it if you so wish:

"Supposing there was no intelligence behind the universe, no creative mind. In that case, nobody designed my brain for the purpose of thinking. It is merely that when the atoms inside my skull happen, for physical or chemical reasons, to arrange themselves in a certain way, this gives me, as a by-product, the sensation I call thought. But, if so, how can I trust my own thinking to be true? It's like upsetting a milk jug and hoping that the way it splashes itself will give you a map of London. But if I can't trust my own thinking, of course I can't trust the arguments leading to Atheism, and therefore have no reason to be an Atheist, or anything else. Unless I believe in God, I cannot believe in thought: so, I can never use thought to disbelieve in God."[8]

I do not agree with the logic of Lewis in concluding for God, well, not entirely, but I do agree with his remarks about trusting his own, or my own, thinking. What is it about the neural activity in my brain that gave rise to the concept of thought, and had me thinking that I was thinking, a process beyond the capability of biology? If evolution is a biological process, it cannot be claimed to have been proven until scientists can explain the process of the material deriving the immaterial, i.e., the conceptual. The explanation: *emergent property of the brain* will simply not do. That some evolutionists argue that there is no way that evolution could result in free will, then even more so, there is no way that evolution could result in the ability to perceive and imagine conceptually. The natural corollary to these arguments should be obvious.

For those looking for a concise overview of the anatomy of the brain, I can highly recommend this online resource[9].

References:

1. Martin, John H., *Neuroanatomy Text and Atlas*, 4th Edition, McGraw-Hill Inc., New York, NY, 2012

2. *Ibid*, p. xv

3. https://extension.umaine.edu/publications/4356e/

4. https://www.frontiersin.org/articles/10.3389/fncir.2016.00023/full

5. https://becominghuman.ai/neural-networks-relation-to-human-brain-and-cognition-b45575359f64

6. https://en.wikipedia.org/wiki/Neural_circuit

7. Pinker, Steven, *How the Mind Works*, Penguin Books, London, UK, 1998, p. 26

8. Lewis, C.S., *The Case for Christianity*, Touchstone Books, Simon & Schuster, UK, 1996, p. 32

9. https://mayfieldclinic.com/pe-anatbrain.htm

Chapter 8-3: How the Brain Works

"The human brain is an incredible pattern-matching machine"

~ Jeff Bezos, American internet and aerospace entrepreneur, founder of Wikipedia ~

If I take *incredible* to mean: *impossible to believe*, then I agree, for the human biological brain may be capable of matching physical patterns in neural networks, and even that I dispute, but it should be obvious that the brain is incapable of matching *conceptual* patterns from symbolic representations. One might be able to argue concerning *physical patterns*, but any representation in the brain has nothing of the physical characteristics of the real-world entity being represented, as we found argued from another perspective in a previous chapter. If the basis of our mental models does not match reality, then any attempt at pattern matching would be futile.

In this context, let me repeat what I believe to be irrefutably true. Patterns cannot interpret themselves (Primary Axiom #1), nor can patterns interpret other patterns, because *pattern is a concept* relating to inactive entities. Patterns, by themselves, are not capable of activity. If the brain's neural networks are organised in patterns, then there is no entity within the brain capable of performing interpretive activities. Any pattern in the brain, which was organised by the physical and chemical properties of the brain's composition, could not be organised around the concepts which the patterns are said to represent.

In this section, we will look at the composition and architecture of the brain, to consider what the brain, a physical organic blob, can and cannot do. Lest I be considered more knowledgeable than I am, understand that this discussion will be driven not from my own scant knowledge of the subject, but by quoting studies from authoritative sources, and interjecting my comments based on my Primary Axioms and earlier arguments. Where possible, I will attempt to differentiate the *Brain* from the *Mind*, the latter to be discussed in later chapters. This is a difficult task as practically all of my reference sources conflate the two, but I will attempt to convince the reader that the truth lies elsewhere.

"Who" is the Ghost?

Recalling Gilbert Ryle's derogatory comments regarding the *Ghost in the Machine*, I would refer the reader to evolutionists' regular usage, when discussing the mind-brain conundrum, of the personal pronouns, "we" and "our". Of course, if we humans are nothing but biological robots, our thoughts and activities subject to the whims of where evolution has taken us, then there ought not be any such personal distinctions. Certainly, we can distinguish between humans as we distinguish between giraffes, cows, and other lower order species, but no more than that. However, that is not what we observe.

If, as Jeff Bezos claims, "*The human brain is an incredible pattern-matching machine*", where do these patterns reside in the brain such that some other part of the brain does the matching and interprets the result of the matching? Pattern is a concept, and is the term we apply to the arrangement of lines or shapes, such that they evidence symmetry or some other conceptual arrangement such circles, turrets, or the famed Fibonacci sequence in nature. The shapes, lines, or sequences do not know that they have significance to human intellects, and so we must ask why the brain, that consists of patterns, thinks that its patterns are of significance to itself. Again, what part of the brain is doing the observing, thinking, or matching of its own internal patterns?

Despite denying dualism, and denying that there is any ghost in the machine, scientists cannot avoid descriptions which infer the ghost. Why is that?

The Brain as an Interface

The mind-brain complex is, as yet, beyond our understanding, yet we know conclusively that there is a connection. Because there is far too much evidence for me to believe that cognitive processes could be purely biological, it seems logical to me to consider that one of the brain's functions would be to serve as an interface between the mind and the body, utilising the faculties inherent in the brain's autonomous processing. I have heard all of the objections, mostly condensed to: the immaterial cannot influence the material. But of course, if you do not believe in the immaterial to start with, why would you offer such an argument anyway. It reminds me of atheists pronouncing that "God would not do it that way", whilst denying the existence of God. I do not, and cannot, offer an explanation of how the immaterial could control the material, as in the sense of "mind over matter", but the evidence suggests that it does. That the primary activities of the mind concern the *conceptual*, rather than the *material*, convinces me that not everything could be material. I am further convinced by the fact that despite significant research, scientists are unable to explain the conceptual in material terms.

Returning to a computer analogy, for many iterations of technology generations, there was an issue with connecting external devices to a computer, especially other than from the OEM (Original Equipment Manufacturer). Even today, there are legacy connectors such as VGA, DVI, HDMI, PS/2, Ethernet, and generations of USB 1- 3. In recent years, Apple has gone from a 30-pin connector to the Lightning connector, and is even considering ditching that. The evolution of connectors was not just physical, but concerned with communication protocols, electrical power, and data transfer speeds. If one applies that experience to what evolution must have had to deal with, even more problems must become apparent. How long would it have taken evolution to resolve the issue of the functional synapses? Most synapse connections are chemical, but others electrical, involving an ion flow. I cannot even begin to imagine the genetic mutations needed to achieve that level of precise functionality.

A diagnostic course of action for TV or computer problems, is to check the connections. Connectors are known to fail, and I have even been told that polarity or resistance issues arise in HDMI cables, but I am unsure of the truth of that. My point is that what appear to be misbehaviours or failures of the mind, may actually be an issue with the interface – the brain. Certainly, there is evidence of failure in the brain, but by analogy, is that a problem with the software, or the hardware; the CPU or an I/O device? Whilst I am unsure, Patricia Churchland is quite sure:

"Consider this: memories exist as modifications to the connectivity of neurons in the brain. Memories come into existence when brain cells – neurons – change how they connect to other neurons by sprouting new structure and pruning back old structure. This changes how one neuron connects to other neurons. Information about events in my life and about what makes *me* is stored in patterns of connections between living brain cells – neurons. Memories of childhood, social skills, the knowledge of how to ride a bicycle and drive a car – all exist in the way neurons connect to each other."[1]

This must be true at some level, especially where the "memory" concerns physical activities, but there is no basis to extrapolating that to the conceptual, other than a rejection of the immaterial. When neuroscientists speak of *the brain reorganising itself* during sleep, who or what is the organiser, and on what basis is such reorganisation predicated? What imperative exists in a biological brain for "pruning back old structure"? Why do some memories disappear for decades, and later reappear, at times stimulated by an external event, and at other times not? Does the brain lose track of its connections, and if so, why? We will shortly come back to the subject of memories.

The above quotation continues, "In dementing diseases and in normal aging, neurons die, brain structures degenerate ... Without the living neurons that embody information, memories perish, personalities change, skills vanish, motives dissipate." Apart from utterly rejecting the reference to information, I have earlier explained my agreement that even a temporary disease can cause *memories to perish, personalities change, skills vanish, and motives dissipate*. (see chapter entitled *An Unfortunate Experience*) As these experiences were temporary, it suggests that no loss of neurons or neuronal connections occurred. Notice the common, but unsubstantiated conflation of activities that belong in the separate domains of the autonomous and volitional.

Persisting with this difficult trek through Patricia Churchland's book, she describes her speciality as *neurophilosophy*, a synthesis of neuroscience, psychology, and evolutionary biology. Churchland is entirely comfortable with this worldview, stating: "Biology reassures me. The connection, via evolution, with all living things gives me a sense of belonging." (p. 20) I, on the other hand, am not at all reassured by biology, for I can find nothing in biology which can explain how the material can give rise to the conceptual, and nothing in any scientific texts which can explain it, or even seriously consider the conundrum. I am inclined to offer, *ignorance is bliss*, but perhaps I am being unkind.

I have no answer to these questions, and many more that arise in my mind, apart from not understanding the "why" of such questions arising – why do we question when evolution already has the answers whether we ask for them or not? Why and how did we evolve to question our own evolution? But evaluating the evidence in as detached a manner as I am able, I am inclined to believe that for volitional activity, the biological brain acts as an interface to its autonomous activities, and is not, itself, the initiator of volitional activity.

Does the Brain Store Memories?

This article[2] claims that memories, which I assert are conceptual, are actually instantiated in the biology of our brains, being formed when synaptic connections are made.

"Each time a memory is recalled, the connection is reactivated and strengthened. The idea that synapses store memories has dominated neuroscience for more than a century, but a new study by scientists at the University of California, Los Angeles, may fundamentally upend it: instead memories may reside *inside* brain cells."

Firstly, what initiates memory recall? The difficulty I have with this idea is that according to other literature, the result of synapsing is "all or none"; that is, the result is either an action potential in the postsynaptic neuron, or not. I have no difficulty with the assertion that neural connections are "reactivated and strengthened" with use, for there is scientific evidence which confirms it. But this article is suggesting that the chemical neurotransmitters themselves represent memory content. That may be true, I cannot know, but this idea is contrary to the results of my research up to this point. Not all neuroscientists agree that memory could be in the synapse, and so have turned their attention to the cell (soma) itself. The tentative conclusion is that the *engram*, or memory trace, is preserved by molecular and chemical changes that persist even when the potential of the membrane returns to its resting state:

"Alternatively, it could be encoded in modifications to the cell's DNA that alter how particular genes are expressed. Glanzman and others favor this reasoning."

According to this website[3],

"an *engram* is a unit of cognitive information inside the brain, theorized to be the means by which memories are stored as biophysical or biochemical changes in the brain (and other neural tissue) in response to external stimuli. The exact mechanism and location of neurologically defined engrams has been a focus of persistent research for many decades."

In other words, they are hypothetical and may not exist at all as real physical entities. Perhaps this is why the persistent search for the location of engrams continues to this day. I would also question the term, "unit of cognitive information". "Unit" may be used as an analogy for data, as what can only be held is an encoded symbolic representation of a conceptual entity, but information is a correlation of those in context. No matter how or where physiological entities which represent neurological entities for the storage of memories are located, we are still left with the issue of interpreting physical biological material in immaterial conceptual terms.

The research mentioned above should yield solutions for some neurological diseases, like PTSD, and I can only marvel at the progress being made for medical reasons, but it gets me no closer to a solution of the mind-brain conundrum.

Soul Searching

In her book, Patricia Churchland has a chapter entitled as above. Her working hypothesis seems to be based on Aristotle's philosophy of hylomorphism and Cartesian Dualism of the 17th century, where the immaterial mind was considered synonymous with the immortal soul of religious belief. Rather than accepting the possibility that the immaterial may exist absent of any immortal connotation, she continues with her presupposition. Churchland wrote:

"In any case, Descartes concluded that all mental functions – perceiving, thinking, hoping, deciding, dreaming, feeling – all are the work of the nonphysical soul and *not* the brain. Where did he suppose the handoff of information between brain and soul take place? He got it wrong because so very little was known about the brain at the time he lived." (p. 48)

Yes, very little was known about the brain back then, but I contend that very little is understood even today about the capability of the mind to conceptualise. Churchland gives a brief account of split-brain experiments, but all of her examples relate to physical activities of the brain and body, not conceptual activities of the mind. Then we have her contention as here:

"The problem is this: if a nonphysical soul causes events to happen in a physical body, or vice versa, then the law of conservation of mass energy is violated ... How can energy be transferred from a completely nonphysical thing to a physical thing ... Once you give slow thought to what sort of thing a non-physical soul might actually be, awkward facts begin to pummel the idea's plausibility." (pp. 50-51)

I contend this to be a misrepresentation, or misunderstanding. The brain gets its energy from internal biological sources, mostly from the oxygen-dependent metabolism of glucose (i.e., blood sugar). The mind does not directly cause events in the physical body - the brain does that. The issue is whether, or how, the immaterial can direct specific activities in the brain that cause the autonomous bodily systems to respond. There is no violation of the law of conservation of mass energy, because the brain is already ticking over courtesy of its biological structure being fed nutrients by the blood. There is no pummelling of the plausibility of the brain being an interface between mind and body, just a lack of explanation. Churchland continues with her blinkered approach by stating examples which relate entirely to physical activities in the body, as if these could somehow refute contentions regarding the conceptual activities of the mind – they cannot. She offers, "Back to my wisdom tooth. Can the dualist match neuroscience's level of explanatory consilience regarding why procaine block's pain? Not even close." (p. 52) No, they cannot, nor would they even attempt to, because that is a straw man argument irrelevant to the contentions of the Cartesian or Substance dualists.

The physical and conceptual belong in two entirely separate domains. It is illogical, and dare I say a sign of desperation, to attempt to refute arguments for the nonphysical by conflating the two. If, as I contend, the mind is immaterial, then there is no violation of the law of conservation of mass energy as I understand it. The obvious question does arise: how can the immaterial influence the material, and vice versa? I have no idea, but my evaluation of the evidence has me concluding that it does. I shall not defer to philosophical materialism just because I don't have the answer.

How the Brain Works

Churchland asks, "Why is it so hard to figure out how the brain works?" (p. 53) and continues with a useful explanation which I have appreciated, notably the discovery by two British physiologists, Hodgkin and Huxley in 1952, that:

"Like all cells, nerve cells (neurons) have an outer membrane, partly constituted by fat molecules, with special protein gates then open and close to allow particular molecules to pass in or out of the cell. In a resting neuron, the inside of the membrane is negatively charged relative to the outside, owing to active pumping out of positive ions such as sodium. Negative ions, such as chloride, are sequestered inside. This voltage difference can change abruptly when the neuron is stimulated. This fast change in voltage across the membrane is what makes neurons special." (p. 54)

So far, so good, but I see nothing in this explanation for how a neuron could imagine its outside world, or contemplate its own chemistry. Perhaps when neurons get together composing a network, they could also compose music when stimulated in just the right way – we will have to see. Using the term "information" loosely, the following is my understanding of the scientific descriptions of how the system works: *information* to and from the brain travels along neurons which are arranged in networks that let them exchange messages between the body and the brain:

- Messages are sent in the form of signals (electrical spikes) called action potentials.

- Action potentials travel down a single neuron cell as an electrochemical cascade, allowing a net inward flow of positively charged ions into the axon.

- Within a cell, action potentials are triggered at the cell body, travel down the axon, and end at the axon terminal.

- The axon terminal has vesicles filled with neurotransmitters ready to be released

One can find other terminology such as:

"An action potential occurs when a neuron sends information down an axon, away from the cell body. Neuroscientists use other words, such as a "spike" or an "impulse" for the action potential. The action potential is an explosion of electrical activity that is created by a depolarizing current."[4]

Now, from studying descriptions from various sources, I have come to the conclusion that many such descriptions reflect the perspective of the author rather than scientific fact. The reference website above is typical. It has been shown that a depolarising current changes the resting potential of the postsynaptic neuron, and that when the depolarisation reaches a certain level, the neuron fires an action potential. Now, consider the following:

"when the threshold level is reached, an action potential of a fixed sized will always fire...for any given neuron, the size of the action potential is always the same. There are no big or small action potentials in one nerve cell - all action potentials are the same size. Therefore, the neuron either does not reach the threshold or a full action potential is fired - this is the "ALL OR NONE" principle.

If the action potential of a given neuron is always of the same size, then what *information* could it possibly convey, other than as a trigger for a neurotransmitter to be released? Neurotransmitters are chemical molecules such as adrenaline, dopamine, and glycine which cross synapses to elicit excitatory, inhibitory or modulatory responses from downstream neurons. These are the steps:

1. The action potential travels down the axon to the presynaptic terminal.

2. Depolarization of this presynaptic terminal opens ion channels allowing calcium (Ca^{2+}) into the cell.

3. The calcium ion triggers the release of the neurotransmitter from the vesicles (small fluid-filled sacs).

4. The neurotransmitter binds to receptor sites on the dendrite membrane of the postsynaptic neuron.

5. Opening and closing of these ion channels cause changes in the postsynaptic membrane potential, whereas it was previously at rest.

6. The sequence continues, steps 1-5, as the action potential propagates to the next neuron.

7. The neurotransmitter at (3) is inactivated or transported back into the presynaptic terminal.

Again, what *data* or *information* is propagated through the neuron pathway? So far, nothing but chemical and electrical activity operating in conformance with what we believe to be the laws of chemistry and physics. In my attempt to visualise what is happening, I considered a string of party lights where they are all off until a current pulse is initiated. As the pulse reaches each light, that light is turned on, but immediately off again as the pulse passes. To an observer, it appears as a moving light. I believe this to be an apt analogy for what happens in a neural pathway. We saw earlier that the action potential is always of the same size for a given neuron, but differences occur in the chemical composition of the neurotransmitters. It is these differences which determine which dendrites are caused to synapse, but the result of the synapse function is always the same: an excitatory response which causes an actional potential to fire in the postsynaptic neuron; an inhibitory response which I can only assume prevents further propagation; and a modulatory response whose relevance escapes me. I have been unable to comprehend neuromodulation other than as quoted here:

"Neuromodulation is the physiological process by which a given neuron uses one or more chemicals to regulate diverse populations of neurons. Neuromodulators typically bind to metabotropic, G-protein coupled receptors (GPCRs) to initiate a second messenger signalling cascade that induces a broad, long-lasting signal. This modulation can last for hundreds of milliseconds to several minutes. Some of the effects of neuromodulators include: alter intrinsic firing activity, increase or decrease voltage-dependent currents, alter synaptic efficacy, increase bursting activity and reconfiguration of synaptic connectivity."[5]

I can only assume that if neuromodulators cause a "long lasting signal", then the membrane potential of a presynaptic neuron does not return to its resting state after the neurotransmitters are released to the postsynaptic neuron. Notice how all of the previous discussions are about neural networks and their connectivity. There is no mention of the where or how, data or memories are stored. Neuron firing is just that: a temporary state when the action potential travels down the axon to trigger the release of neurotransmitters, whereupon the membrane of the presynaptic neuron returns to its resting state. There seems to be no permanence in networks other than when activated, a neural pathway follows the same route, unless reconfigured by neuromodulation. The computer CPU (Central Processing Unit) is a useful analogy, but what is the neural network's comparison with data storage?

The neurosciences have provided great insights into how the brain works in its autonomous functioning, but provide no insights into the initiation of volitional activities. There is no doubt that sensory events precipitate thoughts and memories, but these are reactions – the initiation is external. To understand the mind-brain complex, we would need explanations not just for events initiated in the mind, but also for where the mind overrides the autonomous functioning of the brain.

Patricia Churchland's *soul searching* will never get her there.

A Stationless Network

I have spent a great deal of time, and even sleepless nights, attempting to understand the functioning of the mind-brain from what I have learned about nerves and neural networks. The mind-brain operates in two modes: autonomous and volitional. In the *autonomous* mode, there is a closed-loop system whereby, for example, a stimulus is received by either an *exteroceptor* (one located near a stimulus in the external environment), or an *interoceptor* (one that interprets stimuli from internal organs and tissues). Afferent neurons carry signals via the central nervous system to the brain, whereby the brain responds with efferent neurons that transport signals back to the effector cells. All very neat and tidy.

In *volitional* mode, the processing may or may not result in efferent neurons carrying signals to parts of the body for actioning whatever request or command. Where we are simply thinking, cogitating, pondering, imagining, designing, or otherwise exercising our imagination or creativity talents, the entire process can be restricted to the mind-brain complex. The resultant brain activity can be measured by functional magnetic resonance imaging (fMRI), but unlike in autonomous mode, where entry and exit points of the neural system can be identified, the same is not true of the volitional. The brain's neural network is rather like an underground railway network covering an entire city, but with no stations for people to get on and off. When we have an idea, or choose to remember, what is the initial stimulus to the brain, and where in the network does that occur? Do different subjects have different starting points, and how could that be determined? What determines the exit point of the processing?

I have been unable to locate any literature which attempts to explain this conundrum. That is to say that such does not exist, only that it is not included in the more available resources.

Who is the Conductor?

All organs of the human body can be described as having both active and reactive modes. The active mode describes what happens autonomously, such as the heart beating, and the reactive mode describes how the organ responds to stimuli. What I, and many others, find curious about the brain, is that despite it having the same biological foundations, subject to the same laws of physics and chemistry, its behaviour is unlike other organs. Autonomously, the *reactive* mode of the brain provides a message exchange service for all other organs, interpreting signals initiated by interoceptors through the *active* mode of the organ pertaining to its health or state, and responding with commands regarding what actions may be necessary. As earlier discussed, signals initiated by *exteroceptors*, those located near a stimulus in the external environment, may convey cognitive content in addition to that experienced via *interoceptors*, sensors that monitor our internal biological workings. We have some understanding of the brain's response to the biological signal content, but only speculation regarding the cognitive content.

In a longish discourse which we need not repeat here, Raymond Tallis presents the challenge facing "those who try to be hard line about consciousness and see it as simply an *effect* of the material world on the material brain"[6]. As we discussed in Chapter 8-1, cognitive psychologists believe that "the world we inhabit is *a mental model* that has only a tangential relation to what is *out there*"[7].

In his book, "*Making up the Mind*", Chris Frith argues that the mind is a construct of the brain: "everything that happens in the mind (mental activity) is caused by, or at least depends upon, brain activity"[8], leading him to conclude as above, that the contents of the mind are not real. Tallis has published a review of Frith's book, one that I highly recommend. Quoting from that review, Tallis quotes Frith:

"Those with undamaged brains have no grounds for smugness: 'even if your brain is intact and functioning perfectly normally, what it tells you about the world may still be false' (p. 39). The most pervasive illusion is that we perceive directly what is out there. In fact the object of perception is the conclusion of what Helmholz characterized in the

19th century as 'an unconscious inference' beyond what arrives at our sense endings. We imagine we see a three-dimensional visual scene in vivid detail when in fact such a scene is constructed from a two-dimensional image on the retina whose edges are blurred and colourless."[9]

It should be obvious that Frith (and Helmholz) is deluded in contending that what we think we see may not be real. We live by what we see and experience, and curiously, despite Frith's contention, we get it right practically 100% of the time. The problem with arguments such as this proposed by Frith, is that it is so nonsensical that it is difficult to refute using logical argument. As I type these words, am I to believe that what I see on the computer screen may not be what I have actually typed? Would this be true of Frith's own writings, that they may not be what he meant? When I, or anyone, read his book, are we to believe that we may each be reading something different, because what our brain tells us may still be false?

We live our lives by trusting the reality of what we see and experience; Frith offers no evidence to support his opinion. I would offer that Frith himself lives by trusting reality as he perceives it – why else would he write and publish his writings if he was unsure of their accuracy and unsure of the perception of readers? Let me expose the fallacy, using two sentences offered by those who assert that the mind and the brain are one and the same, in whatever terms they choose to express that belief:

1. "We are our brains"; and

2. "what it tells you about the world may still be false."

In the first statement, "we" and "brains" are conflated as one and the same entity. From the second, we can identify two separate entities: "it" and "you". Now, if (1) is true, (2) could be paraphrased as: *what the brain tells itself about the world may still be false*. But there is still a problem. There is a part of the brain that is doing the telling, and another part which is being told. Logically, two entities have been identified, even though they share a common biological construction. Frith claims: "'By seeing through these illusions created by our brain, we can begin to develop a science that explains how the brain creates the mind (p. 17)", and in doing so, subconsciously admits to there being two entities. Who is it that is doing the "seeing through" the creations of the brain, if not a separate entity capable of cognition independent of the machinations of the brain? No matter where they turn, or how they attempt to justify conflating the mind and the brain, they inevitably speak of them as two separate entities. If the mind is material, whereabouts in the brain does it reside? What is the biological mechanism that has one part of the brain telling a separate part of the brain what to understand about sensory experiences? Has neuroanatomy identified separate logical regions, or are we to believe that somehow, there is a "master and slave" relationship between specific neural circuits? It is as if one part is the conductor, and other parts represent the orchestra, whilst in effect, the orchestra is conducting itself.

Steven Pinker takes the line that *"the mind is not the brain but what the brain does"*[10], which raises more questions than it answers. Chris Frith offers that the mind is the child of the brain, concluding not only that the brain makes up the mind but also that "The contents of the mind are not real" (p. 16). So if the brain causes the activities of the mind, what is it that the mind actually does? And whatever it does, why would we trust it if its contents are not real, although to reword this as I understand is meant, its perception of reality is false? Truly, these people surely dig a hole for themselves.

Libet's experiment

The conclusions of neurophysiologist, Benjamin Libet, in the 1980s from his experiments are not universally accepted, and I will let the reader research this issue on your own, should you be so interested, for there are numerous, easily accessible articles on the internet. I mention the subject here more to make the reader aware that I am aware, but also to inject some of my own thoughts regarding the claim: our brains make decisions to act before our conscious mind is aware of them. Again, quoting Raymond Tallis:

"In a typical experiment, Libet's subjects are instructed to make a simple movement – to bend their right wrist or fingers of their right hand – in their own time. Using EEG, the experimenter records a particular activity in the brain that indicates a readiness to move. This so-called "readiness potential" is seen in the part of the cerebral cortex most closely associated with voluntary movement. The readiness potential occurs about half a second before activity in the relevant muscles of the arm or hand, as recorded by an electromyogram, because it takes time for the neural activity in the cortex to translate into events in the relevant muscles. Nothing worrying there. But Libet made another observation that seemed to raise serious questions. He asked his subjects to recall the position of a spot revolving on a clock face in order to determine the time when they were first aware of their urge or intention to make a movement. To his surprise, he found that the readiness potential occurred a consistent third of a second *before* the time at which the subjects reported being aware of a decision to move. Libet concluded from that the brain (not the subject or the person) "decided" to initiate or at least to prepare to initiate the act before there was any reportable subjective awareness of a decision having been made. Put more simply, the cerebral causes of our actions seem to occur before our conscious awareness of deciding to perform them,"[11]

Other researchers have concluded similarly from similar experiments, but there are logical flaws. Firstly, what caused the "brain" to decide? If researchers are true to the maxim of cause-and-effect that underpins all scientific enquiry, this should have been the first question asked. Time delays can be explained in many ways, and is not an issue of concern to me. But a fundamental flaw of Libet's process is that he included an experiment within an experiment of the same type, as here: "He asked his subjects to recall the position of a spot revolving on a clock face", but he did not record the readiness potential of them making the decision to recall as requested. How could he have been sure that the readiness potential that he did record, was not in fact related to the subject's intention to recall, rather than their intention to move? Researchers, analysts, and the like know all too well that you do not muddy the waters of experiments with additional variables, but this is what Libet has done. An unfortunate aspect of Libet's experiments is that scientists use them, and others similar, to argue that we do not have free will.

I am not at all surprised that Libet "found that the readiness potential occurred a consistent third of a second *before* the time at which the subjects reported being aware of a decision to move". He readily admits that "it takes time for the neural activity in the cortex to translate into events in the relevant muscles", but seems not to admit that it takes time for the subjects to mentally record an observation. Without this measurement, his analysis is flawed.

Clearly, if it was the "brain" that made the decision, and not the subject, then the biological brain, the child of umpteen years of evolution, is in charge, not us. Paraphrased, one might say that we are our brains, but our brains are not us, which leaves the question: Who or what is us? The ghost in the machine whose existence is denied?

Not Libet, nor any other neurophysiologist, can reach a valid conclusion on the issue until they can explain the biological mechanism that caused the brain to decide independently of its owner, although the existence of the owner is denied. There is no evidence from the biology or physiology of how the neural networks of the brain can autonomously perform cognitive tasks.

More Recent Research

I wanted to make a quick note here to acknowledge the work being undertaken, and the creditable results being achieved, especially for readers who may be aware of modern developments. I do not intend to discuss it here, other than to note that from what I can understand, such research is yet to answer my questions. I was encouraged to read of attempts "to evaluate the potential of the end-to-end approach for directly mapping fMRI activity to stimulus space"[12] which might add some railway stations to the network.

Another headline that caught my attention was "This 'mind-reading' algorithm can decode the pictures in your head"[13]. There may be medical uses for this technology, it is certainly impressive, but it doesn't address conceptual issues, which are my primary interest.

Summary

I believe it fair to assert that nobody understands how the brain works. Scientists researching various aspects have made great progress in understanding the physiology of the nervous system, in extraordinary physical detail, but have made no progress in understanding the signalling protocols between neurons. Whilst differentiations in the signals can be understood in terms of physics and chemistry, the *meaning* of the signals remains as elusive as ever. Similarly with the brain: areas of activity have been mapped, and physical transformations of neural circuits subject to repetitive use have been observed. Such research has been of significant benefit to medicine, leading to treatment and even cures of illnesses related to the brain. However, beyond these discoveries, we are no closer to understanding the cognitive aspects of brain activity.

I have yet to find any scientific attempt to explain how the brain *conceptualises*, such that we can, at times, see the concept in our "mind's eye", and at other times instantiate the concepts in terms of poetry, music, and other forms of art. I have yet to find any article attempting to explain that most curious of phenomena, the "mind's eye" – how is it that we "see" without physically seeing. I have yet to find an explanation from the likes of Chris Frith, as to how he manages to overcome what his brain falsely tells him such that he can publicly assert, with confidence, that he is not, himself, succumbing to such falsehoods when he writes as he does. I have yet to encounter any scientific literature on these subjects which avoids reference to two separate entities: we/us and our brains. To my mind, there is a certain failure of logic when scientists speak of others and themselves in contrast to their brains, yet at the same time claiming that their brains are them. It is as if subconsciously, they know that the mind and brain are two separate entities, yet being unable to find the mind in the brain, and unwilling to admit to the immaterial nature of the mind, they have to make illogical attempts at conflating the two.

In an earlier chapter, I commented that we have a very limited understanding of how light waves and photons convey the image of an object that we see. We have some understanding of how the eye transduces the light energy into an electro-chemical signal that faithfully conveys the reality of the object seen, even if some scientists contest the truth of "faithfully". There is no doubt that the object that we see in our "mind's eye" is not the object itself, but is truly a construct of the mind, but that does not argue that the conceptual object in our mind does not match the reality of the object itself. It is also true that our mind can construct a mental image of objects that *do not exist* in reality, but it is probable that such objects are constructed of real components. That is the magic of our imagination, the basis for the creation of art, music, poetry, and the like.

Unless, and until, we accept that the mind is an entirely separate, although interdependent, entity from the brain, we may never understand how the biological brain works. I doubt that we could ever learn to decipher neural activity in conceptual terms, because to do that, we would need to be able to understand the immaterial mind itself – the *we* and *us* that scientists are unable to avoid in their writings.

References:

1. Churchland, Patricia S., *Touching A Nerve: Our Brains, Our Selves*, W.W. Norton & Company, New York, NY, 2013, p. 12

2. https://www.scientificamerican.com/article/memories-may-not-live-in-neurons-synapses/

3. *https://en.wikipedia.org/wiki/Engram_(neuropsychology)*

4. http://faculty.washington.edu/chudler/ap.html#:~:text=An%20action%20potential%20occurs%20when,created%20by%20a%20depolarizing%20current.

5. https://en.wikipedia.org/wiki/Neuromodulation

6. Tallis, Raymond, *Aping Mankind*, Routledge Classics, New York, NY, 2016, p. 110

7. *Ibid*, p. 110

8. Frith, Chris, *Making up the Mind: How the Brain Creates Our Mental World*, Wiley-Blackwell, Hoboken, NJ, 2007, p. 23

9. https://academic.oup.com/brain/article/130/11/3050/333014

10. Pinker, Steven, *How the Mind Works*, Penguin Books, London, UK, 1998

11. Tallis, *Ibid*, p. 56

12. https://www.frontiersin.org/articles/10.3389/fncom.2019.00021/full

13. https://www.sciencemag.org/news/2018/01/mind-reading-algorithm-can-decode-pictures-your-head

Chapter 8-4: The Concept of Mind

"Many people can talk sense with concepts but cannot talk sense about them."

~ Gilbert Ryle, *The Concept of the Mind*[1] ~

That is quite a challenge that Ryle has presented, so let us attempt to rise to it. In brief, a *concept* is an *abstract idea*, i.e., existing in thought or as an idea, but not having a physical or concrete existence. Our minds are whirling with concepts, whether we are aware of them or not. Perhaps dominating our mind is our concept of *self*, which material monists choose to deny as real, but next might be our personal concept of the mind, wherein even that concept is whirring. Earlier we discussed the competing philosophies of dualism, which itself is a concept. The question must be asked: If concepts are abstract ideas with no physical or material existence, how can they exist and be processed in a material organ of the body (brain)? I am looking forward to Gilbert Ryle's explanation of how that can be so, if indeed, he does address that particular issue.

Ryle begins his argument:

"It is one thing to know how to apply ... concepts, quite another to know how to correlate them with one another and with concepts of other sorts ... For certain purposes it is necessary to determine the logical cross-bearings of the concepts which we know quite well how to apply ... Descartes left as one of his main philosophical legacies a myth which continues to distort the continental geography of the subject. A myth is, of course, not a fairy story. It is the presentation of facts belonging to one category in the idioms appropriate to another. To explode a myth is not to deny the facts but to re-allocate them, and this is what I am trying to do.

To determine the logical geography of concepts is to reveal the logic of propositions in which they are wielded, that is to say, to show with what other propositions they are consistent and inconsistent, what propositions follow from them, and what propositions they follow. The logical type or category to which a concept belongs is the set of ways in which it is legitimate to operate with it. The key arguments employed in this book are therefore intended to show why certain sorts of operations with the concepts of mental powers and processes are breaches of logical rules." (p. 10)

I have said before that it is logical to argue consistently within a proposition, but it can only lead to truth if its foundational presuppositions, and the proposition itself, are true. If Ryle seeks to reorganise the concepts related to the mind, following the proposition of philosophical materialism, then he is begging the question. You cannot argue against Cartesian dualism and the proposed immaterial nature of the mind, by beginning with Carl Sagan's assertion that "the Cosmos is all that is or was or ever will be", meaning that everything is composed of matter as we understand it: atoms and molecules. He notes that a myth is not a fairy story, rather, "It is the presentation of facts belonging to one category in the idioms appropriate to another". But that is precisely what evolutionists and other scientists do when they apply anthropomorphisms to physical objects and systems, and mechanisations to biology. In both cases, these transferred epithets *are presentation of facts belonging to one category in the idioms appropriate to another*. Ryle himself is as guilty as any in this regard.

Consider his opening remarks: "It is one thing to know how to apply … concepts, quite another to know how to correlate them with one another and with concepts of other sorts … For certain purposes it is necessary to determine the logical cross-bearings of the concepts which we know quite well how to apply". Implicit in these remarks is the existence of something responsible for the doing, which is what? Does he truly believe that the entity "we" is the brain? How is it that a material brain can "determine the logical cross-bearings of the concepts", when logic itself is a concept of which the material cannot be aware? As with all people of a similar philosophical persuasion, Ryle is unable to refute the *Ghost in the Machine* without continually introducing a ghost of his own – the "I", "we", or "us" as entities separate from the brain.

Ryle proposes a straw man argument when he states: "Holders of the official theory [of mind] tend, however, to maintain that anyhow in normal circumstances a person must be authentically seized of the present state and workings of his own mind … This self-observation is also commonly supposed to be immune from illusion, confusion or doubt … Sense perception can, but consciousness and introspection cannot, be mistaken or confused." (p. 16) In reviewing my own writings, I commonly find mistakes in my own thinking, and even when thinking on what I am about to write, I commonly ask myself if I am mistaken or confused. I do not perceive myself as Robinson Crusoe in that regard, as many of my contemporaries have expressed similar doubts about their own thoughts. I will be unapologetically dogmatic: I do not, and cannot, validate my own thinking as it occurs, which is why I am given to committing my thoughts to paper for review. Ryle must be speaking of other people with whom I am not familiar, or is deluded about himself, believing that his own introspection cannot be mistaken or confused. *An Unfortunate Experience* is my own testament to the confusion that can arise.

In building his argument regarding the geography of concepts, Ryle makes a valid point regarding category mistakes; e.g., seeing an entity in the same category as the entities of which it is composed. He offers the example of someone when visiting a university for the first time, and being shown various features, offices and buildings, then asks: but where is the university? Curiously, philosophers make precisely the same mistakes when discussing the mind, alternately speaking of the mind as if it were an entity, a property, or an activity. Ryle is right to point out this logical error. He remarks: "The theoretically interesting category mistakes are those made by people who are perfectly competent to apply concepts, at least in the situations with which they are familiar, but are still liable in their abstract thinking to allocate those concepts to logical types to which they do not belong." (p. 19) I wholeheartedly agree. In substantiating his abusiveness (his word) against those who believe in the *Ghost in the Machine*, he explains,

"My destructive purpose is to show that a family of radical category mistakes is the source of this double-life theory. The representation of a person as a ghost mysteriously ensconced in a machine derives from this argument. Because, as is true, a person's thinking, feeling and purposive doing cannot be described solely in the idioms of physics, chemistry and physiology, therefore they must be described in counterpart idioms. As the human body is a complex organized unit, so the human mind must be another complex organized unit, though one made of a different sort of stuff and with a different sort of structure. Or, again, as the human body, like any other parcel of matter, is a field of causes and effects, so the mind must be another field of causes and effects, though not (Heaven be praised) mechanical causes and effects." (pp. 19-20)

An *idiom* is defined as a group of words established by usage as having a meaning not deducible from those of the individual words. I am still struggling to discern Ryle's point here, but perhaps it will later become clear. Ryle believed that the origin of these category mistakes can thus be found: "Still unwittingly adhering to the grammar of mechanics, he [Descartes] tried to avert disaster by describing minds in what was merely an obverse vocabulary. The workings of the minds had to be described by the mere negatives of the specific descriptions given to bodies; they are not in space, they are not motions, they are not modifications of matter, they are not accessible to public observation. Minds are not bits of clockwork, they are just bits of not-clockwork." (p. 21) I somewhat agree with Ryle's observation, but fail to understand why he has a problem with this line of thinking. Descartes, and others including myself, used a process of elimination in attempting to discover what the mind "is", but only succeeded in finding what the mind "is not". He searched in all known physical places, and not finding what he was looking for, concluded that the mind was not there, i.e., material. Philosophical materialism cannot accept that conclusion, thus leading scientists to propose solutions which they can neither explain nor substantiate. When they conclude that the mind is an entity, a property, or an activity, in each case they make their own category mistake. The mind cannot be all three categories simultaneously, it can only be one, or none. The fact that scientists cannot agree which one, is suggestive of it being none.

In my own search for truth, I am seldom able to find it, for at best, I can only uncover untruth – I cannot know what some things are, but I can discover what they are not. There is nothing illogical in that approach. The medieval Sephardic Jewish philosopher, Maimonides, took the same approach to understanding God. Accepting the reference to God as *Ein Sof*, the Infinite and Unknowable God, Maimonides asserted that we can only speak of God in terms of what he is not, for every term that we use refers to the finite or the anthropomorphic, neither of which God is. It will be interesting to discover Ryle's alternative - speaking of what the mind truly is.

Ryle rightly asks how minds can influence and be influenced by bodies; I ask the same questions, having experienced those very same phenomena, but I will not conclude without substantiated evidence or a plausible explanation. It could well be that I fail to properly apprehend some of Ryle's points, but as written, some make little sense. For example, in his discussion of the human will, he attempts to demonstrate that the mind is not a separate faculty by selecting examples that accord with his argument, whilst ignoring those that do not. Referring to mentioning or recalling acts of will, as distinct from actions that may result, he asserts:

"If ordinary men never report the occurrence of these acts, for all that, according to the theory, they should be encountered vastly more frequently than headaches; or feelings of boredom; if ordinary vocabulary has no non-academic names for them; if we do not know how to settle simple questions about their frequency, duration or strength, then it is fair to conclude that their existence is not asserted on empirical grounds." (p. 64)

I know not what "ordinary" world Ryle inhabited, but I can provide numerous examples of where acts of will are identified as separate to any subsequent action. The most common of these being New Year's resolutions. We can ask the question: When did you decide to do this or that? If deciding is not an act of will, then what is it? The crime of plotting an illegal act relates to the acts of will preceding the planning of the act, whether the plan was executed or not. Conspiracy similarly refers to an act of will to accomplish something, entirely separate in time to any subsequent activity including the planning of the act. When we decide what to do tomorrow, that is an act of will. There are many other such examples. The words *decide, conspire,* and *plot* are non-academic terms in our ordinary vocabulary, so I cannot agree with what I believe Ryle was asserting. Certainly, many voluntary actions have no appearance of a preceding act of will, largely a matter of timing, and often we can ask the question: Why did you do that, without receiving a useful reply. Regarding Ryle's assertion, "then it is fair to conclude that their existence is not asserted on empirical grounds", I am forced to agree (partially). Acts of will, which are activities of the mind, are immaterial, and thus there can be no empirical grounds. However, neurologists have identified brain activity when subjects are thinking. Whilst the subject of thoughts cannot be discerned from the activated neural circuits, there is no evidence that acts of will are not included.

Ryle comments: "Nor could it be maintained that the agent himself can know that any overt action of his own is the effect of a given volition." (p. 65) Consider yourself onstage before a magician who, presenting to you a fanned deck of cards, asks you to select one. When you make a choice, enacted by selecting a card, can it be asserted that such an overt act is not the effect of a given volition? If it is not, what was the cause of your selecting a card? No doubt this process is a mystery, but for Ryle to assert, as he does, that "transactions between minds and bodies involve links where no links can be" (p. 65) simply testifies to his presuppositions about the mind and body, which at this point in his book, he has failed to disclose. Part of his argument relates to an infinite regression of voluntary thoughts, rejecting the notion of a "first cause". He states: "if ... an act of choosing is describable as voluntary, then, on this suggested showing, it would have in its turn to be the result of a prior choice to choose, and that from a choice to choose to choose ... " (p. 67) I would like to have asked Ryle about Libet's experiment, where the "readiness potential" arose in the brain before the subject was aware of making a decision – if the brain activity was not voluntary, what was the first physical cause? I doubt that he has a coherent answer. Again, at this point in his argument at least, he fails to offer a way out of this infinite regression of choice postulate. He further notes:

"Volitions have been postulated as special acts, or operations, 'in the mind', by means of which a mind gets its ideas translated into facts. I think of some state of affairs which I wish to come into existence in the physical world, but, as my thinking and wishing are unexecutive, they require the mediation of a further executive mental process." (p. 62)

I do not accept his proposition of "the result of a prior choice to choose, and that from a choice to choose to choose". On what basis does he claim that "thinking" is an *unexecutive* requiring mediation by an antecedent *executive* mental process? What evidence, or line of thought, has him asserting that volition is not an executive mental process in its own right? How did he manage to write his book, choosing to write it in the first place, and choosing what words to write, if these processes were subject to an *infinite regression of choice postulate*? I have been unable to find any explanation as to why he considers *volitions* to be of a different mental type than executive mental processes. Perhaps he understands, but he has failed in his attempts to convey that understanding. As he earlier commented, "Misinterpretations are not always due to the inexpertness or carelessness of the spectator [*reader*]; they are due sometimes to the carelessness [*inexpertness*?] ... of the agent or speaker.' (p. 58)

The irony in all of this is that the *concept* of the mind can only be the *product* of an *immaterial mind*, for all concepts are immaterial by definition. Nothing that I have learned in science provides evidence that the material can give rise to an immaterial concept, nor anything else immaterial. The fact that Gilbert Ryle has written as he has, suggests that subconsciously, he knows that the mind is immaterial, but philosophically, he cannot accept what he knows to be true. To repeat the definition of the term, *concept*, the Oxford dictionary gives it as "an abstract idea, a plan or intention, an idea or invention." Synonyms include notion, opinion, conviction, belief, and hypothesis, the latter being a common trigger in the scientific method. I wonder how many research scientists would accept that their hypotheses originate in evolutionary processes?

References:

1. Ryle, Gilbert, *The Concept of Mind*, Penguin Books Ltd, London, England, 1990, p. 9

Chapter 8-5: What IS the Mind?

"The mind is not the brain but what the brain does, and not even everything it does."

~ Steven Pinker, *How the Mind Works*[1] ~

In one way, it is logical that before we can discuss what the *mind* does, or how it works, we ought first to ascertain just what the mind is. On the other hand, perhaps it is more logical to examine the activities of the mind, and then work backward to the properties which could give rise to such activities, leading us to identify the entity. A lack of precision at this point will allow dubious claims to be made, and in truth, this is what I have found. A common claim by scientists is that "the mind is an emergent property of the brain", which enlightens me not at all. In addition to his observation quoted above, Steven Pinker offers:

a. "The mind, like the Apollo spacecraft, is designed to solve many engineering problems, and thus is packed with high-tech systems each contrived to overcome its own obstacles." (p. 4)

b. "The mind, I claim, is not a single organ but a system of organs, which we can think of as psychological faculties or mental modules." (p. 27)

c. "The mind is a system of organs of computation, designed by natural selection to solve the kinds of problems our ancestors faced." (p. 21)

d. "The mind is organised into modules or mental organs, each with a specialized design that makes it an expert in one area of interaction with the world. The module's basic logic is specified by our genetic program." (p. 21)

e. "Modules are defined by the special things they do with the information available to them, not necessarily by the kinds of information they have available … a (mental) organ of the body is a specialised structure tailored to carry out a particular function … our physical organs owe their complex design to the information in the human genome, and so, I believe, do our mental organs." (p. 31)

f. "Learning ... is made possible by innate machinery designed to do the learning." (p. 33)

g. "The ultimate goal that the mind was designed to attain is maximising the number of copies of the genes that created it." (p. 43)

Earlier, I quoted another assertion by Steven Pinker, "A myth is, of course, not a fairy story. It is the presentation of facts belonging to one category in the idioms appropriate to another. To explode a myth is not to deny the facts but to re-allocate them, and this is what I am trying to do." (p. 10) For a start, he could avoid falling into the same error of which he accuses others, i.e., the inappropriate use of idioms appropriate to another category. Stating that evolution *designed* anything is a myth.

Putting aside for the moment, the incongruity of asserting that evolution *designed* anything, or that evolution has a *goal*, ultimate or otherwise, let us try to understand what Pinker and other scientists understand the mind to be. We know that the brain is a biological organ of the body; thus, I contend, the brain should be considered in the same sense as other organs like the skin, heart, liver, kidney, etc. Steven Pinker seems to agree, although he attempts to differentiate mental organs from physical organs, whilst at the same time, acknowledging that mental organs derive their properties from the same source - the human genome. If there is such a physical entity as a mental organ, then we ought to be identifying its cellular properties. I assume that Pinker speaks of mental organs in the sense that they are specialised to perform intellectual activities, rather than just physical activities as performed by other organ types. The obvious problem, for me at least, is that intellectual activities are conceptual in nature, not physical, and I can see no reason why the mental organs, whether defined as mind or brain, can perform activities so radically different to the so-called physical organs.

If the mind "is not a single organ but a system of organs", where are they located in the human brain? What is their composition, and why are they never identified in any neuroanatomical atlas? Would a comparison of a human brain with that of, for example, a chimpanzee, allow neurologists to identify these mental organs which distinguish humans from other species? Or is Pinker suggesting that the sections of the brain already identified are these organs, or perform the functions of these organs? He is not clear on this point.

Organs have *properties*, defined as attributes, qualities, or characteristics. When speaking of what an organ does, its behaviour can be described as autonomously active, and/or reactive, but its properties both define and constrain those activities. Biologically, these properties are defined by the types of cells comprising the organ. Keeping this brief, *properties* are *regulators* of activities. Now, even where activities are described as autonomous, the organ as a composite entity does not initiate its activity. That may seem contradictory, which in a sense it is, but activities which appear to be initiated by an organ are in fact, in response to the activities of its constituent cells. The same is true of any reactions. If we wind back the clock to just before the moment of conception, neither the male sperm nor the female ovum had been doing much by themselves, but there is inner activity which keeps the cells alive. Fertilization occurs when the male sperm breaches the female ovum and life begins. Thereafter, everything that happens does so in accordance with genetic instructions, which can only operate in accordance with the laws of physics and chemistry as we understand them. That is the nature of biology - it can be neither more nor less. We have evidence, in the case of deformed babies, that sometimes the process of gestation does not work as it should – medical science is continuing research into why this is so. But no revelation so far has concluded for processes that are not in accordance with the known laws of science.

If the mind is a property of the brain, as in "the mind is an emergent property of the brain", then the mind is a *regulator*, not a separately identifiable entity like an organ. Perhaps what is meant is not a property *per se*, but a construct or arrangement of the brain itself – the "scientific" text on that subject is ambiguous. If the mind is an arrangement of the brain, or the term *mind* refers to a set of mental organs as Pinker suggests, then we still have a biomass subject to the laws of chemistry and physics. On the other hand, in the quotation at the beginning of this chapter, Pinker asserts that the mind is not a property, but an *activity*. I would argue that there is a substantive difference between *property* and *activity*, and that the mind cannot be both. This confusion needs to be resolved.

The question, the answer to which is foundational to understanding the nature of the mind, is this: Can biological entities be so arranged as to render properties and/or behaviours which are not intrinsic to the entities themselves, nor to any of the components comprising that entity? If individual neurons are neither sentient nor capable of cognitive behaviour, is it possible that such capabilities could emerge from an arrangement of neurons? There are occasions in nature where the whole is more than the sum of its parts, but before accepting this as a generalization applicable to the mind-brain conundrum, we need to understand in what way the whole is different. Deductive reasoning must delve deeper than the superficial.

Spirits, and the like

I ponder why some scientists are reluctant to consider that the immaterial may be part of our "natural" existence. I can think of many reasons, most seemingly having their basis in a reluctance to accept the possibility of a creator, commonly referred to as "God", just as they reject *Intelligent Design* being fearful of the identity of the designer. Throughout the ages, and even today, there are people who believe in a spirit world, without perceiving such spirits as being related to a transcendent god. One perception of the spirit world is that it is like the world in which we inhabit, but different, being in another dimension; another, that it is totally unlike our world, being of no material construction. In most cases, these spirit worlds have the concept of an "afterlife", somewhere we go after this life, but not always. In some religions, the afterlife is a continuum of reincarnations. We should note that the concept of the spirit is entirely antithetical to biological evolution, just as evolutionists opine that *free will* is an illusion, because such could not be a product of evolution. Anti-theists, Richard Dawkins being the doyen of this movement, assert that a belief in gods arose from ignorance of natural phenomena such as lighting and thunder; after all, someone had to be responsible. However, from where the belief in a spirit world, with no connection to gods? If free will, as a cognitive activity, could not be a product of biological evolution as some prominent evolutionists assert, then neither could the concept of the spiritual or immaterial. That we do have free will, despite evolutionists using their free will to opine that we do not, and that we do have the concepts of the spiritual and immaterial, would surely argue against evolutionary theory. Could we truly conceptualise, unlike other organisms on the planet, if we did not have an "organ" of conceptualisation? Nothing that we perceive as a concept could arise without such a conceptualising organ, which must by definition be immaterial, not material. A concept that I continually grapple with, and to be honest, keeps me awake at night, is "infinity", or an infinite - something without boundaries or constraints. No doubt, my curiosity concerning the attributes of an infinite god has taken me down this path, but I mention this only to evidence that we can conceptualise, dimly, that which may not even exist. I would argue that it is imperative for evolutionists to answer this question, if their hypothesis is to be plausible: How could a mind which is material, and thus finite, have a concept of what it is not?

Materialists might offer explanations, but they could only be speculations in defence of materialism.

I think it was C.S. Lewis who coined the term, *chronological snobbery*, a phenomenon where moderns believe that ancients were less intelligent, because that has been the path of evolution. Of course, there is no evidence for that, it is an unavoidable inference by those believing in evolution. One cannot even say that the ancients were less knowledgeable than moderns – they just had different knowledge, much of which has been lost. The ancients were closer to the land, and closer to each other, in ways that moderns cannot be, surrounded as we are by technology and other distractions. If the ancients believed in a spirit world, could it be because of their real experience of it, undistracted in their simpler lives? I cannot know, but neither can anyone else, most especially scientists beholden to philosophical materialism. If you refuse to believe that a phenomenon exists, then it is illogical to even ponder it. For the deists who refuse to accept a spirit world absent of God, you might wonder whether that was just a stage of God's revelation to humankind, always provided that you are prepared to let go of Young Earth Creationism.

The issue, as I see it, is a failure to differentiate two possible phenomena: a spirit world with God, and a spirit world without God. Likewise, a spirit world representing an afterlife, and a spirit world without an afterlife. Reading texts that reject the notion of a separate immaterial mind, the argument always seems to come back to a concept of the "soul". An assumed property of this soul is that it lives on after our physical death, and perhaps even preceded our birth, such as is said to occur with reincarnation, although just how the first iteration came to be is a subject best avoided. The argument is similar to that of those who reject the concept of god saying, "I cannot believe in a god who … [fill in your personal disbelief]". The fallacy is to assign a property, or activity, to an entity without substantiation. It is illogical to entirely reject the existence of an entity, because you reject some assumed property: the entity may still exist without that property. Such is how strawman arguments are formulated.

In my other field of research, gods and religions, I have found that predominantly, peoples' perception of "God" stems from one or more religions which influences their acceptance or rejection. But they fail to understand that their focus is on properties and activities, not the entity, if it exists at all. If a transcendent God does exist, I entirely agree with Maimonides, that we can only speak of that entity in negation: what God is not, rather than what God is. Thus, to reject the entity called God because you reject properties which he could not possibly have, is illogical.

Naturalistic Nonmaterialism

Some scientists do believe in the phenomenon of a non-material existence, which is nevertheless intrinsic to our natural world, but one which does not rely on a transcendent god for its ongoing presence. Some accept that maybe there was/is a god who set the whole thing rolling, but that is as far as they will go. This is similar to the philosophy of *deism*: God created the universe, but thereafter chose to ignore his creation (perhaps he had moved on to some other project). Thomas Nagel is an American philosopher, and Emeritus Professor of Philosophy and Law, at New York University. In his book, "Mind and Cosmos", he admits that he is a layman, not a scientist, but nevertheless understood sufficiently to question the materialists view of the world. He wrote:

"My preference for an immanent, natural explanation is congruent with my atheism. But even a theist who believes God is ultimately responsible for the appearance of conscious life could maintain that this happens as part of a natural order that is created by God, but that it does not require further divine intervention. A theist not committed to dualism in the philosophy of mind could suppose that the natural possibility of conscious organisms resides already in the character of the elements out of which those organisms are composed, perhaps supplemented by laws of psychophysical emergence. To make the possibility of conscious life a consequence of the natural order created by God while ascribing its actuality to subsequent divine intervention would then seem an arbitrary complication. Some form of teleological naturalism should for these reasons seem no less credible than an interventionist explanation, even to those who believe that God is ultimately responsible for everything."[2]

I applaud Nagel's lateral thinking, for he offers a solution which had not occurred to me: that of consciousness being instantiated in the very elements from which we are constituted. But such a proposition does not deny the existence of the non-material, for how else could consciousness exist in such primitive materials as atoms and molecules? Moving on, his apparent accommodation with deists, theists, and intelligent design proponents drew the ire of evolutionists, as one would expect. I would commend this article[3] in the Chronicle of Higher Education for your review. Quoting from that article,

"His latest book, *Mind and Cosmos* (Oxford University Press, 2012), has been greeted by a storm of rebuttals, ripostes, and pure snark. "The shoddy reasoning of a once-great thinker," Steven Pinker tweeted. *The Weekly Standard* quoted the philosopher Daniel Dennett calling Nagel a member of a "retrograde gang" whose work "isn't worth anything—it's cute and it's clever and it's not worth a damn."

The critics have focused much of their ire on what Nagel calls "natural teleology," the hypothesis that the universe has an internal logic that inevitably drives matter from nonliving to living, from simple to complex, from chemistry to consciousness, from instinctual to intellectual.

This internal logic isn't God, Nagel is careful to say. It is not to be found in religion. Still, the critics haven't been mollified. According to orthodox Darwinism, nature has no goals, no direction, no inevitable outcomes."

Fortunately for me, this book that you are reading will not attract the attention of the evolutionists (I am unknown), for otherwise I would be subjected to even more scorn and derisory comments, being far less qualified than Thomas Nagel. As the article mentions, "Joan Roughgarden, an ecologist and evolutionary biologist at the Hawaii Institute of Marine Biology, agrees that evolutionary biologists can be nasty when crossed. I mean, these guys are impervious to contrary evidence and alternative formulations, she says. What we see in evolution is stasis—conceptual stasis, in my view—where people are ardently defending their formulations from the early 70s." Indeed, they do, most vociferously, but whilst the evolutionists reject philosophy and more importantly, epistemology, they have no answers to the questions Thomas Nagel and others ask.

I find it somewhat amusing that evolutionists ardently defend their science, yet an examination of that science provides little reason for them to be so committed to their beliefs. I would suggest that they are more committed to ideology than to science.

Tides of the Mind

I now return to a book of that title by David Gelernter, sub-titled "*Uncovering the Spectrum of Consciousness*"[4]. My review here is deliberately brief, and I only introduce this particular subject on the off-chance that a reader may be interested in learning more. I apologise if the discussions in my book appear to jump around (they actually do), but there is so much overlap that it is difficult to keep all my ducks in a row, so to speak. The author describes his book thus:

"My task in this book is to assemble the fundamental facts about subjective reality, show you what they are, and try to convince you that they amount to a picture of the mind in motion, to the spectrum of consciousness."[5]

The Preface to this book we have earlier reviewed on the subject of consciousness, but here we will briefly review, as Gelernter describes it, the *spectrum of consciousness*. The topic does not directly address the mind-brain conundrum, but provides food for thought, most especially because of the author's acceptance of a *subjective reality*. As earlier argued, the material can only behave in an objective manner, never subjective – that is the realm of the immaterial. The author's proposition is that the state of the mind ebbs and flows much as oceans do, from a state of intense concentration to one of almost complete idleness. As an aside, that is my own experience, although that lower state is not one in which I can persist, other than when I am asleep. Watching people, sadly often in poverty, who can be observed sitting on a bench for hours not engaged in any physical activity, I wonder whether they are engaged in any mental activity. In my younger days, I could start work around 7:00 am and be so concentrated on a task that it would be close to 11:00 am before I realised how much time had passed.

Gelernter describes the spectrum of consciousness from high focus to low focus. In high focus, we use our memory in a disciplined way: our thoughts are rational, and our reflections and self-awareness are strong. In medium focus, our memory use ranges freely and occasionally wanders; he posits that at this level our thoughts seek experience, with emotions emerging, and daydreaming beginning. At low focus, our memory takes off on its own: our thoughts drift, reflections and self-awareness are weak, emotions are blooming, and we dream[6].

From this brief overview, one can offer many questions and criticisms, but the author does explain the variations from these generalisations. I found it quite absorbing and I might return to it one day, but for now that is all that I wanted to convey.

References:

1. Pinker, Steven, *How the Mind Works*, Penguin Books, London, UK, 1998

2. Nagel, Thomas, *Mind and Cosmos: Why the Materialist Neo-Darwinian Conception of Nature is Almost Certainly False*, Oxford University Press, New York, NY, 2012, p. 95

3. https://www.chronicle.com/article/Where-Thomas-Nagel-Went-Wrong/139129

4. Gelernter, David, *The Tides of Mind: Uncovering the Spectrum of Consciousness*, Liveright Publishing Corporation, New York, NY, 2016

5. Ibid, p. 54

6. Ibid, p. 3

Chapter 8-6: How the Mind Works

"The mind has to be built out of specialized parts because it has to solve specialized problems ... Each of our mental modules solves its unsolvable problem by a leap of faith about how the world works, by making assumptions that are indispensable but indefensible – the only defense being that the assumptions worked well enough in the world of our ancestors."

~ Steven Pinker, *How the Mind Works*[1] ~

Unless otherwise stated, where page numbers are quoted in this chapter, they refer to the book referenced above.

I must confess that I was very disappointed with this book by a Professor of Psychology, because his terminology was so inappropriate when speaking of evolution. I cannot know whether it was simply his choice of words, or whether he truly misunderstands the fundamentals of evolution, but already by page 43, I doubted that I could get through all 565 pages. As an analysis project, it was both tiresome and frustrating. Here are some examples so that you can judge for yourself, emphases mine:

• "Natural selection is not a puppetmaster that pulls the strings of behaviour directly. It acts by **designing** the generator of behaviour. The package of information-processing and **goal pursuing** mechanisms called the mind. Our minds are **designed to generate behaviour** that would have been adaptive." (p. 42)

• "Though the process of **natural selection** itself **has no goal**, it has evolved entities that are **highly organized to bring about certain goals** and sub-goals." (p. 43)

• "The **ultimate goal** that the **mind was designed** to attain is maximising the number of genes that created it." (p. 43)

Evolutionary biologists refute any suggestion of teleology in the evolution process, and thus there are three words that should never be used when referring to evolution: *design*, *goals*, and *purpose*, yet Pinker continually infuses them into his descriptions, contrary to his assertion that one should not use inappropriate idioms. On that basis, he contradicts his own arguments. In another study, he noted, "Logicians tell us that a single contradiction can corrupt a set of statements and allow falsehoods to proliferate through it."[2] Does he not recognise that very failing in his own works? In the same study, but in a different context to the one I am arguing here, he also notes, "The problem is not just that these claims are preposterous but that the writers did not acknowledge they were saying things that common sense might call into question."[3] Curiously, he does not acknowledge the same issues with his own writings on the mind. Pinker writes about reverse-engineering the mind to understand its component parts, working backwards from its design intent, but his premise is utterly wrong: if there is no design, there can be no design intent. Pinker acknowledges, "Reverse -engineering is possible only when one has a hint of what the device was designed to accomplish" (page 42), but ignores the absence of design and a designer. On the other hand, if he does truly sense design, why does he not acknowledge it as do the proponents of *Intelligent Design*?

Similarly, if you cannot identify the entity, how can you begin to deconstruct it? The activities of an entity can suggest some properties of the entity, and perhaps even how those properties interrelate, but cannot take you much further. I am reminded of science fiction scenarios where a structure suddenly arrives from outer space, and scientists attempt to understand what it is, and of what it is composed. They know it can fly, it is made of some unknown material, and is likely the product of an intelligent agency, because only intelligent agencies are capable of design. Beyond that, nothing much can be known until an external agent explains it. Probing the construction of the mind is similar, except that there is no external agent to do the explaining.

It is also illogical to claim that a process which has no goal, can evolve entities that are highly organised to *achieve* goals. Here, he is accepting that the mind is purely biological, and thus only capable of behaving according to the intrinsic chemical properties of the cell, none of which can ever have goals or express intent. Chemistry and biology do not achieve goals – they simply interact according to their properties without purpose: purpose is conceptual, not physical. Of course, the last quotation demonstrates just how confused he truly is, contradicting himself by stating that the mind was *designed* by evolution *for an ultimate goal*. This is fantasy, not science, especially when he speaks of *a leap of faith*, as he does in the opening quotation. Let us have another look at that.

"Each of our mental modules solves its unsolvable problem by a leap of faith about how the world works, by making assumptions that are indispensable but indefensible". I suspect that Pinker is acknowledging what I contend: that the brain cannot learn about reality when its only sources are electro-chemical signals conveying encoded symbolic representations of that reality. There is nothing to enlighten the brain as to the meaning conveyed in the semantic layer of the signals. Pinker's solution is that the mind-brain complex, aka *mental modules*, makes *assumptions* which over time, just prove to be right. Pinker does not explore the implications for evolution of getting the assumptions wrong. He also does not substantiate his assumption that the embryonic mental modules, which are nothing but biological arrangements, were capable of making assumptions in the first place. An assumption requires conceptualisation of possible solutions, pre-knowledge of what something might be. When the brain was in its developing state, it was empty of any concepts about which it could conceptualise and thus make assumptions, valid or otherwise. Here I reprise Primary Axiom #2 – All knowledge is built upon prior knowledge.

Setting aside my criticisms above, I have found Steven Pinker's *"The Blank Slate"* to be a noble and well considered study on the psychological impact of genetic and cultural influences. There are many "experts" who contend that nurture is all, and that genetics play no part. There are others who contend the opposite. Pinker makes a genuine attempt to resolve the conflict, leaning on modern studies in genetics, and whilst I accept that genetics certainly do play a significant role, I remain sceptical of his attempts to reverse-engineer "what is", into "what was", and how it came to be, purely via biological evolution.

Putting a stake in the ground, so to speak, I do NOT know how the *mind* works, and based on reading numerous books and articles on the subject, I contend that neither does anyone else, even when they claim that they do. In this study, I offer the evidence of why all current scientific explanations are false, and why in all likelihood, any scientific endeavour based on *philosophical materialism* to explain the mind, will inevitably fail. Neuroscience has made amazing progress over the past half-century: identifying where in the brain specific functions are performed; developing drugs to counter adverse behaviour; excising tumours, and so on. However, as a human geneticist, Gerard Verschuuren observes, *"brain surgeons are not mind surgeons"*[4]. Some brain surgery does impact the *outputs* of the mind, but not necessarily the mind itself. There are highly qualified specialists in the neurosciences on both sides of the debate, over whether the mind is physical property of the brain, and emergent property of the brain, the activity of the brain, or whether it is some immaterial entity – a mystery beyond our comprehension. I side with the latter. From the outset, I would ask the reader to put aside any suspicion that such a conclusion derives from a belief in God: it does not. As this study attempts to explain, there are solid scientific and logical reasons for rejecting any explanation based entirely on philosophical materialism and evolution.

Returning to Steven Pinker's book, *"How the Mind Works"*, it is disappointing for three primary reasons. Firstly, from my own background in information processing, and from the well-publicised experience of commercial enterprises, most especially the aviation industry, his computing analogies are little better than handwaves, and suggest a very shallow comprehension of the science. For example, he states, "Minds are probably easier to revamp than bodies because software is easier to modify than hardware" (p. 41). He clearly has no proper comprehension of that field, and for readers familiar with the Boeing 737-MAX saga, you would understand this. Secondly, his explanations of how the genome works are out of date. We can forgive him for that, as his book was published some twenty years before my own study, and his knowledge of genomics likely stems from some decades even before that. Nevertheless, such false knowledge detracts from the authority of his arguments, as he himself has acknowledged concerning the works of others. Thirdly, his explanations based on evolutionary theory are entirely unacceptable, and I suspect that they would horrify most evolutionists. As before, there are concepts that should never be suggested in evolutionary theory: *purpose*, *goals* and *design*, yet Pinker's explanations, as previously evidence, continually invoke such concepts. It could be that he does truly understand, but was struggling to find the right words, I will grant him that. However, as written, his book is more fantasy than science. Let me repeat some examples:

- "The mind, like the Apollo spacecraft, is designed to solve many engineering problems, and thus is packed with high-tech systems each contrived to overcome its own obstacles." (p. 4)

- "The mind is a system of organs of computation, designed by natural selection to solve the kinds of problems our ancestors faced." (p. 21)

- "Learning ... is made possible by innate machinery designed to do the learning." (p. 33)

- "The ultimate goal that the mind was designed to attain is maximising the number of copies of the genes that created it." (p. 43)

I suspect from that last point that Pinker has been unfortunately influenced by Richard Dawkins', "*The Selfish Gene*"[5], which again to be fair, was based on science later found to be incomplete. Ironically, despite his attempts to explain how the mind works, Pinker seems to understand that the mind is incomprehensible to itself; he notes: "The faculty (mind) with which we ponder the world has no ability to peer inside itself or our other faculties to see what makes them tick" (p. 4). In this instance, he agrees with Primary Axiom #1, that *nothing can explain itself*. To be fair, Pinker does offer considerable useful commentaries, but so often, contradicts himself, as when he says that the mind both *evolved* and *was designed*: a contradiction in terms. He also offers explanations which I cannot sensibly parse, as found in his numerous descriptions of the mind as quoted above.

Many of his arguments appear to agree with the proponents of *Intelligent Design*, contrary to evolutionary theory, but again, it could just be his choice of words that lead to that impression. He also demonstrates an understanding of some of the fundamentals of information processing when he states, "Information and computation reside in patterns of data and in relations of logic", but then confounds the issue, continuing, "... that are independent of the physical medium that carries them." (p. 24) Pinker is correct in the latter: the semantic (meaning) layer of messaging is independent of the media, but thus arises a problem for the biological brain. If they *are independent of the physical medium that carries them*, i.e., the brain, where are they? How can the patterns be independent of the physical network of neurons, axons, dendrites, and synapses, when it is these physical entities that are said to instantiate the patterns? The neuronal network is claimed to be the physical medium, the pattern, and the process which creates the pattern! If the mind, an independent entity, creates the pattern, the mind must also be independent of the pattern which it creates. This is the cardinal fallacy in Pinker's explanation, and is demanding of emphasis.

Patterns can be created by the material in which they appear, but only subject to their intrinsic physical properties. For example, exempting some exceptions for the moment, crystals are of solid material where the molecules fit together in a repeating pattern; they have flat surfaces called facets, which form geometric shapes such as triangles, rectangles, and squares. The shapes are a direct result of the type of molecules and atoms that make up the crystal. The molecular structure of crystals disallows curves, circles, and spheres. A diamond is a unique form of crystal, actually compressed carbon, and cannot grow as other crystalline structures do. There are many other such examples in nature, but all are subject to the same source of limitations – their chemical and cellular properties.

Biology takes a different path, but under the same rules of chemistry and physics. Keep in mind that a foundational principle of evolution is the complete absence of philosophical purpose or goals. I have yet to understand the origin of the purpose of replication and/or survival, other than that this is what organisms do. Cells replicate, but what was the chemical impetus for the arrangements that led to the initial replication, chance alone? More especially, where did the *purpose* of survival come from? Survival has always been a struggle, as implied by the term, survival of the fittest, but why bother?

Cognitive Ability

"Cognitive psychology has shown that the mind best understands facts when they are woven into a conceptual fabric, such as a narrative, mental map, or intuitive theory. Disconnected facts in the mind are like unlinked pages on the Web. They might as well not exist."

~ Steven Pinker ~

An interesting observation by a Professor of Psychology, and one with which I entirely agree. But the question we must ask is: by what process can *material* neurons, or neural pathways, be woven into a *conceptual* fabric? Does Pinker believe that the *conceptual* is material? The definition of conceptual is: something having to do with the mind, or with mental concepts, or philosophical or imaginary ideas. Conceptual thinking involves abstract ideas, which I contend, are immaterial, not material, which would be the case if the mind is a property or activity of the brain. The behaviour of the material brain is entirely contingent upon the laws of chemistry and physics. It *is* the mind that does the weaving, as the Pinker says, but can the mind be of the same physical material as the brain, subject to the same constraints, or must it be immaterial to be able to process the immaterial conceptual, as I contend? Pinker seems to agree when he states, "Correlation is a mathematical and logical concept; it is not defined in terms of the stuff that the correlated entities are made of" (page 65). This is what I have been contending from the outset. The *stuff that the correlated entities are made of* is biological (material), as in neuronal networks, so how can correlations, which are conceptual, be made in and by the material brain? What, and on what basis, is intuition initiated?

A definition of the *cognitive* includes: relating to, being, or involving conscious intellectual activity such as thinking, reasoning, and remembering. Another says that the cognitive is based on, or is capable of, being reduced to empirical factual knowledge. Knowledge, synonymous with information, is the product of *"facts ... woven into a conceptual fabric"*, such facts being derived from the interpretation of sensory perceptions. Knowledge can be true or false, depending on the validity of the facts, and how other conceptual factors contribute to the weaving. Existing "knowledge", worldview, and other factors, will determine the outcome.

Consciousness, in the sense of being awake or even aware, is not a necessity for intellectual activity, especially reasoning and remembering. A phenomenon that I have observed, and often used, is that of "sleeping on it". Problems that I struggle to solve during the day, whilst actively working on them, are sometimes magically resolved in my mind the next morning. The only conclusion that I can arrive at is that overnight, my mind was chugging away in the background working on my behalf, or perhaps its own.

Intelligence, Intellect and Rationality

"If atheism is true, there can be no absolute or objective standard of right and wrong, for there is no eternal heaven that would make values objective and universal."

~ Jean Paul Sartre, atheistic philosopher ~

Sartre was commenting on morality, but I contend that his observation about the lack of absolutes is applicable to other issues about which we rationalise, including rationalisation itself. If all intellectual activity is subjectively based, with no absolute objectives as a foundation, then we ought not believe in anything that anyone thinks, and that includes my use of the word, *ought*, because there can be no sense of correctness without objective absolutes. As famously quipped by American politician, Daniel Patrick Moynihan, "Everyone is entitled to his own opinion, but not to his own facts". Facts are objective truths built on absolutes, not subjectivity.

Professor of Psychology, Gerard Verschuuren, advises:

"Rationality is not a matter of intelligence but of intellect. Rationality is our capacity for abstract reasoning and having reasons for our thoughts, thus giving us access to the 'unseen' world of thoughts, laws, and truths – allowing us to be masters of our own actions. Reasoning is pondering realities beyond what we experience through our senses. Animals, for their part, seem to live their lives entirely in the present, without having any thoughts about the past or the future."[6]

Animals clearly have a level of intelligence, and perhaps even intellect, as they can be trained to perform various tasks based on both human and other stimuli. On the other hand, it is doubtful that they can decide to *not perform*. Just why, from time to time, they appear to so choose is beyond my understanding, but I am not confident that it is based on reason. I doubt that an animal chooses to exercise or not, or chooses to not eat because it wants to lose weight. When my cat sits on the window sill looking out on the world, likely it is with a sense of curiosity, but I doubt that it is pondering the meaning of life, or whether the lawn needs mowing. Does an animal ponder the rights or wrongs of what it did yesterday, or what it should do tomorrow to improve itself? When a sausage falls off the plate onto the ground, does a pet dog, like Newton, think: "Ah! Gravity!"? Likely not.

It is doubtful that any organism, other than humans, ponders the *how* or *why* of anything. For them, the world just "is", as they respond to the needs of the body, not the mind. This is where humans are so very different – our rationality has us asking questions about all manner of things, even when our thoughts are not considered rational based on the rules of logic. I remember reading a comment by Albert Einstein, a doyen of intellect and reasoning, that the greatest mystery about the world is that it is comprehensible. If it were not, one could not do science. However, it is no so much that the world is comprehensible, as that we, alone on Planet Earth and maybe the Cosmos, are able to comprehend it. We are composed of the same atoms and molecules as everything else, sentient or otherwise, yet we alone exhibit intellect and rationality. We are said to share 99% of our DNA with chimpanzees, but as earlier discussed, that 1% difference is not what accounts for the vast disparity in intelligence, intellect, and rationality, meaning no offence to chimpanzees. Quite frankly, anybody who thinks that they are not much different to a chimpanzee probably isn't, which should be a matter of concern.

The question that evolutionists must answer is this: How can differences in molecular arrangements account for the workings of the human mind - how can biological matter arrange itself in conceptual patterns, and even more fundamentally, how can it devise conceptual patterns for it which it has no previous experience? If the brain is the orchestra, who wrote the music, and who/what is the conductor?

As Dr. Verschuuren questions: "Why does the rationality as present in our minds correspond with the rationality present in the world?"[7] Reasoning from the position that "rationality is not a matter of intelligence but of intellect", and that even intelligence is not an intrinsic property of physical matter, he makes the point that without absolutes, there can be no cause of the order that we know exists, and without absolutes to ground reality, "even science would be a shaky, problematic, and irrational enterprise". He then supports my thinking when he notes that our Universe need not be the way it is, "In other words, our universe is neither necessary nor absolute, but finite and dependent instead; the more philosophical term is *contingent*. However, if there is no inherent necessity for the universe to exist, then the universe is not self-explaining, and therefore must find an explanation outside itself." (Primary Axiom #1)

Thinking

Gilbert Ryle has a chapter in his book entitled, *The Intellect*. As with the whole of the book, I struggle to understand his point, as he spends almost all of his words on asserting what is not, rather than what is. Philosophically, I cannot fault him for that – I have supported that approach myself when we observe a phenomenon that we cannot explain. Reviewers have labelled him an *empiricist* - a person who supports the theory that all knowledge is based on experience derived from the senses. I disagree with this form of empiricism, because I contend that *instinct* is a form of knowledge which exists from birth, prior to any sensory experience other than in the mother's womb (or a test tube). On the other hand, I also argue that all knowledge is built upon prior knowledge (Primary Axiom #2), which raises the question: How did knowledge originate?

Ryle is right to say that, "In one sense, the English verb 'think' is a synonym of 'believe' and 'suppose'; so it is possible for a person, in this sense, to think a great number of silly things, but, in another sense, to think very little. Such a person is ... intellectually idle."[8] I would agree, the modern idiom being, *"just saying"*. In my perception of thinking, and what I attempt to practice, is that it is a process of "gathering my thoughts" to remain focused on the context. Sometimes, I can be found staring into space as I channel my thoughts into recalling salient facts or experiences. At one stage of my career, I was regularly accused of slacking off, as my boss would not accept my explanation for my apparent lack of activity. I sometimes wondered whether my boss had never truly thought as I did, but I digress. Mostly, I think with pen (or keyboard) in hand to order my disordered thoughts into coherence. Many of my ideas come to me unbidden, but others I tease out, although I cannot explain how.

I recall hearing from a Cognitive Psychologist, with whom I once worked, a theory that a human mind can retain a maximum of seven concepts simultaneously, with most people capable of only two. When I worked as a consultant for a software company, whose product included modules for purchasing, sales, inventory, manufacturing, scheduling, and finance, I was often asked by a developer the implications of making a change in just one area. I was very conversant with all software modules, and could answer with the implications for costing, inventory, planning, scheduling, finance, and whatever other business function might be affected. It might, perhaps, be inaccurate to say that I held all of those concepts in mind simultaneously, but there had to be a degree of parallel processing in my mind to be able to articulate the effects in one verbal delivery. Similarly, as I researched for this book, I encountered subjects which I knew I intended to cover, and in which chapter this new material would belong. How that relates to thinking and a theory of mind I will let others ponder – I just know that it happens.

A challenging publication on this subject is *The Mind is Flat: The*

Illusion of Mental Depth and The Improvised Mind by psychologist Nick Chater[9]. Quoting from a review:

"His thesis is simple if stark; we find it hard to plumb our mental depths not because they are so deep and murky, but because there are no mental depths to plumb ... Chater goes on to propose his new theory of the 'cycle of thought'. He notes that our brains need cooperation across a vast number of slow neural processing units, across whole networks or even entire regions of the brain ... and so he concludes that any given problem must be split into tiny fragments to be dealt with in parallel across the entire, densely interconnected network."[10]

Again, I must disagree, firstly with his mention of *"slow neural processing units"*. According to one narrative, there are between 18 and 640 trillion signals zipping around the brain every second. His thought, "it is hard to see how so many interconnected neurons can coordinate on processing more than one thing at a time" is problematic, most especially when we see athletes and musicians at work. In a sport I played in my youth, rugby league, I was envious of players who could run, swerve, vary their pace, identify potential threats and opportunities, and pass the football, all in just a few seconds. Now maybe, just maybe, their minds only worked on one thing at a time, but the mind would need to have been very quick to switch between one activity to another, whilst remembering to switch back to something else of significance. You see, whilst the mind was focused on one specific process, there was a background process running continuously which amongst other things, was reminding the player of the rules of the game. A similar phenomenon is found in the behaviour of, for example, a fighter pilot, whose situational awareness must encompass multiple domains even whilst under stress. Now, it may just be his choice of terminology, but Chater contradicts himself when he doubts the ability to "coordinate on processing more than one thing at a time", yet in the next breath refers to parallel processing across a network. You see, the reason why parallel processing was developed in computing was – to work on more than one problem at a time. From my own experience, my mind does execute parallel processing, because whilst deliberately working on one issue, another one comes to mind, sometimes complementing my work, sometimes distracting. It could be argued that my mind instantly switched from one subject to another and back again, but why would one so argue when a more obvious solution is to hand – the mind is capable of parallel processing multiple subjects simultaneously. Can we not walk and talk at the same time? Can a soldier not watch what is happening in front of him, evaluate threats, talk on his radio, and signal to other troops, all at the same time?

Imagination and Conception

I cannot be certain that animals do not have imagination, I can only rely on the conclusions of animal behaviourists who say that they do not. However, humans most certainly do, and in their imagination, are able to combine concepts which in their experience, have never been combined before. Here is an extract from the works of Maimonides which illustrates the point:

"Imagination is the power by which one retains impressions that have been received by the mind, after they have vanished from the receptor senses. By means of his imagination, man connects some impressions and separates others, in doing so, he creates new ideas out of images that are stored in his memory, and dreams up visions he has never heard of and could not have possibly perceived. For example, a person may imagine: an iron ship sailing through the air, a man whose head reaches up to heaven and whose feet stand on the ground, an animal with a thousand eyes. Many other such impossible things may all be inventions of man's imagination or fantasies of his dreams."[11]

Maimonides (1138 – 1204) lived in a period long before iron ships were developed (1830s), and even longer than before the first flight (1903), let alone when metal aircraft were invented. To what do we attribute Maimonides imagination and creativity? Could a biological brain, composed of neuronal networks, have accidentally created just the right patterns for Maimonides to visualise *an iron ship sailing through the air*?

Creativity

"Creativity is intelligence having fun"

~ Albert Einstein ~

According to Wikipedia, a sufficiently erudite source for our purpose here:

"Creativity is a phenomenon whereby something new and somehow valuable is formed. The created item may be intangible (such as an idea, a scientific theory, a musical composition, or a joke) or a physical object (such as an invention, a printed literary work, or a painting). Creativity is a combined power, and over the years, individuals are able to expand internal reservoir of means as intelligence, wisdom, imagination and all the fragments of mind which have acquired from being conscious, alive and wakeful to the universe in an incredibly new manner."[12]

In general, creativity is purposeful, with a goal in mind. Evolutionists deny teleological processes, i.e., relating to or involving the explanation of phenomena in terms of the purpose they serve, rather than of the cause by which they arise. As we have seen, evolutionists also deny that we have free will, because evolution could not give rise to free will. Similarly then, we ought not have the attributes of imagination and creativity, yet experience demonstrates that we do have such attributes. Evolution theory and philosophical materialism continually founder on the realities of the human condition, and thus their contentions must rest on very sandy foundations. I acknowledge that not all evolutionists consciously believe as I have written above, but when pressed, they invariably do, or offer an alternative narrative which is equally indefensible.

My point here is to further evidence that the mind could not be material, and could not be an emergent property of the brain.

Attention, Comprehension, and Worldview

I conducted a simple experiment on social media to demonstrate that one's worldview invariably distracts attention from a subject, leading to a lack of comprehension. I opened a question on how the mind works by first mentioning God. This was a deliberate red herring to note whether atheists would miss the point, and not unexpectedly, they did. Here is the discussion, with a sample of two of the answers received:

- Question: Atheists will claim that God is but a figment of the human imagination, failing to understand that they are in fact, arguing against themselves. Let me explain. According to the philosophy of materialism which underpins evolutionary theory, the mind-brain is a purely biological entity, the mind being an emergent property of the brain. The workings of the brain depend entirely on the laws of chemistry and physics, as applicable to the structure of the brain, which developed depending on the biology of cells and the genome. My question is: What properties of the material world subject to the laws of nature, could give rise to the concept of a non-material being? How could the material envisage the non-material?

- Answer 1: You seem to be offering a binary choice while of course there is a multitude of possibilities, none of which can be proven. You propound a theory of materialism (which may be accurate, I wouldn't know) which precludes any notion of imagination. Your fallacious binary choice seems to be that 'God' (whatever that means) exists or that this and no alternative theory is possible. You seem to base all this on the odd assertion that without 'God', imagination can't exist. Novel, for sure.

- Answer 2: Gods came into the imagination of numerous ancient communities because they could not deduce many properties of the material world. Examples are the diurnal rising and setting of the sun, the tides, the weather, diseases, etc. A supernatural being (or several of them) were invented to explain the unknowns. The more that mankind discovered the less the need for god(s). Today non-believers are possible the majority in educated societies. Religions are desperately trying to keep and increase the number of their believers.

I was accused of not phrasing my question sufficiently clearly, even though, I would have thought, the question was specific and unambiguous as written - "My question is: What properties of the material world subject to the laws of nature, could give rise to the concept of a non-material being?" The atheist respondents focused on the preamble to the question, rather than on the question itself. Being atheists, the mention of God distracted their attention from the main issue, as I suspected that it would.

My interest is always piqued by scientists observing behaviour in others, but seemingly unaware of the same behaviour in themselves. I am all too aware of this phenomenon, and take steps to ensure that I do not similarly succumb (at least in my own writings). The most common example is where scientists assert that there is no such faculty as free will, but cannot explain themselves, or their own lives, without it. In discussing the history of knowledge and beliefs, and how beliefs have changed over time, Patricia Churchland opined that new evidence "challenged a whole framework of thinking ... that had been taken for granted since ..."[13] Churchland makes a valid point, but does not apply that to her own thinking. Darwinism has influenced scientific thinking for over 150 years, without ever resulting in a coherent explanation of how evolution could account for the whole human condition. Resistance to this thinking is powerful, often resulting in personal confrontations, threats, insulting language, demonisation, and the subversion of careers. As Steven Pinker puts it, "The analysis of ideas is commonly replaced by political smears and personal attacks. The poisoning of the intellectual atmosphere has left us unequipped to analyze pressing issues about human nature just as new scientific discoveries are making them more acute."[14] I entirely agree, and contend that unquestioned acceptance of philosophical materialism and Darwinism has left scientists unequipped to analyze the mind-brain conundrum. Adherents to evolution theory are remarkably similar to the devotees of Aristotelian science in the days of Galileo, but cannot see themselves in that light, because they see themselves as more enlightened. But of course, every generation has thought that way, and likely always will.

Dreams

The following is from this online resource[15]:

"There are several theories about why we dream. Are dreams merely part of the sleep cycle, or do they serve some other purpose? Possible explanations include:

1. representing unconscious desires and wishes.

2. interpreting random signals from the brain and body during sleep.

3. consolidating and processing information gathered during the day; and

4. working as a form of psychotherapy.

From evidence and new research methodologies, researchers have speculated that dreaming serves the following functions:

5. offline memory reprocessing, in which the brain consolidates learning and memory tasks and supports and records waking consciousness.

6. preparing for possible future threats.

7. cognitive simulation of real life experiences, as dreaming is a subsystem of the waking default network, the part of the mind active during daydreaming.

8. helping develop cognitive capabilities.

9. reflecting unconscious mental function in a psychoanalytic way.

10. a unique state of consciousness that incorporates experience of the present, processing of the past, and preparation for the future; and

11. a psychological space where overwhelming, contradictory, or highly complex notions can be brought together by the dreaming ego, notions that would be unsettling while awake, serving the need for psychological balance and equilibrium.

Much that remains unknown about dreams. They are by nature difficult to study in a laboratory, but technology and new research techniques may help improve our understanding of dreams."

The referenced website has a lot more detail on the subject of dreams, which may be of interest, but additional points only serve to confirm my opinion on the subject. That is, that it is the immaterial mind which is the driver, not the brain, although the brain is clearly involved. The offered opinions suggest either a superintending function, or the brain operating as a controlling function of itself. The latter seems unlikely unless there is a separate region in the brain that contains the controlling "firmware": I have not encountered such a region in any of the works that I have studied. Both the "Neuroanatomy Text and Atlas"[16] and "The Brain Atlas"[17] are very detailed, with no suggestion of the brain having any areas reserved for monitoring or controlling the brain itself.

The persistent conundrum here is that of a physical organ being capable of recognising and correlating conceptual facts which it could not possibly understand. Consider (3), *"consolidating and processing information"* – as I have argued, the entire neuronal structure of the brain deals in encoded electro-chemical signals, which need to be decoded by an entity which is separate from the messaging network.

I have no doubt that in some manner, the functions suggested above are performed in a dream state, but it is the nature of those functions which convince me that the active agent is the immaterial mind, not the material brain.

Summary

One author asserted that *"The mind is a system of organs of computation"*[18], and I would offer that at a fundamental level, that analogy does work. However, the term "computation" suggests both mathematical and non-mathematical calculations, whereas computers work via logic alone, "switches" turned on an off, as instructed by the controlling software. The insufficiency of the analogy is the absence of an explanation for how to get from switches to logical conclusions on complex conceptual subjects. The preceding narrative is just one example of the complexity, especially where logic fails entirely.

It is worth revisiting two observations by Steven Pinker, both of which I wholeheartedly support, and both of which bring into question the veracity of his conclusions of the mind:

- "Logicians tell us that a single contradiction can corrupt a set of statements and allow falsehoods to proliferate through it."; and

- "The faculty (mind) with which we ponder the world has no ability to peer inside itself or our other faculties to see what makes them tick."

In Pinker's books, he unconsciously contradicts himself, and studiously writes narratives purporting to peer inside the mind, seeking to explain that which he says, the mind cannot. What is doing the peering if not the mind?

Whilst I cannot know how the mind works, my understanding of how computers work has me not able to accept the *"computational theory of mind"*. On the other hand, if this theory is correct, as Pinker and numerous others claim, then a functional analysis of information processing would be the best way to approach the subject of how the mind ought to work. This we have reviewed in Part 4: *Principles of Information Processing*, wherein it was concluded that the theory was false.

References:

1. Pinker, Steven, *How the Mind Works*, Penguin Books, London, UK, 1998, p. 30

2. Pinker, Steven, *The Blank Slate: The Modern Denial of Human Nature*, Penguin Books, London, UK, 2002, p. ix

3. *Ibid*, p. x

4. Verschuuren, Gerard, *What Makes You Tick? A New Paradigm for Neuroscience*, SOLAS Press, Antioch, CA, 2012, p. xv

5. Dawkins, Richard, *The Selfish Gene*, Oxford University Press, Oxford, UK, 2006

6. Verschuuren, *Ibid*, pp. 77-78

7. *Ibid*, p. 79

8. Ryle, Gilbert, *The Concept of Mind*, Penguin Books Ltd, London, England, 1990, pp. 265-266

9. Chater, Nick, *The Mind is Flat: The Illusion of Mental Depth and The Improvised Mind*, Allen Lane / Penguin, London, UK, 2018

10. https://www.psychologytoday.com/us/blog/ten-zen-questions/201911/can-you-have-two-thoughts-once

11. Finkel, Avraham Yaakov, *The Essential Maimonides: Translations of the Rambam*, Jason Aronson, Inc., Lanham, MD, 1996, p. 171

12. https://en.wikipedia.org/wiki/Creativity

13. Churchland, Patricia S., *Touching A Nerve: Our Brains, Our Selves*, W.W. Norton & Company, New York, NY, 2013, p. 16

14. Pinker, Steven, *The Blank Slate*, p. x

15. https://www.medicalnewstoday.com/articles/284378.php#causes

16. Martin, John H., *Neuroanatomy Text and Atlas*, 4th Edition, McGraw-Hill Inc., New York, NY, 2012

17. Hanaway, Joseph, *The Brain Atlas: A Visual Guide to the Human Nervous System*, Fitzgerald Science Press, Bethesda, MD, 1998

18. Pinker, Steven, *How the Mind Works*

Chapter 8-7: The Mind's Eye

"The eyes are useless when the mind is blind"

~ Anonymous ~

This quotation has multiple meanings, from the philosophical, to the pragmatic, and even to the physiology of our sense of sight. As I mentioned in an earlier chapter, whilst the eye "sees", the mind-brain can only "visualise", for only the eye deals with light – thereafter through the optic nerve to the visual cortex to the upper reaches of the brain, all we have are electro-chemical signals being transmitted. The semantic layer, that of meaning, is hidden in encoded symbolic representations of the reality of what is seen by the eye. There is a process in the mind-brain complex that decodes these signals and reconstructs what the eye sees for the mind-brain to visualise. The question becomes: can the brain alone, a biological organ composed of a complex network of neuronal pathways, transcend itself to the immaterial visualisation of external realities?

I contend not.

The Merriam-Webster dictionary defines the *mind's eye* as *the mental faculty of conceiving imaginary or recollected scenes*. They have it partly right: they have ignored the physiological fact that the mind's eye is also the mental faculty that allows the system of sight to conceive or visualise external realities before us, not just imaginary or recollected scenes. Of course, mention of *imaginary scenes* removes us from the physical reality to another dimension. If this activity be true, then it rests the case for the immateriality of the human mind.

There is not much that I can say on this subject, as I have no idea of how the mind's eye works, other than to contend that it cannot be an activity based solely on the physical properties and contents of the brain. As I sit here contemplating the subject, I can close my eyes and recollect the scene before me, the one that I just saw with my eyes, or alternatively, whatever I choose to imagine, always provided that I have the free will to choose, contrary to the contentions of the evolutionists. Some historical events in my life, and some people, I can recollect in exquisite detail, but others are just a blur. No doubt there are reasons for this phenomenon, perhaps related to the impact that such events or people had on my life, but that is as far as I can venture.

For me, the mind's eye presents the most compelling evidence for duality – that whilst the brain is material, the mind is immaterial, yet both interact with each other to complete the nature of human existence. I would again offer that I am not venturing into the concept of an immortal soul or a life hereafter. Whether the immaterial mind dies with the death of the brain is not something about which I can comment – it might, or it might not, but we should not let one's beliefs about this issue determine whether or not we believe the mind to be immaterial.

What happens to the mind with the death of the brain I will leave to others to express their opinions: I have none, principally because there is no evidence before me.

But here, let me confess a conundrum which has me stumped. If it requires an immaterial mind to perform visualisation, both from memory and as a function of our sense of sight, does that not say that all creatures with a visual sensory system must have an immaterial mind? I know not where to go from there, not even back to the drawing board.

Something to Ponder

I am ever searching for better explanations for what I am attempting to convey, knowing all too well that the thought processes of all of us vary in one way or another. Thus, I offer yet another thought to ponder.

In a book, concepts are symbolically represented by ink blots, with neither the page not the ink blots comprehending their meaning. When light is shone on a page, the process of reflection transforms the symbols into characteristics of light waves and photons in a way for which I have no understanding – I just know that it must happen. When received by the eye, the characteristics of the light are transduced by the multiple layers of neurons in the retina, deconstructing the characteristics into individual components and transmitting them via thousands of dedicated nerve fibres in the optic nerve to the visual cortex, again with no understanding. The visual cortex then performs the amazing task of recombining these individual neural signals into a composite symbolic representation of the external reality, and transforms it into neuronal patterns in the grey matter of our brain, again with no comprehension of the conceptual meaning originally instantiated by symbolic ink blots. The neuronal patterns are in a way, no different to the other stages in the transmissions in that they are physical symbolic representations of the original concepts laid down by whoever wrote the book.

The questions become: Where does the comprehension arise? Who or what is the "we" that does the comprehending? How can the neuronal patterns interpret themselves?

Part 9: Language

"But if thought corrupts language, language can also corrupt thought."

~ George Orwell, 1984 ~

I have chosen this quotation from the works of George Orwell, chiefly because it is so apt for earlier chapters (perhaps it should have been there) concerning the language of scientists, and the "transferred epithet"[1] concept introduced by Raymond Tallis. There can be no doubt that at times, and especially in the context of the mind-brain conundrum, the language used in explanations by scientists does indeed corrupt thought.

What little I review in this section, and it is deliberately not much, comes largely from three published studies:

1. *Language In Our Brain* by Angela Friederici[2],

2. *The First Word: The Search for the Origins of Language* by Christine Kenneally[3], and

3. *The Language Instinct* by Stephen Pinker[4].

I had considered discussing language in more detail, but was alarmed at the ever-rising word/page count of this tome, and concluded that enough is enough. Perhaps I will return to this subject at a later date, as I do find it fascinating, but for now, a quick review will do (I can hear your sighs of relief from here).

From the outset, I am not about to contend with these scientists on the details of their studies, for such would be far beyond my knowledge and competence. My goal in studying these published works is to understand the foundational presuppositions upon which such "scientific" assumptions are made. For example, Friederici admits, *"In Language and* (sic) *Our Brain"*, I start from the assumption that language is a biological system that evolved through phylogeny"[5]. As earlier, if you are as unfamiliar with scientific terms as I am, *phylogeny* is the evolution of a genetically related group of organisms via the study of protein or gene evolution by involving the comparison of homologous sequences. My issue is with the *assumption* of phylogeny without even a rational explanation.

I also wish to understand the rigour in the terminology, believing that without adequate definition of terms, even the best of scientists can be led into error. The most common oversight is the acceptance that information can be stored biologically, when in truth, the nerve cells of the brain cannot do so. We have covered some of this subject before, but here it is useful to review some principles in a specific context. The brain contains two types of cells, but only the nerve cells are of interest. The others, called glia cells, provide neurons with nourishment, protection, and structural support. Nerve cells are described thus:

"There are many sizes and shapes of neurons, but all consist of a cell body, dendrites and an axon. The neuron conveys information through electrical and chemical signals. Try to picture electrical wiring in your home. An electrical circuit is made up of numerous wires connected in such a way that when a light switch is turned on, a light bulb will beam. A neuron that is excited will transmit its energy to neurons within its vicinity.

Neurons transmit their energy, or "talk", to each other across a tiny gap called a synapse. A neuron has many arms called dendrites, which act like antennae picking up messages from other nerve cells. These messages are passed to the cell body, which determines if the message should be passed along. Important messages are passed to the end of the axon where sacs containing neurotransmitters open into the synapse. The neurotransmitter molecules cross the synapse and fit into special receptors on the receiving nerve cell, which stimulates that cell to pass on the message."[6]

A small correction, not all neurons have both dendrites and an axon, but that is not important here. Another issue is that neural circuits are not like "electrical wiring in your home ". Such wiring conducts electricity as a continuous stream, whereas neural circuits conduct "pulses" in the form of action potentials, with postsynaptic neurons returning to their resting state as they transition to being the presynaptic neuron. Consider the words used: electrical and chemical *signals*, *information*, and *messages*. These are treated as synonyms, but as I have argued in earlier chapters, they are not. I would refer the reader back to Chapter 6-2 Neuron Communications, if you would like to refresh your understanding in this area. The most critical for our understanding is that neither messages nor information is passed, for both have a semantic layer. All that occurs is a rippling of action potentials (electrical spikes) which have no meaning in and of themselves.

Many parts of the brain are involved in language processing:

"Neurolinguistics – one of the many neurosciences ... has shown how different parts of the brain seem to be specifically involved in different aspects of verbal *reception* (such as breaking up the sounds and translating them into speech components, connecting them with meaning, parsing the grammar of sentences, detecting the different total tonal envelopes of questions and statements); in equally numerous aspects of verbal *expression* (the selection of words, stringing them together in a grammatical form, dealing with different components of articulation); in linking sight and sound and meaning in reading; in connecting sight, sound, action, and meaning in writing, and so on."[7]

I doubt whether genetic mutation could be directly involved in the positioning and structuring of the neural networks that account for these differentiations, although perhaps epigenetics played a role. But the mystery remains: how could neural networks, based on physics and chemistry, self-organise around the conceptual of which it could not be aware? As for tonal expressions, Queenslanders have developed the habit of ending statements with a rising intonation usually reserved for questions, yet somehow, we understand that they are not asking a question (we hope). Another mystery for neurolinguistics to solve.

Returning to Angela Friederici's book, "*Language In Our Brain*", "the author then admits that the different language-relevant brain regions alone cannot explain language". Indeed, they cannot, and searching through this resource, I have been unable to find a coherent explanation of what does explain language processing in the brain. If we deconstruct the brain, all we find is a network of biological cells which are incapable of subjective activity. If we disassemble a computer network, all we find are physical components which, for the same reason, are incapable of autonomous subjective activity – all such components are comprised of chemicals arranged in different ways.

Evolution of Language

I have encountered a very curious "chicken and egg" conundrum in the scientific studies on the evolution of language, one which I have not found to be adequately addressed. Philosophical approaches date at least as far back as Plato, 5th century BCE, but a biological approach appears to date from the days of Charles Darwin, as discussed in his *"Descent of Man"*, 1871. Many theories have been offered,

"For centuries, philosophers and other scientists have thought about the origins of language, formulating a variety of different theories. Only a few of these theories have survived the time. Historically, one of the first essays on the topic stems from Johann Gottfried Herder (1772) ... *Treatise on the Origin of Language* ... In his essay, he claims that language is a human creation. Humans' first words were the origin of consciousness. The phonological forms of these words were primarily close to the sounds of objects they denominated, they were onomatopoetic. As humans developed, words and language developed and became more and more complex. In his view grammar is not at the origin of language – as 'grammar is no more than a philosophy about language' – but rather it is a 'method of use' that evolved only as the need to talk about abstract entities increased. Herder's theory stands in direct contrast to the modern views such as those of Chomsky, who considers syntax to be the core of language."[8]

I have some sympathy with Herder's theory, as it parallels the development of written language. These started with pictograms – symbols representing concepts. Over time, these were transformed to disambiguate the concepts so that writings could be understood without detailed explanation of concepts. From my studies on the Hebrew language, I learned that early versions of Paleo-Hebrew consisted largely of pictograms which could be ambiguous, i.e., interpreted differently depending on context. The holy Scriptures, written in this form, could only be read correctly by those educated in the final detail, one of the reasons why the script continued to develop. From an evolutionary or developmental perspective, I am not confident that Chomsky is right, for if syntax is the core of language, why do different languages have different syntax if the brain evolution was the same? Certainly, the syntax of individual languages is key to understanding, but that is based on learning, not on biological evolution.

"In the context of brain-related theory of the evolution of language,

Rizzolatti and Arbib (1998) based their theory on experimental work with non-human and human primates. Their claim is that language evolved from motoric gestures. It is based on the finding that in macaque monkeys viewing a grasping movement activated neurons [sic] not only in the visual cortex but also neurons in the hand-motor cortex, which fired in response to the visual stimulus (Rizzolatti et al, 1996). These neurons were subsequently called *mirror neurons*. Similar effects were then observed in humans at the level of the brain systems (rather than single neurons) and were called the *mirror system* (Iacoboni et al., 1999)."[9]

Whilst not discounting the scientific evidence of *mirroring*, there is a logical flaw in the extrapolation to a theory of language development. Language, for the most part, is conceptual, and is structured using arbitrary symbols and grammar, whereas mirroring of gestures is not. There is a cognitive link between gestures and language, else the meaning of a gesture could not be correctly interpreted. I am reminded of a humorous anecdote concerning Winston Churchill's "V for Victory" hand gesture. With the palm facing outward, it connotes peace, but the obverse is usually considered offensive. According to the anecdote, Churchill started with the offensive gesture, and only when informed of its meaning, turned it around but still considered it appropriate to be displayed toward the enemy.

The Language Instinct

The title of this part is taken from a book of the same name by Stephen Pinker[10]. What you may have already guessed, from my earlier expression of interest in origins, is from where did such an instinct derive? What evolutionary processes could give rise to an instinct to communicate, given that such an instinct is teleological, i.e., relating to or involving the explanation of phenomena in terms of the purpose they serve, rather than of the cause by which they arise. We communicate for a reason, well, mostly at least, but what must be obvious is that an instinct to communicate involves purpose. Pinker, at one time Professor in the Department of Brain and Cognitive Sciences at MIT, is undoubtedly a man of intellect and learning, and I mean no disparagement, but as with so many people of science, he starts in the middle of the story, not at the beginning. I have thoroughly enjoyed his book, and learned much from it, but it does not answer my primary question: How, and perhaps more important, why, did the instinct of language arise? Again, the instinct is teleological, the antithesis to evolution theory.

The sub-title of the book also piqued my interest: *The New Science of Language and Mind*. Was I about to learn something new about the science of the mind, or was I to be disappointed to discover that as with practically all other studies, the brain and mind have been conflated, both physically and conceptually? Would I find two associated presuppositions: evolution, and the mind being an emergent property of the brain?

Unfortunately, this volume whilst being very interesting, took me no closer to plausible explanations.

References:

1. Tallis, Raymond, *Why the Mind is Not a Computer*, Imprint Academic, Exeter, UK, 2004, p. 34

2. Friederici, Angela D., *Language In Our Brain: The Origins of a Uniquely Human Capacity*, The MIT Press, Cambridge, MA, 2017

3. Kenneally, Christine, *The First Word: The Search for the Origins of Language*, Viking Penguin, London, England, 2007

4. Pinker, Stephen, *The Language Instinct: The New Science of Language and Mind*, Penguin Books, London, England, 1994

5. Friederici, *Ibid*, p. 2

6. https://mayfieldclinic.com/pe-anatbrain.htm

7. Tallis, Raymond, *Aping Mankind*, Routledge Classics, New York, NY, 2016, p. 24

8. Friederici, Angela D., *Language In Our Brain: The Origins of a Uniquely Human Capacity*, The MIT Press, Cambridge, MA, 2017, p. 203

9. *Ibid*, p. 204

10. Pinker, Stephen, *The Language Instinct: The New Science of Language and Mind*, Penguin Books, London, England, 1994

Part 10: Postscript

"Cognitive psychology tells us that the unaided human mind is vulnerable to many fallacies and illusions because of its reliance on its memory for vivid anecdotes rather than systematic statistics."

~ Steven Pinker ~

This postscript concerns the fallibility of the human mind, and has me pondering whether Steven Pinker ever considered that his too is as susceptible as any other.

My Primary Axiom #1 asserts that *nothing can explain itself*, and as the mind is the only engine of explanation that we possess, it follows logically that the mind cannot explain itself. Fortunately, we are able to examine some behaviours of the mind, but which one should note, are external to the mind itself. Behaviours are activities based on properties, but if we are unable to identify the properties and the substances giving rise to those properties, we cannot diagnose the nature of the entity itself. We can discern its interaction with the brain, and whilst we can get ever closer, we must never lose sight of our fallibility. In other works, I have expressed my disappointment with what I term, the *Cult of Celebrity*. No doubt, hero worship has always been a human trait, but in modern times, with the increasing reach of communications and social media, the noise has reached a crescendo. Now, achievers in practically any field, with the possible exception of the intellectual, are looked upon for guidance, even where there is no reason for such recognition, other than achievements in sport, music, acting, or other forms of entertainment. That is not to denigrate the achievers, but to wonder at the worshippers. Even in sport, in Australia at least, professional players are now not just expected, but required to be role models for younger generations. In a sense, that is a good thing, but in another way, not so, for these people are themselves generally too young and immature, lacking in wisdom, to be surrogate elders.

The *Cult of Celebrity* is the ugly step-sister of the *Cult of Authority*. Where once education and learning were pursued for their own sake, today they are but steps to achieve success in commerce and financial security. In importance, the piece of paper has replaced the learning supposedly acquired to obtain it. Facts quickly crammed into minds with little appreciation of their application, just as quickly leak out again through lack of use. Education has become an industry, focused on revenue and success rates, encouraging a reduction in academic standards. Universities regularly complain of the lack of numeracy and literacy in applicants, but then succumb to intellectual poverty just to maintain the inflow of cash and students. As Nobel Prize winning author, J.M. Coetzee described his students in his book, *Disgrace* (p. 32), "post-historical and post-literate, they might as well have been hatched from eggs yesterday." In that same vein, I remember this quote from the mind of Calvin Coolidge, American politician and lawyer, the 30th president of the United States (1923 – 1929):

"Nothing in this world can take the place of persistence. Talent will not; nothing is more common than unsuccessful men with talent. Genius will not; unrewarded genius is almost a proverb. Education will not; the world is full of *educated derelicts*." (italics mine)

Educated derelicts indeed! One could also ponder the materialists' explanation for the human trait of persistence.

That is not to suggest that all students are of that nature, but increasingly, according to reports from across the Western world, that is the trend. Thus, as academic credentials are acquired in ever increasing numbers, their value as a commodity trends in the opposite direction. Nevertheless, the sheer volume has accorded them respectability that they may not deserve, encouraging this *Cult of Authority*, whereby as long as one has the requisite credentials, they must be listened to. Accordingly, the increasingly credentialled masses will only listen to the credentialled, accepting their words to be authoritative, just as they accept the words of celebrities. Of course, if all are similarly credentialled, then none can be differentiated as particularly authoritative. Discernment is a long-lost talent, as much through disinterest as any other reason.

The human mind, always fallible, is inclining ever more so.

The chapters in this book, from which you have evidenced persistence by getting this far, by an author (me) lacking academic credentials of any kind, are offered for consideration. Although containing some authoritative quotations, my own words are not authoritative, merely thoughtful. I trust that they represent some value to the reader.

Chapter 10-1: A Law Hypothesis

"Because there is a law such as gravity, the universe can and will create itself from nothing. Spontaneous creation is the reason there is something rather than nothing."

~ Stephen Hawking, *The Grand Design*[1] ~

Now, before you think that I am being unfair to the late Stephen Hawking, here I am referring to what people would generally understand from his statements, rather than what I believe he actually meant. However, only Hawking can be blamed for his lack of clarity in his last published work, *"The Grand Design"*. Apart from the misleading title referring to *design*, which Hawking and all other atheists and evolutionists declare is not a function of evolution, nor any other naturalistic processes, his choice of words is very poor in his main theme. Raymond Tallis describes this as transferred epithets, whilst Steven Pinker terms it as "the presentation of facts belonging to one category in the idioms appropriate to another". I do believe that a scientist of Hawking's calibre would understand the concept of law better than his statement above would indicate.

No law, not even a scientific law, can *cause* anything to happen. *Causality* is a metaphysical concept that describes the relationship between cause and effect. Scientific laws are statements of causality. For any cause and effect to occur, there must be something to initiate the cause, and something upon which to have an effect. If nothing exists, then nothing happens. If nothing exists, there is no activity for scientific laws to describe. Thus, Hawking's statement is the most deplorable nonsense, if taken as written, which I will grant, he did not mean to happen. Gravity describes a relationship between two or more masses at any distance from each other, and if there are no masses, i.e., nothing, then gravity does not exist. Without an intelligence, i.e., nothing, to observe an interaction, there can be no formulation of a scientific law.

If one believes Stephen Hawking to be right as written, one must ask: Why do we not observe spontaneous creation happening even now? Why just once, so long ago? As one Jewish sage observed, *some people seem to occasionally wander off the reservation*. Albert Einstein commented that "the man of science is a poor philosopher"[2], but in this instance, we have a particular man of science who would appear to have a very poor grasp of logic, let alone philosophy. But to be fair, Hawking did not mean what his words meant in the ordinary sense. Without going into a full explanation, in the context of Hawking's statement, "nothing" is a quantum vacuum, a seething mass of energy and Higgs bosons upon which gravity can operate as particles pop in and out of existence. To understand scientific pronouncements, one must understand the meta-language which explains the language being used. One must always be aware that scientists do not necessarily mean what you think they mean.

Introduction

I am unsatisfied by the paradigm of philosophical materialism. The theme of this study has been the mind-brain conundrum: Is the *mind* of a material nature, as is the *brain*, or is the mind non-material? Is the mind an emergent property of the brain, or does it have a separate existence, perhaps even spiritual? Why are there so many issues in our existence which seemingly do not fit within the paradigm of materialism? Why have scientists, in the main, chosen to ignore that an answer to many unsolved mysteries may be found by looking beyond the material? Stephen Hawking, an accomplished scientist and theoretician, was unsatisfied with the Big Bang model, as scientists had to go outside established science to forge their explanations. Thus, he found himself seeking alternate explanations for the origin of our Universe. In doing so, he appealed to multiple dimensions, and a mathematical concept - imaginary time. It was as if not liking the imagined singularity which science cannot explain, he proposed imaginary time to avoid explaining it.

What is it about the non-material that has scientists imagining anything and everything but?

Secular and theistic scientists have been using the term, *information*, in numerous unspecific ways, as elastic in nature as the term, *evolution*. Whilst to some extent, discipline has been imposed on other terms, such as speculation, model, paradigm, hypothesis, and theory, I contend that the same has not applied to the concept of *law*. The multiverse theory assumes that particular laws and constants are built into each separate universe at the moment of its origin, such as the Big Bang. The suggestion is that somehow, they are "imprinted" in each universe, which raises the questions: What determines this set of laws, where are they, and how are they remembered? How does an individual universe "know" what laws and constants are governing it, as opposed to the different laws and constants of other universes? Cosmologist Martin Rees suggested that "The physical laws were themselves 'laid down' in the Big Bang", but admitted, "The mechanisms that might 'imprint' the basic laws and constants in a new universe are obviously far beyond anything we understand"[3] ... but obviously not beyond one's imagining. In my own imagining, the allocation of laws and constants is a function of an intelligent agent. If not, however, that is the case, with the allocation being simply random, selected from a wider set of laws and constants, what is the source of that set?

Rupert Sheldrake, a biologist and researcher in the field of parapsychology, has some challenging theories on the nature of reality and evolution. He notes that the founding fathers of modern science, Copernicus, Galileo, Descartes, Kepler, and Newton, were all essentially Platonists or Pythagoreans, advocates of the theory that there are mathematical patterns underly the natural world. He notes,

"Physicists who reject the multiverse theory have a variety of alternative suggestions. Some pin their faith on what they call a 'final theory', a unique mathematical formula that would predict every detail of our *present* universe, including all the so-called constants of nature. The uniqueness of the universe would then be a necessary consequence of mathematics. This ultimate Platonic dream is far from becoming true. But suppose that physicists really did one day discover *The Formula*. The next question would be: where did it come from? And why did it exist in the first place? The answer would probably be a superformula. But where did *that* come from?"[4]

The principle of causality suggests the need for a *First Cause*, but at the same time, cannot find the cause of that beginning. It cannot be explained within philosophical materialism, as indeed, mathematics cannot be. All along, scientists believe in the existence of mathematics, and even attribute ultimate authority to mathematics, whilst acknowledging that mathematics does not have a material existence, and have no idea of the source of mathematics.

This suggests to me that the search should not so much proceed on finding a *theory* of everything, but on finding the *authority* for everything. This I shall attempt to do with my hypothesis that follows.

The Nature of Law

"It is one thing for the human mind to extract from the phenomena of nature, the laws which it has itself put into them; it may be a far harder thing to extract laws over which it has no control."

~ Arthur Eddington (1882-1944) English astronomer, physicist, and mathematician ~

Scientific research is predicated on a fundamental metaphysical law: *causality*. For every action, there is a reaction; for every cause, there is an effect; for every phenomenon observed, the cause is sought. Essential to this method of working is the acceptance that given the same circumstances, a certain cause will always have the same effect. Scientists speak of natural law, or laws of nature, these having been derived based on experience and experimentation, as if such laws were the discoveries of man. But of course, they are not – such laws govern existence irrespective of whether we correctly understand them or not. It is also generally accepted that natural laws are universally valid; apply equally to both inanimate matter and living organisms; should have universal application in all fields of scientific research; and are immutable.

I offer that the concept of *immutability* is where the acceptance of scientific law being synonymous with *natural law* departs from reality, for as we know from history, scientific laws are falsifiable and subject to correction, as new phenomena are experienced. But the intrinsic laws of nature that govern our existence are *not* falsifiable – it is just our understanding of them which is so.

My Hypothesis

In this chapter, I contend that by defining "natural" as only pertaining to the physical or material, we have established an arbitrary *rule of evidence* that militates against a full understanding of existence, and the laws that govern it. This definition of *natural* has been allowed to hijack the agenda, initially perhaps unintentionally, but subsequently used as a blunt instrument to bludgeon those who would dare to consider that there is more to existence that what we can demonstrate with the scientific method. In the context of the mind-brain conundrum, there are aspects which clearly belong in the domain of the scientific definition of natural law, but as I have attempted to explain, there are some which are not so clearly categorised.

Struggling to avoid terms which have established connotations, e.g., spiritual, supernatural, etc., I am going to use the term, *formless*, for that which science cannot, at least as yet, grapple with using the tools of the scientific method.

Defining Science

In its broadest sense, the term "science" refers to any systematic knowledge-base or prescriptive practice that is capable of resulting in a prediction or predictable type of outcome. It pertains to reliable knowledge in almost any field, including philosophy - the study of general and fundamental problems such as those connected with existence, knowledge, values, reason, mind, and language. Today, we still use the term broadly as in *political science*. In its more restricted contemporary sense, science refers to a system of acquiring knowledge of the physical or material world based on the *scientific method*, and to the organized body of knowledge gained through such research. The scientific method is a way of validating, modifying, or falsifying a hypothesis that arose from the observation of some phenomenon. It usually proceeds in this sequence: observation, hypothesis, prediction, and testing the predictions. Though the terminology can vary with usage, in general the *hypothesis* becomes a *theory* when it has been sufficiently validated for the scientific community to accept it as THE explanation. The theory of evolution, and here I refer to the narrative that all life on earth arose from a single common ancestor which itself arose from an inorganic form, is an exception, as no predictions have been made or verified: the proposed mechanisms have not been replicated, and thus cannot be tested, and as we have seen, the explanatory powers of random mutation and natural selection are even now being questioned by evolutionists themselves.

Scientific laws describe the *what*, but not necessarily the *how* or *why*; are usually expressed in mathematical formulae; and can be used in everyday scientific endeavours. The Law of Gravity tells us what happens, but not why, nor does it explain what gravity is. Similarly, Boyle's Law establishes a relationship for pressure, temperature, and volume, but it does not tell us why a gas behaves as it does. This is an entirely rational approach to the material or physical world, but it is reasonable to ask whether this approach is at all useful in seeking to understand whether there is an existence beyond the physical – what I term, *formless* - being without physical form.

I contend that it is not.

Relating to Nature

The term "natural" has long been equated with the physical environment in which we live, including earth, water, air, flora, fauna, and so forth. The non-material or spirit world is termed *supernatural*, for those who believe in such things. Having narrowed the scope of "science" to just the physical world, the terms *scientific law* and *natural law* have been conflated to be synonymous. It is my belief that this conflation of terms has done a disservice to mankind, perhaps inadvertently, excluding as it has, any understanding of the broader aspects of our existence. The difficulty arises when people begin to believe that because science, so defined, deals only with the material, then only the material exists. The silly season starts when people then declare that there is no scientific evidence for the *formless*, seemingly unaware that given the modern usage of the term "science", that there cannot be, by definition.

Another View

We speak of "natural justice", but this is a statement of morality, having nothing to do with the physical world. We speak of a person's "nature", meaning personality, but again it can be argued that this is not material in essence. In this chapter, I want to continue the case that in the context of investigating the nature of the mind-brain complex, we need to return to the broader understanding of science, and allow the possibility that nature extends beyond the physical or material. If we arbitrarily constrain our thinking to just the material, then we automatically disallow the possibility of the *formless*. While many consider scientific laws and natural laws to be equivalent, I contend that they are not. Understand that this is a *philosophical* discussion relating to *existence*, and can only be addressed by logic and deduction, with little if any physical evidence.

Terminology

There are excellent articles that address some issues in this chapter, but most are written from an academic perspective, using a great deal of academic terminology that makes comprehension, for the less familiar, rather difficult. This essay[5] is of that nature, but I would recommend it as a useful foundation for my proposition. It discusses, at length, terms such as methodological naturalism, metaphysical naturalism, Humean categories and the like, but those unfamiliar with such terms need to be patient as they read. I am going to use just a few terms with which the reader may be more familiar: natural[6]; supernatural[7]; materialism[8]; science[9]; and scientism[10], definitions below. To avoid presuppositions concerning the word, *supernatural*, I will use a new word, *extra-natural*, meaning being external to the physical forms considered by scientists to be natural.

Notice that in all cases, these are arbitrary *philosophical* considerations. That all phenomena can be explained in only physical terms, or that physical matter is the only reality, is not something that has, or can, be proven. Similarly, science has chosen, for the most part, to define itself in terms of materialism, an arbitrary definition which one can allow provided that one understands the arbitrariness of it. The important point to bear in mind is that these definitions are arbitrary, and despite the advances in science, may not map to reality.

The Point of Contention

My contention here is that those who argue for materialism, or scientism, have chosen to establish the boundaries and extents of knowledge, but have no reason for doing so other than personal preference. I contend that this is a philosophical position, not even a scientific one in the broader sense, and one that cannot be validated by "science" as the materialists define it. I will further argue that this philosophical position is not even logical, and that we need to take a much broader view of what we consider *natural,* if we are to research the origins and nature of existence.

From here on, I will use the terms "natural law" and "scientific law" as two separate and distinct domains of law.

What is reality?

We cannot know the totality of reality. We have our own reality, that which we experience, and we have the reality of which others tell us. A child in the Australian Outback may have no personal experience of rain, snow, or the oceans, but he or she may nevertheless believe in them. We sense and experience the world in which we live, but our experience of it does not define it. We perceive only what our senses convey to us, but as I have evidenced in earlier chapters, our sensory experiences are cognitive interpretations of external realities, and often, the same external reality can result in varied cognitive interpretations in different people, because such are *subjective*, not objective. We also know that our sensory perceptions are limited, and need to be augmented by intelligence to reveal through scientific discoveries, that there is so much more than what we experience, such as an electromagnetic spectrum beyond the visual, sub-atomic particles, and weird quantum behaviour. The point that I want to establish is that in science, as in everyday life, the definition of reality is arbitrary, being based on experience when we cannot know that humanity has experienced all of reality or more importantly, has properly interpreted it.

For the purposes of this discussion, I will coin yet another new term to avoid stumbling over our different perceptions of what is natural - *The Truths of Existence*, and offer the following definition:

Existence must be subject to a set of laws which determine both the origin, and characteristics, of that which exists, whether material or formless, whether proven or unproven, whether experienced or not. This set of laws stands as the super-set to all other laws which define our material existence, and the formless if such exists, and is both causative and deterministic, not merely descriptive.

Natural law, as I have defined it, just "is" and is immutable. Because scientific law is incomplete and still developing, it is logically a sub-set of natural law. Because natural law is conflated with scientific law in modern parlance, in our thinking it too is a sub-set of whatever super-set must exist. The acceptance of the common usage of the term 'natural', as being synonymous with the material or physical, has led people into muddled thinking. We shall seek to un-muddle.

What is Existence?

Herewith the essence of the argument: despite the assertions by some in the scientific community, in fact we know neither the cause, nor the full extent, of existence. Fortunately, we can logically infer some things, and this is where we shall start, but note from the outset that the extent of existence is entirely independent of our knowledge of it. Scientific Law defines our knowledge of it, but *existence* just is, which suggests a line of enquiry beyond where I wish to go in this study. However, if we are afraid of where our journey may lead us, we will miss the opportunity to learn along the way. We can always turn back when out comfort level is exceeded, but that is no reason to not start.

Scientific law and *natural law* are not equivalent. Those wedded to philosophical materialism can be excused for conflating the two, because they have actually defined it thus, so in fact they have no room to believe anything different. Natural laws exist in reality, control our reality, and insofar as we can ascertain, are immutable, i.e. they cannot change themselves, nor be changed, because if they are not immutable, such changes would be subject to a pre-existing superset of laws. If it is proven, as some research suggests, that the natural laws obtaining in our Universe are not constant throughout the Cosmos, then there is reason to accept the likelihood of an even higher set of laws. Modern science has yet to propose a plausible hypothesis for the origin of natural laws. Some say they arrived with the Big Bang or some prior event, and were created out of nothing, even though amazingly coherent and symmetrical once here. It is admitted that for the singularity of the Big Bang model to exist, it must have been subject to as yet, unknown laws that explain its existence.

What are Scientific Laws?

One definition is: "an independently and sufficiently verified **description** of a direct link between cause and effect of a phenomenon, deduced from experiments and/or observations". A somewhat contradictory definition is: "a law generalizes a body of observations. At the time it is made, no exceptions have been found to a law. Scientific laws **explain** things, but they **do not describe** them". Finally, this one: a scientific law is "a phenomenon of nature that has been proven to invariably occur whenever certain conditions exist or are met; also, a formal statement about such phenomenon, also called **natural law**" [emphasis mine]. I have lost track of where I found these definitions but I am sure you can find them or something similar.

It is unclear from these definitions whether scientific laws describe or explain, but in my view, scientific laws describe what happens, but not why, other than as a first-order cause and effect. At an observed or superficial level, they may describe how something happens, but always stop short of explaining the very beginning of a chain of causally dependent events. For example, random quantum variation – what is that? Can we be sure that it is random, or do we declare it random because as yet, we understand neither the cause nor the rules? Planets continue in their motion due to gravity, but whilst science understands gravity in terms of mass and distance, the nature of gravity continues to elude our understanding. Radioactive elements decay into daughter elements, but what causes a particular atom to decay into a lower energy state? It is said to be random, but is that true, or does it just appear to be random because we have yet to discern a pattern or cause?

How do Scientific and Natural Laws relate?

Scientific laws are not real entities: they are not deterministic; they do not control anything; nor cause anything to happen. All they can say is that when an event occurs, it will always occur in a particular way given the same set of conditions; this allows us to make predictions and utilise behaviours. Scientific laws are descriptions and conclusions drawn from observations and experimentation. They are an attempt to describe natural laws, and as we know, are subject to revision as greater knowledge is acquired (perhaps they should be called science *descriptors*). Scientific law is but a knowledge *sub-set* of natural law, but if we conflate natural law with scientific law, then it in turn is a sub-set of a yet to be uncovered super-set. For convenience in this discussion, I am going to call this super-set *Existence Law*. Ponder how far our knowledge of the natural world has progressed over even the last two hundred years, and it should be obvious that at any point in time, scientific law trails our understanding of natural law by a significant margin.

I would contend that it is impossible for scientific law to fully describe natural law, or Existence Law. The current thinking of so many scientists (and atheists) is constrained by their arbitrary definition of the scope of science. Such constraints, I suspect, are the inevitable child of the confusion between natural law and scientific law, and an unwillingness to seek beyond the material. It is entirely illogical to claim that a sub-set, arbitrarily defined, can explain the super-set on which it is predicated. It is logically equivalent to a national sports team claiming to be the best in the world by defining their nation as the totality of the world.

Natural laws *govern* or *prescribe* reality – all reality, both material and formless. Scientific laws attempt to *describe* material reality only, as currently understood. Many attempt to address the formless by refusing to accept its existence, yet their attempts to describe the formless in materialistic terms inevitably founder on unsubstantiated and unprovable assertions. A fundamental error often encountered, perhaps due to a lack of semantic rigour, is the claim that scientific laws cause events to happen: they do not. Scientific laws are *descriptive*, not *prescriptive*, they cannot cause anything to happen: they simply describe our observations and conclusions of what we see happening, and when science gets it right, these laws can be used to predict behaviour. In physics for example, we have seen the progression from Newtonian laws, to Einstein's Relativity Theories, to Quantum Mechanics. While natural laws do not change, scientific laws do change, and regularly, as they should. It is prudent to be aware that science cannot prove itself on its own terms: any explanation of science must lie outside itself (Primary Axiom #1). Science can attempt to describe natural law, but cannot explain it, as science itself is predicated on natural law. Logically, in any domain or field of endeavour, any explanation must exclude the fact or term being explained.

Numerous scientists have attempted a unified Theory of Everything: a hypothetical single, all-encompassing, coherent theoretical framework of physics that fully explains and links together all physical aspects of the universe. Attempts so far have been foiled by numerous phenomena, such as the proposed singularity preceding the Big Bang not yielding to accepted scientific laws; the speed of light possibly not being constant everywhere; and the observation that Newton's laws work with large objects, but not with the very small, whilst quantum physics works with the very small, but not the large. Whilst we do not have a *scientific law* of everything, I would contend that there must be a *natural law* of everything, or perhaps natural law refers only to our Cosmos, and is a sub-set of the super-set, *Existence Law*.

Hans Sachsse, a German chemist, philosopher of technology, and an influential author on cybernetics, described natural science as *"a census of observational relationships which cannot say anything about first causes or the reasons for things being as they are; it can only establish the regularity of the relationships"*[11]. The literature is replete with examples of scientists admitting to the descriptive nature of scientific laws, but it is a source of intrigue, to me at least, why so many scientists seem to baulk at taking the next step: that of attempting to understand the essence and source of what we call scientific laws. It is into this intellectual vacuum that I shall venture, others can correct my error.

The Truths of Understanding Existence

In an attempt to disambiguate natural and scientific law, and to provide a framework for what may be, hypothetically, beyond the material, i.e., *formless*, I thus offer the first two of what I shall describe as the *Laws of Existence*. I am not much in favour of this term, but I have struggled to find one more apt, so please excuse my lack of scholarship if this term rankles.

1.   Scientific law is a sub-set of natural law, and is predicated upon natural law.

2.   Natural law is a sub-set of Existence Law, and is predicated upon Existence Law.

For many years, the hard sciences, those now most commonly meant by the term *science*, had the luxury of indulging in materialism, or physicalism, that was open to perception by our senses. As knowledge developed, this luxury was eroded by atomic theory, identification of sub-atomic particles, Einstein's Relativity theories, Quantum Mechanics, String Theory, multi dimensions, and so on. Many scientific hypotheses are based on concepts that are not even in evidence; for example, that there are eleven dimensions rather than just the four we can experience. As I have read from some particle physicists, the deeper we delve into this area, the less we know. Quantum physics has revealed weird phenomena such as quantum entanglement, quantum tunnelling, and quantum non-locality, suggestive of another dimension independent of our space-time reality. Without belabouring the point, it is generally understood by scientists that our current scientific knowledge is perhaps but a fraction of the knowledge yet to be obtained. Thus, any commitment to materialism must be on dubious foundations as we have yet to fully understand what matter is, or whether our understanding of physical phenomena is accurate. This is an important point and bears restating, even if does not represent the opinion of the experts. Here I put my contention in italics to emphasise the point:

The philosophy of materialism, or physicalism, cannot be said to be logical, because by definition, it rests on an ill-defined foundation. We do not yet know what "physical" is, nor do we know the cause of it. Thus, the philosophy is presumptive at best, built as it is on inadequate knowledge and the shifting sands of understanding.

The sub-set of natural law, known as scientific law, has been further defined into disciplines such as physics, chemistry, astronomy, cosmology, engineering, etc. All are inter-related and are informed by one another. The divisions are arbitrary. The boundaries are blurred, and perhaps in time the definitions may be redefined to reflect newer scientific understanding. A most curious science is mathematics. Is it a part of scientific law, or more likely as it appears to transcend the physical sciences, a part of natural law? If it belongs in the domain of natural law, what determined the rules? Nobody knows where mathematics comes from, but we use it to describe all of the hard sciences, and it forms the basis of investigation of most physical phenomena that we do not understand, like relativity, gravity, and quantum mechanics. Now it matters not that I may be not entirely accurate with my scientific definitions, a correction will not invalidate the point of arbitrariness. Returning to natural law, are there such divisions as we have decided on for science? As we do not know the scope, source, or cause of natural law, we cannot say with any certainty. This then leads to the most essential points of the argument: the scope and source of natural law.

We have seen that science has defined itself, just as materialists have defined their own philosophical position, such as it is. The history of science reveals that no generation of scientists ever know the full extent of scientific knowledge that may yet be acquired; each generation knows only a sub-set of what future generations will know. Thus, no generation of scientists can know the scope, boundaries, or limits of the physical sciences. Science, as a sub-set of natural law, in turn a sub-set of the super-set Existence Law, cannot know the scope, boundaries, or limits of natural law, nor Existence Law. Science can only add to its own sub-set as new knowledge is obtained. In a sense, scientific endeavour is blind: it knows not where it is heading, or what it may find, even though this is untrue of some specific research. Scientific research may know the next step, and may have a destination in mind, but progress and direction will be ultimately determined by what is discovered along the way. Logically, this gives rise to the next Laws of Existence:

3.   The ultimate scope, boundaries, and limits of both natural law and Existence Law are beyond explanation by scientific law.

4.   The ultimate scope, boundaries, and limits of these parent law sets being beyond explanation, one can never know when one has discovered all of natural law, let alone Existence Law.

The most perplexing of all the questions facing science and the humanities is this: the origin of existence. Let us define existence, for this part of the discussion at least, as being space, time, energy, and matter - where did it come from? The most commonly accepted working model for the origin of our universe is the Big Bang.

Deliberately avoiding a scientific definition of the Big Bang model, so as to avoid nit-picking by those who actually know better (ambiguity intended), the model says that there was an infinitely dense and hot singularity that for whatever reason, chose to expand rapidly, followed by a second expansion phase, to create the initial conditions of the universe. The trigger might have been a quantum fluctuation, we do not know, but many have speculated, and some have even tried to apply scientific principles. Of course, if there was such a singularity, it must have complied with some law-set, but which? The singularity is not consistent with current scientific law, may even be outside natural law, but if real, must be subject to Existence Law, however defined. This issue has occupied the best scientific minds without resolution, giving rise to questions as phrased by Rupert Sheldrake[12]:

a.   If the laws of nature existed before the Big Bang, and governed the Big Bang from the first instant, where were they?

b.   If the laws and constants of nature all came into being at the moment of the Big Bang, how does the universe remember them? Where are they imprinted?

Let us hypothesize that the Big Bang was not the beginning of existence, nor was the singularity which preceded it: that maybe something preceded the singularity, something we shall call the *First*

Event. If we say that there was nothing prior to the *First Event*, define *nothing*. Nothing is a total absence of everything - no space, no time, no energy, no matter, no laws scientific or natural. Thus, existence came to be, uncaused, contrary to the first principle of science. If there was truly nothing in terms of *material* existence, and we accept the scientific assertion that natural law relates only to the material, then Existence Law was the *formless* entity operating prior to the First Event, the Big Bang, or any of its predecessors, and is independent of material existence.

If you prefer any one of the many theories such as oscillating universes, multi-verses, or whatever, through any number of regressions, you cannot avoid running up against another fundamental question: where did the energy running through all existence come from, not that we truly know what energy *is*? We know what it does, but there is no scientific definition of energy, absent of function or relationship to the material. So far, that question has proven unanswerable. Let me put a stake in the ground: we do not know the origin of existence. I have read of supposedly mathematical proof that the material itself is eternal, always was, and had no beginning. I could not even begin to understand the mathematics of that, so I will leave it there, other than to note the irony that *formless* mathematics is being used to explain an eternity of material form.

Just like physicists exploring the realm of sub-atomic particles and quantum mechanics, the deeper we explore the origin of existence, the deeper we must go to find an answer, as yet unfound. It is the very nature of this subject, and my Primary Axiom #1, that *nothing can describe itself*, which reveals the Fifth Law of Existence:

5. The source of natural Law cannot be found in natural Law.

As a logical consequence of the preceding laws, we can deduce the Sixth Law:

6. The source of existence cannot be found by the physical sciences.

One further point before we summarize. Science has arbitrarily defined itself, and corralled itself within its own definition of 'natural', deliberately and arbitrarily excluding what I have termed, *formless*. Natural law has been constrained by that same definition, despite circumstantial evidence that such is not warranted. For natural law to be so constrained, any such constraint must be contained in a super-set to natural law, and would be the source of natural law itself. What could that be? That is the basis of the Seventh Law, which takes us where those beholden to philosophical materialism truly do not want to go.

To summarize, let us restate the seven *Laws of Existence* as I have formulated them:

1. Scientific law is a sub-set of natural law and is predicated upon natural law.

2. Natural law is a sub-set of Existence law and is predicated upon Existence Law.

3. The ultimate scope, boundaries, and limits of both natural law and Existence Law are beyond explanation by scientific law.

4. The ultimate scope, boundaries, and limits of these parent law sets being beyond explanation, one can never know when one has discovered all of natural law, let alone Existence Law.

5. The source of natural law cannot be found in natural law.

6. The source of existence cannot be found by the physical sciences.

7. Existence Law as the source of natural law must transcend natural law and determine it, or is itself, Existence Law.

Summary

Scientific law is falsifiable by definition, but whatever law set that superintends scientific law must itself be immutable, otherwise even scientific law is predicated upon a sandy foundation. Scientific law is but a sub-set of natural law, and it is logical to assert that we cannot discern the totality of natural law from our knowledge of scientific law: thus, it is impossible for us to assert with any level of confidence, what it is in natural law that we are yet to discover. Not knowing the totality of natural law, we cannot comprehend its source, Existence Law, and what that does or does not encompass, including entities or phenomena beyond those we term material or physical.

The most puzzling question is the existence of formless laws, which are conceptual, and how they are imprinted in physical material such that even from before the Big Bang, they were imprinted somewhere.

Existence Law #7 is no doubt perplexing and an apparent tautology: *Existence Law is itself Existence Law*. I have termed it such to allude to my belief that the existence of a transcendent *First Cause* is more probable than not. If that be true, then the properties and activities are not separately discernible from the entity itself – it is "one" in a way that we cannot possibly comprehend.

References:

1. Hawking, Stephen, and Mlodinow, Leonard, *The Grand Design*, Bantam Books, New York, NY, 2010, p. 180

2. Einstein, Albert, *Out of My Later Years*, Thames & Hudson, London, UK, 1950, p. 58

3. Rees, Martin, *Before the Beginning: Our Universe and Others*, Simon & Schuster, London, UK, 1997, p.3

4. Sheldrake, Rupert, *The Science Delusion*, Coronet, London, UK, 2013, p. 96

5. Doyle, Shaun, *Defining arguments away - the distorted language of secularism*, Journal of Creation, Vol. 26(2), 2012, pp. 120-127

6. *Natural*: Of nature, relating to nature, relating to the physical world not the spiritual.

7. *Supernatural*: Pertains to being above or beyond what is natural, unexplainable by natural law or phenomena.

8. *Materialism*: The theory that physical matter is the only reality and that everything, including thought, feeling, mind, and will, can be explained in terms of matter and physical phenomena.

9. *Science*: In its broadest sense, any systematic knowledge-base or prescriptive practice that is capable of resulting in a prediction or predictable type of outcome. In its more restricted contemporary sense, science refers to a system of acquiring knowledge based on the *scientific method*, and to the organized body of knowledge gained through such research.

10. *Scientism*: The belief that the methods of natural science, or the categories and things recognized in natural science, form the only proper elements in any philosophical or other enquiry.

11. Sachsse, H. Die Stellung des Menschen im Kosmos in der Sicht der Naturwissenschaft, Herrenalber Texte HT33, "Mensch und Kosmos", 1981, p. 93-103, as quoted in "In the Beginning was Information", Werner Gitt, p. 27

12. Sheldrake, *Ibid*, p. 108

Chapter 10-2: Reflections on Science

"A paradigm shift is a fundamental change in the basic concepts and experimental practices of a scientific discipline."

~ Thomas Kuhn, *The Structure of Scientific Revolutions* ~

I have left this subject to last, as I am reluctant to be openly critical of human minds, most especially those of scientists and academics, notwithstanding my previous remarks about the late Stephen Hawking, but with a sense of humility for I greatly admired the man. I would not want a reader to dismiss my arguments simply because I was perceived as anti-science. A limitation of the human intellect is *pride*, accepting that pride in individual achievements is often legitimate. In the scientific and academic world, reputations and careers are built on acquired and demonstrated knowledge. At times, some knowledge is proven to be false, and that is when the trouble starts. With reputations and careers at stake, intellectuals will vigorously defend their positions, even to the point of denigrating those who challenge them. The history of science is littered with such incidents, notably in the persecution of Galileo Galilei by his academic peers when he contradicted the Aristotelian science of his day. Religious people display the same limitation, especially those of Christian denominations, who in a last ditch attempt to defend their beliefs based on their interpretation of some bible version, assert that their scripture is the inerrant Word of God - those who believe otherwise lack the guidance of the Spirit of God, aka., the Holy Spirit. I have written extensively on that subject, having been subjected to such criticism myself.

Curiously, it was a belief in an ordered universe created by a transcendent God which gave rise to many principles of modern science. Theoretical physicist, J. Robert Oppenheimer, reportedly a non-observant Ashkenazi Jew, acknowledged the role of Catholic scholars when he offered, "Christianity was needed to give birth to modern science"[1]. There are claims, much disputed, that Bishop Robert Grosseteste (1175-1273) was the first in the Latin West to develop an account of an experimental method in science; that he "seems to have been the first writer to make systematic use of a method of experimental verification and falsification"; and that he gave a "special importance to mathematics in attempting to provide scientific explanations of the physical world"[2]. Dr. Verschuuren noted, "The Franciscan friar, Roger Bacon, established concepts such as using hypotheses, experimentation, and verification, so science would be free from foreign authorities and habitual bias. In other words, what some consider a period of darkness was actually a blaze of light and reason."[3] Roger Bacon (1214-1292) lived during latter stages of the so-called *"Dark Ages"*, a historical periodization traditionally referring to the Middle Ages, $5^{th} - 15^{th}$ century, asserting that a demographic, cultural, and economic deterioration occurred in Western Europe, following the decline of the Roman Empire. The persistence of this characterization owes much to the thinking in the eighteen century *"Age of Enlightenment"*, although in my opinion, such a period is better viewed as a *false dawn.*"

I would remind readers that it was during these so called, Dark Ages, that philosophers such as Maimonides, whom I often quote, enlightened us with his thoughts. It was in this same period that the earliest, still existing, universities were founded: Paris (1045), Bologna (1088), Oxford (1096), and Cambridge (1209). Sometimes, written history was formulated to deceive us.

My objective in this chapter is to have the reader ponder the observations of scientists about their own colleagues, and the disciplines they pursue. It is not my intention to denigrate anyone, but simply to offer what I believe is a useful perspective, which may assist in the discernment of truth in science. I would ask the reader to accept the following in the spirit that it is offered.

Science is Self-Correcting

In their better moments, scientists admit that science is "self-correcting", but then seem to ignore that in some cases, they have no way of knowing where in the cycle of truth their current beliefs belong - in the *corrected phase*, or still *in need of correction*. A great deal of scientific theory is reliable, demonstrated in the implementation of technology which has been proven over time. Other theories, or more precisely, hypotheses, are clearly speculative even when claimed to be proven. Quoting from a book[4] by evolutionary biologist, Professor Douglas Futuyma, he appears to concur: "At its best, science challenges not only non-scientific views but established scientific view as well. This, in fact, is the wellspring of progress in science. Our knowledge can progress only if we can find errors and learn from them. Thus, much of the history of science consists of a rejection or modification of views that were once widely held." (p. 163) But, then, he seeks to lead us astray, when on the same page he asserts, "Science is science only if it limits itself to determining the nature of reality". How, I ask, can it truly be science, if it limits itself to its own arbitrary, and unprovable, *definition* of reality? As we noted earlier, Robert Wesson, evolution proponent, political scientist, and Senior Research Fellow at the Hoover Institute, takes a different view, effectively refuting Futuyma's perspective of reality: "The contention that reality consists of only material particles and their modes of interaction is not even a clear-cut theory. It implies a narrow definition of reality, making the thesis true by definition."[5]

In a chapter on scientific knowledge, Professor Futuyma wrote of how in Communist Russia, politics was allowed to override scientific fact. He concluded,

"A grim story indeed, but what do we learn from it? That reality stubbornly refuses to be bent to our desires or ideologies. Genes cannot be altered to suit our ends, as devoutly as we may wish them to be. Truth cannot be established by the Communist Party, nor by the vote of a democratic society or by a board of education. Reality does not yield to wishful thinking." (p. 162)

With the greatest respect to the good Professor, there is a certain irony here regarding *reality stubbornly refuses to be bent to our desires or ideologies*, for in defining science the way he has, he is stubbornly bending his understanding of science to the desires and ideologies of himself and his peers! Similarly, *genes cannot be altered to suit our ends*. I am confident that he was referring to our understanding of genes, not the genes themselves, but with the benefit of hindsight, given that his book was based on the science of some four decades ago, I would contend that evolutionary biologists of his time did precisely that. Scientific understanding of the genome has advanced significantly in recent times, such that its role in replication is now seen in a very different light. I fully accept Futuyma's integrity as he states, "Science is not the acquisition of truth; it is the quest for truth" (p. 164), quoting Stephen J. Gould[6]. I am not a scientist, *per se*, but pursuit of truth is also my goal; I have no preference for what it should be, I just want to know what it is. Futuyma admitted:

"The picture I have just painted is, of course, a somewhat idealistic one. In fact, scientists are just as human as anyone else. They believe that one or another hypothesis is most likely to be true, and they engage in sometimes bitter battles to defend their ideas. Scientists' beliefs are also shaped by their political, social, and religious environment." (p. 164)

He concluded his discussion on this subject by commenting on how people can be misled:

"Thus, the common image of scientists as abstracted, unbiased, detached intellects has no foundation in reality. Scientists are often highly opinionated, even in the face of contrary evidence; and they are often not particularly intelligent either. The spectrum of scientists, as of any other group of people, runs from the brilliant to the fairly stupid."

An Honest Apology

In a remarkable display of humility, Dr Frances Arnold, who won the Nobel prize in chemistry in 2018, sharing the prize with researchers from Harvard and Cambridge Universities, came forward herself to let her followers know that a 2019 paper of hers had been retracted. She tweeted: "For my first work-related tweet of 2020, I am totally bummed to announce that we have retracted last year's paper on enzymatic synthesis of beta-lactams," she began, before accepting responsibility. "It is painful to admit, but important to do so. I apologize to all. I was a bit busy when this was submitted, and did not do my job well."[7] Her comments were also reported in The Guardian[8], which went on to comment:

"In doing so, Arnold has mastered the art of a good apology ... (one) that gives context where relevant; but accepts full responsibility and moves on. In return, she has been thanked for her honesty."

Other scientists, such as Dr Betül Kacar, an astrobiologist from the University of Arizona, and Jonathan P Dowling, an academic from Louisiana State University, tweeted their admiration of Arnold for her humility. On another good note, an official statement in the Journal of Science concluded that Arnold's oversight of missing data, later found in a lab notebook, was not deliberate, and no scientific misconduct could be attributed to her.

Summary

Perhaps before continuing, you might care to reread under the heading, *Paradigms*, in Chapter 5-2, as the comments by eminent scientists recorded there are relevant here. Science pursues the truth of the laws of nature, yet scientific research is limited to arbitrary definitions of its own discipline, and fails to acknowledge that what is known about the laws of nature is what scientists have learned from their own limited scientific approach. Sadly, secular scientists are not (yet?) prepared to acknowledge that their pursuit of science cannot be explained by their own discoveries in science. In the minds of many, scientific law and natural law are one and the same. However, as I sought to explain in the previous chapter, this cannot be true. The generally accepted definition of natural law is unnecessarily restrictive, as is philosophical materialism: *the contention that reality consists of only material particles and their modes of interaction*. Professor Futuyma asserts that *reality does not yield to wishful thinking*, and I agree, but I suspect that he is being disingenuous given his commitment to philosophical materialism. As has been proven, time and time again throughout science history, accepted science does yield to wishful thinking, and continues to do so. However, even more significant I would offer, is that *reality does not yield to arbitrary definitions or accepted presumptive paradigms*.

If the quest of science is truly truth, then it should acknowledge the possibility of realities beyond its reach, especially when its reach is self-constrained. It is my contention that a paradigm shift is needed if science is ever to find the truth of existence, and most especially, the origins of existence. As for the evolutionists' take on the mind-brain complex, this would require them to honestly accept the evidence for the formless, and how the mind is not subject to any laws pertaining to physics and chemistry. Whilst the material brain is constrained by the properties of its components, the mind *has a mind of its own*, as it were, and can operate freely, unconstrained, and even contrary to what the material would recommend.

This deserves an explanation.

References:

1. Oppenheimer, J. Robert, Reflections on Science and Culture, University of Alberta, 1961

2. https://plato.stanford.edu/entries/grosseteste/#SciMet

3. Verschuuren, Gerard, *What Makes You Tick? A New Paradigm for Neuroscience*, SOLAS Press, Antioch, CA, 2012, p. 85

4. Futuyma, Douglas J., *Science on Trial – The Case for Evolution*, Pantheon Books, New York, 1982

5. Wesson, Robert, *Beyond Natural Selection*, A Bradford Book, The MIT Press, Cambridge Massachusetts, 1991, p. 4

6. Gould, Stephen J., *The Mismeasure of Man*, W. W. Norton & Co Inc, New York, NY, 1981

7. https://twitter.com/francesarnold/status/1212796190711959552

8. https://www.theguardian.com/science/2020/jan/06/nobel-prize-winner-demonstrates-best-way-apologize-chemist-frances-arnold

Chapter 10-3: A Final Word on Evolution

"I would like that to be known; these facts are in the summary which I think is a very good one."

~ John Sherman Cooper (1901-1991), American politician, jurist, and diplomat ~

Of course, I think that *my summary* also is a very good one, but you are at liberty to disagree. In a future decade or so when I have learned so much more, I may well decide that it is not so good after all.

This final chapter summarises the issues which prevent me from so far accepting undirected evolution as true. It has been a difficult conclusion to reach, as I have no starting point, no presuppositions, no loyalty to any perspective, and no preference for what the truth should be. I have concluded based on my understanding of the science, with dollops of logic thrown in. I readily admit to possible misunderstandings of the science, but I have analysed as comprehensively as I have been able. The following, in no particular order, are the primary stumbling blocks for evolution as I perceive them.

1. Inheritance: Epigenetic inheritance aside, two cell types are relevant: gamete (reproductive) and somatic (non-reproductive). Only gamete cells are inherited by offspring - sperm from the male and ova from the female. Thus, only mutations which occur to the germ line from conception onward can contribute to evolution; there is no known biological mechanism for mutations in somatic cells, i.e., the rest of the body, to make their way into the germ line.

2. Unit of Inheritance: is the gene, which in humans, has an average composition of 3,500 base pairs (nucleotides).

3. Mutations: are permanent alterations to the DNA sequence of a gene, caused by insertions, deletions, or substitutions of nucleotides. The external causes of mutations, called mutagens, can affect anywhere from a single base pair, to multiple genes in a chromosome.

4. Distribution: mutations are mostly deleterious, with neutral being next in frequency and beneficial the least common. Given the unit of inheritance being the gene, the frequency of inherited mutations would favour deleterious over beneficial, suggesting devolution (genetic entropy) rather than evolution.

5. Complexity: of numerous features in humans is so finely detailed, precise and interrelated that the probability of ultimately favourable genetic mutations happening in precisely the right sequence is extremely low. There is no check-pointing of a sequence in anticipation of subsequent mutations to continue to process for a purpose or goal. One can only assume that, in the majority of cases, a partial mutation accumulation would offer no survival or reproductive benefit, resulting in that line of an organism's development ceasing through deselection.

6. Systems: are what enable us to tick. Any propositions which attempt to describe how evolution might have occurred must consider the implications for systems, not just individual organisms. In the literary sources that I have researched, I find very little evidence of *systems thinking*.

7. Definitions: I contend that scientists pay insufficient attention to the fundamentals of their calling. Physical reality is composed of entities, having properties that determine their activities, and their state. Mental activities have none of the restrictions of physical activities, and thus must be of different properties and form.

8. Axioms: Despite their commitment to the disciplines of their respective scientific pursuits, it would appear that scientists lack a "code of conduct" for evaluating hypotheses and conclusions, such as I have proposed with my ten Primary Axioms.

9. Mind-Brain: Chemicals can only do what they *must*, as determined by their properties and those of other entities with which they interreact in a specified environment. Chemicals, and entities composed of chemicals, always behave *objectively*, whereas the mind always behaves *subjectively*.

10. The Mind: is conflated by scientists as related to the brain, in being a "mental organ" of the brain itself, an emergent property of the brain, or an activity of the brain. It cannot be all three, and given that its activities relate to the conceptual, it cannot be any of those three.

11. Visual Acuity: The eye "sees" because it interacts with light; the mind-brain complex "visualises" because it constructs a mental image of an external reality, allowing the cognitive to enhance visualisation by the application of prior learnings.

12. Imagination: is the source of human creativity, yielding music, poetry, literature, the arts, and yes, even scientific hypotheses, most especially those not suggested by experienced reality. The brain, through evolution, cannot suggest anything not so experienced, leaving only the immaterial mind as the cause of truly original thought.

13. Infinite: How could a mind which is material, and thus finite, have a concept of what it is not?

This list could be expanded significantly, drawing on various topics discussed in this book. I will leave the reader to refute my list, to compose their own, or to simply follow whatever worldview with which they are most comfortable.

Bibliography

1. Alexander, Denis, *Creation or Evolution: Do We Have To Choose?* Monarch Books, Oxford, UK, 2008

2. Andrews, Professor E.H., *What is Man? Adam, Alien, or Ape?* Thomas Nelson Publishers, Nashville, TN, 2018

3. Andrews, Professor E.H., *Who Made God? Searching for a Theory of Everything*, EP Books, Darlington, England, 2009

4. Angliss, Sarah, and Connell, Tom, *All About Your Brain, Senses and Nervous System (Human Machine)*, Belitha Press, London, UK, 1999

5. Barr, Murray L., and Kiernan, John A., *Barr's the Human Nervous System: An Anatomical Viewpoint (Periodicals)*, Lippincot-Raven, Philadelphia, PA, 1998

6. Berlinski, David, *The Devil's Delusion: Atheism and its Scientific Pretensions*, Basic Books, New York, NY, 2009

7. Biological Information - New Perspectives, Proceedings of the Symposium *Cornell University, USA, 31 May – 3 June 2011*, Edited by: Robert J Marks II (*Baylor University, USA*), Michael J Behe (*Lehigh University, USA*), William A Dembski (*Discovery Institute, USA*), Bruce L Gordon (*Houston Baptist University, USA*), John C Sanford (*Cornell*)

8. Bray, Dennis, Wetware: *A Computer in Every Living Cell*, Yale University Press, New Haven, CT, 2009

9. Brentano, Franz, *Psychology from an Empirical Standpoint*, edited by Linda L. McAlister, Routledge Publishing, London, UK, 1995

10. Cardozo, Nathan Lopes, *Jewish Law as Rebellion: A Plea for Religious Authenticity and Halachic Courage*, Urim Publications, Jerusalem, Israel, 2018

11. Chater, Nick, *The Mind is Flat: The Illusion of Mental Depth and The Improvised Mind*, Allen Lane / Penguin, London, UK, 2018

12. Churchland, Patricia S., *Touching A Nerve: Our Brains, Our Selves*, W.W. Norton & Company, New York, NY, 2013

13. Cohen, Gillian, editor with Robert Johnson & Kim Plunkett, *Exploring cognition: Damaged Brains and neural networks*, Psychology Press, Hove, UK, 2000

14. Collins, Francis S., *The Language of God*, Free Press, Simon & Schuster, New York, 2006

15. Craig, William Lane, *Time and Eternity*, Crossway Books, Wheaton, Illinois, 2001

16. Critchlow, Hannah, *The Science of Fate: Why Your Future is More Predictable Than You Think*, Hodder & Stoughton, London, UK, 2019

17. Darwin, Charles, *The Descent of Man*, Wordsworth Editions, London, UK, 2013

18. Darwin, Charles, *The Origin of Species by Means of Natural Selection*, Penguin Books, London, 1985

19. Davies, Paul, *The Fifth Miracle: The Search for the Origin and Meaning of Life*, Touchstone, New York, NY, 1999

20. Dawkins, Richard, *The Greatest Show on Earth – the Evidence for Evolution*, Bantam Press, London, 2009

21. Dawkins, Richard, *The Selfish Gene*, Oxford University Press, Oxford, UK, 2006

22. Dembski, William, *Being as Communion - A Metaphysics of Information*, Ashgate Publishing Company, Burlington, VT, 2014

23. Dennett, Daniel C., *Consciousness Explained*, Penguin Press, London, England, 1991

24. Eccles, John C., *How the SELF Controls Its BRAIN*, Springer-Verlag, Berlin, Germany, 1994

25. Einstein, Albert, and Infeld, Leopold, *The Evolution of Physics*, Cambridge University Press, Cambridge, UK, 1938

26. Etkin, David, *Disaster Theory: An Interdisciplinary Approach to Concepts and Causes*, Butterworth-Heinemann, Oxford, UK, 2016

27. Feldman, Rabbi Yaakov, *The 8 Chapters of the Rambam: A Classic Work on the Fundamentals of Jewish Ethics and Character Development*, Targum Press, Southfield, MI, 2008

28. Finkel, Avraham Yaakov, *The Essential Maimonides: Translations of the Rambam*, Jason Aronson, Inc., Lanham, MD, 1996

29. Fodor, Jerry and Piatelli-Palmarini, Massimo, *What Darwin Got Wrong*, Picador, New York, 2011

30. Frith, Christopher D., *Making up the Mind: How the Brain Creates our Mental World*, Wiley-Blackwell, Oxford, UK, 2007

31. Fromm, Jochen, *The Emergence of Complexity*, kassel university press, Kassel, Germany, 2004

32. Futuyma, Douglas J., *Science on Trial – The Case for Evolution*, Pantheon Books, New York, 1982

33. Gelernter, David, *The Tides of Mind: Uncovering the Spectrum of Consciousness*, Liveright Publishing Corporation, New York, NY, 2016

34. Gilder, George, *The Israel Test*, Richard Vigilante Books, USA, 2009

35. Giles, Kimberly, *Choosing Clarity: The Path to Fearlessness*, Thomas Noble Books, Wilmington, DE, 2014

36. Gitt, Dr. Werner, *In the Beginning was Information*, First Master Books, Green Forest, AR, 2007

37. Gitt, Dr. Werner, *Without Excuse*, Creation Book Publishers, Atlanta, GA, 2011

38. Gluck, Mark A., and Myers, Catherine E., *Gateway to Memory: An Introduction to Neural Network Modeling*, The MOT press, Cambridge, MA, 2001

39. Gould, Stephen Jay, *Rocks of Ages: Science and Religion in the Fullness of Life*, Ballantine Books (Random House), New York, NY, 1999

40. Grisel, Judith, *Never Enough: The Neuroscience and Experience of Addiction*, Scribe Publications, London, UK, 2019

41. Hanaway, Joseph, *The Brain Atlas: A Visual Guide to the Human Nervous System*, Fitzgerald Science Press, Bethesda, MD, 1998

42. Harris, Sam, *Free Will*, Free Press, New York, NY, 2012

43. Hawking, Stephen W., *A Brief History of Time*, Bantam Press, Great Britain, 1988

44. Hawking, Stephen, and Mlodinow, Leonard, *The Grand Design*, Bantam Books, New York, NY, 2010

45. Heidegger, Martin, *An Introduction to Metaphysics*, Anchor Books, New York, 1961

46. Hitchens, Christopher and Wilson, Douglas, *Is Christianity Good for the World*, Canon Press, Moscow, Indiana, 2009

47. Hitchens, Peter, *The Rage Against God – How Atheism Led Me to Faith*, Zondervan, Grand Rapids, Michigan, 2010

48. Kasparov, Garry, *Deep Thinking: Where Machine Intelligence Ends and Human Creativity Begins*, John Murray, London, UK, 2017

49. Kauffman, Stuart A., *The Origins of Order: Self-Organization and Selection in Evolution*, Oxford University Press, New York, NY, 1993

50. Lewis, C.S., *The Case for Christianity*, Touchstone Books, Simon & Schuster, UK, 1996

51. Maimonides, Moses, *The Guide for the Perplexed*, Digireads Publishing, 2018, translation from the original text with annotations by M. Friedlander, Trubner & Co., London, UK, 1881

52. Marquez, Gabriel Garcia, *Love in the Time of Cholera*, Penguin Books, London, UK, 1989

53. Martin, John H., *Neuroanatomy Text and Atlas*, 4th Edition, McGraw-Hill Inc., New York, NY, 2012

54. Mazur, Suzan, *The Altenberg 16: An Expose of the Evolution Industry*, North Atlantic Books, Berkeley, CA, 2010

55. McGinn, Colin, *The Mysterious Flame: Conscious Minds in a Material World*, Basic Books, New York, NY, 1999

56. Meyer, Stephen C., *Darwin's Doubts: The Explosive Origin of Animal Life and The Case for Intelligent Design*, HarperCollins, New York, 2013

57. Meyer, Stephen C., *Signature in the Cell: DNA and the Evidence for Design*, HarperCollins, New York, 2009

58. Murray, Charles, and Herrnstein, Richard J., *The Bell Curve: Intelligence and Class Structure in American Life*, Free Press, New York, NY, 1994

59. Nagel, Thomas, *Mind and Cosmos: Why the Materialist Neo-Darwinian Conception of Nature is Almost Certainly False*, Oxford University Press, New York, NY, 2012

60. Nagel, Thomas, *The Last Word*, Oxford University Press, New York, NY, 1997

61. Naylor, Carma, *A Mormon's Unexpected Journey - Volumes I & II*, WinePress Publishing, Enumclaw, WA, 2006

62. Noble, Denis, *The Music of Life: Biology Beyond Genes*, Oxford

University Press, Oxford, UK, 2006

63. Parker, Steve, *Brain and Nervous System (Human Body)*, Carlton Books, London, UK, 1996

64. Phillips, Christopher, *Six Questions of Socrates*, W.W. Norton & Company, New York, NY, 2004

65. Phillips, Christopher, *Socrates Café: A Fresh Taste of Philosophy*, Norton Paperback, New York, NY, 2001

66. Popper, Karl, *Realism and the Aim of Science*, Routledge Press, New York, NY, 1999

67. Popper, Karl, *The Logic of Scientific Discovery*, Routledge Press, London, UK, 1992

68. Popper, Karl, and Eccles, John C., *The Self and Its Brain: An Argument for Interactionism*, Routledge & Kegan Paul, London, England, 1983

69. Popper, Karl, *The Two Fundamental Problems of the Theory of Knowledge*, Routledge Classics, New York, NY, 2012

70. Rees, Martin, *Before the Beginning: Our Universe and Others*, Simon & Schuster, London, UK, 1997

71. ReMine, Walter James, *The Biotic Message: Evolution Versus Message Theory*, St. Paul Science, Inc., St. Paul, MN, 1993

72. Ryle, Gilbert, *The Concept of Mind*, Penguin Books Ltd, London, England, 1990

73. Sanford, Dr John C., *Genetic Entropy & The Mystery of the Genome*, FMS Publications, Waterloo, New York, 2008

74. Searle, J., *Intentionality: An Essay in the Philosophy of Mind*, Cambridge University Press, Cambridge, UK, 1983

75. Seeskin, Kenneth, *Searching for a Distant God: The Legacy of Maimonides*, Oxford University Press, New York, NY, 2000

76. Shannon, C.E., and Weaver, W., *The Mathematical Theory of Communication*, University of Illinois Press, Urbana, IL, 1949

77. Shapiro, James A., *Evolution: A View from the 21st Century*, FT Press Science, Upper Saddle River, NJ, 2013

78. Sheldrake, Rupert, *The Science Delusion*, Coronet, London, UK, 2013

79. Shoebat, Walid, *Why I Left Jihad*, Top Executive Media, USA, 2005

80. Sobel, Dava, Longitude, *Fourth Estate*, Harper Collins, UK, 1998

81. Spetner, Dr. Lee, *The Evolution Revolution - Why People are Rethinking the Theory of Evolution*, Judaica Press, Brooklyn, NY, 2014

82. Spillane, J.D., *The Doctrine of the Nerves*, Oxford University Press, Oxford UK

83. Talbot, Wayne, *Information, Knowledge, Evolution and Self: A question of origins*, Xlibris, Bloomington, IN, 2016

84. Talbot, Wayne, *Religion? Of God or Man?* Xlibris, Bloomington, IN, 2019

85. Talbot, Wayne, *The Dawkins Deficiency – Why Evolution is Not the Greatest Show on Earth*, Deep River Books, Sisters, OR, 2011

86. Tallis, Raymond, *Aping Mankind*, Routledge Classics, New York, NY, 2016

87. Tallis, Raymond, *Why the Mind is Not a Computer*, Imprint Academic, Exeter, UK, 2004

88. Twersky, Isadore, *A Maimonides Reader*, Behrman House, Springfield, NJ, 1972

89. Verschuuren, Gerard, *What Makes You Tick?* A New Paradigm for Neuroscience, SOLAS Press, Antioch, CA, 2012

90. Walton, John H., *The Lost World of Genesis One: Ancient Cosmology and the Origins Debate*, IVP Academic, Downers Grove, Illinois, 2017

91. Walton, John, *The Origin of Life, Should Christians Embrace Evolution*, Inter-Varsity Press, Nottingham, England, 2009

92. Wesson, Robert, *Beyond Natural Selection*, A Bradford Book, The MIT Press, Cambridge Massachusetts, 1991

93. Woodward, Thomas, *Darwin Strikes Back: Defending the Science of Intelligent Design*, Baker Books, Ada, MI, 2006

www.ingramcontent.com/pod-product-compliance
Lightning Source LLC
Chambersburg PA
CBHW071425070526
44578CB00001B/1